ECSTAT

Student Version

The IBM compatible, 360K diskette on the facing page contains a copy of ECSTAT Student Version, a powerful, easy-to-use, and self-contained econometrics tool. The diskette also includes the ECSTAT manual and data files for all the data included in the *Using Econometrics* text. (The use of the text does not require ECSTAT; the examples and exercises can be completed using a wide variety of statistics packages and computers.)

The complete 37-page ECSTAT manual is in the disk file entitled ECSTAT.DOC. The manual can be printed by copying the file to a printer. To do this, boot (start) your computer with DOS, place the ECSTAT disk in the A: drive, and then enter:

A:

PRINTMAN

(This instruction will print out the manual as long as your printer is connected to LPT1, as most printers are.)

To run the ECSTAT program from the disk, start up your computer with DOS, place the ECSTAT disk in the A: drive, and then enter:

A:

ECSTAT

As shipped, the program displays white on black. If you have a color screen, you can change the colors by reconfiguring ECSTAT using Option 19 (Housekeeping). To install ECSTAT on a hard disk or a floppy disk containing DOS, see the instructions on pages 6-7 of the manual. Before you start using your ECSTAT Student Version program, be sure to make a backup copy of the original disk. (The program is not copy-protected.)

USING ECONOMETRICS

A Practical Guide

SECOND EDITION

A. H. STUDENMUND

Occidental College

with the assistance of

HENRY J. CASSIDY

Federal Home Loan Mortgage Corporation

HarperCollins*Publishers*

Dedicated to

Scott Richard Studenmund

Sponsoring Editor: John Greenman
Project Coordination, Text and Cover Design: Ruttle, Shaw & Wetherill, Inc.
Production Manager: Michael Weinstein
Compositor: Ruttle, Shaw & Wetherill, Inc.
Printer and Binder: R. R. Donnelley & Sons Company
Cover Printer: New England Book Components, Inc.

Library of Congress Cataloging-in-Publication Data

Studenmund, A. H.
Using econometrics : a practical guide / A. H. Studenmund, with the assistance of Henry J. Cassidy. —2nd ed.
p. cm.
Includes bibliographical references and index.
ISBN 0-673-52125-7
1. Econometrics. 2. Regression analysis. I. Cassidy, Henry J. II. Title.
HB139.S795 1992
330'.01'5195—dc20 91-29439
CIP

91 92 93 94 9 8 7 6 5 4 3 2 1

Acknowledgments
Table B-1: Reprinted with permission of Hafner Press, a division of Macmillan Publishing Company from *Statistical Methods for Research Workers*, 14th Edition, by Ronald A. Fisher. Copyright © 1970 by University of Adelaide.

Tables B-2 and B-3: Abridged from M. Merrington and C.M. Thompson, "Tables of percentage points of the inverted beta (*F*) distribution," *Biometrica*, Vol. 38, 1951, pp. 159-77. By permission of the Biometrica Trustees.

Table B-4: From N. E. Savin and Kenneth White, "The Durbin-Watson Test for Serial Correlation with Extreme Sample Sizes or Many Regressors," *Econometrica*, Nov. 1977, p. 1994. Reprinted with permission.

Tables B-5, and B-6: From J. Durbin and G. S. Watson, "Testing for serial correlation in least squares regressions," *Biometrika*, Vol. 38, 1951, pp. 159-77. By permission of the Biometrica Trustees.

Tables B-7 and B-8: Based on *Biometrica Tables for Statisticians*, Vol. 1, 3rd ed. (1966). By permission of the Biometrica Trustees.

Contents

Preface

"Econometric education is a lot like learning to fly a plane; you learn more from actually doing it than you learn from reading about it."

Using Econometrics represents a new approach to the understanding of elementary econometrics. It covers the topic of single-equation linear regression analysis in an easily understandable format that emphasizes real-world examples and exercises. As the subtitle, *A Practical Guide,* implies, the book is aimed not only at beginning econometrics students but also at regression users looking for a refresher and at experienced practitioners who want a convenient reference.

The material covered by this book is traditional, but there are four specific features that we feel distinguish *Using Econometrics:*

1. Our approach to the learning of econometrics is simple, intuitive, and easy to understand. We do not use matrix algebra, and we relegate proofs and calculus to the footnotes.
2. We include numerous examples and example-based exercises. We feel that the best way to get a solid grasp of applied econometrics is through an example-oriented approach.
3. Although most of this book is at a simpler level than other econometrics texts, Chapters 6 and 7 on specification choice are among the most complete in the field. We think that an understanding of specification issues is vital for regression users.
4. We include a new kind of learning tool, called an interactive regression learning exercise, that helps students simulate econometric

analysis by giving them feedback on various kinds of decisions without relying on computer time or much instructor supervision.

The formal prerequisites for using this book are few. Most important, readers are assumed to have some familiarity with macroeconomic and microeconomic theory; in addition, the book is easier to use if readers have had a statistics course (even if they have forgotten most of it), or if readers are not afraid of working with mathematical functions. While all the statistical concepts necessary for econometric study are covered in the text, they are covered only to the extent needed for an understanding of regression analysis. Because the prerequisites are few and the statistics material is self-contained, *Using Econometrics* can be used not only in undergraduate econometrics courses but also in MBA-level courses in quantitative methods. We also have been told that the book is a helpful supplement for graduate-level econometrics courses.

What's New in the Second Edition?

The first edition of *Using Econometrics* was the most-adopted new econometrics textbook of the 1980s. As a result, we've been careful to write a second edition that adds more coverage and fifty percent more data sets, exercises, and examples *without* changing the clarity and practicality that made the first edition a success. Twice as many interactive regression learning exercises also have been added. Most of the new materials we added were suggested by users of the text, although some were recommended by reviewers intent on keeping us up with changes in standard practices in applied econometrics.

Specific additions include:

1. Chapter 12 on distributed lag models
2. Chapter 13 on dummy dependent variable techniques
3. an appendix on additional specification criteria in Chapter 6
4. a section on Variance Inflation Factors in Chapter 8
5. a section on the Breusch-Pagan and White tests in Chapter 10

Also, our expanded instructor's manual includes answers to odd numbered exercises, lecture notes, sample examinations, and an additional interactive exercise.

Finally, we're pleased to be able to include the student version of ECSTAT with each copy of the text at virtually no additional cost.

ECSTAT is the extremely user-friendly PC-based regression software package that we use to produce the regression results in the text. While the book is not tied to the use of ECSTAT in any way, we think students will learn to appreciate ECSTAT's accuracy and simplicity. The ECSTAT diskette also includes the data sets published in the text and a manual for using the program.

Acknowledgments

If this book has a spiritual father, it's Henry Cassidy of the Federal Home Loan Mortgage Corporation. It was Henry who saw the need for a follow-on to Rao and Miller's legendary *Applied Econometrics* and who coauthored the first edition of *Using Econometrics* as an expansion of his own work of the same name. Henry also contributed quite a bit to this second edition, as he reviewed all the new material and provided a healthy dose of good humor and inspiration.

From the point of view of this edition, the contributions of two superb economists stand out. Carolyn Summers of the National Education Association continued her invaluable first-edition role as editorial consultant, galley proofreader, indexer, and general gadfly-about-town. Mary Hirschfeld here at Occidental College reviewed the entire text, suggested a number of crucial improvements in its substance, and provided me with extremely perceptive feedback from the point of view of teachers of econometrics.

The quality of the outside reviewers of this edition was also quite high. In particular, I owe a real debt of gratitude to Rob Engle of UCSD for taking the time to give me in-depth feedback on the new material. Others in this excellent group of reviewers were Dennis Byrne (University of Akron), William Brown (CSU Northridge), Edward Coulson (Penn State), William Dawes (SUNY Stony Brook), Cliff Huang (Vanderbilt), Elia Kacapyr (Ithaca College), John Warner (Clemson), Tom Witt (West Virginia), and Phanindra Wunnava (Middlebury College).

Especially helpful in the editorial and production process were Paula Cousin, Jack Greenman, Dave Murphy, Ilana Scheiner, and Robert Dohner (the author of ECSTAT). Others who provided timely and useful assistance in this edition were Sonmez Atesoglu (Clarkson University), Sandra Chadwick, James Keeler (Kenyon College), Bruce Gensemer (Kenyon College), Marlene Penfold, and Justin Meyer (Silver Oak). Finally, but perhaps most important, I'd like to thank my superb Occidental College colleagues and students for their feedback and encouragement. These especially included Ned Cull, Richard Falkenrath, Greg Gordon, Eric Hilt, Nichole Becker Kamman, Jim Kee, Mark Lieberman, Manuel Pastor, Bob Sego, Bruce Stimpson, and Chris Tregillis.

A. H. Studenmund

1

An Overview of Regression Analysis

1.1 What Is Econometrics?

"Econometrics is too mathematical; it's the reason my best friend isn't majoring in economics."

"There are two things you don't want to see in the making—sausage and econometric research."[1]

"Econometrics may be defined as the quantitative analysis of actual economic phenomena."[2]

"It's my experience that 'economy-tricks' is usually nothing more than a justification of what the author believed before the research was begun."

1. Attributed to Edward E. Leamer.

2. Paul A. Samuelson, T. C. Koopmans, and J. R. Stone, "Report of the Evaluative Committee for *Econometrica*," *Econometrica*, 1954, p. 141.

Believe it or not, these are all actual quotations. Obviously, econometrics means different things to different people. To beginning students, it may seem as if econometrics is an overly complex obstacle to an otherwise useful education. To skeptical observers, econometric results should be trusted only when the steps that produced those results are completely known. To professionals in the field, econometrics is a fascinating set of techniques that allows the measurement and analysis of economic phenomena and the prediction of future economic trends.

You're probably thinking that such diverse points of view sound like the statements of blind people trying to describe an elephant based on what they happen to be touching, and you're partially right. Econometrics has both a formal definition and a larger context. While you can easily memorize the formal definition, you'll get the complete picture only by understanding the many uses of and alternative approaches to econometrics.

That said, we need a formal definition. **Econometrics,** literally "economic measurement," is the quantitative measurement and analysis of actual economic and business phenomena. It attempts to quantify economic reality and bridge the gap between the worlds of economic theory and actual business activity. To many students, these worlds may seem far apart. On the one hand, economists theorize equilibrium prices based on carefully conceived marginal costs and marginal revenues; on the other, many firms seem to operate as though they have never heard of such concepts. Econometrics allows us to examine data from real-world firms and to quantify the actions of these firms and other factors, such as the actions of consumers and governments. Such measurements have a number of different uses, and an examination of these uses is the first step to understanding econometrics.

1.1.1 Uses of Econometrics

Econometrics has three major uses:

1. the description of economic reality
2. the testing of hypotheses about economic theory
3. the forecasting of future economic activity.

The simplest use of econometrics is **description.** We can use econometrics to quantify economic activity; econometrics allows us to put numbers in equations that previously contained only abstract symbols. For example, consumer demand for a particular commodity often can be thought of as a relationship between the quantity demanded (C) and the commodity's price (P), the price of a substitute good (P_s), and disposable income (Y_d). For most goods, the relationship between consumption and disposable income is expected to be positive, because an increase in disposable income will be associated with an increase in the consumption of the good. Econometrics actually allows us to estimate that relationship based upon past consumption, income, and prices. In other words, a general and purely theoretical functional relationship like:

$$C = f(P, P_s, Y_d) \tag{1.1}$$

can become explicit:

$$C = -60.5 - 0.45P + 0.12P_s + 12.2Y_d \tag{1.2}$$

This technique gives a much more specific and descriptive picture of the function.[3] Let's compare Equations 1.1 and 1.2. Instead of expecting consumption merely to "increase" if there is an increase in disposable income, Equation 1.2 allows us to expect an increase of a specific amount (12.2 units for each unit of increased disposable income). The number 12.2 is called an estimated regression coefficient, and it is the ability to estimate these coefficients that makes econometrics valuable.

The second and perhaps the most common use of econometrics is **hypothesis testing,** the testing of alternative theories with quantitative evidence. Much of economics involves building theoretical models and testing them against evidence, and hypothesis testing is vital to that scientific approach. For example, you could test the hypothesis that the product in Equation 1.1 is what economists call a normal good (one for which the quantity demanded increases when disposable income increases). You could do this by applying various statistical tests to the estimated coefficient (12.2) of disposable income (Y_d) in Equation 1.2. At first glance, the evidence would seem to support this hypothesis because the coefficient's sign is positive, but the "statistical significance" of that estimate would have to be investigated before such a conclusion could be justified. Even though the estimated coefficient is positive, as expected, it may not be sufficiently different from zero to imply that the coefficient is indeed positive instead of zero. Unfortunately, statistical tests of such hypotheses are not always easy, and there are times when two researchers can look at the same set of data

3. The results in Equation 1.2 are from a model of the demand for chicken that we will examine in more detail in Section 6.1.

and come to different conclusions. Even given this possibility, the use of econometrics in testing hypotheses is probably its most important function.

The third and most difficult use of econometrics is to **forecast** or predict what is likely to happen next quarter, next year, or further into the future. For example, economists use econometric models to make forecasts of variables like sales, profits, Gross National Product (GNP), and the inflation rate. The accuracy of such forecasts depends in large measure on the degree to which the past is a good guide to the future. Business leaders and politicians tend to be especially interested in this use of econometrics because they need to make decisions about the future, and the penalty for being wrong (bankruptcy for the entrepreneur and political defeat for the candidate) is high. To the extent that econometrics can shed light on the impact of their policies, business and government leaders will be better equipped to make decisions. For example, if the president of a company that sold the product modeled in Equation 1.1 wanted to decide whether to increase prices, forecasts of sales with and without the price increase could be calculated and compared to help make such a decision. In this way, econometrics can be used not only for forecasting but also for policy analysis.

1.1.2 Alternative Econometric Approaches

There are many different approaches to econometrics. For example, the fields of biology, psychology, and physics all face quantitative questions similar to those faced in economics and business. However, these fields tend to use somewhat different techniques for analysis because the problems they face aren't the same. Different approaches also make sense within the field of economics. The kind of econometric tools used to quantify a particular function depends in part on the uses to which that equation will be put. A model built solely for descriptive purposes might be different from a forecasting model, for example.

To get a better picture of these approaches, let's look at the steps necessary for any kind of quantitative research:

1. specifying the models or relationships to be studied
2. collecting the data needed to quantify the models
3. quantifying the models with the data.

Steps 1 and 2 are similar in all quantitative work, but the techniques used in step 3, the quantification of models, differ widely between and within disciplines. Choosing among techniques for the quantification of a model given a particular set of data is often referred to as the "art" of econometrics. There are many different alternative approaches to quantifying the same equation, and each approach may give somewhat different results. The choice of approach is left to the individual econometrician (the researcher using econometrics), but each researcher should be able to justify that choice.

This book will focus primarily on one particular econometric approach: single-equation linear *regression analysis*. The majority of this book will thus discuss only regression analysis, but it is important for every econometrician to remember that regression is only one of many approaches to econometric quantification.

The importance of critical evaluation cannot be stressed enough; a good econometrician is one who can diagnose faults in a particular approach and figure out how to repair them. The limitations of the regression analysis approach must be fully perceived and appreciated by anyone attempting to use regression analysis or its findings. The possibility of missing or inaccurate data, incorrectly formulated relationships, poorly chosen estimating techniques, or improper statistical testing procedures implies that the results from regression analyses need to be viewed with some caution.

1.2 What Is Regression Analysis?

Econometricians use regression analysis to make quantitative estimates of economic relationships that previously have been completely theoretical in nature. After all, anybody can claim that the quantity of compact discs demanded will increase if the price of those discs decreases (holding everything else constant), but not many people can actually put numbers into an equation and estimate *by how many* compact discs the quantity demanded will increase for each dollar that price decreases. To predict the *direction* of the change, you need a knowledge of economic theory and the general characteristics of the product in question. To predict the *amount* of the change, though, you need a sample of data, and you need a way to estimate the relationship. The most frequently used method to estimate such a relationship in econometrics is regression analysis.

1.2.1 Dependent Variables, Independent Variables, and Causality

Regression analysis is a statistical technique that attempts to "explain" movements in one variable, the **dependent variable,** as a function of movements in a set of other variables, called the **independent (*or* explanatory) variables,** through the quantification of a single equation. For example, in Equation 1.1:

$$C = f(P, P_s, Y_d) \tag{1.1}$$

C is the dependent variable and P, P_s, and Y_d are the independent variables. Regression analysis is a natural tool for economists because most economic propositions can be stated in such single-equation[4] functional forms. For example, the quantity demanded (dependent variable) is a function of price, income, and the prices of substitutes (independent variables).

Much of economics and business is concerned with cause-and-effect propositions: If the price of a good increases by one unit, then the quantity demanded decreases on average by a certain amount, depending on the price elasticity of demand (defined as the percentage change in the quantity demanded that is caused by a one percent change in price). Similarly, if the quantity of capital employed increases by one unit, then output increases by a certain amount, called the marginal productivity of capital. Propositions such as these pose an if-then, or causal, relationship that logically postulates a dependent variable having movements that are causally determined by movements in a number of specified independent variables.

Don't be deceived by the words dependent and independent, however. While many economic relationships are causal by their very nature, a regression result, no matter how statistically significant, cannot prove causality. All regression analysis can do is test whether a significant quantitative relationship exists. Judgments as to causality must also include a healthy dose of economic theory and common sense. For example, the fact that the bell on the door of a flower shop rings just before a customer enters and purchases some flowers by no means implies that the ringing of the bell causes the purchase! If events A and B are related statistically, it may be that A causes B, that B causes A, that some omitted factor causes both, or that a chance correlation exists between the two.

The cause and effect relationship is often so subtle that it fools

4. Often there are several related propositions that, when taken as a group, suggest a *system* of regression equations. An example is a two-equation model of supply and demand. Usually, these two equations must be considered simultaneously instead of separately. The estimation of such simultaneous systems will be discussed in Chapter 14.

even the most prominent economists. For example, in the late nineteenth century, English economist Stanley Jevons hypothesized that sunspots caused an increase in economic activity. To test this theory, he collected data on national output (the "dependent" variable) and sunspot activity (the "independent" variable) and showed that a significant positive relationship existed. This result led him, and some others, to jump to the conclusion that sunspots did indeed cause output to rise. Such a conclusion was unjustified because regression analysis cannot confirm causality; it can only test the strength and direction of the quantitative relationships involved.

1.2.2 Single-Equation Linear Models

The simplest single-equation linear regression model is:

$$Y = \beta_0 + \beta_1 X \tag{1.3}$$

Equation 1.3 states that Y, the dependent variable, is a single-equation linear function of X, the independent variable. The model is a single-equation model because no equation for X as a function of Y (or any other variable) has been specified. The model is linear because if you were to plot Equation 1.3 on graph paper, it would be a straight line rather than a curve.

The βs are the **coefficients** (or **parameters**) that determine the coordinates of the straight line at any point. β_0 is the **constant** or **intercept** term; it indicates the value of Y when X equals zero. β_1 is the **slope coefficient,** and it indicates the amount that Y will change when X changes by one unit. Figure 1.1 illustrates the relationship between the coefficients and the graphical meaning of the regression equation. As can be seen from the diagram, Equation 1.3 is indeed linear.

The slope, β_1, shows the response of Y to a change in X. Since being able to explain and predict changes in the dependent variable is the essential reason for quantifying behavioral relationships, most of the emphasis in regression analysis is on slope coefficients such as β_1. In Figure 1.1 for example, if X were to increase from X_1 to X_2, the value of Y in Equation 1.3 would increase from Y_1 to Y_2. For linear (i.e., straight-line) regression models, the response in the predicted value of Y due to a change in X is constant and equal to the slope coefficient β_1:

$$\frac{(Y_2 - Y_1)}{(X_2 - X_1)} = \frac{\Delta Y}{\Delta X} = \beta_1$$

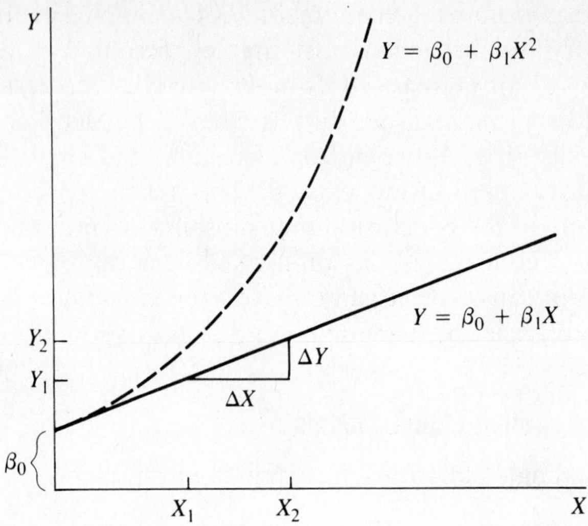

Figure 1.1 GRAPHICAL REPRESENTATION OF THE COEFFICIENTS OF THE REGRESSION LINE
The graph of the equation $Y = \beta_0 + \beta_1 X$ is linear with a constant slope equal to $\beta_1 = \Delta Y/\Delta X$. The graph of the equation $Y = \beta_0 + \beta_1 X^2$, on the other hand, is nonlinear with an increasing slope (if $\beta_1 > 0$).

where Δ is used to denote a change in the variables. Some readers may recognize this as the "rise" (ΔY) divided by the "run" (ΔX). For a linear model, the slope is constant over the entire function.

We must distinguish between an equation that is linear in the variables and one that is linear in the coefficients (or parameters). This distinction is necessary because while linear regressions need to be linear in the coefficients, they do not necessarily need to be linear in the variables. An equation is **linear in the variables** if plotting the function in terms of X and Y generates a straight line. For example, Equation 1.3:

$$Y = \beta_0 + \beta_1 X \tag{1.3}$$

is linear in the variables, but Equation 1.4:

$$Y = \beta_0 + \beta_1 X^2 \tag{1.4}$$

is not linear in the variables because if you were to plot Equation 1.4 it would be a quadratic, not a straight line. This difference can be seen in Figure 1.1.

An equation is **linear in the coefficients** (or parameters) only if the coefficients (the βs) appear in their simplest form—they are not raised to any powers (other than one), are not multiplied or divided by other coefficients, and do not themselves include some sort of function (like logs or exponents). For example, Equation 1.3 is linear in the coefficients, but Equation 1.5:

$$Y = \beta_0 + X^{\beta_1} \quad (1.5)$$

is not linear in the coefficients β_0 and β_1. Equation 1.5 is not linear because there is no rearrangement of the equation that will make it linear in the βs of original interest, β_0 and β_1. In fact, of all possible equations for a single explanatory variable, *only* functions of the general form:

$$f(Y) = \beta_0 + \beta_1 f(X) \quad (1.6)$$

are linear in the coefficients β_0 and β_1. In essence, any sort of configuration of the Xs and Ys can be used and the equation will continue to be linear in the coefficients. However, even a slight change in the configuration of the βs will cause the equation to become nonlinear in the coefficients. For example, Equation 1.4 is not linear in the variables but *is* linear in the coefficients. The reason that Equation 1.4 is linear in the coefficients is that if you define $f(X) = X^2$, Equation 1.4 fits into the general form of Equation 1.6.

All this is important because if linear regression techniques are going to be applied to an equation, that equation *must be* linear in the coefficients. Linear regression analysis can be applied to an equation that is nonlinear in the variables if the equation can be formulated in a way that is linear in the coefficients. Indeed, when econometricians use the phrase "linear regression," they usually mean "regression that is linear in the coefficients." The application of regression techniques to equations that are nonlinear in the coefficients will be discussed in Section 7.6.

1.2.3 The Stochastic Error Term

Besides the variation in the dependent variable (Y) that is caused by the independent variable (X), there is almost always variation that comes from other sources as well. This additional variation comes in part from omitted explanatory variables (e.g., X_2 and X_3). Even if these extra variables are added to the equation, there still is going to be some variation in Y that simply cannot be explained by the model.[5] This variation probably comes from sources such as omitted influences, measurement error, incorrect functional form, or purely random and totally unpredictable occurrences.

Econometricians admit the existence of such inherent unexplained variation ("error") by explicitly including a stochastic (or random) error term in their regression models. A **stochastic error term** is a term that is added to a regression equation to introduce all of the variation in Y that cannot be explained by the included Xs. It is, in effect, a symbol of the econometrician's ignorance or inability to model all the movements of the dependent variable. The error term is usually referred to with the symbol epsilon (ϵ), although other symbols (like u or v) are sometimes used.

The addition of a stochastic error term (ϵ) to Equation 1.3 results in a typical regression equation:

$$Y = \beta_0 + \beta_1 X + \epsilon \tag{1.7}$$

Equation 1.7 can be thought of as having two components, the *deterministic* component and the *stochastic*, or random, component. The expression $\beta_0 + \beta_1 X$ is called the *deterministic* component of the regression equation because it indicates the value of Y that is determined by a given value of X, which is assumed to be nonstochastic. This deterministic component can also be thought of as the **expected value** of Y given X, the mean value of the Ys associated with a particular value of X. For example, if the average height of all 14-year-old girls is 5 feet, then 5 feet is the expected value of a girl's height given that she is 14. The deterministic part of the equation may be written:

$$E(Y|X) = \beta_0 + \beta_1 X \tag{1.8}$$

5. An exception would be the case where the data can be explained by some sort of physical law and are measured perfectly. Here, continued variation would point to an omitted independent variable. A similar kind of problem is often encountered in astronomy, where planets can be discovered by noting that the orbits of known planets exhibit variations that can only be caused by the gravitational pull of another heavenly body. Absent these kinds of physical laws, researchers in economics and business would be foolhardy to believe that *all* variation in Y can be explained by a regression model because there are always elements of error in any attempt to measure a behavioral relationship.

which states that the expected value of Y given X, denoted as $E(Y|X)$, is a linear function of the independent variable (or variables if there are more than one).[6]

Unfortunately, the value of Y observed in the real world is unlikely to be exactly equal to the deterministic expected value $E(Y|X)$. After all, not all 14-year-old girls are 5 feet tall. As a result, the stochastic element (ϵ) must be added to the equation:

$$Y = E(Y|X) + \epsilon = \beta_0 + \beta_1 X + \epsilon \tag{1.9}$$

The stochastic error term must be present in a regression equation because there are at least four sources of variation in Y other than the variation in the included Xs:

1. Many minor influences on Y are *omitted* from the equation (for example, because data are unavailable).

2. It is virtually impossible to avoid some sort of *measurement error* in at least one of the equation's variables.

3. The underlying theoretical equation might have a *different functional form* (or shape) than the one chosen for the regression. For example, the underlying equation might be nonlinear in the variables for a linear regression (or vice-versa).

4. All attempts to generalize human behavior must contain at least some amount of unpredictable or *purely random* variation.

To get a better feeling for these components of the stochastic error term, let's think about a consumption function (aggregate consumption as a function of aggregate disposable income). First, consumption in a particular year may have been less than it would have been because of uncertainty over the future course of the economy, causing consumers to save more and to consume less than they would if the uncertainty had not existed. Since this uncertainty is hard to measure, there might be no variable measuring consumer uncertainty in the equation. In such a case, the impact of the omitted variable (consumer uncertainty)

6. This property holds as long as $E(Y|\epsilon) = 0$ [read as "the expected value of Y, given ϵ" equals zero], which is true as long as the Classical Assumptions (to be outlined in Chapter 4) are met. It's easiest to think of $E(Y)$ as the mean of Y, but the expected value operator E technically is a summation of all the values that a function can take weighted by the probability of each value. The expected value of a constant is that constant, and the expected value of a sum of variables equals the sum of the expected values of those variables.

would likely end up in the stochastic error term. Second, the observed amount of consumption may have been different from the actual level of consumption in a particular year due to an error (such as a sampling error) in the measurement of consumption in the National Income Accounts. Third, the underlying consumption function may be nonlinear, but a linear consumption function might be estimated. To see how this incorrect functional form would cause errors, see Figure 1.2. Fourth, the consumption function attempts to portray the behavior of people, and there is always an element of unpredictability in human behavior. At any given time, some random event might increase or decrease aggregate consumption in a way that might never be repeated and couldn't be anticipated.

Any or all of these possibilities may explain the existence of a difference between the observed values of Y and the values expected from the deterministic component of the equation, E(Y|X). These sources of error will be covered in more detail in the following chapters, but for now it is enough to recognize that in economics there will

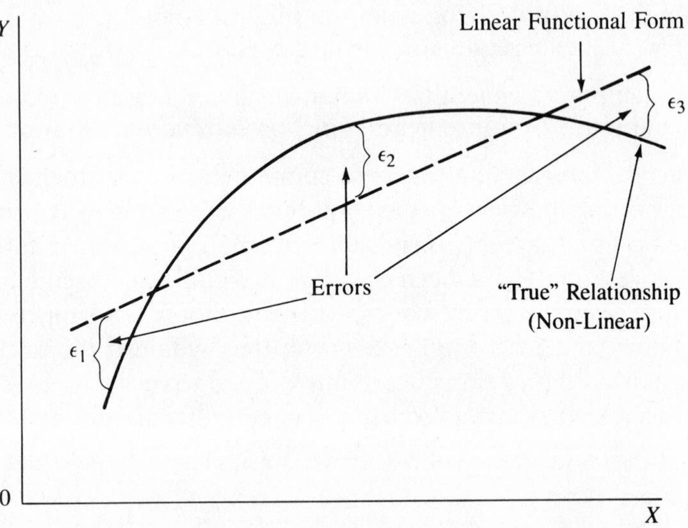

Figure 1.2 ERRORS CAUSED BY USING A LINEAR FUNCTIONAL FORM TO MODEL A NONLINEAR RELATIONSHIP

One source of stochastic error is the use of an incorrect functional form. For example, if a linear functional form is used when the underlying relationship is nonlinear, systematic errors (the ϵs) will occur. These nonlinearities are just one component of the stochastic error term. The others are omitted variables, measurement error, and purely random variation.

always be some stochastic or random element, and, for this reason, an error term must be included in all regression equations.

1.2.4 Extending the Notation

Our regression notation needs to be extended to include reference to the number of observations and to allow the possibility of more than one independent variable. If we include the specific reference to the observations, the single-equation linear regression model may be written as:

$$Y_i = \beta_0 + \beta_1 X_i + \epsilon_i \qquad (i = 1,2,\ldots,n) \qquad (1.10)$$

where: Y_i = the *i*th observation[7] of the dependent variable
X_i = the *i*th observation of the independent (or explanatory) variable
ϵ_i = *i*th observation of the stochastic error (or disturbance) term
β_0, β_1 = the regression coefficients (or parameters)
n = the number of observations

This equation is actually n equations, one for each of the n observations:

$$Y_1 = \beta_0 + \beta_1 X_1 + \epsilon_1$$
$$Y_2 = \beta_0 + \beta_1 X_2 + \epsilon_2$$
$$Y_3 = \beta_0 + \beta_1 X_3 + \epsilon_3$$
$$\vdots$$
$$Y_n = \beta_0 + \beta_1 X_n + \epsilon_n$$

That is, the regression model is assumed to hold for each observation. The coefficients do not change from observation to observation, but the values of Y, X, and ϵ do.

The second notational addition allows for more than one indepen-

7. A typical observation is an individual person, year, or country. For example, a series of annual observations starting in 1950 would have Y_1 = Y for 1950, Y_2 for 1951, etc.

dent variable. Since more than one independent variable is likely to have an effect on the dependent variable, our notation should allow these additional explanatory Xs to be added. If we define:

X_{1i} = the *i*th observation of the first independent variable
X_{2i} = the *i*th observation of the second independent variable
X_{3i} = the *i*th observation of the third independent variable

then all three variables can be expressed as determinants of Y in a **multivariate** (more than one independent variable) linear regression model:

$$Y_i = \beta_0 + \beta_1 X_{1i} + \beta_2 X_{2i} + \beta_3 X_{3i} + \epsilon_i \qquad (1.11)$$

The *meaning of the regression coefficient* β_1 is now the effect of a one unit change in X_1 on the dependent variable Y, holding constant the other included independent variables (X_2 and X_3). Similarly, β_2 gives the effect of X_2 on Y, holding X_1 and X_3 constant. These multivariate regression coefficients (which are parallel in nature to partial derivatives in calculus) serve to isolate the impact on Y of a change in one variable from the impact on Y of changes in the other variables. This is possible because multivariate regression takes the movements of X_2 and X_3 into account when it estimates the coefficient of X_1. The result is quite similar to what we would obtain if we were capable of conducting controlled laboratory experiments in which only one variable at a time was changed.

In the real world, though, it is almost impossible to run controlled experiments, because many economic factors change simultaneously, often in opposite directions. Thus the ability of regression analysis to measure the impact of one variable on the dependent variable, holding constant the influence of the other variables in the equation, is a tremendous advantage. Note that if a variable is not included in an equation, then its impact is *not* held constant in the estimation of the regression coefficients. This will be discussed further in Chapter 6.

The general multivariate regression model with K independent variables is written as:

$$Y_i = \beta_0 + \beta_1 X_{1i} + \beta_2 X_{2i} + \cdots + \beta_K X_{Ki} + \epsilon_i \qquad (1.12)$$
$$(i = 1,2,\ldots,n)$$

If the sample consists of a series of years or months (called a **time series**), then the subscript i is usually replaced with a t to denote time.

It also does not matter if X_{1i}, for example, is written as X_{i1} as long as the appropriate definitions are presented. Often the observational subscript (i or t) is deleted, and the reader is expected to understand that the equation holds for each observation in the sample.

1.2.5 The Estimated Regression Equation

Once a specific equation has been decided upon, it must be quantified. This quantified version of the "true" regression equation[8] is called the **estimated regression equation** and is obtained from a sample of Xs and Ys. While the true equation is purely theoretical in nature:

$$Y_i = \beta_0 + \beta_1 X_i + \epsilon_i \tag{1.13}$$

the estimated regression equation has actual numbers in it:

$$\hat{Y}_i = 103.40 + 6.38X_i \tag{1.14}$$

The observed values of X and Y are used to determine the coefficient estimates 103.40 and 6.38. These estimates are used to determine $\hat{Y}$ (read as "Y-hat"), the *estimated* or *fitted* value of Y_i. To allow this section to focus on the general characteristics of the estimated regression concept, we'll postpone an explanation of how these estimates are obtained until the next chapter.

Let's look at the differences between a true regression equation and an estimated regression equation. First, the theoretical regression coefficients β_0 and β_1 in Equation 1.13 have been replaced with *estimates* of those coefficients like 103.40 and 6.38 in Equation 1.14. We can't actually observe the values of the true regression coefficients, so instead we calculate estimates of those coefficients from the data. The **estimated regression coefficients,** more generally denoted by $\hat{\beta}_0$ and $\hat{\beta}_1$ (read as "beta-hats"), are empirical best guesses of the true regression coefficients and are obtained from data from a sample of the Ys and Xs. The expression

$$\hat{Y}_i = \hat{\beta}_0 + \hat{\beta}_1 X_i \tag{1.15}$$

8. Our use of the word "true" throughout the text should be taken with a grain of salt. Many philosophers argue that the concept of truth is useful only relative to the scientific research program in question. Many economists agree, pointing out that what is true for one generation may well be false for another. To us, the true coefficient is the one that you'd obtain if you could run a regression on the entire relevant population. Thus readers who so desire can substitute the phrase "population coefficient" for "true coefficient" with no loss in meaning.

is the general applied empirical counterpart of the true regression Equation 1.13. The calculated estimates in Equation 1.14 are examples of estimated regression coefficients $\hat{\beta}_0$ and $\hat{\beta}_1$. For each sample we calculate a different set of estimated regression coefficients.

$\hat{Y}_i$ is the *estimated value* of Y_i, and it represents the value of Y calculated from the estimated regression equation for the *i*th observation. As such, $\hat{Y}_i$ is our prediction of $E(Y_i|X_i)$ from the regression equation. The closer $\hat{Y}_i$ is to Y_i, the better the fit of the equation. (The word fit is used here much as it would be used to describe how well clothes fit.)

The difference between the estimated value of the dependent variable ($\hat{Y}_i$) and the actual value of the dependent variable (Y_i) is defined as the **residual:**

$$e_i = Y_i - \hat{Y}_i \tag{1.16}$$

Note the distinction between the residual in Equation 1.16 and the error term:

$$\epsilon_i = Y_i - E(Y_i|X_i) \tag{1.17}$$

The *residual* is the difference between the observed Y and the estimated regression line ($\hat{Y}$), while the *error term* is the difference between the observed Y and the true regression equation (the expected value of Y). In other words, the error term is a theoretical value that can never be observed, but the residual is a real-world value that is calculated for each observation every time a regression is run. Indeed, most regression techniques not only observe the residual but also attempt to select values of $\hat{\beta}_0$ and $\hat{\beta}_1$ that keep the residual as low as possible. The smaller the residual, the better the fit, and the closer the $\hat{Y}$s will be to the Ys.

All these concepts are shown in Figure 1.3. The (X,Y) pairs are shown as points on the diagram, and both the true regression equation (which cannot be seen in real applications) and an estimated regression equation are included. Notice that the estimated equation is close to but not equivalent to the true line. This is a typical result. As can be seen in the figure, $\hat{Y}_6$, the computed value of Y for the sixth observation, lies on the estimated (dashed) line, and it differs from Y_6, the actual observed value of Y for the sixth observation. The difference between the observed and estimated values is the residual, denoted by e_6. In addition, while we usually would not be able to observe an error

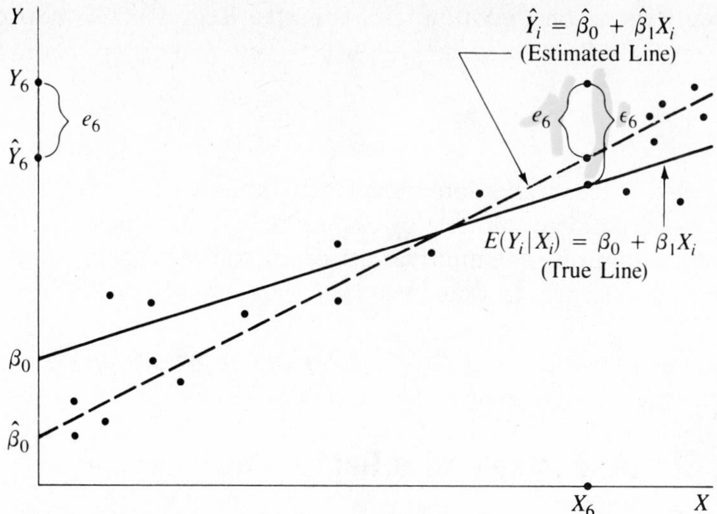

Figure 1.3 TRUE AND ESTIMATED REGRESSION LINES
The true relationship between X and Y (the solid line) cannot typically be observed, but the estimated regression line (the dotted line) can. The difference between an observed data point (for example, $i = 6$) and the true line is the value of the stochastic erorr term (ϵ_6). The difference between the observed Y_6 and the fitted value from the regression line ($\hat{Y}_6$) is the value of the residual e_6 for this observation.

term, we have drawn the assumed true regression line here (the solid line) to see the error term for the sixth observation, ϵ_6, which is the difference between the true line and the observed value of Y, Y_6.

Another way to state the estimated regression equation is to combine Equations 1.15 and 1.16, obtaining:

$$Y_i = \hat{\beta}_0 + \hat{\beta}_1 X_i + e_i \tag{1.18}$$

Compare this equation to Equation 1.13. When we replace the theoretical regression coefficients with estimated coefficients, the error term must be replaced by the residual, because the error term, like the regression coefficients β_0 and β_1, can never be observed. Instead, the residual is observed and measured whenever a regression line is estimated with a sample of Xs and Ys. In this sense, the residual can be thought of as an estimate of the error term, and e could have been denoted as $\hat{\epsilon}$.

The chart on page 18 summarizes the notation used in the true and estimated regression equations:

True Regression Equation	Estimated Regression Equation
β_0	$\hat{\beta}_0$
β_1	$\hat{\beta}_1$
ϵ_i	e_i

The estimated regression model can be extended easily to more than one independent variable by adding the additional Xs to the right side of the equation. The multivariate estimated regression counterpart of Equation 1.12 would thus be:

$$\hat{Y}_i = \hat{\beta}_0 + \hat{\beta}_1 X_{1i} + \hat{\beta}_2 X_{2i} + \cdots + \hat{\beta}_K X_{Ki} \tag{1.19}$$

1.3 A Simple Example of a Regression

Let's look at a fairly simple example of regression analysis. Suppose you've accepted a summer job as a weight guesser at the local amusement park, Magic Flag Farm. Customers pay 50 cents each, which you get to keep if you guess their weight within ten pounds. If you miss by more than ten pounds, then you have to give the customer a small prize that you buy from Magic Flag Farm for 60 cents each. Luckily, the friendly managers of Magic Flag Farm have arranged a number of marks on the wall behind the customer so that you are capable of measuring the customer's height accurately. Unfortunately, there is a five-foot wall between you and the customer, so you can tell little about the person except for height and (usually) gender.

On your first day on the job, you do so poorly that you work all day and actually lose two dollars, so on the second day you decide to collect data to run a regression to formally estimate the relationship between weight and height. Since most of the participants are male, you decide to limit your sample to males. You hypothesize the following relationship:

$$Y_i = f(\overset{+}{X_i}) = \beta_0 + \beta_1 X_i + \epsilon_i \tag{1.20}$$

where: Y_i = weight in pounds of the *i*th customer
X_i = height in inches (above 5 feet) of the *i*th customer
ϵ_i = the stochastic error term

In this case, the sign of the theoretical relationship between height and weight is believed to be positive (signified by the positive sign above

X_i in the general theoretical equation), but you want to actually quantify that relationship to be able to estimate weights given heights. To do this, you need to collect a data set, and you need to apply regression analysis to the data.

The next day you collect the data summarized in Table 1.1 and run your regression on the Magic Flag Farm computer, obtaining the following estimates:

$$\hat{\beta}_0 = 103.40 \qquad \hat{\beta}_1 = 6.38$$

This means that the equation

$$\text{Estimated Weight} = 103.40 + 6.38 \cdot \text{Height (inches above five feet)} \tag{1.21}$$

is worth trying as an alternative to just guessing the weights of your customers. Such an equation estimates weight with a constant base of

TABLE 1.1 DATA FOR AND RESULTS OF THE WEIGHT-GUESSING EQUATION

Observation i (1)	Height Above 5′ X_i (2)	Weight Y_i (3)	Predicted Weight $\hat{Y}_i$ (4)	Residual e_i (5)	$ Gain or Loss (6)
1	5.0	140.0	135.3	4.7	+.50
2	9.0	157.0	160.8	−3.8	+.50
3	13.0	205.0	186.3	18.7	−.60
4	12.0	198.0	179.9	18.1	−.60
5	10.0	162.0	167.2	−5.2	+.50
6	11.0	174.0	173.6	0.4	+.50
7	8.0	150.0	154.4	−4.4	+.50
8	9.0	165.0	160.8	4.2	+.50
9	10.0	170.0	167.2	2.8	+.50
10	12.0	180.0	179.9	0.1	+.50
11	11.0	170.0	173.6	−3.6	+.50
12	9.0	162.0	160.8	1.2	+.50
13	10.0	165.0	167.2	−2.2	+.50
14	12.0	180.0	179.9	0.1	+.50
15	8.0	160.0	154.4	5.6	+.50
16	9.0	155.0	160.8	−5.8	+.50
17	10.0	165.0	167.2	−2.2	+.50
18	15.0	190.0	199.1	−9.1	+.50
19	13.0	185.0	186.3	−1.3	+.50
20	11.0	155.0	173.6	−18.6	−.60
				TOTAL =	$6.70

103.40 pounds and adds 6.38 pounds for every inch of height over 5 feet. Note that the sign of $\hat{\beta}_1$ is positive, as you expected.

How well does the equation work? To answer this question, you need to calculate the residuals (Y_i minus $\hat{Y}_i$) from Equation 1.21 to see how many were greater than ten. As can be seen in the last column in Table 1.1, if you had applied the equation to these 20 people you wouldn't exactly have gotten rich, but at least you would have earned \$6.70 instead of losing \$2.00. Figure 1.4 shows not only Equation 1.21 but also the weight and height data for all 20 of the customers used as the sample.

Equation 1.21 would probably help a beginning weight guesser, but it could be improved by adding other variables or by collecting a larger sample. Such an equation is realistic, though, because it's likely that every successful weight guesser uses an equation like this without consciously thinking about that concept.

Our goal with this equation was to quantify the theoretical weight/

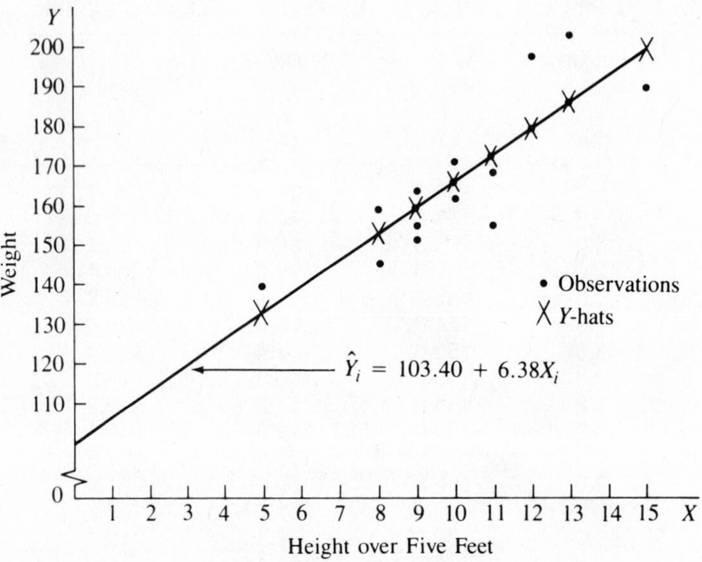

Figure 1.4 A WEIGHT-GUESSING EQUATION
If we plot the data from the weight-guessing example and include the estimated regression line, we can see that the fitted $\hat{Y}$s come fairly close to the observed Ys for all but three observations. Find a male friend's height and weight on the graph; how well does the regression equation work?

height equation, Equation 1.20, by collecting data (Table 1.1) and calculating an estimated regression (Equation 1.21). Although the true equation, like the actual stochastic error terms, can never be observed, we were able to come up with an estimated equation that had the sign we expected for $\hat{\beta}_1$ and that helped us in our job. Before you decide to quit school or your job and try to make your living guessing weights at Magic Flag Farm, there is quite a bit more to learn about regression analysis, so we had better move on.

1.4 Using Regression to Explain Housing Prices

As much fun as guessing weights at an amusement park might be, it's hardly a typical example of the use of regression analysis. For every regression run on such an off-the-wall topic, there are literally hundreds run to *describe* the reaction of GNP to an increase in the money supply, to *test* an economic theory with new data, or to *forecast* the effect of a price change on a firm's sales.

As a more realistic example, let's look at a model of housing prices. The purchase of a house is probably the most important financial decision in an individual's life, and one of the key elements in that decision is an appraisal of the house's value. If you overvalue the house, you can lose thousands of dollars by paying too much; if you undervalue the house, someone might outbid you.

All this wouldn't be much of a problem if houses were homogeneous products, like corn or gold, that have generally known market prices with which to compare a particular asking price. Such is hardly the case in the real estate market. Consequently, an important element of every housing purchase is an appraisal of the market value of the house, and many real estate appraisers use regression analysis to help them in their work.

Suppose your family is about to buy a house in Southern California, but you're convinced that the owner is asking too much money. The owner says that the asking price of $230,000 is fair because a larger house next door sold for $230,000 about a year ago. You're not sure it's reasonable to compare the prices of different-size houses that were purchased at different times. What can you do to help decide whether to pay the $230,000?

Since you're taking an econometrics class, you decide to collect data on all local houses that were sold within the last few weeks and to build a regression model of the sales prices of the houses as a function of their sizes.[9] Such a data-set is called **cross-sectional** in nature because all of the observations are from the same point in time and represent different individual economic entities (like countries or, in this case, houses). Cross-sectional models avoid having to measure the impact of elapsed time by forcing all the observations to come from the same time period.

To measure the impact of size on price, you include the size of the house as an independent variable in a regression equation that has the price of that house as the dependent variable. You expect a positive sign for the coefficient of size, since big houses cost more to build and tend to be more desirable than small ones. Thus the model is:

$$P_i = f(\overset{+}{S_i}) = \beta_0 + \beta_1 S_i + \epsilon_i \tag{1.22}$$

where: P_i = the price (in thousands of \$) of the *i*th house
S_i = the size (in square feet) of that house
ϵ_i = a stochastic error term

You collect the records of all recent real estate transactions, find that 43 local houses were sold within the last 4 weeks, and estimate the following regression on those 43 observations:

$$\hat{P}_i = 40.1 + 0.138S_i \tag{1.23}$$

What do these estimated coefficients mean? The most important coefficient is $\hat{\beta}_1 = 0.138$, since the reason for the regression is to find out the impact of size on price. This coefficient means that if size increases by 1 square foot, price will increase by 0.138 thousand dollars (\$138). $\hat{\beta}_1$ thus measures the change in P_i associated with a one-unit change in S_i. It's the slope of the regression line in a graph like Figure 1.5.

What does $\hat{\beta}_0 = 40.1$ mean? $\hat{\beta}_0$ is the estimate of the constant or intercept term. In our equation, it means that price equals 40.1 when size equals zero. As can be seen in Figure 1.5, the estimated regression line intersects the price axis at 40.1. While it might be tempting to say that the average price of a vacant lot is \$40,100, such a conclusion

9. It's unusual for an economist to build a model of price without including some measure of quantity on the right-hand side. Such models of the price of a good as a function of the attributes of that good are called *hedonic* models and will be discussed in greater depth in Section 11.7. The interested reader is encouraged to skim the first few paragraphs of that section before continuing on with this example.

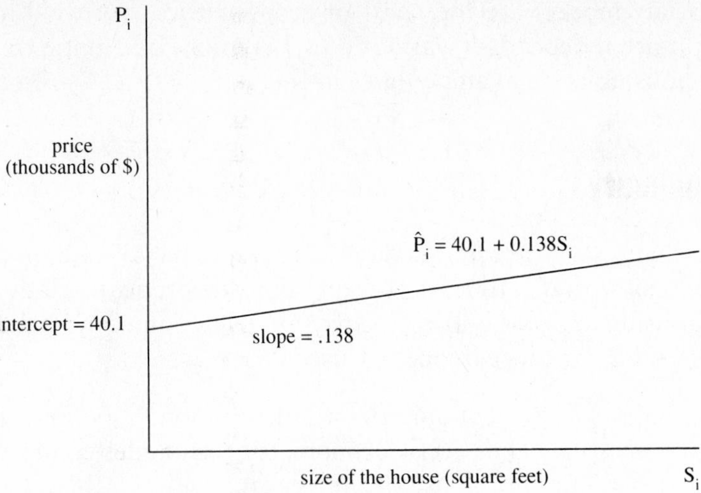

Figure 1.5 A CROSS-SECTIONAL MODEL OF HOUSING PRICES
A regression equation that has the price of a house in Southern California as a function of the size of that house has an intercept of 40.1 and a slope of 0.138, using Equation 1.23.

would be unjustified for a number of reasons, which will be discussed in later chapters. It's much safer to interpret $\hat{\beta}_0 = 40.1$ as nothing more than the value of the estimated regression when $S_i = 0$ or to not interpret $\hat{\beta}_0$ at all.

How can you use this estimated regression to help decide whether to pay $230,000 for the house? If you calculate a $\hat{Y}$ (predicted price) for a house that is the same size (1,600 square feet) as the one you're thinking of buying, you can then compare this $\hat{Y}$ with the asking price of $230,000. To do this, substitute 1600 for S_i in Equation 1.23, obtaining:

$$\hat{P}_i = 40.1 + 0.138(1600) = 40.1 + 220.8 = 260.9$$

The house seems to be a good deal. The owner is asking "only" $230,000 for a house when the size implies a price of $260,900! Perhaps your original feeling that the price was too high was a reaction to the steep housing prices in Southern California in general and not a reflection of this specific price.

On the other hand, perhaps the price of a house is influenced by more than just the size of the house. (After all, what good's a house in Southern California unless it has a pool or air conditioning?) Such multivariate models are the heart of econometrics, but we'll hold off adding more independent variables to Equation 1.23 until we return to this housing price example later in the text.

1.5 Summary

1. Econometrics, literally "economic measurement," is a branch of economics that attempts to quantify theoretical relationships. Regression analysis is only one technique used in econometrics, but it is by far the most frequently used.

2. The major uses of econometrics are description, hypothesis testing, and forecasting. The precise econometric techniques employed may vary depending on the use of the research.

3. While regression analysis specifies that a dependent variable is a function of one or more independent variables, regression analysis alone cannot prove or even imply causality.

4. Linear regression can only be applied to equations that are *linear in the coefficients,* which means that the regression coefficients are in their simplest possible form. For an equation with two explanatory variables, this form would be:

$$f(Y_i) = \beta_0 + \beta_1 f(X_{1i}) + \beta_2 f(X_{2i}) + \epsilon_i$$

5. A stochastic error term must be added to all regression equations to account for variations in the dependent variable that are not explained completely by the independent variables. The components of (and reasons for) this error term include:
   a. omitted or left-out variables
   b. measurement errors in the data
   c. an underlying theoretical equation that has a different functional form (shape) than the regression equation
   d. purely random and unpredictable events.

6. An estimated regression equation is an approximation of the true equation that is obtained by using data from a sample of actual Ys and Xs. Since we can never know the true equation, econometric analysis focuses on this estimated regression equation and the estimates of the regression coefficients. The difference between a

particular observation of the dependent variable and the value estimated from the regression equation is called the residual.

Exercises

(Answers to even-numbered exercises are in Appendix A.)

1. Write the meaning of each of the following terms without reference to the book (or your notes), and compare your definition with the version in the text for each:
   a. stochastic error term
   b. regression analysis
   c. linear in the variables
   d. slope coefficient
   e. multivariate regression model
   f. expected value
   g. residual

2. Decide whether you would expect relationships between the following pairs of dependent and independent variables (respectively) to be positive, negative, or ambiguous. Explain your reasoning.
   a. Aggregate net investment in the U.S. in a given year and GNP in that year.
   b. The amount of hair on the head of a male professor and the age of that professor.
   c. The number of acres of wheat planted in a season and the price of wheat at the beginning of that season.
   d. Aggregate net investment and the real rate of interest in the same year and country.
   e. The growth rate of GNP in a year and the average hair length in that year.
   f. The quantity of canned heat demanded and the price of a can of heat.

3. Review the four components of (reasons for) a stochastic error term.
   a. Can you think of other possible reasons for adding ϵ to the equation? What would they be? Why do you think they were omitted from the discussion in Section 1.2.3?
   b. If you were to run a regression on each of the pairs of dependent and independent variables in Question 2 above, for which regressions would an error term not be necessary? Why?

4. Look over the following equations and decide whether they are linear in the variables, linear in the coefficients, both, or neither.
   a. $Y_i = \beta_0 + \beta_1 X_i^3 + \epsilon_i$
   b. $Y_i = \beta_0 + \beta_1 \log X_i + \epsilon_i$
   c. $\log Y_i = \beta_0 + \beta_1 \log X_i + \epsilon_i$
   d. $Y_i = \beta_0 + \beta_1 X_i^{\beta_2} + \epsilon_i$
   e. $Y_i^{\beta_0} = \beta_1 + \beta_2 X_i^2 + \epsilon_i$
   f. $Y_i = 1 + \beta_0(1 - X_i^{\beta_1}) + \epsilon_i$
   g. $Y_i = \beta_0 + \beta_1 X_i/10 + \epsilon_i$

5. Your friend estimates a simple equation of bond prices in different years as a function of the interest rate that year (for equal levels of risk) and obtains:

$$\hat{Y}_i = 101.40 - 4.78X_i$$

where: Y_i = U.S. Government bond prices (per \$100 bond) in the *i*th year

X_i = the Federal funds rate (percent) in the *i*th year

   a. Carefully explain the meanings of the two estimated coefficients. Are the estimated signs what you would have expected?
   b. Why is the left-hand variable in your friend's equation $\hat{Y}$ and not Y?
   c. Didn't your friend forget the stochastic error term in the estimated equation?
   d. What is the economic meaning of this equation? What criticisms would you have of this model? (Hint: The Federal funds rate is a rate that applies to overnight holdings in banks.)

6. Recall the weight-guessing regression in Section 1.3:
   a. Go back to the data set and identify the three customers who seem to be quite a distance from the estimated regression line. Would we have a better regression equation if we dropped these customers from the sample?
   b. Measure the height of a male friend and plug it into Equation 1.21. Does the equation come within ten pounds? If not, do you think you see why? Why does the estimated equation predict the same weight for all males of the same height when it is obvious that all males of the same height don't weigh the same?

c. Look over the sample with the thought that it might not be randomly drawn. Does the sample look abnormal in any way? (Hint: Are the customers who choose to play such a game a random sample?) If the sample isn't random, would this have an effect on the regression results and the estimated weights?
d. Think of at least one other factor besides height that might be a good choice as a variable in the weight/height equation. How would you go about obtaining the data for this variable? What would the expected sign of your variable's coefficient be if the variable were added to the equation?

7. Continuing with the height/weight example, suppose you collected data on the heights and weights of 29 more customers and estimated the following equation:

$$\hat{Y}_i = 125.1 + 4.03X_i \tag{1.24}$$

where: Y_i = the weight (in pounds) of the *i*th person
X_i = the height (in inches over five feet) of the *i*th person

a. Why aren't the coefficients in Equation 1.24 the same as those we estimated previously (Equation 1.21)?
b. Compare the estimated coefficients of Equation 1.24 with those in Equation 1.21. Which equation has the steeper estimated relationship between height and weight? Which equation has the higher intercept? At what point do the two intersect?
c. Use Equation 1.24 to "predict" the 20 original weights given the heights in Table 1.1. How many weights does Equation 1.24 misguess by more than ten pounds? Does Equation 1.24 do better or worse than Equation 1.21? Could you have predicted this result beforehand?
d. Suppose you had one last day on the weight-guessing job. What equation would you use to guess weights? (Hint: There are more than two possible answers.)

8. Not all regression coefficients have positive expected signs. For example, a recent *Sports Illustrated* article by Jaime Diaz reported on a study of golfing putts of various lengths on the Professional Golfers Association (PGA) Tour.[10] The article included data on the percentage of putts made (P_i) as a function of the length of the putt

10. Jaime Diaz, "Perils of Putting," *Sports Illustrated*, April 3, 1989, pp. 76-79.

in feet (L_i). Since the longer the putt, the less likely even a professional is to make it, we'd expect L_i to have a negative coefficient in an equation explaining P_i. Sure enough, if you estimate an equation on the data in the article, you obtain:

$$\bar{P}_i = f(L_i) = 83.6 - 4.1L_i + e_i \qquad (1.25)$$

a. Carefully write out the exact meaning of the coefficient of L_i.
b. What percent of the time would you expect a PGA golfer to make a ten-foot putt? Does this seem realistic? How about a 1-foot putt or a 25-foot putt? Do these seem as realistic?
c. Your answer to part b should suggest that there's a problem in applying a linear regression to this data. What is that problem? (Hint: If you're stuck, first draw the theoretical diagram you'd expect for P_i as a function of L_i, then plot Equation 1.25 onto the same diagram.)
d. Suppose someone else took the data from the article and estimated the following equation:

$$\hat{P}_i = 83.6 - 4.1L_i$$

Is this the same result as that in Equation 1.25? If so, what definition do you need to use to convert this equation back to Equation 1.25?

9. Return to the housing price model of Section 1.4 and consider the following equation:

$$\hat{S}_i = 72.3 + 5.77P_i$$

where: S_i = the size (in square feet) of the *i*th house
P_i = the price (in thousands of \$) of that house

a. Carefully explain the meaning of each of the estimated regression coefficients.
b. Suppose you're told that this equation explains a significant portion (more than 80 percent) of the variation in the size of a house. Have we shown that high housing prices cause houses to be large? If not, what have we shown?
c. What do you think would happen to the estimated coefficients of this equation if we had measured the price variable in dollars instead of in thousands of dollars? Be specific.

10. Housing price models can be time-series in nature as well as cross-sectional. If you study aggregate time-series housing prices (see Table 1.2 for data and sources), you have:

$$\hat{P}_t = f(\overset{+}{GNP}) = 4363.1 + 21.4Y_t$$
$$n = 26 \qquad \text{(annual 1964-89)}$$

where: P_t = the nominal median price of new single family houses in the U.S. in year t

Y_t = U.S. GNP in year t (billions of current \$)

TABLE 1.2 DATA FOR THE TIME-SERIES MODEL OF HOUSING PRICES.

t	Year	Price (P_t)	GNP(Y_t)
1	1964	18,900.	649.8
2	1965	20,000.	705.1
3	1966	21,400.	772.0
4	1967	22,700.	816.4
5	1968	24,700.	892.7
6	1969	25,600.	963.9
7	1970	23,400.	1015.5
8	1971	25,200.	1102.7
9	1972	27,600.	1212.8
10	1973	32,500.	1359.3
11	1974	35,900.	1472.8
12	1975	39,300.	1598.4
13	1976	44,200.	1782.8
14	1977	48,800.	1990.5
15	1978	55,700.	2249.7
16	1979	62,900.	2508.2
17	1980	64,600.	2732.0
18	1981	68,900.	3052.6
19	1982	69,300.	3166.0
20	1983	75,300.	3405.7
21	1984	79,900.	3772.2
22	1985	84,300.	4014.9
23	1986	92,000.	4231.6
24	1987	104,500.	4524.3
25	1988	112,500.	4880.6
26	1989	118,900.	5233.2

P_t = the nominal median price of new single family houses in the U.S. in year t. (Source: *The Statistical Abstract of the U.S.*)

Y_t = U.S. GNP in year t (billions of current dollars). (Source: *The Economic Report of the President*)

a. Carefully interpret the economic meaning of the estimated coefficients.
b. What is Y_t doing on the right side of the equation? Shouldn't it be on the left side?
c. Both the price and GNP variables are measured in nominal (or current, as opposed to real, or inflation-adjusted) dollars. Thus a major portion of the excellent explanatory power of this equation (more than 99 percent of the variation in P_t can be explained by Y_t alone) comes from capturing the huge amount of inflation that took place between 1964 and 1988. What could you do to eliminate the impact of inflation in this equation?
d. GNP is included in the equation to measure more than just inflation. What factors in housing prices other than inflation does the GNP variable help capture? Can you think of a variable that might do a better job?

11. The distinction between the stochastic error term and the residual is one of the most difficult concepts to master in this chapter.
a. List at least four differences between the error term and the residual.
b. Usually, we can never observe the error term, but we can get around this difficulty if we assume values for the true coefficients. Calculate error terms and residuals for each of the following six observations given that the true β_0 equals 0.0, the true β_1 equals 1.5, and the estimated regression equation is $\hat{Y}_i = 0.48 + 1.32X_i$:

Y_i	2	6	3	8	5	4
X_i	1	4	2	5	3	4

2

Ordinary Least Squares

The bread and butter of regression analysis is the estimation of the coefficients of econometric models with a technique called Ordinary Least Squares (OLS). The first two sections of this chapter summarize the reasoning behind and the mechanics of OLS. Regression users usually rely on computers to do the actual OLS calculations, so the emphasis here is on understanding what OLS attempts to do and how it goes about doing it.

How can you tell a good equation from a bad one once it has been estimated? One factor is the extent to which the estimated equation fits the actual data. The rest of the chapter is devoted to developing an understanding of the most commonly used measures of this fit, R^2, the adjusted R^2 ($\bar{R}^2$, called R-Bar-Squared), and r, the simple correlation coefficient. The use of $\bar{R}^2$ is not without perils, however, and the chapter concludes with an example of the misuse of this statistic.

2.1 Estimating Two-Variable Models with OLS

Once the variables have been specified and the data collected, the next step in regression analysis is to estimate the coefficients of the equation.

In a sense, this estimation is what regression analysis is all about. The purpose of regression analysis is to take a purely theoretical equation like:

$$Y_i = \beta_0 + \beta_1 X_i + \epsilon_i \qquad (2.1)$$

and use a set of data to create an estimated equation like:

$$\hat{Y}_i = \hat{\beta}_0 + \hat{\beta}_1 X_i \qquad (2.2)$$

where each "hat" indicates a sample estimate of the true population value. (In the case of Y, the "true population value" is E[Y|X].) The purpose of the estimation technique is to obtain numerical values of the coefficients of an otherwise completely theoretical regression equation.

The most widely used method of obtaining these estimates is Ordinary Least Squares (OLS). OLS has become so standard that its estimates are presented as a point of reference even when results from other estimation techniques are used. **Ordinary Least Squares** is a regression estimation technique that calculates the $\hat{\beta}$s so as to minimize the sum of the squared residuals, thus:[1]

$$\text{OLS minimizes } \sum_{i=1}^{n} e_i^2 \qquad (i = 1,2,\ldots,n) \qquad (2.3)$$

Since these residuals (e_is) are the differences between the actual Ys and the estimated Ys produced by the regression (the $\hat{Y}$s in Equation 2.2), it's equivalent to saying that OLS minimizes $\Sigma(Y_i - \hat{Y}_i)^2$.

2.1.1 Why Use Ordinary Least Squares?

While OLS is the most-used regression estimation technique, it's not the only one. Indeed, econometricians have invented what seems like

1. The summation symbol, Σ, means that all terms to its right should be added (or summed) over the range of the *i* values attached to the bottom and top of the symbol. In Equation 2.3, for example, this would mean adding up e_i^2 for all integer values between 1 and n:

$$\sum_{i=1}^{n} e_i^2 = e_1^2 + e_2^2 + \cdots + e_n^2$$

Often the Σ notation is simply written as $\sum_i$ as in Equation 2.5, and it is assumed that the summation is over all observations from i = 1 to i = n. Sometimes, the i is omitted entirely, as in Equation 2.8, and the same assumption is made implicitly.

zillions of different estimation techniques, a number of which we'll discuss later in this text.

There are at least three important reasons for using OLS to estimate regression models:

1. OLS is relatively easy to use.
2. The goal of minimizing Σe_i^2 is quite appropriate from a theoretical point of view.
3. OLS estimates have a number of useful properties.

The first reason for using OLS is that it's the simplest of all econometric estimation techniques. Most other techniques involve complicated nonlinear formulas or iterative procedures, many of which are extensions of OLS itself. In contrast, OLS estimates are simple enough that, if you had to, you could compute them without using a computer or a calculator.

The second reason for using OLS is that minimizing the summed squared residuals is an appropriate theoretical goal for an estimation technique. To see this, recall that the residual measures how close the regression equation comes to the actual observed data:

$$e_i = Y_i - \hat{Y}_i \qquad (i = 1,2,...,n) \tag{1.16}$$

Since it's reasonable to want our estimated regression equation to be as close as possible to the observed data, it's reasonable to want to minimize these residuals. The main problem with simply totaling the residuals and choosing that set of $\hat{\beta}$s that minimizes them is that e_i can be negative as well as positive. Thus, negative and positive residuals might cancel each other out, allowing a wildly inaccurate equation to have a very low Σe_i. For example, if $Y = 100{,}000$ for two consecutive observations and if your equation predicts 1.1 million and $-900{,}000$, respectively, your residuals will be $+1$ million and -1 million, which add up to zero!

We could get around this problem by minimizing the sum of the absolute values of the residuals, but this approach has problems as well. Absolute values are difficult to work with mathematically, since some mathematical operations provide ambiguous answers when applied to absolute numbers. In addition, summing the absolute values of the residuals gives no extra weight to extraordinarily large residuals. That is, it often doesn't matter if a number of estimates are off by a small amount, but it is important if one estimate is off by a huge amount. For example, recall the weight-guessing equation of Chapter

1; you lost only if you missed the customer's weight by ten or more pounds. In such a circumstance, you'd want to avoid large residuals.

Minimizing the summed squared residuals gets around these problems. Squared functions pose no unusual mathematical difficulties in terms of manipulations, and the technique avoids canceling positive and negative residuals because squared terms are always positive. In addition, squaring gives greater weight to big residuals than it does to smaller ones because e_i^2 gets relatively larger as e_i increases. For example, one residual equal to 4.0 has a greater weight than two residuals of 2.0 when the residuals are squared ($4^2 = 16$ vs. $2^2 + 2^2 = 8$).

The final reason for using OLS is that its estimates have at least three desirable characteristics:

1. The estimated regression line (Equation 2.2) goes through the means of Y and X. That is, if you substitute $\overline{Y}$ and $\overline{X}$ into Equation 2.2, the equation holds *exactly.*
2. The sum of the residuals is exactly zero.
3. OLS can be shown to be the "best" estimator possible under a set of fairly restrictive assumptions.

While we won't discuss the third property until Chapter 4, we'll provide examples of the first two of these properties later in this chapter.

2.1.2 How Does OLS Work?

How would OLS estimate a two-variable regression model like Equation 2.1?

$$Y_i = \beta_0 + \beta_1 X_i + \epsilon_i \tag{2.1}$$

OLS selects those estimates of β_0 and β_1 that minimize the squared residuals, summed over all the sample data points:

$$\sum_{i=1}^{n} e_i^2 = \sum_{i=1}^{n} (Y_i - \hat{Y}_i)^2 \qquad (i = 1,2,\ldots,n) \tag{2.4}$$

However, $\hat{Y}_i = \hat{\beta}_0 + \hat{\beta}_1 X_{1i}$, so OLS actually minimizes

$$\sum_i e_i^2 = \sum_i (Y_i - \hat{\beta}_0 - \hat{\beta}_1 X_i)^2 \tag{2.5}$$

by choosing the $\hat{\beta}$s that do so. In other words, OLS yields the $\hat{\beta}$s that minimize Equation 2.5. For an equation with just one independent variable, these coefficients are:

$$\hat{\beta}_1 = \frac{\sum_{i=1}^{n} [(X_i - \overline{X}) \cdot (Y_i - \overline{Y})]}{\sum_{i=1}^{n} (X_i - \overline{X})^2} \tag{2.6}$$

and, given this estimate of β_1,

$$\hat{\beta}_0 = \overline{Y} - \hat{\beta}_1\overline{X} \tag{2.7}$$

where $\overline{X}$ = the mean of X or $\Sigma X/n$ and $\overline{Y}$ = the mean of Y or $\Sigma Y/n$.

What do these equations mean? Equation 2.6 sets $\hat{\beta}_1$ equal to the joint variation of X and Y (around their means) divided by the variation of X around its mean. It thus measures the portion of the variation in Y that is associated with variations in X. Equation 2.7 defines $\hat{\beta}_0$ to ensure that the regression equation does indeed pass through the means of X and Y. In addition, it can be shown that Equations 2.6 and 2.7 provide $\hat{\beta}$s that minimize the summed squared residuals.[2]

For each different data set, we'll get different estimates of β_1 and β_0. So it's of interest to look at how these estimated coefficients vary depending on the sample. One measure of this variability is the **standard error of the estimated coefficients,** which is an estimate of the square root of the variance of the distribution of the $\hat{\beta}$s. The larger the standard error, the more the $\hat{\beta}$ estimates will vary. OLS provides estimates of these standard errors; the standard error of the slope[3] coefficient, $SE(\hat{\beta}_1)$, is:

$$SE(\hat{\beta}_1) = \frac{SEE}{\sqrt{\Sigma(X_i - \overline{X})^2}} \tag{2.8}$$

SEE is the *standard error of the estimate,* a measure of the quality of fit of the overall equation that is useful in its own right:

$$SEE = \sqrt{\frac{\Sigma e_i^2}{n - 2}} \tag{2.9}$$

2. For those with a moderate grasp of calculus and algebra, this proof is informative. First, differentiate Equation 2.5 with respect to $\hat{\beta}_0$ and $\hat{\beta}_1$. Second, set these two derivatives equal to zero. Third, simultaneously solve the two resulting equations (called "normal equations") for $\hat{\beta}_0$ and $\hat{\beta}_1$. The results are the OLS estimators in Equations 2.6 and 2.7. This same general minimization procedure is applied in multivariate models as well.

3. The standard error of $\hat{\beta}_0$ is calculated automatically by most computer regression programs, but we don't include an equation for $SE(\hat{\beta}_0)$ because we discourage hypothesis tests of $\hat{\beta}_0$ (as will be discussed in future chapters).

While we won't use the standard errors of the coefficients until Chapter 5, they are extremely important in hypothesis testing and model specification.

2.1.3 Total, Explained, and Residual Sums of Squares

Before going on, let's pause to develop some measures of how much of the variation of the dependent variable is explained by the estimated regression equation. A comparison of the estimated values with the actual values can help the researcher get a feeling for the adequacy of the hypothesized regression model.

Various statistical measures can be used to assess the degree to which the $\hat{Y}$s approximate the corresponding sample Ys, but all of them are based on the degree to which the regression equation estimated by OLS explains the values of Y better than a naive estimator, the sample mean, denoted by $\bar{Y}$. That is, econometricians use the squared variations of Y around its mean as a measure of the amount of variation to be explained by the regression. This computed quantity is usually called the **total sum of squares,** or TSS, and is written as:

$$\text{TSS} = \sum_{i=1}^{n}(Y_i - \bar{Y})^2 \qquad (2.10)$$

If we rewrite the equation for the residual (Equation 1.16) as $Y_i = \hat{Y}_i + e_i$, we can see that Y is equal to the estimated value of Y plus the residual (because the residual is defined as the difference between Y and $\hat{Y}$). Thus Equation 2.10 can be rewritten as:

$$\text{TSS} = \sum_i (Y_i - \bar{Y})^2 = \sum_i [(\hat{Y}_i - \bar{Y}) + e_i]^2 \qquad (2.11)$$

$$= \sum_i (\hat{Y}_i - \bar{Y})^2 + \sum_i e_i^2 + 2\sum_i (\hat{Y}_i - \bar{Y})e_i$$

For the estimation techniques dealt with in this text, it can be shown that the last term of Equation 2.11 equals zero. Thus, a few manipulations have allowed us to decompose the total sum of squares into two components, that which can be explained by the regression and that which cannot:

$$\sum_i (Y_i - \bar{Y})^2 = \sum_i (\hat{Y}_i - \bar{Y})^2 + \sum_i e^2 \qquad (2.12)$$

Total Sum of Squares (TSS)	=	Explained Sum of Squares (ESS)	+	Residual Sum of Squares (RSS)

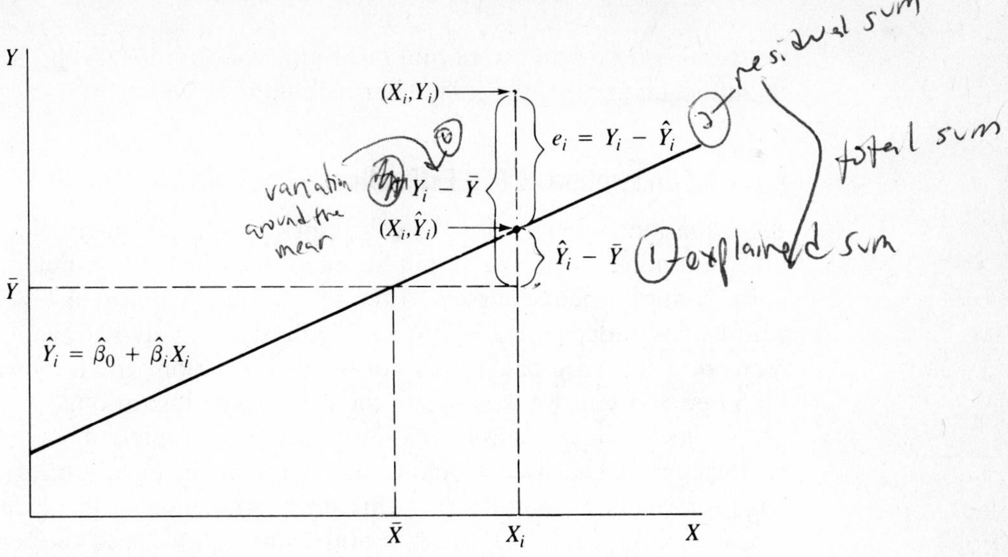

Figure 2.1 DECOMPOSITION OF THE VARIANCE IN Y
The variation of Y around its mean $(Y_i - \bar{Y})$ can be decomposed into two parts: 1. $(\hat{Y}_i - \bar{Y})$, the difference between the estimated value of Y $(\hat{Y})$ and the mean value of Y $(\bar{Y})$, and 2. $(Y_i - \hat{Y}_i)$, the difference between the actual value of Y and the estimated value of Y.

This decomposition is usually called the "decomposition of variance" or the decomposition of the squared deviations of Y_i from its mean.

Figure 2.1 illustrates the decomposition of variance for the simple regression model. All estimated values of Y_i lie on the estimated regression line $\hat{Y}_i = \hat{\beta}_0 + \hat{\beta}_1 X_i$. The sample mean of Y is denoted by the horizontal line at $\bar{Y}$. The total deviation of the actual value of Y_i from its sample mean value is decomposed into two components, the deviation of $\hat{Y}_i$ from the mean and the deviation of the actual value of Y_i from the fitted value $\hat{Y}_i$. Thus, the first component of Equation 2.12 measures the amount of the squared deviation of Y_i from its mean that is explained by the regression line. In Figure 2.1 the fitted value $\hat{Y}_i$ lies closer to the value of Y_i than does $\bar{Y}$, thus explaining in a purely empirical sense a portion of the squared deviation of Y_i from its mean. This component of the total sum of the squared deviations, called the **explained sum of squares,** or ESS, is attributable to the fitted regression line.

The ESS is the explained portion of the TSS. The unexplained portion, that is, unexplained in an empirical sense by the estimated regression equation, is called the **residual sum of squares,** or RSS.

We can see from Equation 2.12 that the smaller the RSS is relative to the TSS, the better the estimated regression line appears to fit the data. Thus, given the TSS, which no estimating technique can alter, researchers desire an estimating technique that minimizes the RSS and therefore maximizes the ESS. That technique is OLS.

2.1.4 An Illustration of OLS Estimation

The equations specified for the calculation of regression coefficients by OLS computer programs might seem a little forbidding, but it is not hard to apply them yourself to data-sets that have only a few observations and independent variables. While you usually will want to use regression software packages to do your estimation, you'll understand OLS better if you work through the following illustration.

To keep things simple, let's attempt to estimate the regression coefficients of the height and weight data given in Section 1.3. For your convenience in following this illustration, the original data are reproduced in Table 2.1. As was noted in Section 2.1.2, the formulas

TABLE 2.1 THE CALCULATION OF ESTIMATED REGRESSION COEFFICIENTS FOR THE WEIGHT/HEIGHT EXAMPLE

Raw Data			Required Intermediate Calculations					
i (1)	Y_i (2)	X_i (3)	$(Y_i-\bar{Y})$ (4)	$(X_i-\bar{X})$ (5)	$(X_i-\bar{X})^2$ (6)	$(X_i-\bar{X})\cdot(Y_i-\bar{Y})$ (7)	$\hat{Y}_i$ (8)	$e_i=Y_i-\hat{Y}_i$ (9)
1	140	5	−29.40	−5.35	28.62	157.29	135.3	4.7
2	157	9	−12.40	−1.35	1.82	16.74	160.8	−3.8
3	205	13	35.60	2.65	7.02	94.34	186.3	18.7
4	198	12	28.60	1.65	2.72	47.19	179.9	18.1
5	162	10	−7.40	−0.35	0.12	2.59	167.2	−5.2
6	174	11	4.60	0.65	0.42	2.99	173.6	0.4
7	150	8	−19.40	−2.35	5.52	45.59	154.4	−4.4
8	165	9	−4.40	−1.35	1.82	5.94	160.8	4.2
9	170	10	0.60	−0.35	0.12	−0.21	167.2	2.8
10	180	12	10.60	1.65	2.72	17.49	179.9	0.1
11	170	11	0.60	0.65	0.42	0.39	173.6	−3.6
12	162	9	−7.40	−1.35	1.82	9.99	160.8	1.2
13	165	10	−4.40	−0.35	0.12	1.54	167.2	2.8
14	180	12	10.60	1.65	2.72	17.49	179.9	0.1
15	160	8	−9.40	−2.35	5.52	22.09	154.4	5.6
16	155	9	−14.40	−1.35	1.82	19.44	160.8	−5.8
17	165	10	−4.40	−0.35	0.12	1.54	167.2	−2.2
18	190	15	20.60	4.65	21.62	95.79	199.1	−9.1
19	185	13	15.60	2.65	7.02	41.34	186.3	−1.3
20	155	11	−14.40	0.65	0.42	−9.36	173.6	−18.6
sum	3388	207	0.0	0.0	92.50	590.20	3388.3	−0.3
mean	169.4	10.35	0.0	0.0			169.4	0.0

for OLS estimation for a regression equation with one independent variable are Equations 2.6 and 2.7:

$$\hat{\beta}_1 = \frac{\sum_{i=1}^{n}[(X_i - \overline{X}) \cdot (Y_i - \overline{Y})]}{\sum_{i=1}^{n}(X_i - \overline{X})^2} \tag{2.6}$$

$$\hat{\beta}_0 = \overline{Y} - \hat{\beta}_1\overline{X} \tag{2.7}$$

If we undertake the calculations outlined in Table 2.1 and substitute them into Equations 2.6 and 2.7, we obtain these values:

$$\hat{\beta}_1 = \frac{590.20}{92.50} = 6.38$$

$$\hat{\beta}_0 = 169.4 - (6.38 \cdot 10.35) = 103.4$$

If you compare these estimates, you'll find that the manually calculated coefficient estimates are the same as the computer regression results summarized in Section 1.3.

Table 2.1 can also be used to exemplify some of the properties of OLS estimates. For instance, the sum of the $\hat{Y}$s (column 8) equals the sum of the Ys (column 2), so the sum of the residuals (column 9) does indeed equal zero (except for rounding errors) as predicted. Another property of OLS estimates, that the estimated regression line goes through the means of Y and X, can be shown by substituting $\overline{Y}$ and $\overline{X}$ from Table 2.1 into the estimated regression equation. (Of course, this is hardly a surprise, since OLS calculates $\hat{\beta}_0$ so as to ensure that this property holds.)

The figures in Table 2.1 can also be used to derive the total sum of squares (TSS), the explained sum of squares (ESS), and the residual sum of squares (RSS). The TSS equals $\Sigma(Y_i - \overline{Y})^2$, or the sum of the squares of the values in column four, which equals 5,065. The ESS equals $\Sigma(\hat{Y}_i - \overline{Y})^2$, or the sum of the squared differences between the values in column eight and $\overline{Y}$, which equals 3,765. The RSS, Σe_i^2, is the sum of the squares of the values in column nine, which equals 1,305. Note that TSS = ESS + RSS except for rounding errors.[4] For practice in the use of these concepts, see Exercise 4.

4. If there is no constant term in the equation, TSS will not necessarily equal ESS + RSS, nor will Σe necessarily equal zero. This topic is covered in more detail in Section 7.4. Also, note that some authors reverse the meanings of TSS, RSS, and ESS (defining ESS as Σe^2), and other authors reverse the order of the letters, as in SSR.

2.2 Estimating Multivariate Regression Models with OLS

Let's face it, only a few dependent variables can be explained fully by a single independent variable. A person's weight, for example, is influenced by more than just that person's height. What about bone structure, percent body fat, exercise habits, or diet?

As important as additional explanatory variables might seem to the height/weight example, there's even more reason to include a variety of independent variables in economic and business applications. While the quantity demanded of a product is certainly affected by price, that's not the whole story. Advertising, aggregate income, the prices of substitutes, the influence of foreign markets, the quality of customer service, possible fads, and changing tastes all are important in real-world models. As a result, we feel that it's vital to move from two-variable regressions to *multivariate regression models,* equations with more than one independent variable.

2.2.1 The Meaning of Multivariate Regression Coefficients

The general multivariate regression model with K independent variables can be represented by Equation 1.12:

$$Y_i = \beta_0 + \beta_1 X_{1i} + \beta_2 X_{2i} + \cdots + \beta_K X_{Ki} + \epsilon_i \qquad (1.12)$$

where *i,* as before, goes from 1 to n and indicates the observation number. Thus, X_{1i} indicates the *i*th observation of independent variable X_1, while X_{2i} indicates the *i*th observation of another independent variable, X_2.

The biggest difference between a two-variable regression model and a multivariate regression model is in the interpretation of the latter's slope coefficients. These coefficients, often called *partial*[5] regression coefficients, are defined to allow a researcher to distinguish the impact of one variable on the dependent variable from that of other variables. Specifically, a **multivariate regression coefficient** indicates the change in the dependent variable associated with a one-unit increase in the independent variable in question *holding constant all the other independent variables in the equation.*

This last italicized phrase is a key to understanding multiple regres-

5. The term "partial regression coefficient" will seem especially appropriate to those readers who have taken calculus, since multivariate regression coefficients correspond to partial derivatives. Indeed, in Equation 1.12 the partial derivative of Y_i with respect to X_{1i} is β_1, etc.

sion (as multivariate regression is often called). The coefficient β_1 measures the impact on Y of a one-unit increase in X_1 holding constant X_2, X_3, . . . and X_K, but *not* holding constant any relevant variables that might have been omitted from the equation (e.g., X_{K+1}). The coefficient β_0 is the value of Y when all the Xs and the error term equal zero. As we'll learn in Section 7.4, you should always include a constant term in a regression equation, but you should not rely on estimates of β_0 for inference.

For example, consider the following annual model of the per capita demand for beef in the U.S. from 1960 to 1987:

$$\hat{\beta}_t = 37.53 - 0.88P_t + 11.9Yd_t \tag{2.13}$$

where: B_t = per capita consumption of beef in year t (in pounds per person)
P_t = price of beef in year t (in cents per pound)
Yd_t = per capita disposable income in year t (in thousands of dollars)

The estimated coefficient of income, 11.9, tells us that beef consumption will increase by 11.9 pounds per person if per capita disposable income goes up by $1,000, holding constant the price of beef. The ability to hold price constant is crucial because we'd expect such a large increase in per capita income to fuel demand, therefore pushing up prices and making it hard to distinguish the effect of the income increase from the effect of the price increase. The multivariate regression estimate allows us to focus on the impact of the income variable by holding the price variable constant.

Note, however, that the equation does not hold other possible variables (like the price of a substitute) constant because these variables are not included in Equation 2.13. Before you move on to the next section, take the time to think through the meaning of the estimated coefficient of P in Equation 2.13; do you agree that the sign and relative size fit with economic theory?

2.2.2 OLS Estimation of Multivariate Regression Models

The application of OLS to an equation with more than one independent variable is quite similar to its application to a two-variable model. To see this, let's follow the estimation of the simplest possible multivariate model, one with just two independent variables:

$$Y_i = \beta_0 + \beta_1 X_{1i} + \beta_2 X_{2i} + \epsilon_i \tag{2.14}$$

The goal of OLS is to choose those $\hat{\beta}$s that minimize the summed squared residuals, but now these residuals are from a multivariate model. For Equation 2.14, OLS would minimize:

$$\Sigma e_i^2 = \Sigma(Y_i - \hat{Y}_i)^2 = \Sigma(Y_i - \hat{\beta}_0 - \hat{\beta}_1 X_{1i} - \hat{\beta}_2 X_{2i})^2 \quad (2.15)$$

While OLS estimation of multivariate models is identical in general approach to that of two-variable models (see footnote 2), the equations themselves are more cumbersome. For Equation 2.14, the estimated coefficients are:

$$\hat{\beta}_1 = \frac{(\Sigma yx_1)(\Sigma x_2^2) - (\Sigma yx_2)(\Sigma x_1 x_2)}{(\Sigma x_1^2)(\Sigma x_2^2) - (\Sigma x_1 x_2)^2} \quad (2.16)$$

$$\hat{\beta}_2 = \frac{(\Sigma yx_2)(\Sigma x_1^2) - (\Sigma yx_1)(\Sigma x_1 x_2)}{(\Sigma x_1^2)(\Sigma x_2^2) - (\Sigma x_1 x_2)^2} \quad (2.17)$$

$$\hat{\beta}_0 = \overline{Y} - \hat{\beta}_1 \overline{X}_1 - \hat{\beta}_2 \overline{X}_2 \quad (2.18)$$

where lower case variables indicate deviations from the mean, as in $y = Y_i - \overline{Y}$; $x_1 = X_{1i} - \overline{X}_1$; and $x_2 = X_{2i} - \overline{X}_2$.

Also of interest are equations for the standard errors of the estimated slope coefficients:

$$SE(\hat{\beta}_1) = \sqrt{\frac{\Sigma e_i^2/(n-3)}{\Sigma(X_{1i} - \overline{X}_1)^2(1 - r^2_{12})}} \quad (2.19)$$

$$SE(\hat{\beta}_2) = \sqrt{\frac{\Sigma e_i^2/(n-3)}{\Sigma(X_{2i} - \overline{X}_2)^2(1 - r^2_{12})}} \quad (2.20)$$

where r^2_{12} is the square of the *simple correlation coefficient* between X_1 and X_2. We will define the simple correlation coefficient, r, in Section 2.3.3.

For the reader who is just about to throw this book away because of the complexity of the previous five equations, there's both bad and good news. The bad news is that Equations 2.16 through 2.20 are for a regression model with only two independent variables; with three or more, the situation really gets out of hand! The good news is that numerous user-friendly computer packages can calculate all of the above in less than a second of computer time. Indeed, only someone lost in time or stuck on a desert island would bother estimating a multivariate regression model without a computer. The rest of us will use SHAZAM, SAS, TSP, RATS, BIOMED, MINITAB, ECSTAT, or any of the other commercially available regression packages. The purpose of presenting these equations is to help you understand what multivariate estimation involves, not to teach you how to do it without a computer.

2.2.3 An Example of a Multivariate Regression Model

As an example of the estimation of a multivariate regression model, let's return to the beef demand equation of the previous section, Equation 2.13:

$$\hat{\beta}_t = 37.53 - 0.88P_t + 11.9Yd_t \tag{2.13}$$

where: B_t = per capita consumption of beef in year t (in pounds per person)
P_t = price of beef in year t (in cents per pound)
Yd_t = per capita disposable income in year t (in thousands of dollars)

These coefficients were calculated by a computer program using Equations 2.16–2.18 on the data in Table 2.2.

The standard errors of the estimated coefficients were calculated by the same computer program, using Equation 2.19 and 2.20:

$$SE(\hat{\beta}_P) = 0.16 \qquad SE(\hat{\beta}_{Yd}) = 1.76$$

As we'll see in future chapters, these standard errors will be useful in testing hypotheses about the $\hat{\beta}$s.

How would we go about graphing a multivariate regression result? We could use a three-dimensional diagram to graph Equation 2.13, as can be seen in Figure 2.2, but any additional variables would push us into four or more dimensions. What can we do? The answer is to draw a diagram of the dependent variable as a function of one of the independent variables *holding the other independent variable(s) constant.* In geometric terms, this means restricting the diagram to just one slice (or plane) of its actual multidimensional space.

To illustrate, look at Figures 2.3 and 2.4. These figures contain two different views of Equation 2.13. Figure 2.3 is a diagram of the effect of P on B, holding Yd constant, and Figure 2.4 shows the effect of Yd on B, holding P constant. These two figures are graphical rep-

TABLE 2.2 DATA FOR THE DEMAND FOR BEEF EXAMPLE

Year	B	Yd	P
1960	85.1	6.036	20.40
1961	87.8	6.113	20.20
1962	88.9	6.271	21.30
1963	94.5	6.378	19.90
1964	99.9	6.727	18.00
1965	99.5	7.027	19.90
1966	104.2	7.280	22.20
1967	106.5	7.513	22.30
1968	109.7	7.728	23.40
1969	110.8	7.891	26.20
1970	113.7	8.134	27.10
1971	113.0	8.322	29.00
1972	116.0	8.562	33.50
1973	108.7	9.042	42.80
1974	115.4	8.867	35.60
1975	118.9	8.944	32.20
1976	127.4	9.175	33.70
1977	123.5	9.381	34.40
1978	117.9	9.735	48.50
1979	105.4	9.829	66.10
1980	103.2	9.722	62.40
1981	104.2	9.769	58.60
1982	103.7	9.725	56.70
1983	105.7	9.930	55.50
1984	105.5	10.419	57.30
1985	106.5	10.625	53.70
1986	107.3	10.905	52.60
1987	103.3	10.970	61.10

resentations of multivariate regression coefficients, since they measure the impact on the dependent variable of a given independent variable, holding constant the other variables in the equation.

2.3 Describing the Overall Fit of the Estimated Model

Once the computer estimates of the coefficients have been obtained, it's time to evaluate the results. We usually ask at least two general questions in this evaluation process:

1. How well do the estimated coefficients correspond to the expectations developed by the researcher before the data were collected?
2. How well does the regression as a whole fit the data?

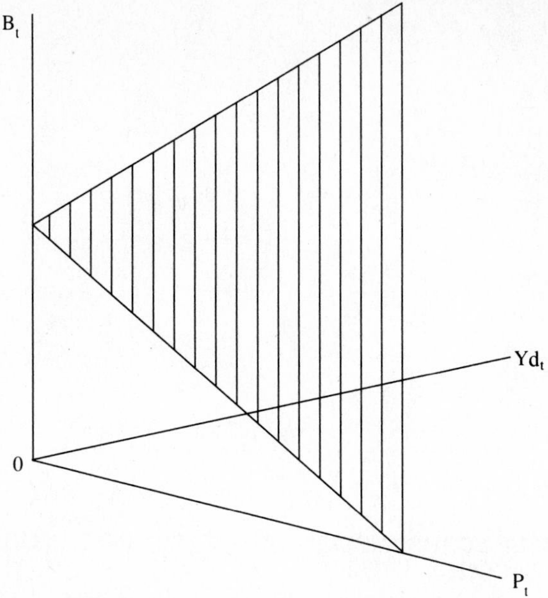

Figure 2.2 BEEF CONSUMPTION AS A FUNCTION OF PRICE AND INCOME
A three-dimensional rendering of Equation 2.13 is a plane that rises as per capita disposable income rises but falls as the price of beef rises.

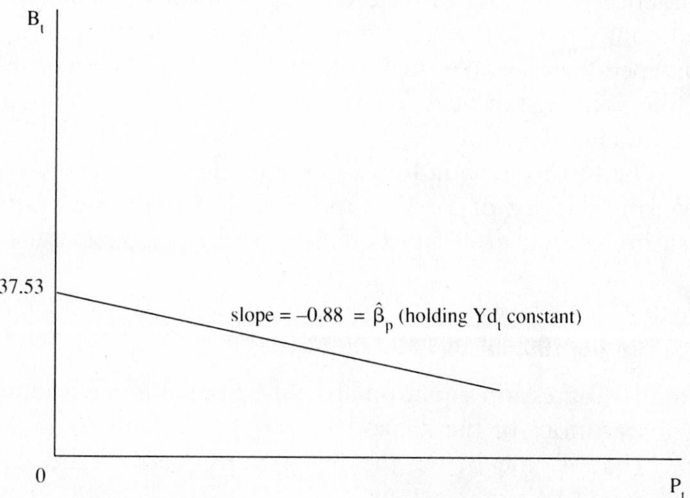

Figure 2.3 BEEF CONSUMPTION AS A FUNCTION OF THE PRICE OF BEEF
In Equation 2.13, an increase in the price of beef by a penny decreases per capita beef consumption by 0.88 pounds, holding disposable income (per capita) constant.

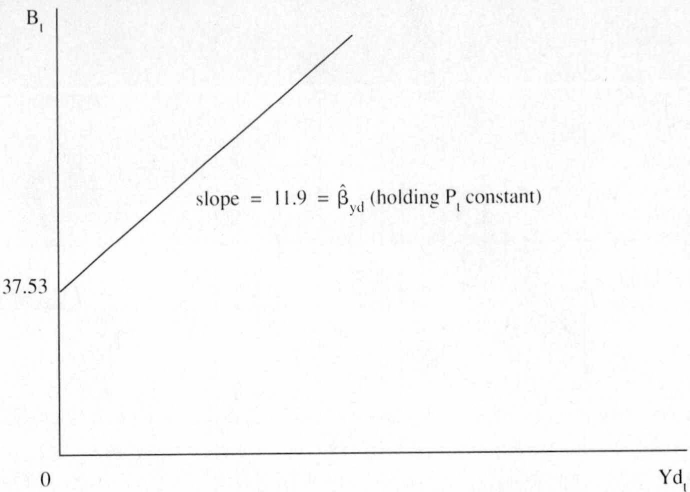

Figure 2.4 BEEF CONSUMPTION AS A FUNCTION OF PER CAPITA DISPOSABLE INCOME
In Equation 2.13, an increase in per capita disposable income of a thousand dollars increases per capita beef consumption by 11.9 pounds, holding the price of beef constant.

The purpose of this section is to deal with the second of these general topics, the overall fit of the estimated model.

In practice, a number of different regression models may be specified and compared. These may differ from one another with respect to the independent variables included, the functional form of the equation, or the data-set sampled. We can never be sure that one estimated model represents the truth any more than another, but evaluating the quality of the fit of the equation is one ingredient in a choice between different formulations of a regression model. The simplest commonly used measure of that fit is the coefficient of determination, R^2.

2.3.1 R^2, The Coefficient of Determination

An estimated regression equation should be capable of explaining the sample observations of the dependent variable Y with some degree of accuracy. That is, the better the fit of the equation, the closer the estimated Y will be to the actual Y.

The **coefficient of determination** is the ratio of the explained sum of squares to the total sum of squares:

$$R^2 = \frac{ESS}{TSS} = 1 - \frac{RSS}{TSS} = 1 - \frac{\Sigma e_i^2}{\Sigma (Y_i - \bar{Y})^2} \tag{2.21}$$

The higher R^2, the closer the estimated regression equation fits the sample data. Measures of this type are called "goodness of fit" measures. Since TSS, RSS, and ESS are all non-negative (being squared deviations), and since ESS $\leq$ TSS, R^2 must lie in the interval

$$0 \leq R^2 \leq 1 \tag{2.22}$$

A value of R^2 close to one shows a "good" overall fit, whereas a value near zero shows a failure of the estimated regression equation to explain the values of Y_i better than could be explained by the sample mean $\bar{Y}$. In words, R^2 can be defined as the percentage of the variation of Y around $\bar{Y}$ that is explained by the regression equation. Since OLS selects the parameter estimates that minimize RSS, OLS provides the largest possible R^2, given the linear specification of the model.

Figures 2.5 through 2.7 demonstrate some extremes. Figure 2.5 shows an X and Y that are unrelated. Thus, the fitted regression line might as well be $Y = \bar{Y}$, the same value it would have if X were omitted. As a result, the estimated linear regression is no better than the sample mean as an estimate of Y_i. The explained portion ESS $= 0$, and the unexplained portion RSS equals the total squared deviations TSS; thus, $R^2 = 0$. In this case, the residuals are large relative to the deviations in Y from its mean, implying that a regression line is not useful in describing the relationship between X and Y.

Figure 2.6 shows a relationship between X and Y that can be adequately "explained" by a linear regression equation: the value of R^2 is .95. This kind of result is typical of a time-series regression with a "good fit." Much of the variation has been explained, but there still remains a portion of the variation that is essentially random or unexplained by the model. "Goodness of fit" is relative to the topic being studied. If the sample is *cross-sectional,* that is, consisting of observations of many individuals, countries, or states for the same time period, an R^2 of .50 might be considered a good fit. In other words, there is no simple method of determining how high R^2 must be for the fit to be considered satisfactory. Instead, knowing when R^2 is relatively large or small is a matter of experience.

Figure 2.7 shows a perfect fit of $R^2 = 1$. Since a fit implies that no estimation is required. The relationship is completely deterministic,

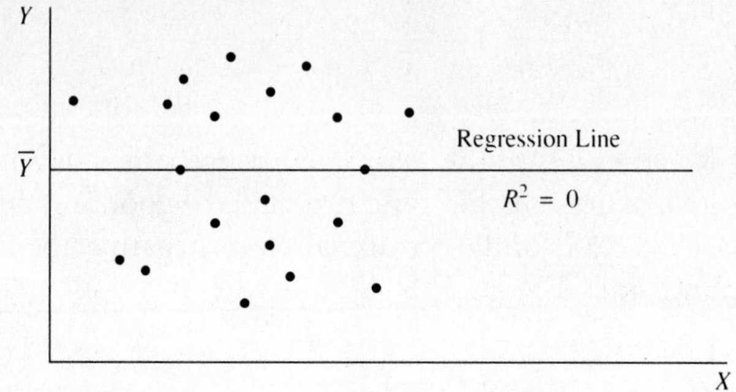

Figure 2.5
X and Y are not related; in such a case, R^2 would be 0.

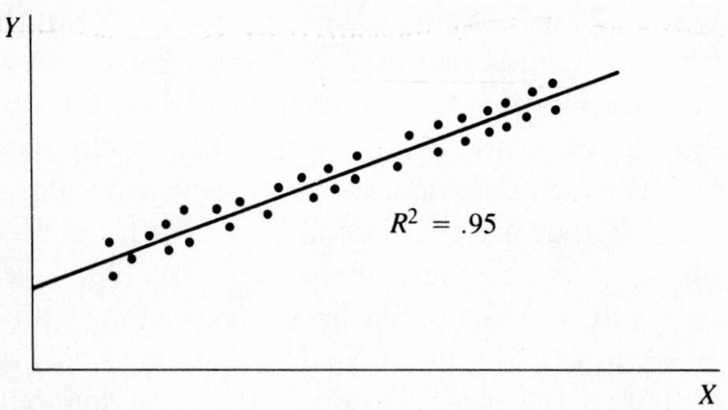

Figure 2.6
A set of data for X and Y that can be adequately explained with a regression line ($R^2 = .95$).

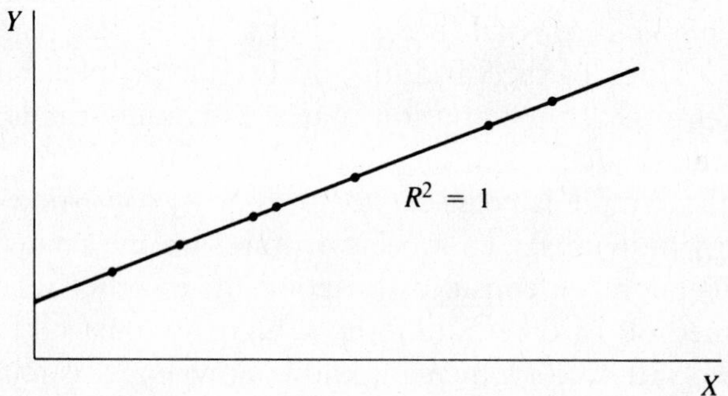

Figure 2.7
A perfect fit: all the data points are on the regression line, and the resulting R^2 is 1.

and the slope and intercept can be calculated from the coordinates of any two points. In fact, reported equations with R^2s equal to (or very near) one should be viewed with suspicion; they very likely do not explain the movements of the dependent variable Y in terms of the causal proposition advanced, even though they explain them empirically.

2.3.2 $\bar{R}^2$, The Adjusted R^2

A major problem with R^2 is that adding another independent variable to a particular equation can never decrease R^2. That is, if you compare two identical regressions (same dependent variable and independent variables), except that one has an additional independent variable, the equation with the greater number of independent variables will always have a better (or equal) fit as measured by R^2. To see this, remember the equation for R^2, Equation 2.21:

$$R^2 = 1 - \frac{RSS}{TSS} = 1 - \frac{\Sigma e_i^2}{\Sigma (Y_i - \bar{Y})^2} \tag{2.21}$$

Since the dependent variable has not changed, TSS is still the same. Also, since OLS estimating techniques ensure that adding a variable will not increase the summed squared residuals, RSS will only decrease or stay the same. If RSS decreases, RSS/TSS will also decrease, and 1 − RSS/TSS will increase. Thus adding a variable to an equation virtually guarantees that R^2 will increase.[6]

Perhaps an example will make this clear. Let's return to our weight guessing regression:

$$\text{Estimated weight} = 103.40 + 6.38 \cdot (\text{height over five feet})$$

The R^2 for this equation is .74. If we now add a completely nonsensical variable to the equation (say, the campus post office box number of each individual in question), then it turns out that the results become:

6. You know that RSS will never decrease because the OLS program could always set the coefficient of the added variable equal to zero, thus giving the same fit as the previous equation. The coefficient of the newly added variable being zero is the only circumstance in which R^2 will stay the same when a variable is added. Otherwise, R^2 will always increase when a variable is added to an equation.

Estimated weight = 102.35 + 6.36 (height > five feet) + 0.02 (box#)

but the R^2 for this equation is .75! Thus an individual using R^2 alone as the measure of the quality of the fit of the regression would choose the second version as better fitting.

The inclusion of the campus post office box variable not only adds a nonsensical variable to the equation, but it also requires the estimation of another coefficient. This lessens the **degrees of freedom,** or the excess of the number of observations (n) over the number of coefficients (including the intercept) estimated (K + 1). For instance, when the campus box number variable is added to the weight/height example, the number of observations stays constant at 20, but the number of estimated coefficients increases from 2 to 3, so the number of degrees of freedom falls from 18 to 17. This decrease has a cost, since the lower the degrees of freedom, the less reliable the estimates are likely to be. Thus the increase in the quality of the fit caused by the addition of a variable needs to be compared to the decrease in the degrees of freedom before a decision can be made with respect to the statistical impact of the added variable.

To incorporate the impact of changes in the number of independent variables, it's necessary to use $\bar{R}^2$ (pronounced R-Bar-Squared), which is R^2 adjusted for degrees of freedom:

$$\bar{R}^2 = 1 - \frac{RSS/(n-K-1)}{TSS/(n-1)} = 1 - \frac{\Sigma e_i^2/(n-K-1)}{\Sigma(Y_i - \bar{Y})^2/(n-1)} \qquad (2.23)$$

Notice that the only difference between R^2 and $\bar{R}^2$ is that the latter has been adjusted to take account of the K degrees of freedom that were lost in the calculations of the estimated slope coefficients. As a result, it's no surprise to learn that one can be expressed in terms of the other. If we substitute Equation 2.21 into Equation 2.23, it turns out that $\bar{R}^2$ can be expressed as a function of R^2:

$$\bar{R}^2 = 1 - (1 - R^2) \cdot \frac{(n-1)}{(n-K-1)} \qquad (2.24)$$

$\bar{R}^2$ will increase, decrease, or stay the same when a variable is added to an equation, depending on whether the improvement in fit caused by the addition of the new variable outweighs the loss of the degree of freedom. Indeed, the $\bar{R}^2$ for the weight guessing equation *decreases* to .72 when the mail box variable is added. The mail box variable, since it has no theoretical relation to weight, should not have been included in the equation, and the $\bar{R}^2$ measure supports this conclusion. Empirically, $\bar{R}^2$ can be used to compare the fits of equations with the same dependent variable and different numbers of independent variables. Because of this property, most researchers automatically use $\bar{R}^2$ instead of R^2 when evaluating the fit of their estimated regression equations; in fact, $\bar{R}^2$ has become so popular that it replaces R^2 in most reported regression results.

Finally, a warning is in order. Always remember that the quality of fit of an estimated equation is only one measure of the overall quality of that regression. As mentioned above, the degree to which the estimated coefficients conform to economic theory and the researcher's previous expectations about those coefficients are just as important as the fit itself. For instance, an estimated equation with a good fit but with an implausible sign for an estimated coefficient might give implausible predictions and thus not be a very useful equation. Other factors, such as theoretical relevance and usefulness, also come into play. We'll return to this issue in Section 2.4.

2.3.3 r, The Simple Correlation Coefficient

The **simple correlation coefficient, r,** is a measure of the strength and direction of the linear relationship between two variables. The simple correlation coefficient between X_1 and X_2 is:

$$r_{12} = \frac{\Sigma[(X_{1i} - \bar{X}_1)(X_{2i} - \bar{X}_2)]}{\sqrt{\Sigma(X_{1i} - \bar{X}_1)^2 \Sigma(X_{2i} - \bar{X}_2)^2}} \tag{2.25}$$

If two variables are perfectly positively correlated, then $r = +1$. To see this, assume that $X_{1i} = X_{2i}$ and substitute into Equation 2.25:

$$r_{12} = \frac{\Sigma[(X_{1i} - \bar{X}_1)(X_{1i} - \bar{X}_1)]}{\sqrt{\Sigma(X_{1i} - \bar{X}_1)^2 \Sigma(X_{1i} - \bar{X}_1)^2}} \tag{2.26}$$

$$= \frac{\Sigma[(X_{1i} - \bar{X}_1)^2]}{\Sigma[(X_{1i} - \bar{X}_1)^2]} = +1$$

If two variables are perfectly negatively correlated, then $r = -1$. To see this, substitute $X_{2i} = -X_{1i}$ into Equation 2.25:

$$r_{12} = \frac{\Sigma[(X_{1i} - \bar{X}_1)(-X_{1i} + \bar{X}_1)]}{\sqrt{\Sigma(X_{1i} - \bar{X}_1)^2\Sigma(-X_{1i} + \bar{X}_1)^2}} \tag{2.27}$$

$$= \frac{-\Sigma[(X_{1i} - \bar{X}_1)^2]}{\Sigma[(X_{1i} - \bar{X}_1)^2]} = -1$$

If two variables are totally uncorrelated, then $r = 0$.

The major econometric use of the simple correlation coefficient is as a measure of the correlation between two explanatory variables, and we'll discuss this use in Sections 5.4 and 8.3. It's interesting to note, however, that for a two-variable regression the square of the simple correlation coefficient, r^2, is identical to the coefficient of determination, R^2. Thus a simple correlation coefficient between Y and X can be used as a measure of the overall fit of a linear regression that contains only one independent variable. For more on this relationship, see Exercise 6. However, the equivalence of r^2 and R^2 does *not* hold for a multiple regression; therefore it is of limited use to econometric researchers.

2.4 An Example of the Misuse of $\bar{R}^2$

Section 2.3 implies that the higher the overall fit of a given equation, the better. Unfortunately, many beginning researchers assume that if a high $\bar{R}^2$ (or R^2 or r) is good, then maximizing $\bar{R}^2$ is the best way to maximize the quality of an equation. Such an assumption is dangerous because a good overall fit is only one measure of the quality of an equation.

Perhaps the best way to visualize the dangers inherent in maximizing $\bar{R}^2$ without regard to the economic meaning or statistical significance of an equation is to look at an example of such misuse. This is important because it is one thing for a researcher to agree in theory that "$\bar{R}^2$ maximizing" is bad, and it is another thing entirely for that researcher to avoid subconsciously maximizing $\bar{R}^2$ on projects. It is easy to agree that the goal of regression is not to maximize $\bar{R}^2$, but many researchers find it hard to resist that temptation.

As an example, assume that you have been hired by the State of California to help the legislature evaluate a bill to provide more water to Southern California. This issue is important because a decision must be made whether or not to ruin, through a system of dams, one of the state's best trout fishing areas. On one side of the issue are Southern

Californians who claim that their desert-like environment requires more water; on the other side are outdoors-lovers and environmentalists who want to retain the natural beauty for which California is famous. Your job is to forecast the amount of water demanded in Los Angeles County, the biggest user of water in the state.

Because the bill is about to come before the state legislature, you are forced to choose between two regressions that have already been run for you, one by the state econometrician and the other by an independent consultant. You will base your forecast on one of these two equations:[7]

The state econometrician's equation:

$$\hat{W} = 24{,}000 + 48{,}000PR + 0.40P - 370RF \qquad (2.28)$$
$$\bar{R}^2 = .859 \quad DF = 25$$

The independent consultant's equation:

$$\hat{W} = 30{,}000 + 0.62P - 400RF \qquad (2.29)$$
$$\bar{R}^2 = .847 \quad DF = 26$$

where: W = the total amount of water consumed in L.A. County in a given year (measured in millions of gallons)
PR = the price of a gallon of water that year (measured in real dollars)
P = the population in L.A. County that year
RF = the amount of rainfall that year (measured in inches)
DF = degrees of freedom, which equal the number of observations (n = 29, since the years in the sample are 1950 through 1978) minus the number of coefficients estimated

Review these two equations carefully before going on with the rest of the section. What do you think the arguments of the state econometrician were for using his equation? What case did the independent econometrician make for her work?

The question is whether or not the increased $\bar{R}^2$ is worth the unexpected sign in the price of water coefficient in Equation 2.28. The

7. The principle involved in this section is the same one that was discussed during the actual research, but these coefficients are hypothetical because the complexities of the real equation are irrelevant to our points.

state econometrician argued that given the better fit of his equation, it would do a better job of forecasting water demand. The independent consultant argued that it did not make sense to expect that an increase in price in the future would, holding the other variables in the equation constant, increase the quantity of water demanded in Los Angeles. Furthermore, given the unexpected sign of the coefficient, it seemed much more likely that the demand for water was unrelated to price during the sample period or that some important variable (such as real per capita income) had been left out of both equations. Since the amount of money spent on water was fairly low compared with other expenditures during the sample years, the consultant pointed out, it was possible that the demand for water was fairly price-inelastic. The economic argument for the positive sign observed by the state econometrician is difficult to justify; it implies that as the price of water goes up, so does the quantity of water demanded.

Was this argument simply academic? The answer, unfortunately, is no. If a forecast is made with Equation 2.28, it will tend to over-forecast water demand in scenarios that foresee rapidly rising prices and under-forecast water demand otherwise. In essence, the equation with the better fit would do a worse job of forecasting.[8]

Thus, a researcher who uses $\bar{R}^2$ as the sole measure of the quality of an equation (at the expense of economic theory or statistical significance) increases the chances of having unrepresentative or misleading results. This practice should be avoided at all costs. No simple rule of econometric estimation is likely to work in all cases. Instead, a combination of technical competence, theoretical judgment, and common sense makes for a good econometrician.

To help avoid the natural urge to maximize $\bar{R}^2$ without regard to the rest of the equation, you might find it useful to imagine the following conversation:

YOU: Sometimes I wish I could play the game of maximizing $\bar{R}^2$, that is, choosing the model that gives the highest $\bar{R}^2$.

YOUR CONSCIENCE: But that would be wrong.

YOU: I know that the goal of regression analysis is to obtain

8. A couple of caveats to this example are in order. First, the purpose of the rainfall variable in both equations was to explain past behavior. For forecasting purposes, average rainfall figures would likely be used because future rainfall would not be known. Second, the income variable suggested by the independent consultant turned out to have a relatively small coefficient. This is because water expenditure is so minor in relation to the overall budget that the demand for water turned out to be fairly income-inelastic as well as fairly price-inelastic.

dependable estimates of the true population coefficients and not to get a high $\bar{R}^2$, but my results "look better" if my fit is good.

YOUR CONSCIENCE: Look better to whom? It's not at all unusual to get a high $\bar{R}^2$ but find that some of the regression coefficients have signs that are contrary to theoretical expectations.

YOU: Well, I guess I should be more concerned with the logical relevance of the explanatory variables than with the fit, huh?

YOUR CONSCIENCE: Right! If in this process we obtain a high $\bar{R}^2$, well and good, but if $\bar{R}^2$ is high, it doesn't mean that the model is good.

YOU: Amen.

2.5 Summary

1. Ordinary Least Squares (OLS) is the most frequently used method of obtaining estimates of the regression coefficients from a set of data. OLS chooses those $\hat{\beta}$s that minimize the summed squared residuals (Σe_i^2) for a particular sample.

2. The coefficient of determination, R^2, is the simplest measure of the degree of statistical fit of an estimated equation. It can be thought of as the percentage of the variation of Y around its mean that has been explained by a particular regression equation and is defined as the explained sum of squares (ESS) divided by the total sum of squares (TSS). A major fault of R^2 is that it always increases (technically, never decreases) when a variable is added to an equation.

3. R-bar-squared ($\bar{R}^2$) is the coefficient of determination (R^2) adjusted for degrees of freedom. $\bar{R}^2$ increases when a variable is added to an equation only if the improvement in fit caused by the addition of the new variable more than offsets the loss of the degree of freedom that is used up in estimating the coefficient of the new variable. As a result, most researchers will automatically use $\bar{R}^2$ instead of R^2 when evaluating the fit of their estimated regression equations.

4. Always remember that the quality of fit of an estimated equation is only one of the measures of the overall quality of that regression. The degree to which the estimated coefficients conform to economic theory and expectations developed by the researcher before the data were collected are at least as important as the size of $\bar{R}^2$ itself.

Exercises

(Answers to even-numbered exercises are in Appendix A.)

1. Write the meaning of each of the following terms without reference to the book (or your notes), and compare your definition with the version in the text for each:
   a. ordinary least squares
   b. the meaning of a multivariate regression coefficient
   c. total, explained, and residual sums of squares
   d. coefficient of determination
   e. degrees of freedom
   f. $\bar{R}^2$
   g. simple correlation coefficient

2. In a two-variable model, what is the relationship between $\hat{\beta}_0$ and $\hat{\beta}_1$? More specifically, if $\hat{\beta}_1$ is known to be "too high" in a given equation, would you expect $\hat{\beta}_0$ to be too high, too low, or unaffected? Why?

3. Suppose you estimate equations A and B on the same data and find that $\hat{\beta}_1 = \hat{\alpha}_1$. What values for $\hat{\beta}_0$, $\hat{\alpha}_0$ and/or $\hat{\alpha}_2$ does this result imply?

$$A: Y_i = \beta_0 + \beta_1 X_{1i} + \epsilon_{ai}$$
$$B: Y_i = \alpha_0 + \alpha_1 X_{1i} + \alpha_2 X_{2i} + \epsilon_{bi}$$

4. Just as you are about to estimate a regression project (due tomorrow), massive sunspots cause magnetic interference that ruins all electrically powered machines (e.g., computers). Instead of giving up and flunking, you decide to calculate estimates from your data (on per capita income in thousands of U.S. dollars as a function of the percent of the labor force in agriculture in 10 developed countries in 1981) using methods like those used in Section 2.1.4 *without* a computer. Your data are:

Country	A	B	C	D	E	F	G	H	I	J
Income	6	8	8	7	7	12	9	8	9	10
Percent on farms	9	10	8	7	10	4	5	5	6	7

a. Calculate $\hat{\beta}_0$ and $\hat{\beta}_1$.
b. Calculate R^2, $\bar{R}^2$, and the r (between income and percent on farms).
c. If the percentage of the labor force in agriculture in another developed country was 8 percent, what level of per capita income (in thousands of U.S. dollars) would you guess that country had?

5. Consider the following two least-squares estimates[9] of the relationship between interest rates and the federal budget deficit in the U.S.:

Model A: $\hat{Y}_1 = 0.103 - 0.079X_1 \qquad R^2 = .00$

where: Y_1 = interest rate on Aaa corporate bonds
X_1 = the federal deficit as a percentage of GNP
(quarterly model: 1970–1983)

Model T: $\hat{Y}_2 = 0.089 + 0.369X_2 + 0.887X_3 \qquad R^2 = .40$

where: Y_2 = interest rate on 3-month Treasury bills
X_2 = federal budget deficit in billions of dollars
X_3 = rate of inflation in percent
(quarterly model: April 1970–September 1979)

a. What does "least squares estimates" mean? What is being estimated? What is being squared? In what sense are the squares "least"?
b. What does it mean to have an R^2 of .00? Is is possible for an R^2 to be negative?
c. Calculate $\bar{R}^2$ for both equations. Is it possible for $\bar{R}^2$ to be negative?
d. Compare the two equations. Which model has estimated signs that correspond to your prior expectations? Is Model T automatically better because it has a higher $\bar{R}^2$? If not, which model do you prefer and why?

9. These estimates are simplified versions of results presented in the June/July 1984 issue of the *Review* of the Federal Reserve Bank of St. Louis (Model A) and the Summer 1983 issue of the *Review* of the Federal Reserve Bank of San Francisco (Model T).

6. Show that in a two-variable regression equation, the square of the simple correlation coefficient, r^2, does indeed equal the coefficient of determination, R^2.

7. Suppose that you have been asked to estimate an econometric model to explain the number of people jogging a mile or more on the school track to help decide whether to build a second track to handle all the joggers. You collect data by living in the press box for the spring term, and you run two possible explanatory equations.

$$\text{A:} \quad \hat{Y} = 125.0 - 15.0X_1 - 1.0X_2 + 1.5X_3 \qquad \bar{R}^2 = .75$$

$$\text{B:} \quad \hat{Y} = 123.0 - 14.0X_1 + 5.5X_2 - 3.7X_4 \qquad \bar{R}^2 = .73$$

where: Y = the number of joggers on a given day
X_1 = inches of rain that day
X_2 = hours of sunshine that day
X_3 = high temperature for that day (in degrees F)
X_4 = the number of classes with term papers due the next day

a. Which of the two (admittedly hypothetical) equations do you prefer? Why?
b. How is it possible to get different estimated signs for the coefficient of the same variable using the same data?

8. David Katz[10] studied faculty salaries as a function of their "productivity" and estimated a regression equation with the following coefficients:

$$\hat{S}_i = 11{,}155 + 230B_i + 18A_i + 102E_i + 489D_i + 189Y_i + \cdots$$

where: S_i = the salary of the *i*th professor in 1969-70 in dollars per year
B_i = the number of books published, lifetime
A_i = the number of articles published, lifetime
E_i = the number of "excellent" articles published, lifetime

10. David A. Katz, "Faculty Salaries, Promotions, and Productivity at a Large University," *American Economic Review*, June 1973, pp. 469–477. Katz' equation included other variables as well, as indicated by the "+ · · ·" at the end of the equation.

D_i = the number of dissertations supervised since 1964
Y_i = the number of years teaching experience

a. Do the signs of the coefficients match your prior expectations?
b. Do the relative sizes of the coefficients seem reasonable?
c. Suppose a professor had just enough time (after teaching, etc.) to write a book, write two excellent articles, or supervise three dissertations, which would you recommend? Why?
d. Would you like to reconsider your answer to Part b above? Which coefficient seems out of line? What explanation can you give for that result? Is the equation in some sense invalid? Why or why not?

9. What's wrong with this kind of thinking? "I understand that R^2 is not a perfect measure of the quality of a regression equation because it always increases when a variable is added to the equation. Once we adjust for degrees of freedom by using $\bar{R}^2$, though, it seems to me that the higher the $\bar{R}^2$, the better the equation."

10. In 1985, Charles Lave[11] published a study of driver fatality rates. His overall conclusion was that the variance of driving speed (the extent to which vehicles sharing the same highway drive at dramatically different speeds) is important in determining fatality rates. As part of his analysis, he estimated an equation with cross-state data from two different years:

$$1981: \hat{F}_i = \hat{\beta}_0 + 0.176\,V_i + 0.0136C_i - 7.75H_i$$
$$\bar{R}^2 = .624 \qquad n = 41$$

$$1982: \hat{F}_i = \hat{\beta}_0 + 0.190\,V_i + 0.0071C_i - 5.29H_i$$
$$\bar{R}^2 = .532 \qquad n = 44$$

where: F_i = fatalities on rural interstate highways (per 100 million vehicle miles traveled) in the *i*th state
$\hat{\beta}_0$ = an unspecified estimated intercept
V_i = driving speed variance in the *i*th state
C_i = driving citations per driver in the *i*th state
H_i = hospitals per square mile (adjusted) in the *i*th state

11. Charles A. Lave, "Speeding, Coordination, and the 55MPH Limit," *American Economic Review*, December 1985, pp. 1159–1164.

a. Think through the theory behind each variable, and develop expected signs for each coefficient. (Hint: Be careful with C.) Do Lave's estimates support your expectations?
b. Should we attach much meaning to the differences between the estimated coefficients from the two years? Why or why not? Under what circumstances might you be concerned about such differences?
c. The 1981 equation has the higher $\bar{R}^2$, but which equation has the higher R^2? (Hint: You can calculate the R^2s given the information above, or you can attempt to figure the answer theoretically.)

11. In Exercise #7 in Chapter One, we estimated a height/weight equation on a new data-set of 29 male customers, Equation 1.24:

$$\hat{Y}_i = 125.1 + 4.03X_i \tag{1.24}$$

where: Y_i = the weight (in pounds) of the *i*th person
X_i = the height (in inches above five feet) of the *i*th person

Suppose that a friend now suggests adding F_i, the percent bodyfat of the *i*th person, to the equation.

a. What is the theory behind adding F_i to the equation? How does the meaning of the coefficient of X change when you add F?
b. Assume you now collect data on the percent bodyfat of the 29 males and estimate:

$$\hat{Y}_i = 120.8 + 4.11X_i + 0.28F_i \tag{2.30}$$

Do you prefer Equation 2.30 or Equation 1.24? Why?
c. Suppose you learn that the $\bar{R}^2$ of Equation 1.24 is .75 and the $\bar{R}^2$ of Equation 2.30 is .72. Which equation do you prefer now? Explain your answer.
d. Suppose that you learn that the mean of F for your sample is 12.0. Which equation do you prefer now? Explain your answer.

3

Learning To Use Regression Analysis

From a quick reading of Chapter 2, it would be easy to conclude that regression analysis is little more than the mechanical application of a set of equations to a sample of data. Such a notion would be similar to deciding that all there is to golf is hitting the ball well. Golfers will tell you that it does little good to hit the ball well if you've used the wrong club or have hit the ball towards a trap, tree, or pond. Similarly, experienced econometricians spend much less time thinking about the OLS estimation of an equation than they do about a number of other factors. Our goal in this chapter is to introduce some of these "real world" concerns.

The first section focuses on economic data. Data collection, though usually frustrating and time-consuming, is seldom theoretically difficult, but a few generalizations about frequently occurring problems will save the reader time and effort later. The second section introduces a new type of variable, a dummy variable, that is useful in quantifying concepts that usually can't be measured.

This is followed by an overview of the steps typically taken by researchers engaged in applied regression analysis. The purpose of suggesting these steps is not to discourage the use of innovative or unusual approaches but rather to develop in the reader a sense of how regression ordinarily is done by professional economists and business analysts.

The chapter concludes with a more complete example of applied regression analysis, a location analysis for the "Woody's" restaurant chain that is based on actual company data and to which we will return in future chapters to apply new ideas and tests.

A few professors may want to skip some or all of this chapter and come back to the material later, perhaps after covering Chapter 5. We feel that the earlier you begin thinking about the topics in this chapter the better, but we've tried to write the material so that most of it can be delayed without much loss in continuity (except for Section 3.2).

3.1 Economic Data

Before any quantitative analysis can be done, data must be defined, collected, organized, and entered into a computer. Usually, this is a time-consuming and frustrating task because of the difficulty of finding data, the existence of definitional differences between theoretical variables and their empirical counterparts, and the high probability of keypunching or typographical errors. In general, though, time spent thinking about and collecting the data is well-spent, since a researcher who knows the data sources and definitions is much less likely to make mistakes using or interpreting regressions run on that data.

3.1.1 What Data to Look For

When choosing a research topic, it's good advice to make sure that data for your dependent variable and all relevant independent variables are available. However, checking for data availability means deciding what specific variables you want to study. Half of the time that beginning researchers spend collecting data is wasted by looking for the wrong variables in the wrong places. A few minutes thinking about what data to look for will save hours of frustration later.

For example, if the dependent variable is the quantity of television sets demanded per year, then most independent variables should be measured annually as well. It would be inappropriate and possibly misleading to define the price of TVs as the price from a particular month. An average of prices over the year (usually weighted by the number of TVs sold per month) would be more meaningful. If the dependent variable includes all TV sets sold regardless of brand, then the price would appropriately be an aggregate based on prices of all brands. Calculating such aggregate variables, however, is not straightforward. Researchers typically make their best efforts to compute the respective aggregate variables and then acknowledge that problems still remain. For example, if the price data for all the various brands are not available, a researcher may be forced to compromise and use the prices of one or a few of the major brands as a substitute for the proper aggregate price.

Another issue is suggested by the TV example. Over the years of the sample, it is likely that the market shares of particular kinds of TV sets have changed. For example, 19-inch color TV sets might have made up a majority of the market in one decade, while 13-inch black and white sets might have been the favorite ten years before. In cases where the composition of the market share, the size, or the quality of the various brands have changed over time, it would make little sense to measure the dependent variable as the number of TV sets because a "TV set" from one year has little in common with a "TV set" from another. The approach usually taken to deal with this problem is to measure the variable in dollar terms, under the assumption that value encompasses size and quality. Thus we would work with the dollar sales of TVs rather than the number of sets sold.

A third issue, whether to use nominal or real variables, usually depends on the underlying theory of the research topic. Nominal (or money) variables are measured in current dollars and thus include increases caused by inflation. If theory implies that inflation should be filtered out, then it's best to state the variables in real (constant dollar) terms by selecting an appropriate price deflator, such as the Consumer Price Index, and adjusting the money (or nominal) value by it.

As an example, the appropriate price index for Gross National Product is called the GNP deflator. Real GNP is calculated by multiplying nominal GNP by the ratio of the GNP deflator from the base year to the GNP deflator from the current year:

$$\text{Real GNP} = \text{nominal GNP} \cdot (\text{base GNP deflator}/\text{current GNP deflator})$$

In 1981, U.S. nominal GNP was \$2925 billion and the GNP deflator (for a base year of 1972 = 100) was 193.7, so real GNP was:

$$\text{Real GNP} = \$2925 \cdot (100/193.7) = \$1510 \text{ billion}$$

That is, the goods and services produced in 1981 were worth $2925 billion if 1981 prices were used but only $1510 billion if 1972 prices were used.

Fourth, recall that all economic data are either time-series or cross-sectional in nature. Since time-series data are for the same economic entity from different time periods while cross-sectional data are from the same time period but for different economic entities, the appropriate definitions of the variables depend on whether the sample is a time-series or a cross-section.

To understand this, consider the TV set example once again. A time-series model might study the sales of TV sets in the U.S. from 1950 to 1990, while a cross-sectional model might study the sales of TV sets by state for 1990. The time-series data set would have 41 observations, each of which would refer to a particular year. In contrast, the cross-sectional data set would have 50 observations, each of which would refer to a particular state. A variable that might be appropriate for the time-series model might be completely inappropriate for the cross-sectional model and vice versa; at the very least, it would have to be measured differently. National advertising in a particular year would be appropriate for the time-series model, for example, while 1990 advertising in or near each particular state would make more sense for the cross-sectional one.

Finally, learn to be a critical reader of the descriptions of variables in econometric research. For instance, most readers breezed right through Equation 2.13 on the demand for beef (and the accompanying data in Table 2.2) without asking some vital questions. Where did the data originate? Are prices and income measured in nominal or real terms? Is the price of beef wholesale or retail? A careful reader would want to know the answers to these questions before analyzing the results of Equation 2.13. (For the record, Yd measures real income, P measures real wholesale prices, and the data come from various issues of *Agricultural Statistics,* published in Washington D.C. by the U.S. Department of Agriculture.)

3.1.2 Where to Look for Economic Data

While some researchers generate their own data through surveys or other techniques, the vast majority of regressions are run on publicly available data. The best sources for such data are government publications. In fact, the U.S. government has been called the most thorough statistics-collecting agency in history.

Excellent sources of U.S. data include the annual *Statistical Abstract of the U.S.*, the annual *Economic Report of the President,* the *Handbook of Labor Statistics,* and *Historical Statistics of the U.S.* (published in 1975). One of the best places to start with U.S. data is the annual *Census Catalog and Guide,* which provides overviews and abstracts of data sources and various statistical products as well as details on how to obtain each item.[1]

Consistent international data are harder to come by, but the United Nations publishes a number of useful compilations of figures from various nations. The best of these are the *U.N. Statistical Yearbook* and the *U.N. Yearbook of National Account Statistics.*

Recently, more and more researchers have started using on-line computer databases to find data instead of plowing through stacks of printed volumes. These on-line databases, available through most college and university libraries, contain complete series on literally thousands of possible variables. For example, one database, "U.S. Econ.," contains more than 25,000 time-series variables for the U.S. economy. The *Directory of Online Databases* (New York: Quadra/Elsevier, 1991) contains a complete listing of all available on-line databases. Two of the most useful of these are the "Economic Literature Index," which is an on-line summary of the *Journal of Economic Literature,* and "Dialog," which provides on-line access to a large number of datasets at a lower cost than many alternatives.

3.1.3 Proxy Variables

Suppose the data aren't there? What happens if you choose the perfect variable and look in all the right sources and can't find the data?

The answer to this question depends on how much data are missing. If a few observations have incomplete data in a cross-sectional study, you usually can afford to drop these observations from the sample. If the incomplete data are from a time-series, you can sometimes estimate the missing value by interpolating (taking the mean of adjacent values). Similarly, if one variable is only available annually in an otherwise quarterly model, you may want to consider quarterly interpolations of that variable. In either case, interpolation can be justified only if the variable moves in a slow and smooth manner. Extreme caution should always be exercised when "creating" data in such a way (and full documentation is required).

1. To obtain the *Census Catalog and Guide,* published each June, write: Superintendent of Documents, Government Printing Office, Washington, D.C., 20402–9325, or call 202–783–9325 and ask for stock/catalog #C3–163/3.

If no data exist for a theoretically relevant variable, then the problem worsens significantly. Omitting a relevant variable runs the risk of biased coefficient estimates, as we'll learn in Chapter 6. After all, how can you hold a variable constant if it's not included in the equation? In such cases, most researchers resort to the use of proxy variables.

Proxy variables substitute for theoretically desired variables when data on variables are incomplete or missing altogether. For example, the value of net investment is a variable that is not measured directly in a number of countries. As a result, a researcher might use the value of gross investment as a proxy, the assumption being that the value of gross investment is directly proportional to the value of net investment. This proportionality (which is similar to a change in units) is all that is required because the regression analyzes the relationship between changes among variables, rather than the absolute levels of the variables.

In general, a proxy variable is a "good" proxy when its movements correspond relatively well to movements in the theoretically correct variable. Since the latter is unobservable whenever a proxy must be used, there is usually no easy way to examine a proxy's "goodness" directly. Instead, the researcher must document as well as possible why the proxy is likely to be a good or bad one. In some cases, proxies are admittedly of poor quality, but they are the only quantitative measures available and are used with the appropriate caveats being stated. Poor proxies and variables with large measurement errors constitute "bad" data. We'll discuss the consequences of such "errors in the variables" in Section 14.6. The degree to which the data are bad is a matter of judgment by the individual researcher.

In fact, a good regression project may be thwarted by the lack of adequate data. In many cases, even the simplest of regression techniques may not be appropriate because the information is inaccurate. Sometimes it is measured with so much error that the researcher should only compose tables or graphs or make general inferences. By the way, such tables and graphs are generally useful adjuncts to the documentation of regression equations.

3.1.4 The Use of Lags in Economics and Econometrics

Most of the regressions studied so far have been "instantaneous" in nature. In other words, they have included independent and dependent variables from the same time period as in:

$$Y_t = \beta_0 + \beta_1 X_{1t} + \beta_2 X_{2t} + \epsilon_t \tag{3.1}$$

The subscript t is used to refer to a particular point in time; if all variables have the same subscript value, then the equation is instantaneous. However, not all economic or business situations imply such instantaneous relationships between the dependent and independent variables. In many cases we must allow for the possibility that time might elapse between a change in the independent variable and the resulting change in the dependent variable. The length of this time between cause and effect is called a **lag.** Many econometric equations include one or more *lagged independent variables* like X_{1t-1}, where the subscript $t-1$ indicates that the observation of X_1 is from the time period previous to time period t, as in the following equation:

$$Y_t = \beta_0 + \beta_1 X_{1t-1} + \beta_2 X_{2t} + \epsilon_t \tag{3.2}$$

In this equation, X_1 has been lagged by one time period, but the relationship between Y and X_2 is still instantaneous.

For example, think about the process by which the supply of an agricultural product is determined. Since agricultural goods take time to grow, decisions on how many acres to plant or how many eggs to let hatch into egg-producing hens (instead of selling them immediately) must be made months if not years before the product is actually supplied to the consumer. Any change in an agricultural market, such as an increase in the price that the farmer can earn for providing the product, has a lagged effect on the supply of that product:

$$\text{Quantity Supplied}_t = f(\text{Price}_{t-1}, \text{etc.})$$

Similarly, many macroeconomic theories have explicit lag structures built into them. The length of time between the decision to undertake a macroeconomic policy (like a cut in government spending or an increase in the money supply) and the impact of that policy on GNP, employment, or prices is usually measured in years. An increase in the money supply stimulates GNP in part by stimulating investment; but investment cannot be increased overnight because decisions need to be made, plans need to be designed, additional workers need to be hired, and so on. Indeed, noted economist Milton Friedman once estimated that it takes between 6 and 30 months for a monetary policy change to be felt fully in the economy.

If a *simple lag* is hypothesized, there is little difficulty in using lags in econometric equations. The lagged independent variable is added to the equation just like any other variable. For example, an equation for the supply for cotton might look like:

$$C_t = f(\overset{+}{PC_{t-1}}, \overset{-}{PF_t}) = \beta_0 + \beta_1 PC_{t-1} + \beta_2 PF_t + \epsilon_t \qquad (3.3)$$

where: C_t = the quantity of cotton supplied in year t
PC_{t-1} = the price of cotton in year $t-1$
PF_t = the price of farm labor in year t

Note that this equation hypothesizes a lag between the price of cotton and the production of cotton, but not between the price of farm labor and the production of cotton. It's reasonable to think that if cotton prices change, farmers won't be able to react immediately because it takes a while for cotton to be planted and to grow. On the other hand, if the price of farm labor changes, farmers can react immediately, since the main use of labor in cotton production is in the picking of cotton just before it goes to market. As a result, PF is not lagged even though PC is.

The meaning of the regression coefficient of a lagged variable is not the same as the meaning of the coefficient of an unlagged variable. The estimated coefficient of a lagged X measures the change in *this year's* Y attributed to a one-unit change in *last year's* X (holding constant the other Xs in the equation). Thus β_1 in Equation 3.3 measures the extra number of units of cotton that would be produced this year as a result of a one-unit change in last year's price of cotton, holding this year's price of farm labor constant.

If the lag structure is hypothesized to take place over more than one time period, or if a lagged dependent variable is specified, the question becomes significantly more complex. Such cases, called *distributed lags,* will be dealt with in Chapter 12.

3.2 Dummy Variables

Some variables (for example, gender) defy explicit quantification and can only be expressed in a qualitative manner. One of the most common ways such variables are quantified is with binary or dummy variables. **Dummy variables** take on the values one or zero depending on whether some condition does or does not hold. The basic use of these variables is presented here; some frequently used extensions will be discussed in Chapters 7 and 14.

As an illustration, suppose that Y_i represents the salary of the *i*th high school teacher, and that the salary level depends primarily on the type of degree earned and the experience of the teacher. All teachers have a B.A., but some also have an M.A. An equation representing the relationship between earnings and the type of degree might be:

$$Y_i = \beta_0 + \beta_1 X_{1i} + \beta_2 X_{2i} + \epsilon_i \tag{3.4}$$

$$\text{where: } X_{1i} = \begin{cases} 1 \text{ if the } i\text{th teacher has an M.A.} \\ 0 \text{ otherwise} \end{cases} \tag{3.5}$$

$$X_{2i} = \text{number of years teaching experience of the } i\text{th teacher}$$

The variable X_1 only takes on values of zero or one, so X_1 is called a dummy variable, or just a dummy. Needless to say, the term has generated many a pun. In this case, the dummy variable represents the condition of having a master's degree.

The coefficients of this regression equation are interpreted as follows:

(1) If the teacher has a B.A. only, X_1 is zero, and

$$E(Y_i|X_1) = \beta_0 + \beta_2 X_{2i} \tag{3.6}$$

(2) If the teacher has an M.A., X_1 equals one, and

$$E(Y_i|X_1) = \beta_0 + \beta_1 + \beta_2 X_{2i} \tag{3.7}$$

Comparing these two equations, β_1 represents the additional average earnings gained by having the master's degree compared to the bachelor's degree, holding experience constant. This interpretation is important because it allows the researcher to formulate an expectation concerning the sign of β_1 prior to estimation. The ability to postulate the signs of coefficients is essential to regression analysis. Such a dummy variable is called an *intercept dummy* because it actually changes the intercept of the regression depending on whether the teacher in question has a master's degree. Figure 3.1 shows this relationship graphically.

An alternative formulation of the regression model (Equation 3.4) would be to define X_{1i} as:

$$X_{1i} = \begin{cases} 0 \text{ of the } i\text{th teacher has an M.A.} \\ 1 \text{ otherwise} \end{cases} \tag{3.8}$$

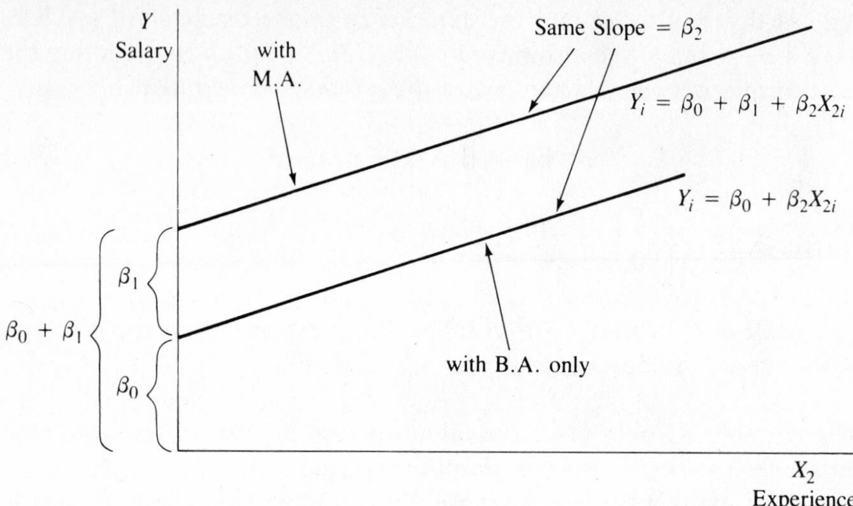

Figure 3.1 GRAPHICAL INTERPRETATION OF AN INTERCEPT DUMMY VARIABLE (X_1)
An intercept dummy changes the intercept of the regression equation depending on whether the qualitative condition specified by the dummy variable is met. The difference between the two intercepts is the estimated coefficient of the dummy variable.

That is, the conditions are turned around. In this case, β_1 would be interpreted as the difference between the average earnings of a B.A. and those of an M.A., and its sign would be expected to be negative. While the β_1 of the redefined variable would have the opposite sign, it would have the same absolute magnitude as the β_1 of the original variable in Equation 3.4. The reason is that the two βs are measuring exactly the same thing (but in opposite directions). The definitions of dummy variables are, in this sense, arbitrary. Once they have been defined, though, only one interpretation can be placed on them. That is, in Equation 3.5, β_1 is expected to be positive while in Equation 3.8, β_1 is expected to be negative, although they have the same absolute magnitude. This works for estimated as well as theoretical models.

Note that in this example only one dummy variable was used even though there were two conditions. This is because one fewer dummy variable is constructed than conditions. The event not explicitly represented by a dummy variable, the **omitted condition,** forms the basis against which the included conditions are compared. Thus, for dual situations (e.g., B.A. and M.A.), only one dummy variable is entered as an independent variable; the coefficient is interpreted as the effect of the included condition relative to the omitted condition. If a third condition (such as having a Ph.D.) is to be included, then only two dummy variables should be used.

Beginners often err by including as many dummy variables as conditions, but such a model is useless becaue the dummies add up to a constant, which is perfectly multicollinear with the intercept term already in the equation. **Perfect multicollinearity** is defined as an exact linear relationshp among some or all of the explanatory variables. This occurs because the sum of the dummy variables equals one for each observation in the data set. Computer regression programs will produce no output under perfect multicollinearity unless there is rounding error. We'll discuss perfect multicollinearity in more depth in Chapter 8.

A dummy variable that has only a single observation with a value of one while the rest of the observations are zeroes (or vice versa) is to be avoided even if the variable is suggested by theory. Such a "one-time dummy" acts merely to eliminate that observation from the data set, improving the fit artificially by setting the dummy's coefficient equal to the residual for that observation. One would obtain exactly the same estimates of the other coefficients if that observation were deleted, but the deletion of an observation is rarely if ever appropriate.

Sometimes dummy variables are used to account for seasonal variation in the data in time-series models. For example, if:

$$X_{1t} = \begin{cases} 1 \text{ in quarter } 1 \\ 0 \text{ otherwise} \end{cases}$$

$$X_{2t} = \begin{cases} 1 \text{ in quarter } 2 \\ 0 \text{ otherwise} \end{cases}$$

$$X_{3t} = \begin{cases} 1 \text{ in quarter } 3 \\ 0 \text{ otherwise} \end{cases}$$

then:

$$Y_t = \beta_0 + \beta_1 X_{1t} + \beta_2 X_{2t} + \beta_3 X_{3t} + \beta_4 X_{4t} + \epsilon_t \qquad (3.9)$$

where X_4 is a non-dummy independent variable and t indexes the quarterly observations. Notice that only three dummy variables are required to represent four seasons. In this formulation, β_1 shows the extent to which the expected value of Y in the first quarter differs from its expected value in the fourth quarter, the omitted condition. β_2 and β_3 can be interpreted similarly.

This procedure may be used as long as Y and X_4 are not "seasonally adjusted" prior to estimation. Inclusion of a set of seasonal dummies "deseasonalizes" Y as well as any other independent variables that are not seasonally adjusted. Many researchers believe that the type of seasonal adjustment done prior to estimation distorts the data in unknown and arbitrary ways, but seasonal dummies have their own limitations such as remaining constant for the entire time period. As a result, there is no unambiguously best approach to deseasonalizing data.

As an example of the meaning of the coefficient of a dummy variable, let's look at a model of the relationship between fraternity/sorority membership and grades. Most noneconometricians would approach this research problem by calculating the mean grades of fraternity/sorority ("Greek") members and comparing them to the mean grades of nonmembers. However, such a technique ignores the possibility that differences in mean grades might be related to characteristics other than Greek membership.

Instead, we'd want to build a regression model that explains college GPA. Independent variables would include not only Greek membership but also other predictors of academic performance such as SAT scores and high school grades. Being a member of a social organization is a qualitative variable, however, so we'd have to create a dummy variable to represent fraternity or sorority membership quantitatively in a regression equation:

$$G_i = \begin{cases} 1 \text{ if the } i\text{th student is an active member of} \\ \quad \text{a fraternity or sorority} \\ 0 \text{ otherwise} \end{cases}$$

If we collect data from all students in our class and estimate the equation implied above, we get:

$$\widehat{CG}_i = 0.37 + 0.81HG_i + 0.00001S_i - 0.38G_i \qquad (3.10)$$
$$\bar{R}^2 = .45 \qquad n = 25$$

where: CG_i = the cumulative college GPA (4-point scale) of the *i*th student

HG_i = the cumulative high school GPA (4-point scale) of the *i*th student

S_i = the sum of the highest verbal and mathematics SAT scores earned by the *i*th student (1600 maximum)

The meaning of the estimated coefficient of G_i in Equation 3.10 is very specific. Stop for a second and figure it for yourself. What is it? The estimate that $\hat{\beta}_G = -0.38$ means that for this sample, the GPA of fraternity/sorority members is 0.38 lower than for nonmembers, holding SATs and high school GPA constant. Thus Greek members are doing about a third of a grade worse than otherwise might be expected.

Before you rush out and quit whatever social organization you're in, note that this sample is quite small and that we've surely omitted some important determinants of academic success from the equation. As a result, we shouldn't be too quick to conclude that the equation shows that Greeks are dummies. To understand this example better, try using Equation 3.10 to predict your own GPA; how close does it come? For an example of the interpretation of the coefficients of seasonal dummy variables, see Exercise 10.

3.3 Steps in Applied Regression

While there are no hard and fast rules for conducting econometric research, most investigators commonly follow a standard method for applied regression analysis. The relative emphasis and effort expended on each step may vary, but normally all the steps are considered necessary for successful research. The researcher must first identify the problem to be studied and select the dependent variable; this choice is determined by the purpose of the research. After that, it is logical to follow this sequence:

1. Review the literature.
2. Specify the model: Select the independent variables and the functional form.
3. Hypothesize the expected signs of the coefficients.
4. Collect the data.
5. Estimate and evaluate the equation.
6. Document the results.

3.3.1 Step 1: Review the Literature

Before developing a theoretical model too deeply (and certainly before estimating it), it's a good idea to review the scholarly literature. Researchers can find out if someone has already examined a topic sufficiently to permit the use of previously generated results. On the other hand, they may disagree with the approach or assumptions used by previous authors, or they may want to apply their theoretical model to a different data set. In either event, researchers should not have to "reinvent the wheel"—they can start their investigations where earlier ones left off. Any research paper should begin with a comment on the extent and quality of previous research.

The most convenient approach to reviewing the literature is to obtain several recent issues of the *Journal of Economic Literature* or a business-oriented publication of abstracts, find and read several recent articles related to the selected topic, and trace appropriate references cited in the literature.

3.3.2 Step 2: Specify the Model: Select the Independent Variables and the Functional Form

The most important step in applied regression analysis is the *specification* of the theoretical regression model. After selecting the dependent variable, the following components should be specified: the independent variables and how they should be measured, the functional (mathematical) form of the equation, and the type of error term in the equation. A regression equation is specified when each of these elements is treated appropriately, but we will focus on the selection of the independent variables until the other two topics can be introduced more completely (in Chapters 7 and 4 respectively).

Each of the elements of specification is determined primarily on the basis of economic theory, rather than on the results of an estimated regression equation. A mistake in any of the three elements results in a **specification error.** Of all the kinds of mistakes that can be made in applied regression analysis, specification error is usually the most disastrous to the validity of the estimated equation. Thus, the more attention paid to economic theory at the beginning of a project, the more satisfying the regression results are likely to be.

The emphasis in this text is on estimating behavioral equations, those that describe the behavior of economic entities. The researcher selects independent variables based on economic theory concerning that behavior. An explanatory variable is chosen because it is a causal determinant of the dependent variable; it is expected to explain at least part of the variation in the dependent variable. Recall that regression gives evidence but does not *prove* economic causality. Just as an example does not prove the rule, a regression result does not prove the theory.

There are dangers in specifying the wrong independent variables. A researcher's goal is to specify only relevant explanatory variables, those expected theoretically to assert a "significant" influence on the dependent variable. Variables suspected of having virtually no effect should be excluded unless their possible effect on the dependent variable is of some particular (e.g., policy) interest.

For example, an equation that explains the quantity demanded of a consumption good might use the price of the product and consumer income or wealth as likely variables. Theory also indicates that complementary and substitute goods are important. Therefore, the researcher may select as independent variables the prices of complements and substitutes, but which complements and substitutes? Of course, selection of the closest complements and/or substitutes is appropriate, but how far should one go? The choice must be based on theoretical judgment.

When researchers decide that, for example, the prices of only two other goods need to be included, they are said to impose their *priors* (i.e., prior information) or their working hypotheses on the regression equation. Imposition of such priors is a common practice that determines the number and kind of hypotheses that the regression equation has to test. The danger is that a prior may be wrong and could diminish the usefulness of the estimated regression equation. Each of the priors therefore should be explained and justified in detail.

3.3.3 Step 3: Hypothesize the Expected Signs of the Coefficients

Once the variables are selected, it is important to hypothesize carefully the signs you expect their regression coefficients to have. For example, in the demand equation for a final consumption good, the quantity demanded (Q_d) is expected to be inversely related to its price (P) and the price of a complementary good (P_c), and positively related to consumer income (Y) and the price of a substitute good (P_s). The first step in the written development of a regression model usually is to express the equation as a general function:

$$Q_d = f(\overset{-}{P}, \overset{+}{Y}, \overset{-}{P_c}, \overset{+}{P_s}) \tag{3.11}$$

The signs above the variables indicate the hypothesized sign of the respective regression coefficient in a linear model. This is a simple convention that has become increasingly popular.

In many cases, the basic theory is general knowledge, so that the reasons for each sign need not be discussed. However, if any doubt surrounds the selection of an expected sign, the researcher should document the opposing forces at work and the reasons for hypothesiz-

ing a positive or negative coefficient. Alternatively, if the theory is unclear about an expected sign, then a question mark may be placed above the respective variable. For example, if the dependent variable in a demand equation is one for which the impact of income on demand is ambiguous, Equation 3.11 may be written as:

$$Q_d = f(\overset{-}{P}, \overset{?}{Y}, \overset{-}{P_c}, \overset{+}{P_s}) \tag{3.12}$$

3.3.4 Step 4: Collect the Data

Data collection may begin after the specification of the regression model. This step entails more than a mechanical recording of data, though, because the type and size of the sample must also be chosen. Often, analysis begins as the researcher examines the data and looks for typographical, conceptual, or definitional errors.

A general rule regarding sample size is the more observations the better. Ordinarily, the researcher takes all the roughly comparable observations that are readily available. Even if a number of computations are necessary to quantify a variable, the burden is on the researcher to explain why as many observations as were available were not used. In regression analysis, all the variables must have the same number of observations. They also should have the same frequency (monthly, quarterly, annual, etc.) and time period. Often, the frequency selected is determined by the availability of data.

The reason there should be as many observations as possible concerns the statistical concept of *degrees of freedom* first mentioned in Section 2.3.2. Consider fitting a straight line to two points on an X,Y coordinate system, as in Figure 3.2. Such an exercise can be done mathematically without error. Both points lie on the line, so there is no estimation of the coefficients involved. The two points determine the two parameters, the intercept and the slope, precisely. Estimation takes place only when a straight line is fitted to three or more points that were generated by some process that is not exact. The excess of the number of observations (three) over the number of parameters to be estimated (in this case two, the intercept and slope) is called the **degrees of freedom.**[2] All that is necessary for estimation is a single

2. Throughout the text, we will calculate the number of degrees of freedom (d.f.) in a regression equation as d.f. = (n − K − 1), where K is the number of independent variables in the equation. Equivalently, some authors will set K′ = K + 1 and define d.f. = (n − K′).

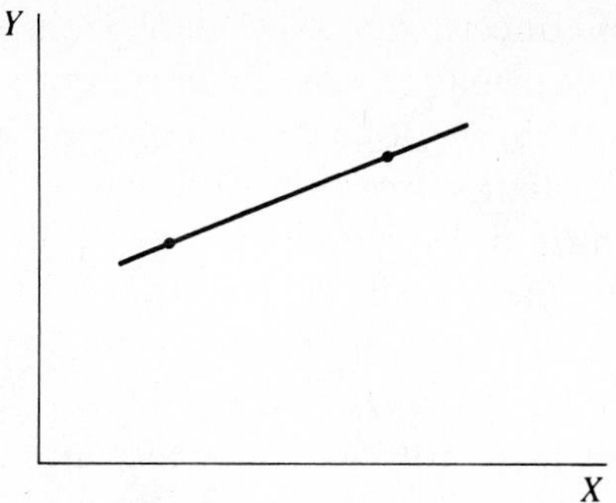

Figure 3.2 MATHEMATICAL FIT OF A LINE TO TWO POINTS
If there are only two points in a data set, as in Figure 3.2, a straight line can be fitted to those points mathematically without error, because two points completely determine a straight line.

degree of freedom, as in Figure 3.3, but the more degrees of freedom there are, the less likely it is that the stochastic or purely random component of the equation (the error term) will affect inferences about the deterministic portion, the portion of primary interest. This is because when the number of degrees of freedom is large, every large positive error is likely to be balanced by a large negative error. With only a few points, the random element is likely to fail to provide such offsetting observations. For example, the more a coin is flipped, the more likely it is that the observed proportion of heads will reflect the true underlying probability (namely 0.5).

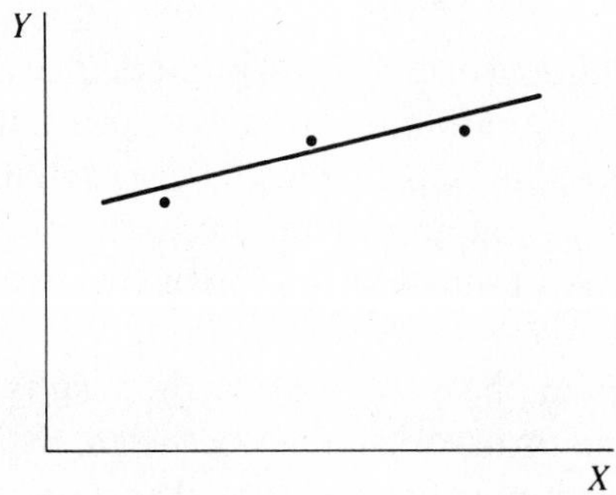

Figure 3.3 STATISTICAL FIT OF A LINE TO THREE POINTS
If there are three (or more) points in a data set, as in Figure 3.3, then the line must almost always be fitted to the points statistically, using the estimation procedures of Section 2.1.

Another area of concern has to do with the units of measurement of the variables. Does it matter if an independent variable is measured in dollars or thousands of dollars? Does it matter if the measured variable differs consistently from the true variable by ten units? Surprisingly, such changes do not matter in terms of regression analysis except in interpreting the scale of the coefficients. All conclusions about signs, significance, and economic theory are independent of units of measurement.[3]

That is, it makes little difference whether an independent variable is measured in dollars or thousands of dollars. The constant term and measures of overall fit remain unchanged. Such a multiplicative factor does change the slope coefficient, but only by the exact amount necessary to compensate for the change in the units of measurement of the independent variable. Similarly, a constant factor added to a variable alters only the intercept term without changing the slope coefficient itself. Recall the weight-height regression of Section 1.3. The explanatory variable was measured as inches above 5 feet, so that 5 feet, or 60 inches, was a constant factor that was subtracted from each observation. If we reestimate the equation with the absolute height in inches (Z) as the independent variable, we get $Y_i = -279.2 + 6.38Z_i$. Since the original equation was $Y_i = 103.4 + 6.38X_i$, only the constant term has changed, and for each height we would obtain the same weight. The essential relationship between Y and Z is the same as between Y and X. That is, adding a constant to a variable will not change the slope coefficients of a linear equation (but it will in most equations that are nonlinear in the coefficients).

3.3.5 Step 5: Estimate and Evaluate the Equation

The Ordinary Least Squares (OLS) technique, discussed in Section 2.1, is the typically used estimating technique. Where alternative techniques might be appropriate, estimates from the alternative techniques should be compared to OLS estimates. The alternatively estimated equations must be evaluated carefully and judgment applied before a choice is made.

Once the model has been estimated, the results should be checked for errors. Data transformations and coefficient estimation are usually done in the same computer program, so it is wise to obtain a printout

3. The units of measurement of the dependent variable also do not alter the interpretation of the regression equation (except, as above, in interpreting the magnitude of the regression coefficients), hypothesis testing, or measures of fit such as R^2.

of the data set exactly as it was used in the regression estimation. Check one or two values of any variables that were transformed. If these values are correct, it may be assumed that the computer did not make any mistakes transforming the rest of the observations. Also obtain a printout or a plot of the data and look for outliers. An **outlier** is an observation that lies outside the range of the rest of the observations. Looking for outliers is a very economical way to look for typing or transcription errors.

After checking for data errors, examine the signs, magnitudes, and significance of the coefficients and the overall measures of fit. Regression results are rarely what one expects. Usually, additional model development is required or alternative estimating techniques are called for. Be sure to reevaluate the model and make any necessary changes before jumping into fancy regression "fix-up" routines. Sometimes these routines improve the overall measures of goodness of fit or some other statistic while playing havoc with the reliability of estimates of the model's parameters. A famous econometrician, Zvi Griliches, has warned that errors in the data coming from their measurement, usually computed from samples or estimates, imply that the fancier estimating techniques should be avoided because they are more sensitive to data errors than is OLS.[4] For a more detailed discussion of such *errors in the variables,* see Section 14.6. Also, when faced with unexpected regression results (which happen all too often), a reexamination of the theoretical basis of the model is in order. However, one should avoid adjusting the theory merely to fit the data, thus introducing researcher bias. The researcher has to walk the fine line between making appropriate and inappropriate adjustments to the model. Choosing proper modifications is one of the artistic elements in applied regression analysis.

3.3.6 Step 6: Document the Results

A standard format usually is used to present estimated regression results:

$$\begin{aligned} \hat{Y}_i &= 103.40 + 6.38X_i \\ &\qquad\qquad (0.88) \\ &\qquad\qquad t = 7.22 \\ n &= 20 \qquad \bar{R}^2 = .73 \end{aligned} \tag{3.13}$$

4. Zvi Griliches, "Data and Econometricians—The Uneasy Alliance," *American Economic Review,* May 1985, p. 199.

The number in parenthesis is the estimated standard error of the estimated coefficient, and the t-value is the one used to test the hypothesis that the true value of the coefficient is different from zero. Other measures of the quality of the regression will be discussed in later chapters. What is important to note is that the documentation of regression results using an easily understood format is considered part of the analysis itself. For time-series data-sets, the documentation also includes the frequency (e.g., quarterly or annual) and the time period of the data.

Most computer programs present statistics to eight or more digits, but it is important to recognize the difference between the number of digits computed and the number of *significant figures,* which may be as low as two or three. The number of significant figures is the number of meaningful digits in a result and is a function of the quality of the measurement of the data. Simply because the computer prints out a result to eight places is not an indication that all those digits have meaning. Indeed, it would be surprising if more than two or three did. Only if data are known to be measured extremely accurately is there justification for an increase in the number of places presented.

One of the important parts of the documentation is the explanation of the model, the assumptions, and the procedures and data used. The written documentation must contain enough information so that the entire study could be replicated[5] exactly (except for rounding errors) by others. Unless the variables have been defined in a glossary or table, short definitions should be presented along with the equations. If there is a series of estimated regression equations, then tables may provide the relevant information for each equation. All data manipulations as well as data sources should be documented fully. When there is much to explain, this documentation usually is relegated to a data appendix. If the data are not available generally or are available only after computation, the data-set itself might be included in this appendix.

3.4 Using Regression Analysis to Pick Restaurant Locations

To solidify our understanding of the six basic steps of applied regression analysis, let's work through a complete regression example. Sup-

5. For example, the *Journal of Money, Credit and Banking* has requested authors to submit their actual data-sets so that regression results can be verified. See W.G. Dewald et al., "Replication in Empirical Economics," *American Economic Review,* Sept. 1986, pp. 587–603.

pose that you've been hired to determine the best location for the next Woody's restaurant, where Woody's is a moderately priced, 24-hour, family restaurant chain. You decide to build a regression model to explain the gross sales volume at each of the restaurants in the chain as a function of various descriptors of the location of each individual branch. If you can come up with a sound equation to explain gross sales as a function of location, then you can use this equation to help Woody's decide which of the locations being considered for the newest eatery will have the highest potential for gross sales. Given data on land costs, building costs, and local building and restaurant municipal codes, the owners of Woody's will be able to make an informed decision.

1. *Review the literature.* You do some reading about the restaurant industry, but your review of the literature consists mainly of talking to various experts within the firm to get their hypotheses, based on experience, as to the particular attributes of a location that contribute to success at selling food at Woody's. The experts tell you that all of the chain's restaurants are identical (indeed, this is sometimes a criticism of the chain) and that all the locations are in what might be called "suburban, retail, or residential" environments (as distinguished from central cities or rural areas, for example). Because of this, you realize that many of the reasons that might help explain differences in sales volume in other chains do not apply in this case because all the Woody's locations are similar. (If you were comparing Woody's to another chain, such variables might be appropriate.)

   In addition, discussions with the people in the Woody's strategic planning department convince you that price differentials and consumption differences between locations are not as important as is the number of customers a particular location attracts. This causes you to be concerned for a while because the variable you had planned to study originally, gross sales volume, would vary as prices changed between locations. Since your company controls these prices, you feel that you would rather have an estimate of the "potential" for such sales. As a result, you decide to specify your dependent variable as the number of customers served (actually, the number of checks or bills that the waiters and waitresses handed out) in a given location in the most recent year for which complete data are available.

2. *Specify the model: Select the independent variables and the functional form.* Your discussions and personal investigations lead to a number of suggested variables that should help explain the attractiveness of a particular site to potential customers. After a while, you realize that there are three major determinants of sales (cus-

tomers) on which virtually everyone agrees. These are the number of people who live near the location, the general income level of the location, and the number of direct competitors close to the location. In addition, there are two other good suggestions for potential explanatory variables. These are the number of cars passing the location per day and the number of months that the particular restaurant has been open. After some serious consideration of your alternatives, you decide not to include the last possibilities. All the locations have been open long enough to have achieved a stable clientele, so the number of months open would not be likely to be important. In addition, data are not available for the number of passing cars for all the locations. Should population prove to be a poor measure of the available customers in a location, you'll have to decide whether to ask your boss for the money to collect complete traffic data.

The exact definitions of the independent variables you decide to include are:

C = Competition – the number of direct market competitors within a two-mile radius of the Woody's location

P = Population – the number of people living within a three-mile radius of the Woody's location

I = Income – the average household income of the population measured in variable P

Since you have no reason to suspect anything other than a linear functional form, that's what you decide to use.

3. *Hypothesize the expected signs of the coefficients.* After thinking about which variables to include, you expect hypothesizing signs will be easy. For two of the variables, you're right. Everyone expects that the more competition, the fewer customers (holding constant the population and income of an area), and also that the more people that live near a particular restaurant, the more customers (holding constant the competition and income). You expect that the greater the income in a particular area, the more people will choose to eat away from home and the more people will choose to eat in a family restaurant instead of in the lower-priced fast-food chains. However, people in especially high-income areas might want to eat in a restaurant that is higher in quality than a family restaurant. Some investigation reveals that it is virtually impossible to get zoning clearance to build a 24-hour facility in a "ritzy" residential neighborhood. You remain slightly worried that the income variable might not be as unambiguous a measure of the appeal of a location as you had thought. To sum, you expect:

$$\overset{}{Y_i} = f(\overset{-}{C}, \overset{+}{P}, \overset{+}{I}) = \beta_0 + \beta_c C_i + \beta_p P_i + \beta_I I_i + \epsilon_i$$

where the signs above the variables indicate the expected impact of that particular independent variable on the dependent variable, holding constant the other two explanatory variables, and ϵ_i is a typical stochastic error term.

4. *Collect the data.* You want to include every restaurant in the Woody's chain in your study, and, after some effort, you come up with data for your dependent variable and all your independent variables for all 33 locations. You're confident that the quality of your data is excellent for three reasons: each manager measured each variable identically, you've included each restaurant in the sample, and all the information is from the same year. (The data-set is at the end of this section, along with a sample computer output for the regression estimated.)

5. *Estimate and evaluate the equation.* You take the data-set and enter it into the computer. You then ask for an OLS regression on the data, but you do so only after thinking through your model once again to see if there are any hints that you've made theoretical mistakes. You end up admitting that, while you cannot be sure you are right, you've done the best you can, and so you estimate the equation, obtaining:

$$\begin{aligned} Y_i = 102{,}192 &- \underset{(2053)}{9075C_i} + \underset{(0.073)}{0.354P_i} + \underset{(0.543)}{1.288I_i} \qquad (3.14) \\ t &= \quad -4.42 \qquad\quad 4.88 \qquad\quad 2.37 \\ n &= 33 \quad \bar{R}^2 = .579 \end{aligned}$$

This equation satisfies your needs in the short run. In particular, the estimated coefficients in the equation have the signs you expected. The overall fit, while not outstanding, seems reasonable for such a diverse group of locations. To predict sales at potential locations, you obtain the values of C, P, and I for each location and then plug them into Equation 3.14. Other things being equal, the higher the predicted Y, the better the location from Woody's point of view.

6. *Document the results.* The results summarized in Equation 3.14 meet our documentation requirements. (Note that we include *t*-values[6] for completeness even though we won't make use of them

6. Throughout the text, the number in parentheses below a coefficient estimate will be the standard error of that estimated coefficient. Some authors put the *t*–score in parentheses, though, so be alert when reading journal articles or other books.

until Chapter 5.) However, it's not easy for a beginning researcher to wade through a computer's regression output to find all the numbers required for documentation. You'll probably have an easier time reading your own computer system's printout if you take the time to "walk through" the sample computer output for the Woody's model on the next two pages. This sample output was produced by the ECSTAT computer program, but it's similar to those produced by SAS, SHAZAM, TSP, and others.

Page one of the computer output summarizes the input data. The first items listed are the actual data. These are followed by two tables that describe these data: The first includes means, standard deviations, and variances; the second lists the simple correlation coefficients between all pairs of variables in the data-set. Numbers followed by "E+06" or "E−01" are expressed in a scientific notation indicating that the printed decimal point should be moved six places to the right or one place to the left, respectively.

The second page summarizes the OLS estimates generated from the data. It starts with a listing of the estimated coefficients, their estimated standard errors, and the associated *t*-values, and follows with R^2, $\bar{R}^2$, the standard error of the regression, RSS, and the *F*-ratio. This is followed by a listing of the observed Ys, the predicted Ys, and the residuals for each observation.

In future chapters, we'll return to this example in order to apply various tests and ideas as we learn them.

3.5 Summary

1. A proxy variable is one that substitutes for a theoretically relevant variable when data on the desired variable are incomplete or missing altogether. The more highly correlated the proxy variable is with the originally specified variable, the better a proxy it is.

2. A dummy variable takes on only the value of one or zero, depending on whether or not some condition is met. An example of a dummy variable would be X equals 1 if a particular individual is female and 0 if the person is male. One fewer dummy variable is constructed than there are qualitative conditions.

3. Six steps typically taken in applied regression analysis are:
   a. Review the literature.
   b. Specify the model: Select the independent variables and the functional form.
   c. Hypothesize the expected signs of the coefficients.
   d. Collect the data.
   e. Estimate and evaluate the equation.
   f. Document the results.

TABLE 3.1 DATA FOR THE WOODY'S RESTAURANTS EXAMPLE

OBSERV.	Y	C	P	I
1	107919.0	3.000000	65044.00	13240.00
2	118866.0	5.000000	101376.0	22554.00
3	98579.00	7.000000	124989.0	16916.00
4	122015.0	2.000000	55249.00	20967.00
5	152827.0	3.000000	73775.00	19576.00
6	91259.00	5.000000	48484.00	15039.00
7	123550.0	8.000000	138809.0	21857.00
8	160931.0	2.000000	50244.00	26435.00
9	98496.00	6.000000	104300.0	24024.00
10	108052.0	2.000000	37852.00	14987.00
11	144788.0	3.000000	66921.00	30902.00
12	164571.0	4.000000	166332.0	31573.00
13	105564.0	3.000000	61951.00	19001.00
14	102568.0	5.000000	100441.0	20058.00
15	103342.0	2.000000	39462.00	16194.00
16	127030.0	5.000000	139900.0	21384.00
17	166755.0	6.000000	171740.0	18800.00
18	125343.0	6.000000	149894.0	15289.00
19	121886.0	3.000000	57386.00	16702.00
20	134594.0	6.000000	185105.0	19093.00
21	152937.0	3.000000	114520.0	26502.00
22	109622.0	3.000000	52933.00	18760.00
23	149884.0	5.000000	203500.0	33242.00
24	98388.00	4.000000	39334.00	14988.00
25	140791.0	3.000000	95120.00	18505.00
26	101260.0	3.000000	49200.00	16839.00
27	139517.0	4.000000	113566.0	28915.00
28	115236.0	9.000000	194125.0	19033.00
29	136749.0	7.000000	233844.0	19200.00
30	105067.0	7.000000	83416.00	22833.00
31	136872.0	6.000000	183953.0	14409.00
32	117146.0	3.000000	60457.00	20307.00
33	163538.0	2.000000	65065.00	20111.00

MEANS, VARIANCES, AND CORRELATIONS
SAMPLE RANGE: 1-33

VARIABLE	MEAN	STANDARD DEV	VARIANCE
Y	125634.6	22404.09	5.01943E+08
C	4.393939	1.919300	3.683712
P	103887.5	55884.51	3.12308E+09
I	20552.58	5141.865	2.64388E+07

	CORRELATION COEFF		CORRELATION COEFF
Y,Y	1.000000	C,Y	-0.144224
C,C	1.000000	P,Y	0.392567
P,C	0.726250	P,P	1.000000
I,Y	0.537022	I,C	-0.031534
I,P	0.245197	I,I	1.000000

TABLE 3.2 ACTUAL COMPUTER OUTPUT (USING THE ECSTAT PROGRAM) FROM THE WOODY'S REGRESSION

```
ORDINARY LEAST SQUARES                    DEPENDENT VARIABLE IS Y
SAMPLE RANGE:              1-33

                         COEFFICIENT        STANDARD ERROR        T-SCORE
      CONST               102192.4            12799.83           7.983891
      C                  -9074.674            2052.674          -4.420904
      P                   0.354668            0.072680           4.879810
      I                   1.287923            0.543293           2.370584

   R-squared               0.618153     Mean of depend var       125634.6
   Adjusted R-squared      0.578652     Std dev depend var       22404.09
   Std err of regress      14542.78     Residual sum            5.02041E-10
   Durbin Watson stat      1.758193     Sum squared resid       6.13328E+09
   F Statistic             15.64894


OUT OF SAMPLE FORECAST: Y

 OBSERVATION       ACTUAL            PREDICTED           RESIDUAL
      1           107919.0           115089.6           -7170.559
      2           118866.0           121821.7           -2955.740
      3           98579.00           104785.9           -6206.864
      4           122015.0           130642.0           -8627.041
      5           152827.0           126346.5            26480.55
      6           91259.00           93383.88           -2124.877
      7           123550.0           106976.3            16573.66
      8           160931.0           135909.3            25021.71
      9           98496.00           115677.4           -17181.36
     10           108052.0           116770.1           -8718.094
     11           144788.0           138502.6            6285.425
     12           164571.0           165550.0           -979.0342
     13           105564.0           121412.3           -15848.30
     14           102568.0           118275.5           -15707.47
     15           103342.0           118895.6           -15553.63
     16           127030.0           133978.1           -6948.114
     17           166755.0           132868.1            33886.91
     18           125343.0           120598.1            4744.898
     19           121886.0           116832.3            5053.700
     20           134594.0           137985.6           -3391.591
     21           152937.0           149717.6            3219.428
     22           109622.0           117903.5           -8281.508
     23           149884.0           171807.2           -21923.22
     24           98388.00           99147.65           -759.6514
     25           140791.0           132537.5            8253.518
     26           101260.0           114105.4           -12845.43
     27           139517.0           143412.3           -3895.303
     28           115236.0           113883.4            1352.599
     29           136749.0           146334.9           -9585.906
     30           105067.0           97661.88            7405.122
     31           136872.0           131544.4            5327.620
     32           117146.0           122564.5           -5418.450
     33           163538.0           133021.0            30517.00
```

Exercises

(Answers to even-numbered exercises are in Appendix A.)

1. Write the meaning of each of the following terms without reference to the book (or your notes), and compare your definition with the version in the text for each:
   a. proxy variable
   b. dummy variable
   c. omitted condition
   d. cross-sectional data-set
   e. specification error
   f. degrees of freedom

2. Contrary to their name, dummy variables are not easy to understand without a little bit of practice:
   a. Specify a dummy variable that would allow you to distinguish between undergraduate students and graduate students in your econometrics class.
   b. Specify a regression equation to explain the grade (measured on a scale of 4.0 for an A) each student in your class received on his or her first econometrics test (Y) as a function of the student's grade in a previous course in statistics (G), the number of hours the student studied for the test (H), and the dummy variable you created above (D). Are there other variables you would want to add? Explain your answer.
   c. What is the hypothesized sign of the coefficient of D? Does the sign depend on the exact way in which you defined D? How?
   d. Suppose that you collected the data and ran the regression and found an estimated coefficient for D that had the expected sign and an absolute value of 0.5. What would this mean in real-world terms? By the way, what would have happened if you had only undergraduates or only graduate students in your class?

3. For each of the following dependent variables, data for a particular independent variable are not available and the choice of a proxy variable must be considered:
   a. In an equation where the dependent variable is total sales (dollars) of Classic Cola (time-series) and the desired independent variable is the amount of advertising by Dr. Popper (Classic Cola's main competitor), it would be hard to get the Dr. Popper folks to give advertising data to their archrivals. Would a good

proxy variable be total advertising in the soft-drink industry? What about total sales of Dr. Popper? Explain your reasoning.

b. In an equation where the dependent variable is the value of exports from the U.S. to Japan (time-series) and the desired independent variable is the amount of bureaucratic red tape making such exports more difficult, the independent variable is so vaguely defined as to make its exact measurement impossible. Would a good proxy be the Japanese average tariff rate? What about the number of laws in Japan that apply to goods imported from the U.S.?

c. In an equation where the dependent variable is the "reservation wage" (the lowest wage that an unemployed worker will accept) and the desired independent variable is the "expected wage" (the wage that the worker expects to earn once a job is landed), data on expected wages are usually very hard to get. Would the worker's last wage be a good proxy? How about the average wage in the worker's profession? Explain your reasoning.

4. Return to the Woody's regression example of Section 3.4.
   a. In any applied regression project there is the distinct possibility that an important explanatory variable has been omitted. Reread the discussion of the selection of independent variables and come up with a suggestion for an independent variable that has not been included in the model (other than the traffic variable already mentioned). Why do you think this variable was not included?
   b. What other kinds of criticisms would you have of the sample or independent variables chosen in this model?
   c. What's wrong with this kind of thinking?: "I know that the coefficients in Equation 3.14 are wrong because if you just divide the average number of checks in the sample (125,635 as can be seen in the computer output at the end of Table 3.1) by the average population in the sample (103,888 from the same source) you get 1.21, a number that is much larger than the computer estimate of the same coefficient (0.3547). Therefore, it's easier and more accurate just to do the calculations myself!"

5. Suppose you were told that, while data on traffic for Equation 3.14 are still too expensive to obtain, a variable on traffic, called T_i, *is* available that is defined as 1 if more than 45,000 cars per day pass the restaurant and 0 otherwise. Further suppose that when the new variable (T_i) is added to the equation the results are:

$$
\begin{aligned}
\hat{Y}_i = 95{,}236 \; - \; & 7{,}307C_i \; + \; 0.320P_i \; + \; 1.28I_i \; + \; 10{,}994T_i \\
& (2{,}153) \qquad (0.073) \qquad (0.51) \qquad (5{,}577) \\
t = \; & -3.39 \qquad\;\; 4.24 \qquad\;\; 2.47 \qquad\;\; 1.97 \\
n = \; & 33 \qquad\qquad \bar{R}^2 = .617
\end{aligned}
$$

a. Is the new variable a dummy variable or a proxy variable? Why?
b. What is the expected sign of the coefficient of the new variable?
c. Would you prefer this equation to the original one? Why?
d. Does the fact that $\bar{R}^2$ is higher in the new equation mean that it is necessarily better than the old one?

6. Suppose that the population variable in Section 3.4 had been defined in different units as in:

P = Population – thousands of people living within a three-mile radius of the Woody's location

a. Given this definition of P, what would the estimated slope coefficients in Equation 3.14 have been?
b. Given this definition of P, what would the estimated slope coefficients in the equation in Exercise 5 (above) have been?
c. Are any other coefficients affected by this change?

7. Develop your own regression equation to explain a dependent variable that relates directly to you and your immediate surroundings or personal interests. Your main focus here should be on choosing appropriate independent variables and on hypothesizing expected signs. Follow through the first three steps in applied regression analysis:
a. Write a review of the relevant theory and research about your chosen dependent variable.
b. Select your independent variables, being specific as to how they are defined and measured.
c. Hypothesize the expected signs of the coefficients of your independent variables and explain your reasoning.

At this point in your econometric career, it's probably a bit premature to attempt to estimate your equation, but if you're still interested in this dependent variable after completing Chapter 5, you might collect data and estimate the equation.

8. In an effort to explain regional wage differentials, you collect wage data from 7338 unskilled workers, divide the country into four

regions (Northeast, South, Midwest, and West), and estimate the following equation (standard errors in parentheses):

$$\hat{Y}_i = 4.78 - \underset{(0.019)}{0.038E_i} - \underset{(0.010)}{0.041S_i} - \underset{(0.012)}{0.048W_i}$$
$$\bar{R}^2 = .49 \qquad n = 7{,}338$$

where: Y_i = the hourly wage (in dollars) of the *i*th unskilled worker
E_i = a dummy variable equal to 1 if the *i*th worker lives in the Northeast and 0 otherwise
S_i = a dummy variable equal to 1 if the *i*th worker lives in the South and 0 otherwise
W_i = a dummy variable equal to 1 if the *i*th worker lives in the West and 0 otherwise

a. What is the omitted condition in this equation?
b. If you add a dummy variable for the omitted condition to the equation without dropping E_i, S_i, or W_i, what will happen?
c. If you add a dummy variable for the omitted condition to the equation and drop E_i, what will the sign of the new variable's estimated coefficient be?
d. Which of the following three statements is most correct? Least correct? Explain your answer.
   I. The equation explains 49 percent of the variation of Y around its mean with regional variables alone, so there must be quite a bit of wage variation by region.
   II. The coefficients of the regional variables are virtually identical, so there must not be much wage variation by region.
   III. The coefficients of the regional variables are quite small compared with the average wage, so there must not be much wage variation by region.
e. If you were going to add one variable to this model, what would it be? Justify your choice.

9. Calculate the real price of a "Whitney GT" automobile for the following years:

	Nominal Price	CPI(current year)	CPI(base year)
a.	\$4,000	95	100
b.	\$10,000	160	100
c.	\$18,000	300	100

10. Consider the following model[7] of the per capita consumption of a particular kind of meat in the U.S. (standard errors in parentheses):

$$\widehat{PK}_t = 17.0 - \underset{(0.5)}{7.7PP_t} + \underset{(0.4)}{4.2PB_t} + \underset{(0.08)}{0.23YD_t} - \underset{(0.2)}{0.9D_{1t}} - \underset{(0.2)}{1.6D_{2t}} - \underset{(0.2)}{1.5D_{3t}}$$

$$\bar{R}^2 = .931 \qquad n = 40 \text{ (quarterly: 1975–1984)}$$

where: PK_t = per capita pounds of a particular kind of meat consumed in quarter t
PP_t = dollar price of a pound of the meat in quarter t
PB_t = dollar price of a pound of a substitute meat in quarter t
YD_t = per capita disposable income (dollars) in quarter t
D_{1t} = a dummy equal to 1 in the first quarter of the year and 0 otherwise
D_{2t} = a dummy equal to 1 in the second quarter of the year and 0 otherwise
D_{3t} = a dummy equal to 1 in the third quarter of the year and 0 otherwise

a. Hypothesize signs for each of the first three slope coefficients in the equation. Which if any of the signs of the estimated coefficients are different from your expectations?
b. Do the sizes of the coefficients of PP_t and PB_t make sense? Explain your reasoning.
c. Carefully interpret the meanings of the estimated coefficients of the three seasonal dummy variables. (Hint: Remember the omitted condition.)
d. The price and income variables in this model are measured in current (nominal) dollars. How would you go about converting this model to real dollars?

11. Michael Lovell[8] estimated the following model of the gasoline mileage of various models of cars (standard errors in parentheses):

$$\hat{G}_i = 22.008 - \underset{(0.001)}{0.002W_i} - \underset{(0.71)}{2.76A_i} + \underset{(1.41)}{3.28D_i} + \underset{(0.097)}{0.415E_i}$$

$$\bar{R}^2 = .82$$

7. This example is explored in more detail in Section 11.6.
8. Michael C. Lovell, "Tests of the Rational Expectations Hypothesis," *American Economic Review*, March 1986, pp. 110–124.

where: G_i = miles per gallon of the *i*th model as reported by Consumers' Union based on actual road tests
W_i = the gross weight (in pounds) of the *i*th model
A_i = a dummy variable equal to 1 if the *i*th model has an automatic transmission and 0 otherwise
D_i = a dummy variable equal to 1 if the *i*th model has a diesel engine and 0 otherwise
E_i = the published U.S. Environmental Protection Agency estimate of the miles per gallon the *i*th model

a. Hypothesize signs for the slope coefficients of W and E. Which, if any, of the signs of the estimated coefficients are different from your expectations?
b. Carefully interpret the meanings of the estimated coefficients of A_i and D_i.
c. Lovell included one of the variables in the model to test a specific hypothesis, but that variable wouldn't necessarily be in another researcher's gas mileage model. What variable do you think Lovell added? What hypothesis do you think Lovell wanted to test?

12. Your boss is about to start production of her newest box office smash-to-be, "The Invasion of the Economists, Part II," when she calls you in and tells you to build a model of the gross receipts of all the movies produced in the last five years. Your regression[9] is (standard errors in parentheses):

$$\begin{aligned}\hat{G}_i = 781 + &\underset{(5.9)}{15.4T_i} - \underset{(674)}{992F_i} + \underset{(800)}{1770J_i} \\ &+ \underset{(1006)}{3027S_i} - \underset{(2381)}{3160B_i} + \cdots \\ \bar{R}^2 &= .485 \qquad n = 254\end{aligned}$$

where: G_i = final gross receipts of the *i*th motion picture (in thousands of dollars)
T_i = the number of screens (theaters) on which the *i*th film was shown in its first week
F_i = a dummy variable equal to 1 if the star of the *i*th film is a female and 0 otherwise

9. This estimated equation (but not the question) comes from a final exam in managerial economics given at the Harvard Business School in Feb. 1982, pp. 18–30.

J_i = a dummy variable equal to 1 if the *i*th movie was released in June or July and 0 otherwise

S_i = a dummy variable equal to 1 if the star of the *i*th film is a superstar (like Tom Cruise or Milton) and 0 otherwise

B_i = a dummy variable equal to 1 if at least one member of the supporting cast of the *i*th film is a superstar and 0 otherwise

a. Hypothesize signs for each of the slope coefficients in the equation. Which if any of the signs of the estimated coefficients are different from your expectations?

b. Milton, the star of the original "Invasion of the Economists," is demanding $4 million from your boss to appear in the sequel. If your estimates are trustworthy, should she say yes or hire Arnold (a nobody) for $500,000?

c. Your boss wants to keep costs low, and it would cost $1.2 million to release the movie on an additional 200 screens. Assuming your estimates are trustworthy, should she spring for the extra screens?

d. The movie is scheduled for release in September, and it would cost $1 million to speed up production enough to allow a July release without hurting quality. Assuming your estimates are trustworthy, is it worth the rush?

e. You've been assuming that your estimates are trustworthy. Do you have any evidence that this is not the case? Explain your answer. (Hint: Assume that the equation contains no specification errors.)

4

The Classical Model

The classical model of econometrics has nothing to do with ancient Greece or even the classical economic thinking of Adam Smith. Instead, the term *classical* refers to a set of fairly basic assumptions required to hold in order for the Ordinary Least Squares (OLS) procedure to be considered the Best (minimum variance) Linear Unbiased Estimator (BLUE) available for regression models. When one or more of these assumptions do not hold, other estimation techniques sometimes may be better than OLS.

As a result, one of the most important jobs in regression analysis is to decide whether the classical assumptions hold for a particular equation. If so, the OLS estimation technique is the best available. Otherwise, the pros and cons of alternative estimation techniques must be weighed. These alternatives are usually adjustments to OLS that take account of the particular assumption that has been violated. In a sense, most of the rest of this book deals in one way or another with the question of what to do when one of the classical assumptions is not met. Since econometricians spend so much time analyzing violations of them, it is crucial that they know and understand these assumptions.

4.1 The Classical Assumptions

The **classical assumptions** must be met in order for OLS estimators to be the best available. Because of their importance in regression analysis, the assumptions are presented here in tabular form and standard statistical notation as well as in words. Subsequent chapters will investigate major violations of the assumptions and introduce estimation techniques that may provide better estimates in such cases.

An error term satisfying Assumptions I through V is called a **classical error term,** and if Assumption VII is added, the error term is called a **classical normal error term.**

The Classical Assumptions

I. *The regression model is linear in the coefficients and the error term.*

II. *The error term has a zero population mean.*

III. *All explanatory variables are uncorrelated with the error term.*

IV. *Observations of the error term are uncorrelated with each other (no serial correlation).*

V. *The error term has a constant variance (no heteroskedasticity).*

VI. *No explanatory variable is a perfect linear function of other explanatory variables (no perfect multicollinearity).*

VII. *The error term is normally distributed (this assumption is optional but usually is invoked).*

I. The regression model is linear in the coefficients and the error term. The regression model is assumed to be linear in the coefficients:

$$Y_i = \beta_0 + \beta_1 X_{1i} + \beta_2 X_{2i} + \cdots + \beta_K X_{Ki} + \epsilon_i \qquad (4.1)$$

On the other hand, the regression model does not have to be linear in the variables because OLS can be applied to equations that are nonlinear in the variables. The good properties of OLS estimators hold regardless of the functional form of the *variables* as long as the form of the equation to be estimated is linear in the *coefficients*. For example, an exponential function:

$$Y_i = e^{\beta_0} X_i^{\beta_1} e^{\epsilon_i} \tag{4.2}$$

where e is the base of the natural log, can be transformed by taking the natural log of both sides of the equation:

$$\ln(Y_i) = \beta_0 + \beta_1 \ln(X_i) + \epsilon_i \tag{4.3}$$

The variables can be relabeled as $Y_i^* = \ln(Y_i)$ and $X_i^* = \ln(X_i)$, and the form of the equation is linear in the coefficients:

$$Y_i^* = \beta_0 + \beta_1 X_i^* + \epsilon_i \tag{4.4}$$

In Equation 4.4, the properties of the OLS estimator of the βs still hold because the equation is linear in the coefficients. Equations that are nonlinear in the coefficients will be discussed in Section 7.6.

II. The error term has a zero population mean. As was pointed out in Section 1.2.3, econometricians add a stochastic (random) error term to regression equations to account for variation in the dependent variable that is not explained by the model. The specific value of the error term for each observation is determined purely by chance. Probably the best way to picture this concept is to think of each observation of the error term as being drawn from a random variable distribution such as the one illustrated in Figure 4.1.

Classical Assumption II says that the mean of this distribution is zero. That is, when the entire population of possible values for the stochastic error term is considered, the average value of that population is zero. For a small sample, it is not likely that the mean would be exactly zero, but as the size of the sample approaches infinity, the mean of the sample approaches zero.

To compensate for the chance that the mean of the population ϵ might not equal zero, the mean of ϵ_i for any regression is forced to be zero by the existence of the constant term in the equation. If the mean of the error term is not equal to zero, then this nonzero amount is implicitly (because error terms are unobservable) subtracted from each error term and added instead to the constant term. This leaves the equation unchanged except that the new error term has a zero mean

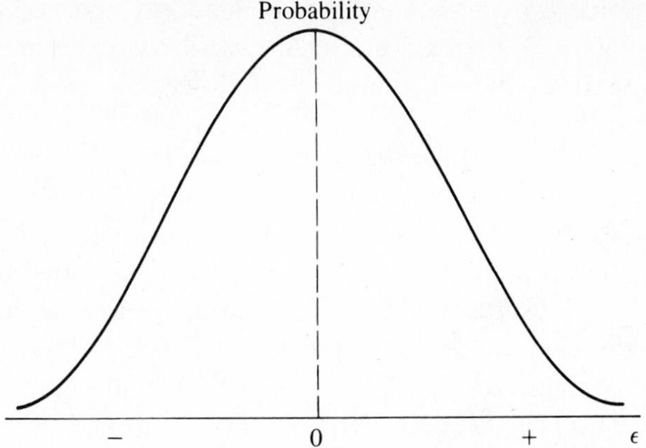

Figure 4.1 A ZERO-MEANED STOCHASTIC ERROR TERM DISTRIBUTION
Observations of stochastic error terms are assumed to be drawn from a zero-meaned random variable distribution. If Classical Assumption II is met, the expected value (the mean) of the error term is zero.

(and thus conforms to Assumption II). In addition, the constant term has been changed by the difference between the sample mean of the error term and zero. Partially because of this difference, it is risky to place much importance on the estimated magnitude of the constant term. In essence, the constant term equals the fixed portion of Y that cannot be explained by the independent variables, while the error term represents the stochastic portion of the unexplained value of Y.

While it's true that the error terms can never be observed, it might be instructive to imagine that we can observe them to see how the existence of a constant term forces the mean of the error term to be zero even in a sample. Consider a typical regression equation:

$$Y_i = \beta_0 + \beta_1 X_i + \epsilon_i \tag{4.5}$$

For example, if the mean of ϵ_i is 3 instead of 0, then[1] $E(\epsilon_i - 3) = 0$. If we add 3 to the constant term and subtract it from the error term, we obtain:

$$Y_i = (\beta_0 + 3) + \beta_1 X_i + (\epsilon_i - 3) \tag{4.6}$$

1. Here, as in Chapter 1, the "E" refers to the expected value (mean) of the item in parentheses after it. Thus $E(\epsilon_i - 3)$ equals the expected value of the stochastic error term epsilon minus 3. In this specific example, since we've defined $E(\epsilon_i) = 3$, we know that $(\epsilon_i - 3) = 0$.

Since Equations 4.5 and 4.6 are equivalent (do you see why?), and since $E(\epsilon_i - 3) = 0$, then Equation 4.6 can be written in a form that has a zero mean for the error term:

$$Y_i = \beta_0^* + \beta_1 X_i + \epsilon_i^* \tag{4.7}$$

where $\beta_0^* = \beta_0 + 3$ and $\epsilon_i^* = \epsilon_i - 3$. As can be seen, Equation 4.7 conforms to Assumption II. This form is always assumed to apply for the true model. Therefore, the second classical assumption is assured as long as there is a constant term included in the equation. This statement is correct as long as all other classical assumptions are met and there are no specification errors in the equation.

III. All explanatory variables are uncorrelated with the error term. It is assumed that the observed values of the explanatory variables are determined independently of the values of the dependent variable and the error term. Explanatory variables (Xs) are considered to be determined outside the context of the regression equation in question.

If an explanatory variable and the error term were instead correlated with each other, the OLS estimates would be likely to attribute to the X some of the variation in Y that actually came from the error term. If the error term and X were positively correlated, for example, then the estimated coefficient would probably be higher than it would otherwise have been (biased upward), because the OLS program would mistakenly attribute the variation in Y caused by ϵ to have been caused by X instead. As a result, it's important to assure that the explanatory variables are uncorrelated with the error term.

A common economic application that violates this assumption is any model that is simultaneous in nature. For example, in a simple Keynesian macroeconomic model, an increase in consumption (caused perhaps by an unexpected change in tastes) will increase aggregate demand and therefore aggregate income. An increase in income, however, will also increase consumption; so, income and consumption are interdependent. Note, however, that the error term in the consumption function (which is where an unexpected change in tastes would appear) and an explanatory variable in the consumption function (income) have now moved together. As a result, Classical Assumption III has been violated; the error term is no longer uncorrelated with all the explanatory variables. This will be considered in more detail in Chapter 14.

IV. Observations of the error term are uncorrelated with each other (no serial correlation). The observations of the error term are drawn independently from each other. If a systematic correlation exists between one observation of an error term and another, then it will be more difficult for OLS to get precise estimates of the coefficients of the explanatory variables. For example, if the fact that the ϵ from one observation is positive increases the probability that the ϵ from another observation also is positive, then the two observations of the error term are said to be positively correlated. Such a correlation would violate Classical Assumption IV.

In economic applications, this assumption is most important in time-series models. In such a context, Assumption IV says that an increase in the error term in one time period (a random shock, for example) does not show up in or affect in any way the error term in another time period. In some cases, though, this assumption is unrealistic, since the effects of a random shock sometimes last for a number of time periods. If, over all the observations of the sample, ϵ_{t+1} is correlated with ϵ_t, then the error term is said to be **serially correlated** (or *autocorrelated*), and this assumption is violated. Violations of this assumption are considered in more detail in Chapter 9.

V. The error term has a constant variance. The variance of the distribution from which the observations of the error term are drawn is constant. That is, the observations of the error term are assumed to be drawn continually from identical distributions (for example, the one pictured in Figure 4.1). The alternative would be for the variance of the distribution of the error term to change for each observation or range of observations. In Figure 4.2, for example, the variance (or dispersion) of the error term is shown to increase as X increases; such a pattern violates Classical Assumption V. Such a violation makes precise estimation difficult, because a particular deviation from a mean (in this case an error term) can be called a statistically large or small deviation only when it is compared with the standard deviation (which is the square root of the variance) of the distribution in question. If you assume that all error term observations are drawn from a distribution with a constant variance when in reality they are drawn from distributions with different variances, then the relative importance of changes in Y is very hard to judge. While the actual values of the error term are not observable directly, the lack of a constant variance for the distribution of the error term still causes OLS to generate imprecise estimates of the coefficients of the independent variables.

In economic applications, Assumption V is most important in cross-sectional data-sets. For example, in a cross-sectional analysis of household consumption patterns, the variance (or dispersion) of the consumption of certain goods might be greater for higher-income house-

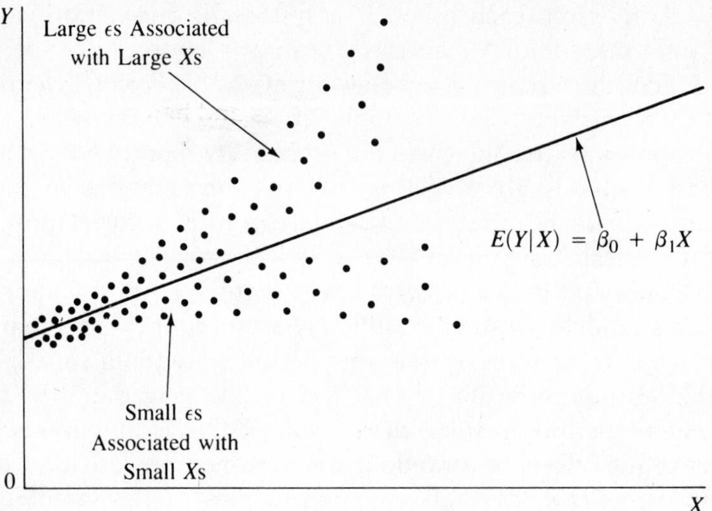

Figure 4.2 AN ERROR TERM WHOSE VARIANCE INCREASES AS X INCREASES (HETEROSKEDASTICITY)

One example of Classical Assumption V not being met is when the variance of the error term increases as X increases. In such a situation (called heteroskedasticity), the observations are on average farther from the true regression line for large values of X than they are for small values of X.

holds because they have more discretionary income than do lower-income households. Thus the absolute amount of the dispersion is greater even though the percentage dispersion is the same. The violation of Assumption V is referred to as **heteroskedasticity** and will be discussed in more detail in Chapter 10.

VI. No explanatory variable is a perfect linear function of other explanatory variables. Perfect **collinearity** between two independent variables implies that they are really the same variable, or that one is a multiple of the other and/or a constant has been added to one of the variables. That is, the relative movements of one explanatory variable will be matched exactly by the relative movements of the other even though the absolute size of the movements might differ. Because every time one of the variables moves, it is matched exactly by a relative movement in the other, the OLS estimation procedure will be incapable of distinguishing one variable from the other.

Many instances of perfect collinearity (or *multicollinearity* if more than two independent variables are involved) are the result of the researcher not accounting for identities (definitional equivalences) among the independent variables. It can be corrected easily by dropping one of the perfectly collinear variables from the equation. Suppose you were attempting to explain home purchases and had included both real and nominal interest rates as explanatory variables in your equation for a time period in which inflation (and expected inflation) was constant. In such an instance, real and nominal interest rates would differ by a constant amount, and the OLS procedure would not distinguish between them. Note that perfect multicollinearity can be caused by an accident in the sample at hand. While real and nominal interest rates would be perfectly multicollinear if inflation were constant in a given sample, they would not be perfectly multicollinear in samples where there was some change in inflation.

Similarly, perfect multicollinearity occurs when two independent variables always sum to a third. For example, the explanatory variables "games won" and "games lost" for a sports team with a constant number of games played and no ties will always sum to that constant, and perfect multicollinearity will exist. In such cases, the OLS computer program (or any other estimation technique) will be unable to estimate the coefficients unless there is a rounding error. The remedy is easy in the case of perfect multicollinearity: just delete one of the two perfectly correlated variables.

Finally, it's also possible to violate Assumption VI if one of the explanatory variables has a variance of zero. In this case, the variable will be perfectly collinear with the constant term, and OLS estimation will be impossible. Luckily, it's quite unusual to encounter perfect multicollinearity, but, as we shall see in Chapter 8, even imperfect multicollinearity can cause problems for estimation.

VII. The error term is normally distributed. While we have already assumed that observations of the error term are drawn independently (Assumption IV) from a distribution that is zero-meaned (Assumption II) and that has a constant variance (Assumption V), we have said little about the shape of that distribution. Assumption VII states that the observations of the error term are drawn from a distribution that is normal, that is, bell-shaped and generally following the symmetrical pattern portrayed in Figure 4.1. This assumption of normality is not required for OLS estimation. Its major use is in **hypothesis testing,** which uses the calculated regression statistics to accept or reject hypotheses about economic behavior. One example of such a test is deciding whether a particular demand curve is elastic or inelastic in a particular range. Hypothesis testing is the subject of Chapter 5, and without the normality assumption, most of the tests in that chapter would be invalid.

4.2 The Normal Distribution of the Error Term

In this section we briefly introduce the concept of the normal distribution and explain why the Central Limit Theorem tends to justify the assumption of normality for a stochastic error term.

The only assumption that is optional to the definition of the classical model is that the error term is normally distributed. It is usually justified and advisable to add the assumption of normality to the other six assumptions for two reasons:

1. The error term ϵ_i can be thought of as the composite of a number of minor influences or errors. As the number of these minor influences gets larger, the distribution of the error term tends to approach the normal distribution. This tendency is called the Central Limit Theorem.

2. The *t* statistic and the *F* statistic, which will be developed in Chapter 5, are not truly applicable unless the error term is normally distributed.

4.2.1 A Description of the Normal Distribution

The normal distribution is a symmetrical, continuous, bell-shaped curve. The parameters that describe normal distributions and allow us to differentiate between various normal distributions are the **mean** (μ, the measure of central tendency) and the **variance** (σ^2, the measure of dispersion). Two such normal distributions are shown in Figure 4.3. In normal distribution #1, the mean is 0 and the variance is 1; in normal distribution #2, the mean is 2 and the variance is 0.5.

A quick look at Figure 4.3 shows how normal distributions differ when the means and variances are different. When the mean is different, the entire distribution shifts. For example, distribution #2 is to the right of distribution #1 because its mean, 2, is greater than the mean of distribution #1. When the variance is different, the distribution becomes fatter or skinnier. For example, distribution #2 is distributed more compactly around its mean than distribution #1 because distribution #2 has a smaller variance. Observations drawn at random from distribution #2 will tend to be closer to the mean than those drawn from distribution #1, while distribution #1 will tend to have a higher likelihood of observations quite far from its mean.

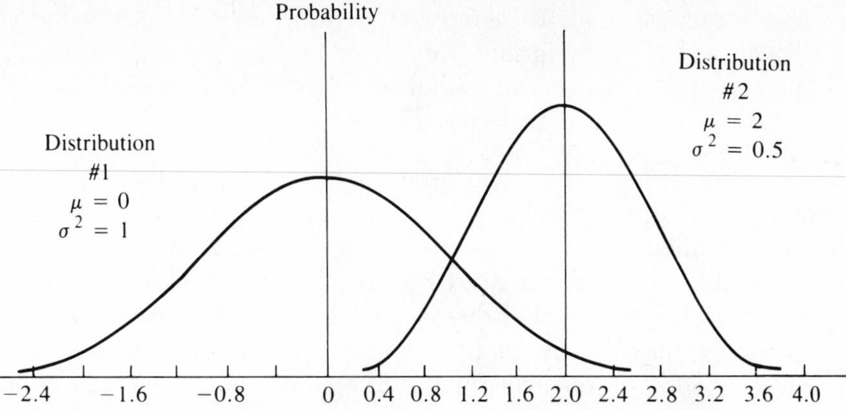

Figure 4.3 NORMAL DISTRIBUTIONS
While all normal distributions are symmetrical and bell-shaped, they do not necessarily have the same mean and variance. Distribution #1 has a mean of 0 and a variance of 1, while distribution #2 has a mean of 2 and a variance of 0.5. As can be seen, the whole distribution shifts when the mean changes, and the distribution gets fatter as the variance increases.

In Figure 4.3, distribution #1 represents what is called the **standard normal distribution** because it is a normal distribution with a mean equal to zero and a variance equal to one. This is the usual distribution given in statistical tables, such as Table B-7 in the back of this book. Often the parameters of a normal distribution will be listed in a compact summary form: $N(\mu, \sigma^2)$. For distribution #1, this notation would be N(0,1) and would stand for a normal distribution with mean zero and variance one.

4.2.2 The Central Limit Theorem and the Normality of the Error Term

As was mentioned in Chapter 1, the error term in a regression equation is assumed to be caused in part by the omission of a number of variables from the equation. These variables are expected to have relatively small individual effects on the hypothesized regression equation, and it is not advisable to include them as independent variables. The error term represents the combined effects of these omitted variables. This component of the error term is usually cited as the justification of the assumption of normality for the error term. In general, a random variable generated by the combined effects of a number of omitted, individually unimportant variables will be normally distributed according to the **Central Limit Theorem** that states:

Central Limit Theorem

> The mean (or sum) of a number of independent, identically distributed random variables will tend to be normally distributed, regardless of their distribution, if the number of different random variables is large enough.

The Central Limit Theorem becomes more valid as the number of omitted variables approaches infinity, but even a few are sufficient to show the tendency toward the normal bell-shaped distribution. The more variables omitted, the more quickly the distribution of the error term approaches the normal distribution because the various omitted variables are more likely to cancel out extreme observations. As a result, it is good econometric practice to assume a normally distributed stochastic error term in a regression that must omit a number of minor unrelated influences. Purposely omitting a few variables to help achieve normality for the error term, however, should *never* be considered.

Let's look at an example of the Central Limit Theorem and how the error term tends to be normally distributed if the number of omitted variables is large enough. First, suppose that only two potentially relevant variables are so minor that they are not included from an equation. Figure 4.4 shows ten computer-generated[2] observations of a stochastic error term that is the sum of two identically distributed variables. Note that while the observations of the error term are near zero, their distribution is hardly normal looking.

Suppose that we now increase the number of omitted variables from two to ten. Figure 4.5 shows ten computer-generated observations of a stochastic error term that is the sum of ten identically distributed variables. As we'd expect from the Central Limit Theorem, the resulting distribution is much more bell-shaped (normal) than is Figure 4.4. If we were to continue to add variables, the distribution would tend to look more and more like the normal distribution.

These figures show the tendency of errors to cancel each other out as the number of omitted variables increases. Why does this occur? Averaging the Xs, each of which is distributed according to the uniform distribution, bunches the observations toward the middle (because extreme values of any X tend to be offset by the others) resulting in a fairly normal distribution. The more Xs to be averaged, the more normal this distribution becomes.

2. In generating these figures, we assumed that all variables were uniformly distributed (every value equally and uniformly likely), that the other components of the error term were small in size compared with the omitted variables, and that the coefficients of the omitted variables were 1.0. These assumptions made the computations easier; the property shown holds even without them.

By the way, the omitted variables do not have to conform to the uniform distribution to produce this result; they can follow *any* probability distribution. Indeed, if the omitted variables were normally distributed, the error term would be normally distributed by definition, since the sum (or average) of normally distributed variables is also a normally distributed variable.

4.3 The Sampling Distribution of $\hat{\beta}$

Just as the error terms follow a probability distribution, so too do the estimates of the true slope βs (the $\hat{\beta}$s or "β-hats") follow such a

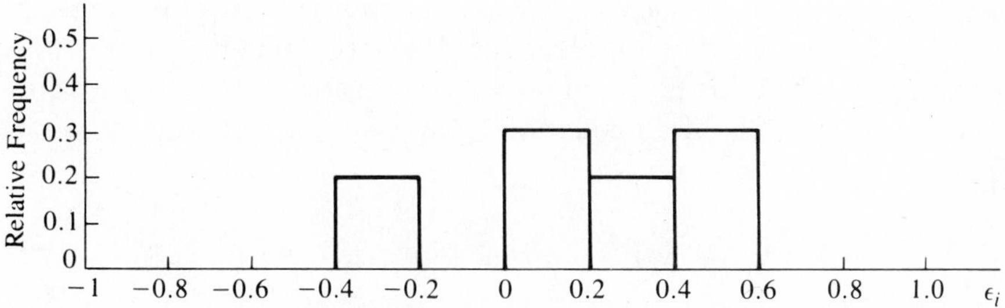

Figure 4.4 RELATIVE FREQUENCY OF THE ERROR TERM AS AN AVERAGE OF 2 OMITTED VARIABLES: 10 OBSERVATIONS

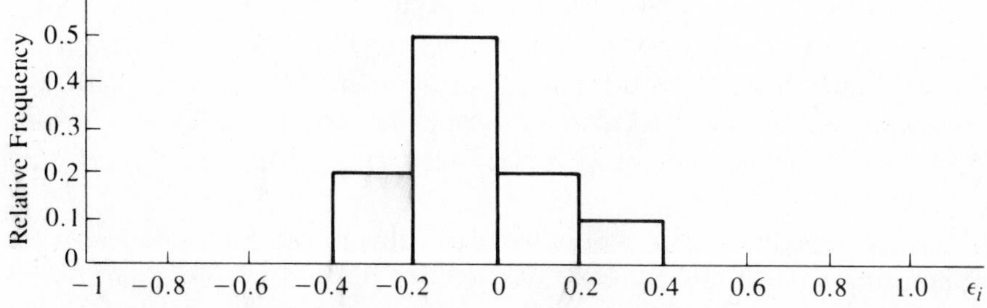

Figure 4.5 RELATIVE FREQUENCY OF THE ERROR TERM AS AN AVERAGE OF 10 OMITTED VARIABLES: 10 OBSERVATIONS

As the number of different omitted variables increases, the distribution of the sum of those variables approaches the normal distribution. This tendency (called the Central Limit Theorem) can be seen in Figures 4.4 and 4.5. As the number of omitted variables increases from 2 to 10, the distribution of their average does indeed become more symmetrical and bell-shaped.

probability distribution. In fact, each different sample of data typically produces a different set of $\hat{\beta}$s. These $\hat{\beta}$s usually are assumed to be normally distributed because the normality of the error term implies that the OLS estimator of the $\hat{\beta}$s in the simple linear regression model is normally distributed as well. The probability distribution of the $\hat{\beta}$s is called a **sampling distribution** because it is based on a number of sample drawings of the error term. To show this, we will discuss the general idea of the sampling distribution of the $\hat{\beta}$s and then use a computer-generated example to demonstrate that such distributions do indeed tend to be normally distributed.

4.3.1 Sampling Distributions of Estimators

Technically, an **estimator** is a formula, such as the OLS formula in Equation 2.16, that tells you how to compute $\hat{\beta}_k$; an **estimate** is the value of $\hat{\beta}_k$ computed by that formula. We have noted that the purpose of regression analysis is to obtain good estimates of the true (or population) coefficients of an equation from a sample of that population. In other words, given an equation like:

$$Y_i = \beta_0 + \beta_1 X_{1i} + \beta_2 X_{2i} + \epsilon_i \tag{4.8}$$

we want to estimate βs by taking a sample of the population and calculating those estimates (typically by OLS if the classical assumptions of Section 4.1 are met). Since researchers usually only have one sample, beginning econometricians often assume that regression analysis can produce only one estimate of the βs. In reality, each different sample from a given population will produce a different set of estimates of the βs. For example, one sample might produce an estimate considerably higher than the true β while another might come up with a $\hat{\beta}$ that is lower. We need to discuss the properties of the distribution of these $\hat{\beta}$s, even though in most real applications we will encounter only a single draw from it.

A simplified example will help clarify this point. Suppose you were attempting to estimate the average age of your class from a sample of the class; let's say that you were trying to use a sample of 5 to estimate the average age of a class of 30. The estimate you would get would obviously depend on the exact sample you picked. If your random sample accidentally included the five youngest or the five oldest people in the class, then your estimated age would be dramatically different from the one you would get if your random sample were more centered. In essence, then, there is a distribution of all the possible estimates that will have a mean and a variance just as the distribution of error terms does. To illustrate this concept, assume that the population is distributed uniformly between 19 and 23. Here are three samples from this population:

sample #1: 19, 19, 20, 22, 23; mean = 20.6
sample #2: 20, 21, 21, 22, 22; mean = 21.2
sample #3: 19, 20, 22, 23, 23; mean = 21.4

Each sample yields an estimate of the true population mean (which is 21), and the distribution of the means of all the possible samples has its own mean and variance. For a "good" estimation technique, we would want the mean of the distribution of sample estimates to be equal to the true population mean. This is called *unbiasedness*. While the mean of our three samples is a little over 21, it seems likely that if we took enough samples, the mean of our group of samples would eventually equal 21.0.

In a similar way, the $\hat{\beta}$s estimated by OLS for Equation 4.8 form a distribution of their own. Each sample of observations of Y and the Xs will produce different $\hat{\beta}$s, but the distribution of these estimates for all possible samples has a mean and a variance like any distribution. When we discuss the properties of estimators in the next section, it will be important to remember that we are discussing the properties of the distribution of estimates generated from a number of samples (a sampling distribution).

Properties of the mean: A desirable property of a distribution of estimates is that its mean equals the true mean of the item being estimated. An estimator that yields such estimates is called an unbiased estimator. An **unbiased estimator** is an estimator whose sampling distribution has as its expected value the true value of β.

$$E(\hat{\beta}_k) = \beta_k \tag{4.9}$$

A single estimate of β obtained for a particular sample from an unbiased estimator is called an unbiased estimate. Only one value of $\hat{\beta}$ is obtained in practice, but the property of unbiasedness is useful because a single estimate drawn from an unbiased distribution is more likely to be near the true value (assuming identical variances of the $\hat{\beta}$s) than one taken from a distribution not centered around the true value. If an estimator produces $\hat{\beta}$s that are not centered around the true β, the estimator is referred to as a *biased estimator.*

We cannot ensure that every estimate from an unbiased estimator is better than every estimate from a biased one because a particular unbiased estimate could, by chance, be further from the true value than a biased estimate might be. This could happen by chance, for example, or because the biased estimator had a smaller variance. Without any other information about the distribution of the estimates, however, we would always rather have an unbiased estimate than a biased one.

Properties of the variance: Just as we would like the distribution of the $\hat{\beta}$s to be centered around the true population β, so too would we like that distribution to be as narrow (or precise) as possible. A distribution centered around the truth but with an extremely large variance might be of very little use because any given estimate would quite likely be far from the true β value. For a $\hat{\beta}$ distribution with a small variance, the estimates are likely to be close to the mean of the sampling distribution. To see this more clearly, compare distributions #1 and #2 (both of which are unbiased) in Figure 4.6. Distribution #1, which has a larger variance than distribution #2, is less precise than distribution #2. For comparison purposes, a biased distribution (distribution #3) is also pictured; note that bias implies that the ex-

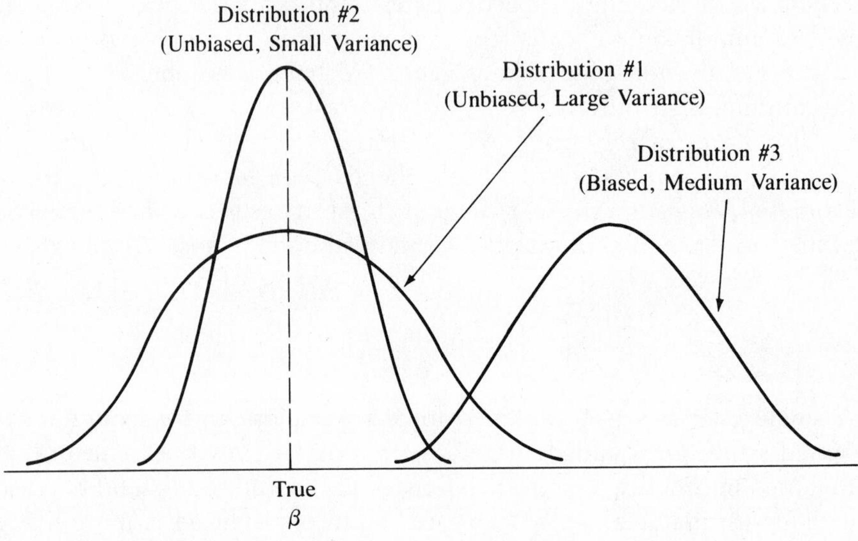

Figure 4.6 DISTRIBUTIONS OF $\hat{\beta}$
Different distributions of $\hat{\beta}$ can have different means and variances. Distributions #1 and #2, for example, are both unbiased, but distribution #1 has a larger variance than does distribution #2. Distribution #3 has a smaller variance than distribution #1, but it is biased.

pected value of the distribution has moved to the right or left of the true β.

The variance of the distribution of the $\hat{\beta}$s can be decreased by increasing the size of the sample. This also increases the degrees of freedom, since the number of degrees of freedom equals the sample size minus the number of parameters estimated. As the number of observations increases, other things held constant, the distribution of $\hat{\beta}$s becomes more centered around its sample mean, and the variance of the sampling distribution (as well as the square root of the variance, called the *standard deviation*) tends to decrease. Although it is not true that a sample of 15 will always produce estimates closer to the true β than a sample of 5, it is quite likely to do so; such larger samples should be sought. Figure 4.7 presents illustrative sampling distributions of $\hat{\beta}$s for 15 and 5 observations for OLS estimators of β when the true β equals one. The larger sample indeed produces a sampling distribution that is more closely centered around β.

In econometrics, general tendencies must be relied on. The element of chance, a random occurrence, is always present in estimating regression coefficients, and some estimates may be far from the true value

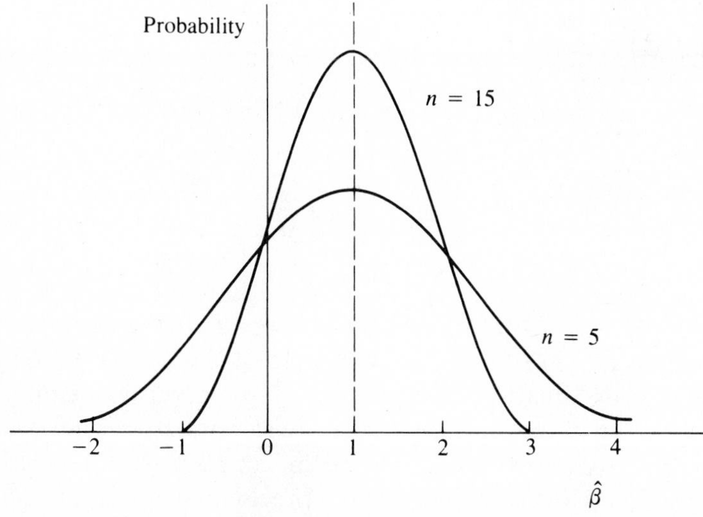

Figure 4.7 SAMPLING DISTRIBUTION OF $\hat{\beta}$ FOR VARIOUS OBSERVATIONS (n)
As the size of the sample increases, the variance of the distribution of βs calculated from that sample tends to decrease. In the extreme case (not shown) a sample equal to the population would yield only an estimate equal to the mean of that distribution, which (for unbiased estimators) would equal the true β, and the variance of the estimates would be zero.

no matter how good the estimating technique. However, if the distribution is centered around the true value and has as small a variance as possible, the element of chance is less likely to induce a poor estimate. If the sampling distribution is centered around a value other than the true β (that is, if $\hat{\beta}$ is *biased*) then a lower variance implies that most of the sampling distribution of $\hat{\beta}$ is concentrated on the wrong value. However, if this value is not very different from the true value, which is usually not known in practice, then the greater precision will still be valuable.

A final item of importance is that as the variance of the error term increases, so too does the variance of the distribution of $\hat{\beta}$. The reason for the increased variance of $\hat{\beta}$ is that, with the larger variance of ϵ_i, the more extreme values of ϵ_i are observed with more frequency, and the error term becomes more important in determining the values of Y_i. Thus, the relative portion of the movements of Y_i explained by the deterministic component βX_i is less, and there are more unexplained changes in Y_i caused by the stochastic element ϵ_i. This implies that empirical inferences about the value of β are more tenuous. The R^2 of the equation will tend to decrease as the variance of the error term increases, symptomatic of this tendency.

Properties of the standard error: Since the standard error of the estimated coefficient, $SE(\hat{\beta})$, is the square root of the estimated variance of the $\hat{\beta}$s, it too is affected by the size of the sample and other factors. To see this, let's look at an equation for $SE(\hat{\beta}_1)$, the standard error of the estimated slope coefficient from a model with two independent variables:

$$SE(\hat{\beta}_1) = \sqrt{\frac{\Sigma e_i^2/(n-3)}{\Sigma(X_{1i} - \bar{X}_1)^2(1 - r^2_{12})}} \tag{2.19}$$

Take a look at Equation 2.19. What happens if the sample size, n, increases? As n increases, so too will Σe_i^2 and $\Sigma(X_{1i} - \bar{X}_1)^2$. While the denominator of Equation 2.19 will rise unambiguously, the numerator will not, because the increase in Σe_i^2 will tend to be offset by the increase in n. As a result, an increase in sample size will cause $SE(\hat{\beta})$ to fall; the larger the sample, the more precise our coefficient estimates will be.

How about when Σe_i^2 increases, holding the sample size constant? In this case, $SE(\hat{\beta}_1)$ will increase. Since Σe_i^2 is an estimate of $VAR[\epsilon_i]$,(the variance of the error term), such a relationship makes sense because a more widely varying error term will make it harder for us to obtain precise coefficient estimates. The more ϵ varies, the less precise the coefficient estimate will be.

Finally, what about that $\Sigma(X_i - \bar{X})^2$ term? When $\Sigma(X_i - \bar{X})^2$ increases, holding the sample size constant, $SE(\hat{\beta}_1)$ decreases. Thus the more X varies around its mean, the more precise the coefficient estimate will be. This makes sense, since a wider range of X will provide more information on which to base the $\hat{\beta}$. As a result, the $\hat{\beta}$s will be more accurate, and $SE(\hat{\beta}_1)$ will fall. While we've used an equation from a three-variable model to show these properties, they also hold for $SE(\hat{\beta})$s from equations with any number of variables.

4.3.2 A Demonstration that the $\hat{\beta}$s Are Normally Distributed

One of the properties of the normal distribution is that any linear function of normally distributed variables is itself normally distributed. Given this property, it is not difficult to prove mathematically that the assumption of the normality of the error terms implies that the $\hat{\beta}$s are themselves normally distributed. This proof is not as important as an understanding of the meaning of such a conclusion, so this section presents a simplified demonstration of that property.

To demonstrate that normal error terms imply normally distributed $\hat{\beta}$s, we'll use a number of computer-generated samples and then calculate $\hat{\beta}$s from these samples in much the same manner as mean ages were calculated in the previous section. All the samples generated will conform to the same arbitrarily chosen true model, and the error term distribution used to generate the samples will be assumed to be normally distributed as is implied by the Central Limit Theorem. An examination of the distribution of $\hat{\beta}$s generated by this experiment not only shows its normality, but is also a good review of the discussion of the sampling distribution of $\hat{\beta}$s.

For this demonstration, assume that the following model is true:

$$Y_i = \beta_0 + \beta_1 X_i + \epsilon_i = 0 + 1X_i + \epsilon_i \tag{4.10}$$

This is the same as stating that, on average, Y = X. If we now assume that the error term is (independently) normally distributed with mean zero and variance of 0.25, and if we further choose a sample size of 5 and a given set of fixed Xs, we can use the computer to generate a large number of random samples (data-sets) conforming to the various assumptions listed above. We then can apply OLS and calculate a $\hat{\beta}$ for each sample, resulting in a distribution of $\hat{\beta}$s as discussed in this section. The sampling distribution for a large number[3] of such computer-generated data-sets and OLS-calculated $\hat{\beta}$s is shown in Figure 4.8.

Two conclusions can be drawn from an examination of this figure:

1. The distribution of $\hat{\beta}$s appears to be a symmetrical, bell-shaped distribution that is approaching a continuous normal distribution as the number of samples of $\hat{\beta}$s increases.
2. The distribution of the $\hat{\beta}$s is unbiased but shows surprising variation. $\hat{\beta}$s from -2.5 to $+4.5$ can be observed even though the true value of β is 1.0. Such a result implies that any researcher who bases an important conclusion on a single regression result runs a severe risk of overstating the case. This danger depends on the variance of the estimated coefficients, which decreases with the size of the sample. Note from Figure 4.7 that as the sample size increases from 5 to 15, the chance of observing a single $\hat{\beta}$ far from its true value falls; this demonstrates the desirability of larger samples.

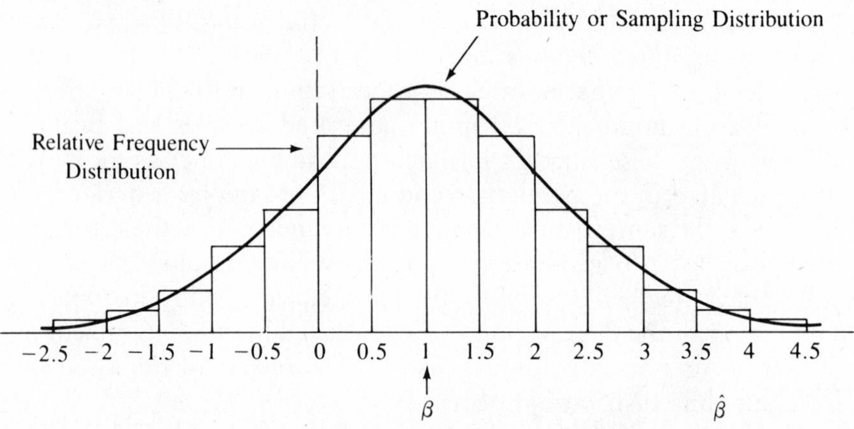

Figure 4.8 SAMPLING DISTRIBUTION OF THE OLS ESTIMATE $\hat{\beta}$ FOR $\beta = 1$ AND $\sigma^2 = 0.25$

To demonstrate that the distribution of $\hat{\beta}$s is indeed normal, we calculated 5,000 estimates of $\hat{\beta}$ from 5,000 samples where the true β was known to be 1.0. As can be seen, the resulting distribution of the $\hat{\beta}$s is not only centered around 1.0, the true β, but it is also symmetrical and bell-shaped, as is the normal distribution.

3. The number of data sets was 5000, but the exact number of $\hat{\beta}$s calculated is less important than the fact that the number was large enough for the sampling distribution to approach the underlying population distribution.

Computer-generated simulations of this kind are usually referred to as **Monte Carlo experiments.** Monte Carlo experiments typically have seven steps:

1. Assume a "true" model with specific coefficient values (for example $Y = 3.0 + 1.2X_1 - 5.3X_2$) and an error term distribution (for example N[0,1.0]).
2. Select values for the independent variables.
3. Select an estimating technique (usually OLS).
4. Create various samples of values of the dependent variable, using the assumed model, by randomly generating error terms from the assumed distribution.
5. Compute the estimates of the βs from the various samples using the estimating technique.
6. Evaluate the results.
7. Return to step #1 and choose other values for the coefficients, independent variables, or error term variance; compare these results with the first set (this step, which is optional, is called *sensitivity analysis*).

4.4 The Gauss-Markov Theorem and the Properties of OLS Estimators

The Gauss-Markov Theorem proves two important properties of OLS estimators. This theorem is proven in all advanced econometrics textbooks, and readers interested in the proof should see Exercise 8 and its answer. For a regression user, however, it's more important to know what the theorem implies than to be able to prove it. The **Gauss-Markov Theorem** states that:

> Given Classical Assumptions I through VI (Assumption VII, normality, is not needed for this theorem), the OLS estimator of β_k is the minimum variance estimator from among the set of all linear unbiased estimators of β_k, for k = 0,1,2,...,K.

The Gauss-Markov Theorem is perhaps most easily remembered by stating that "OLS is BLUE" where **BLUE** stands for "*B*est (meaning minimum variance) *L*inear *U*nbiased *E*stimator." Students who might forget that "best" stands for minimum variance might be better served by remembering "OLS is MvLUE," but such a phrase is hardly catchy or easy to remember.

If an equation's coefficient estimation is unbiased (that is, if each of the estimated coefficients is an unbiased estimate of the true population coefficient), then:

$$E(\hat{\beta}_k) = \beta_k \qquad (k = 0,1,2,...,K)$$

Best, as mentioned above, means that each $\hat{\beta}_k$ has the smallest variance possible (in this case, out of all the linear unbiased estimators of β_k). An unbiased estimator with the smallest variance is called **efficient,** and that estimator is said to have the property of efficiency.

Given all seven classical assumptions, the OLS coefficient estimators can be shown to have the following properties:

1. *They are unbiased.* That is, $E(\hat{\beta})$ is β. This means that the OLS estimates of the coefficients are centered around the true population values of the parameters being estimated.
2. *They are minimum variance.* The distribution of the coefficient estimates around the true parameter values is as tightly or narrowly distributed as is possible for an unbiased distribution. No other linear unbiased estimator has a lower variance for each estimated coefficient than OLS.
3. *They are consistent.* As the sample size approaches infinity, the estimates converge on the true population parameters. Put differently, as the sample size gets larger, the variance gets smaller, and each estimate approaches the true value of the coefficient being estimated.
4. *They are normally distributed;* the $\hat{\beta}$s are $N(\beta, VAR[\hat{\beta}])$. Thus various statistical tests based on the normal distribution may indeed be applied to these estimates, as will be done in Chapter 5.

If the seven classical assumptions are met and if OLS is used to calculate the $\hat{\beta}$s, then it can be stated that an estimated regression coefficient is an unbiased, minimum variance estimate of the impact on the dependent variable of a one-unit change in a given independent variable, holding constant all other independent variables in the equation. Such an estimate is drawn from a distribution of estimates that is centered around the true population coefficient and has the smallest possible variance for such unbiased distributions.

4.5 The Classical Assumptions in Standard Notation

We now introduce the notation that is typically used to represent the measures of central tendency and the degree of dispersion of sampling and probability distributions. Then we present the classical regression assumptions in this standard notation.

4.5.1 Standard Econometric Notation

While Section 4.3 portrayed graphically the notions of central tendency and dispersion, this section presents the standard notation used throughout the econometrics literature for these concepts.

The measure of the central tendency of the sampling distribution of $\hat{\beta}$, which can be thought of as the mean of the $\hat{\beta}$s, is denoted as $E(\hat{\beta})$, read as "the expected value of beta-hat." The expected value of a random variable is the population mean of that variable (with observations weighted by the probability of observation).

The variance of $\hat{\beta}$ is the typical measure of dispersion of the sampling distribution of $\hat{\beta}$. The variance has several alternative notational representations, including $VAR(\hat{\beta})$ and $\sigma^2(\hat{\beta})$. Each of these is read as the "variance of beta-hat" and represents the degree of dispersion of the sampling distribution of $\hat{\beta}$.

Table 4.1 presents various alternative notational devices used to represent the different population (true) parameters and their corresponding estimates (based on samples).

To review Table 4.1, the true coefficient β_k is estimated by $\hat{\beta}_k$. The estimator is stochastic, or random, because its value depends on the values of the stochastic error term in the true function. Thus $\hat{\beta}$ has a sampling distribution based on the distribution of the ϵ_is. If the ϵ_is are normally distributed, as usually is assumed by invoking the Central Limit Theorem, then the $\hat{\beta}$s are normally distributed.

Two population parameters, the mean and the variance, fully describe the $\hat{\beta}$ distribution. The mean, the measure of central tendency of the sampling distribution, is $E(\hat{\beta}_k)$, which equals β_k if the estimator is unbiased. The variance (or, alternatively, the square root of the variance, called the **standard deviation**) is a measure of dispersion in the sampling distribution of $\hat{\beta}_k$. The variance of the estimates is a population parameter that is never actually observed in practice; instead, it is estimated with $\hat{\sigma}^2(\hat{\beta}_k)$, also written as $s^2(\hat{\beta}_k)$. Note, by the way, that the variance of the true β, $\sigma^2(\beta)$, is zero, since there is only

TABLE 4.1 NOTATION CONVENTIONS

Population Parameter (True Values, but Unobserved)		Estimate (Observed from Sample)	
Name	**Symbol(s)**	**Name**	**Symbol(s)**
Regression coefficient	β_k	Estimated regression coefficient	$\hat{\beta}_k$
Expected value of the estimated coefficient	$E(\hat{\beta}_k)$		
Variance of the error term	σ^2 or $VAR(\epsilon_i)$	Estimated variance of the error term	s^2 or $\hat{\sigma}^2$
Standard deviation of the error term	σ	Standard error of the equation (estimate)	s or SEE
Variance of the estimated coefficient	$\sigma^2(\hat{\beta}_k)$ or $VAR(\hat{\beta}_k)$	Estimated variance of the estimated coefficient	$s^2(\hat{\beta}_k)$ or $\widehat{VAR}(\hat{\beta}_k)$
Standard deviation of the estimated coefficient	$\sigma_{\hat{\beta}_k}$ or $\sigma(\hat{\beta}_k)$	Standard error of the estimated coefficient	$\hat{\sigma}(\hat{\beta}_k)$ or $SE(\hat{\beta}_k)$
Error or disturbance term	ϵ_i	Residual (estimate of error in a loose sense)	e_i

one true β_k with no distribution around it. Thus the estimated variance of the estimated coefficient is defined and observed, the true variance of the estimated coefficient is unobservable, and the true variance of the true coefficient is zero. The square root of the estimated variance or the coefficient estimate is the standard error of $\hat{\beta}$, $SE(\hat{\beta}_k)$, which we will use extensively in hypothesis testing.

4.5.2 The Classical Assumptions in Notational Form

Most econometrics texts and articles state the assumptions of the classical model in terms of expectation and variance operators. The classical assumptions typically stated in this notation (Assumptions II through V) are presented here:

$$\text{II.} \quad E(\epsilon_i) = 0$$

Assumption II states that the distribution of the error term has a central tendency or expected value of zero.

$$\text{III.} \quad E(X_{ki} \cdot \epsilon_i) = 0 \qquad (k = 1,2,...,K \text{ and } i = 1,2,...,n)$$

Assumption III states that the error term and each explanatory variable are uncorrelated. This can be summarized by stating that the expected value of the product of an independent variable and the error term from the same observation is zero. This results because the expected value of the error term is zero, and because if the values of the error term and the explanatory variables are indeed uncorrelated, the positive $X_{ki} \cdot \epsilon_i$ products will offset the negative ones. If X_{ki} and ϵ_i are correlated, then they will either be positively or negatively correlated, and the expected value of their product will no longer be zero. For example, a positive correlation will occur if values of any given X are above its mean when ϵ is positive.[4]

$$\text{IV.} \quad E(\epsilon_i \cdot \epsilon_j) = 0 \quad \text{for all } i,j = 1,2,...,n \text{ except } i = j$$

Assumption IV states that the observations of the error term are similarly uncorrelated. If observations of the error term are systematically related to each other, then the expected value of their product will not be zero.

$$\text{V.} \quad VAR(\epsilon_i) = \sigma^2 \quad \text{a constant, for all } i = 1,2,...,n$$

Assumption V states that the variance of the distribution of the error term is constant for all observations. That is, the value of the variance does not change with any observation of the error term. To make this clearer, note that the alternative to Assumption V (heteroskedasticity) would be that the variance of the distribution of the error term *does* depend on exactly which observation is being discussed, or:

$$VAR(\epsilon_i) = \sigma_i^2 \qquad (i = 1,2,...,n) \tag{4.11}$$

4. Only when two variables are uncorrelated is the expected value of their product equal to the product of their expected values. Therefore, even if the expected value of epsilon is zero, the expected value of epsilon *times* X is not equal to zero unless epsilon and X are uncorrelated.

The only difference between Assumption V and Equation 4.11 is that with homoskedasticity there is no subscript and with heteroskedasticity there is a subscript. This subscript implies that the variance of the error term in this situation differs from observation to observation, and Assumption V is violated.

4.6 Summary

1. The seven classical assumptions state that the regression model is linear with an error term that is zero-meaned, is uncorrelated with the explanatory variables and other observations of the error term, has a constant variance, and is normally distributed (optional). In addition, the explanatory variables must not to be perfect linear functions of each other.

2. The two most important properties of an estimator are unbiasedness and minimum variance. An estimator is unbiased when the expected value of the estimated coefficient is equal to the true value. Minimum variance holds when the estimating distribution has the smallest variance of all the estimators.

3. Given the classical assumptions, OLS can be shown to be the minimum variance, linear, unbiased estimator (or BLUE, for best linear unbiased estimator) of the regression coefficients. This is the Gauss-Markov Theorem. When one or more of the classical properties do not hold (excluding normality), OLS is no longer BLUE, although it still may provide better estimates in some cases than the alternative estimation techniques discussed in subsequent chapters.

4. Because the sampling distribution of the OLS estimator of $\hat{\beta}_k$ is BLUE, it has desirable properties. Moreover, the variance, or the degree of dispersion of the sampling distribution of $\hat{\beta}_k$, decreases as the number of observations increases. The rule here is simple: If the cost of additional observations is within reason, obtain them and use them for estimation.

5. There is a standard notation used in the econometric literature. Table 4.1 presents this fairly complex set of notational conventions for use in regression analysis. This table should be reviewed periodically as a refresher.

6. An OLS-estimated regression coefficient from a model that meets the classical assumptions is an unbiased, minimum variance estimate of the impact on the dependent variable of a one-unit change in the independent variable in question, holding constant the other independent variables in the equation.

Exercises

(Answers to even-numbered exercises are in Appendix A.)

1. Write the meaning of each of the following terms without reference to the book (or to your notes), and compare your definition with the version in the text for each:
   a. The Classical Assumptions
   b. classical error term
   c. standard normal distribution
   d. The Central Limit Theorem
   e. unbiased estimator
   f. BLUE
   g. sampling distribution

2. Think back to the assumption of linearity in the coefficients (Assumption I in Section 4.1) and the equation reproduced below:

$$\ln(Y_i) = \beta_0 + \beta_1 \ln(X_i) + \epsilon_i$$

   a. What is the elasticity of Y with respect to X? (Elasticity is defined as the percentage change in Y brought about by a one percent change in X.)
   b. Would the above equation still be linear in the coefficients if it also included an additional (additive) term, $\beta_2 \ln(Z_i)$? Why or why not? What is the economic application of such an equation?

3. Suppose one of your friends said that "Assumptions III and IV don't really mean that much." After all, if ϵ is zero, then ϵ multiplied by something should be zero, and it's no big deal that:

$$E(X_{ki} \cdot \epsilon_i) = 0 \text{ or } E(\epsilon_i \cdot \epsilon_j) = 0$$

Would your friend be right or wrong? Why?

4. Which of the following pairs of independent variables would violate Assumption VI? (That is, which pairs of variables are perfect linear functions of each other?)
   a. right shoe size and left shoe size (of students in your class)
   b. consumption and disposable income (in the U.S. over the last 30 years)
   c. X_i and $2X_i$
   d. X_i and $(X_i)^2$

5. Consider the following estimated regression equation (standard errors in parentheses):

$$\hat{Y}_t = -120 + \underset{(0.05)}{0.10F_t} + \underset{(1.00)}{5.33R_t} \qquad \bar{R}^2 = .50$$

where: Y_t = corn yield (bushels/acre) in year t
F_t = fertilizer intensity (pounds/acre) in year t
R_t = rainfall (inches) in year t

   a. Carefully state the meaning of the coefficients 0.10 and 5.33 in this equation in terms of the impact of F and RS on Y.
   b. Does the constant term of −120 really mean that *negative* amounts of corn are possible? If not, what is the meaning of that estimate?
   c. Suppose you were told that the true value of β_F is *known* to be 0.40. Does this show that the estimate is biased? Why or why not?
   d. Suppose you were told that the equation does not meet all the classical assumptions and therefore is not BLUE. Does this mean that the true β_{RS} is definitely *not* equal to 5.33? Why or why not?

6. Consider a random variable that is distributed N(0,0.5); that is, normally distributed with a mean of zero and a variance of 0.5. What is the probability that a single observation drawn from this distribution would be greater than one or less than minus one? (Hint: To answer this question, you will need to convert this distribution to a standard normal one [with mean equal to zero and standard deviation equal to one] and then refer to Table B-7 in the back of the book. That table includes a description of how to make such a transformation.)

7. Bowen and Finegan[5] estimated the following regression equation for 78 cities (standard errors in parentheses):

$$\hat{L}_i = \underset{}{94.2} - \underset{(0.08)}{0.24U_i} + \underset{(0.06)}{0.20E_i} - \underset{(0.16)}{0.69I_i} - \underset{(0.18)}{0.06S_i}$$

$$+ \underset{(0.03)}{0.002C_i} - \underset{(0.53)}{0.80D_i}$$

$$n = 78 \qquad R^2 = .51$$

where: L_i = percent labor force participation (males ages 25 to 54) in the *i*th city
U_i = percent unemployment rate in the *i*th city
E_i = average earnings (hundreds of dollars/year) in the *i*th city
I_i = average other income (hundreds of dollars/year) in the *i*th city
S_i = average schooling completed (years) in the *i*th city
C_i = percent of the labor force that is nonwhite in the *i*th city
D_i = dummy equal to 1 if the city is in the South and 0 otherwise

a. Interpret the estimated coefficients of C and D. What do they mean?

b. How likely is perfect collinearity in this equation? Explain your answer.

c. Suppose that you were told that the data for this regression were from 1950 and that estimates on the data from 1960 yielded a much different coefficient of the dummy variable. Would this imply that one of the estimates was biased? If not, why not? If so, how would you determine which year's estimate was biased?

d. Comment on the following statement. "I know that these results are not BLUE because the average participation rate of 94.2 percent is way too high." Do you agree or disagree? Why?

8. A typical exam question in a more advanced econometrics class is to prove the Gauss-Markov Theorem. How might you go about starting such a proof? What is the importance of such a proof?

5. W.G. Bowen and T.A. Finegan, "Labor Force Participation and Unemployment," in Arthur M. Ross (ed.), *Employment Policy and Labor Markets,* (Berkeley: University of California Press, 1965), Table 4–2.

(Hint: If you're having trouble getting started answering this question, see Appendix A.)

9. Ray Fair[6] built an econometric model of adultery by estimating an equation where the dependent variable was the number of extramarital affairs that the *i*th woman said she had per year (Y_i). The data were obtained from 6366 women in their first marriages who filled out and voluntarily returned a survey that was included in *Redbook* magazine. The results (standard errors in parentheses) were:

$$\hat{Y}_i = 7.18 + \underset{(0.07)}{0.26J_i} - \underset{(0.06)}{1.53H_i} - \underset{(0.03)}{0.12A_i} + \underset{(0.02)}{0.14X_i} - \underset{(0.09)}{0.95R_i}$$

where: J_i = a job index, with 1 = student, 2 = unskilled or semi-skilled worker, 3 = white collar worker, 4 = teacher or skilled worker, 5 = manager, 6 = professional

H_i = the *i*th woman's assessment of the happiness of her marriage (1 = low to 5 = high)

A_i = the age in years of the *i*th woman

X_i = the number of years the *i*th woman has been married

R_i = a scale of the strength of religious belief of the *i*th woman (1 = none to 4 = strong)

a. What do you think about applying econometrics to this topic? Is it sexist? What risks does it run?
b. Which Classical Assumptions do you know hold for this equation? Which do you think might not hold? About which assumption(s) is it virtually impossible to make a judgment?
c. Develop your own hypotheses for the signs (and, if possible, magnitudes) of the coefficients of the independent variables. Compare the estimated coefficients with your expected ones. Are there any differences? Why do you think they happened?
d. What do you think about this data collection technique? Do you think that the responses are representative of the population? Why or why not?
e. Does the estimated constant term of 7.18 indicate that the average married woman has more than 7 extramarital affairs per year? Why or why not?

6. Ray C. Fair, "A Theory of Extramarital Affairs," *Journal of Political Economy*, Feb. 1978, pp. 45–61.

f. Is J_i a dummy variable? Is it a proxy variable? Why do you think this variable was included in the equation? Do you see any problems with the definition of this variable? Explain your answer.

10. The Gauss-Markov Theorem shows that OLS is BLUE, so we of course hope and expect that our coefficient estimates will be unbiased *and* minimum variance. Suppose, however, that you had to choose one or the other.

a. If you had to pick one, would you rather have an unbiased non-minimum variance estimate or a biased minimum variance one? Explain your reasoning.

b. Are there circumstances in which you might change your answer to part a? (Hint: Does it matter *how* biased or less-than-minimum variance the estimates are?)

c. Can you think of a way to systematically choose between estimates that have varying amounts of bias and less-than-minimum variance?

11. For your first econometrics project you decide to model sales at the frozen yogurt store nearest your school. The owner of the store is glad to help you with data collection because she believes that students from your school make up the bulk of her business. After countless hours of data collection and an endless supply of tutti-fruiti frozen yogurt, you estimate the following regression equation (standard errors in parentheses):

$$\hat{Y}_t = 262.5 + \underset{(0.7)}{3.9T_t} - \underset{(20.0)}{46.94P_t} + \underset{(108.0)}{134.3A_t} - \underset{(138.3)}{152.1C_t}$$

$$n = 29 \qquad \bar{R}^2 = .78$$

where: Y_t = the total number of frozen yogurts sold during the *t*th two-week time period

T_t = average high temperature (in degrees F) during period t

P_t = the price of frozen yogurt (in dollars) at the store in period t

A_t = a dummy variable equal to 1 if the owner places an ad in the school newspaper during period t, 0 otherwise

C_t = a dummy variable equal to 1 if your school is in regular session in period t (early September through early December and early January through late May), 0 otherwise

a. This is a demand equation without any supply equation specified. Does this violate any of the Classical Assumptions? What kind of judgments do you have to make to answer this question?

b. What is the real-world economic meaning of the fact that the estimated coefficient of A_t is 134.3? Be specific.

c. You and the owner are surprised at the sign of the coefficient of C_t. Can you think of any reason for this estimate? (Hint: Assume that your school has no summer session.)

d. If you could add one variable to this equation, what would it be? Be specific.

5

Basic Statistics and Hypothesis Testing

The most important use of econometrics for many researchers is in testing their theories with data from the real world, so hypothesis testing is more meaningful to them than are the other major uses of econometrics, description and forecasting. This chapter starts with a brief discussion of statistical inference and then introduces the topic of hypothesis testing. We will then examine the *t*-test, the statistical tool typically used for hypothesis tests of individual regression coefficients, and the *F*-test, the statistical tool typically used for joint tests of more than one coefficient at a time.

We are merely returning to the essence of econometrics—an effort to quantify economic relationships by analyzing sample data—and asking what conclusions we can draw from this quantification. Hypothesis testing goes beyond calculating estimates of the true population parameters to a much more complex set of questions. Hypothesis testing asks what we can learn about the real world from this sample. Is it likely that our result could have been obtained by chance? Can our theories be rejected using the results generated by our sample? If our theory is correct, what are the odds that this particular sample would have been observed?

All approaches to hypothesis testing take into account that any particular estimate comes from a distribution of estimates and must be interpreted with respect to the standard error of that distribution before any importance can be attached to it. A single estimate of a regression coefficient β is only one of many possible estimates, and conclusions drawn from that estimate must take the entire distribution into account. As a result, the classical approach to hypothesis testing, which is what we present in this chapter, consists of three steps:

1. *The development of hypotheses about the truth.* These are usually the result of the application of economic theory or common sense to the details of the particular equation being estimated.
2. *The calculation of estimates of the coefficients and their standard errors.* This has been the subject of the previous four chapters.
3. *The testing of the implications of the estimates obtained in #2 for the hypotheses stated in #1.* This usually involves inferring whether the estimated coefficient is significantly different from the population value that was hypothesized.

Hypothesis testing and the *t*-test should be familiar topics to readers with strong backgrounds in statistics, who are encouraged to skim this chapter and focus only on those applications that seem somewhat new. The development of hypothesis testing procedures is explained here in terms of the regression model, however, so parts of the chapter may be instructive even to those already skilled in statistics.

5.1 Statistical Inference

Throughout this text, we've introduced basic statistical concepts only when they've been required for an understanding of the point at hand (rather than lumping them into an all–too–easily–skipped early chapter). Thus, while the current chapter is about statistics, we've already covered many statistical concepts. For example, we've discussed random variables, samples, the normal probability distribution, estimated means and standard errors, the sampling distribution of the βs, and the properties of estimators.

One topic we've touched on only briefly is **statistical inference,** the drawing of conclusions about a population from a sample of data from that population. Thus far, we've learned how to get the best possible estimates of the true regression coefficients by specifying an equation, collecting a sample, and calculating estimates of those true coefficients. Now it's time to see what our estimates mean so we can infer something about the population from our sample estimates. Our approach will be classical in nature, since we assume that the sample data are our best and only information about the population. An alternative, **Bayesian statistics,** adds prior information to the sample to draw statistical inferences.[1]

Let's face it, there might not be a "truth" for us to discover with regression analysis and statistical inference. Since human institutions are always changing and human behavior often seems inherently random, it may well be that by the time we collect data and analyze an equation, the world around us will have changed once again.

If it's impossible to imagine discovering what the true coefficients are, it's perhaps easier to think of statistical inference in terms of attempting to draw conclusions about the coefficients of the regression that best applies to the population in which we're interested. To get such a **population regression,** we'd have to run a properly specified regression on the entire population; the estimated coefficients of that regression would be as close as we could come to the truth.

Unfortunately for economists (but fortunately for econometricians, since otherwise we'd all be mathematicians or weatherforecasters), collecting and keypunching data for an entire population is an almost impossible task unless the population is defined so narrowly as to be of little importance. Instead, we have to run a regression on a representative sample of the population and then attempt to infer the properties of the population from the sample.

Since we usually have only one estimated regression, how do we decide whether the sample estimate $\hat{\beta}$ is close to the population β? Each sample from the population gives us a different $\hat{\beta}$, resulting in

1. Bayesian econometrics combines estimates generated from samples with estimates based on prior theory or research. For example, suppose you attempt to estimate the marginal propensity to consume (MPC) with the coefficient of income in an appropriately specified consumption regression equation. If your prior belief is that the MPC is 0.9 and if the estimated coefficient from your sample is 0.8, then a Bayesian estimate of the MPC would be somewhere between the two, depending on the strength of your belief. Bayesians, by being forced to state explicitly their prior expectations, tend to do most of their thinking before estimation, which is a good habit for a number of important reasons. For more on this approach, see A. Zellner, *An Introduction to Bayesian Analysis in Econometrics* (New York: Wiley, 1971).

the sampling distribution of the $\hat{\beta}$s discussed in Chapter 4. As a result, it's possible for an estimate drawn from an unbiased distribution to be quite far from the population coefficient. This is where statistical inference comes in. We can decide whether our estimate supports or refutes our prior expectation by running statistical tests on our $\hat{\beta}$s and using measures of dispersion like SE($\hat{\beta}$) to standardize the comparisons. The main technique economists use to achieve such inference is called hypothesis testing, so let's move on to that topic.

5.2 What Is Hypothesis Testing?

Hypothesis testing is used in a variety of settings. The Food and Drug Administration (FDA), for example, requires testing of new products before allowing their sale. If the sample of people exposed to the new product shows some side effect significantly more frequently than would be expected to occur by chance, the FDA is likely to withhold approval of the marketing of that product. Similarly, economists have been statistically testing various relationships between consumption and income for half a century; theories developed by John Maynard Keynes and Milton Friedman, among others, have all been tested on macroeconomic and microeconomic data-sets.

Although researchers are always interested in learning whether the theory in question is supported by estimates generated from a sample of real-world observations, it's almost impossible to prove that a given hypothesis is correct. All that can be done is to state that a particular sample conforms to a particular hypothesis. Even though we cannot prove that a given theory is "correct" using hypothesis testing, we *can* often reject a given hypothesis with a certain degree of confidence. In such a case, the researcher concludes that the sample result would be very unlikely to have been observed if the hypothesized theory were correct. If there is conflicting evidence on the validity of a theory, the question is often put aside until additional data or a new approach shed more light on the issue.

Let's begin by investigating three topics that are central to the application of hypothesis testing to regression analysis:

1. the specification of the hypothesis to be tested
2. the choice of a particular decision rule for deciding whether to reject the hypothesis in question
3. the kinds of errors that might be encountered if the application of the decision rule to the appropriate statistics yields an incorrect inference.

5.2.1 Classical Null and Alternative Hypotheses

The first step in hypothesis testing is to state explicitly the hypothesis to be tested. To ensure fairness, the researcher should specify the hypothesis *before* the equation is estimated. The purpose of prior theoretical work is to match the hypothesis to the underlying theory as completely as possible. Hypotheses formulated after generation of the estimates are at times justifications of particular results rather than tests of their validity. As a result, most econometricians take pains to specify hypotheses before estimation.

In making a hypothesis, you must state carefully what you think is not true and what you think is true. These reflections of the researcher's expectations about a particular regression coefficient (or coefficients) are summarized in the null and alternative hypotheses. The **null hypothesis** is typically a statement of the range of values of the regression coefficient that would be expected to occur if the researcher's theory were *not* correct. The **alternative hypothesis** is used to specify the range of values of the coefficient that would be expected to occur if the researcher's theory were correct. The word "null" also means "zero," and the null hypothesis can be thought of as the hypothesis that the researcher does *not* believe. The reason it's called a null or zero hypothesis is that a variable would not be included in an equation if its expected coefficient were zero.

We set up the null and alternative hypotheses in this way so we can make rather strong statements when we reject the null hypothesis. It is only when we define the null hypothesis as the result we do *not* expect (the "strawman"[2] result) that we can control the probability of rejecting the null hypothesis accidentally when it is in fact true. The converse does not hold. That is, we can never actually know the probability of agreeing accidentally that the null hypothesis is correct when it is in fact false. As a result, we can never say that we *accept* the null hypothesis; we always must say that we *cannot reject* the null hypothesis, or we put the word "accept" in quotes.

Researchers occasionally will have to switch the null and alternative hypotheses. For instance, some tests of the rational expectations theory have put the preferred hypothesis as the null hypothesis in order to make the null hypothesis a specific value. In such cases of tests of specific nonzero values, the reversal of the null and alternative hy-

2. The word strawman draws its meaning from the name of a debating technique that sets up an easily refuted opposing argument as a "strawman" (that is easily knocked down). Thus the strawman is what we expect to be false and the "pet theory" is what we expect to be true.

potheses is regrettable but unavoidable; an example of this kind of reversal is in Section 5.4. Except for these rare cases, all null hypotheses in this text will be the result we expect not to occur.

The notation used to refer to a null hypothesis is "H_0:," and this notation is followed by a statement of the value or range of values you do *not* expect the particular parameter to take. For example:

$$H_0: \beta = S \quad \text{(where S is the specific value you do not expect)}$$

If the estimated coefficient is significantly different from S, then we reject the null hypothesis. If the estimated coefficient is not significantly different from S, then we cannot reject H_0.

The alternative hypothesis is expressed by "H_A:" followed by the parameter value or values you expect to observe:

$$H_A: \beta = T \quad \text{(where T is the value you expect to be true)}$$

Hypotheses in econometrics usually do not specify particular values, but rather the particular signs that the researcher expects the estimated coefficients to take. We frequently hypothesize that a particular coefficient will be positive (or negative). In such cases, the null hypothesis still represents what is expected not to occur, but that expectation is now a range; the same is true for the alternative hypothesis. If, for example, you expect a negative coefficient, then correct[3] null and alternative hypotheses are:

$$H_0: \beta \geq 0$$
$$H_A: \beta < 0$$

Note that the null hypothesis is the "strawman" and the alternative hypothesis is your "pet theory."

Let's look at an example of making such null and alternative hypotheses. Suppose you are about to estimate an equation to explain aggregate retail sales of new cars (Y) as a function of real disposable income (X_1) and the average retail price of a new car deflated by the consumer price index (X_2):

3. Some researchers prefer to use a two-tailed test around zero for this hypothesis because they feel that the classical approach requires the null hypothesis to contain a single value. As mentioned in the text, we feel that the use of a two–tailed test in such a circumstance is a mistake. However, we have no quarrel with using $\beta = 0$ as the null hypothesis as long as the alternative hypothesis remains the same. As will be explained in footnote 4, these null hypotheses yield identical results in a one-tailed test.

$$Y_i = f(\overset{+}{X_1},\overset{-}{X_2}) = \beta_0 + \beta_1 X_{1i} + \beta_2 X_{2i} + \epsilon_i \qquad (5.1)$$

As was discussed in Section 3.3, the positive sign above X_1 in the functional notation on the left side of Equation 5.1 indicates that you expect income to have a positive impact on car sales, holding prices constant. Because you expect X_1 to have a positive impact on Y, you have already made a hypothesis. That is, you have already stated that you expect β_1 to be greater than zero. Since the null hypothesis contains that which you are hoping to reject, it would be:

$$H_0\text{: } \beta_1 \leq 0$$

The alternative hypothesis (which should correspond to your expected sign) would be:

$$H_A\text{: } \beta_1 > 0$$

To test yourself, go back to Equation 5.1 and attempt to create null and alternative hypotheses for X_2. Do you see that the answer is H_0: $\beta_2 \geq 0$ and H_A: $\beta_2 < 0$? This answer is correct because you have stated in the signs above Equation 5.1 that you expect price to have a negative impact on car sales, holding disposable income constant, and because you should place your expected coefficient values in the alternative hypothesis.[4]

Another way to state the null and alternative hypotheses for the same equation and underlying theory would be to test the null hypothesis that β_1 is not significantly different from zero in either direction. In this second approach, the null and alternative hypotheses would be:

$$H_0\text{: } \beta_1 = 0$$
$$H_A\text{: } \beta_1 \neq 0$$

4. As mentioned in footnote 3, you can use $\beta = 0$ as your null hypothesis for X_1 instead of $\beta \leq 0$ as long as you don't change the alternative hypothesis of H_A: $\beta > 0$. These two versions of the null hypothesis give identical answers because, in order to test a null hypothesis that is a range like $\beta \leq 0$, you must focus on the value in that range which is closest to the range implied by the alternative hypothesis. If you can reject that value, you can reject values that are further away as well. Truncating the range of the null hypothesis in this way has no practical importance because you are really asking on which side of zero is the true parameter.

Since the alternative hypothesis has values on both sides of the null hypothesis, this approach is called a **two-sided test** (or *two-tailed test*) to distinguish it from the **one-sided test** of the previous example (where the alternative hypothesis was only on one side of the null hypothesis).

One difficulty with the two-sided approach in the new car sales model is that you expect β_1 to be positive, but the alternative hypothesis does not distinguish between highly positive values, which support your theory, and highly negative values, which do not. Such difficulties do not apply if there are conflicting theories about which sign the coefficient is expected to take because either dramatically large positive *or* negative values would fit your expectations. Since the expected sign of a coefficient (such as for price or income) is usually predicted by the underlying theory, most null hypotheses involving zero tend to be one-tailed tests in economics and business. Null hypotheses involving nonzero values can be either one-tailed or two-tailed depending on the underlying theory.[5]

5.2.2 Type I and Type II Errors

The typical testing technique in econometrics is to hypothesize an expected sign (or value) for each regression coefficient (except the constant term) and then to determine whether to reject the null hypothesis. Since the regression coefficients are only estimates of the true population parameters, it would be unrealistic to think that conclusions drawn from regression analysis will always be right.

There are two kinds of errors we can make in such hypothesis testing:

Type I. We reject a true null hypothesis.

Type II. We do not reject a false null hypothesis.

We will refer to these errors as **Type I** and **Type II Errors,** respectively.[6]

5. The reason that null hypothesis involving nonzero values can sometimes involve two-tailed tests is that if your "pet theory" is $\beta = 1$ (or any other nonzero value), then you have no alternative but to violate the strawman principle and put $\beta = 1$ in the null hypothesis. For more on this, see Section 5.4.2.

6. Some authors refer to these as α and β errors, respectively, but no matter which titles are used, many beginning students understandably have trouble remembering which error is which. Such students might be helped by thinking of a Type I Error as an error of short–sightedness (not seeing that a null hypothesis is actually true) and a Type II Error as an error of overimagination (imagining that a null hypothesis is true when it actually is false). Those with poor memories but vivid imaginations seem to remember this most easily by recalling that someone with only one eye (one I) would probably be short-sighted.

Let's return to the new car sales model to get a better understanding of these errors. In particular, recall that the null and alternative hypotheses to test the impact of disposable income on new car sales (holding prices constant) are:

$$H_0: \beta_1 \leq 0$$
$$H_A: \beta_1 > 0$$

There are two distinct possibilities. The first is that the true β_1 in the population is equal to or less than zero, as specified by the null hypothesis; this is the same as saying that income does *not* have a positive effect on car sales. When the true β_1 is not positive, unbiased estimates of β_1 will be distributed around zero or some negative number, but any given estimate is very unlikely to be exactly equal to that number. Any single sample (and therefore any estimate of β calculated from that sample) might be quite different from the mean of the distribution. As a result, even if the true parameter β_1 is not positive, the particular estimate obtained by a researcher may be sufficiently positive to lead to the rejection of the null hypothesis that $\beta_1 \leq 0$. This is a Type I Error; we have rejected the truth! For the new car sales model this would mean rejecting the null hypothesis of a nonpositive impact of income on sales even though it is true. A Type I Error is graphed in Figure 5.1.

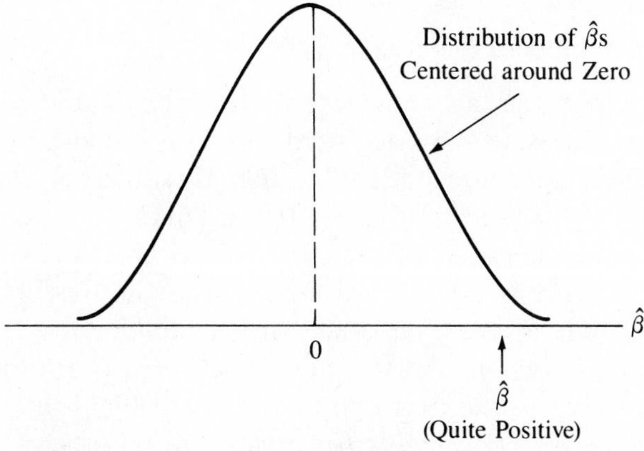

Figure 5.1 REJECTING A TRUE NULL HYPOTHESIS IS A TYPE I ERROR
If $\beta = 0$, but you observe a $\hat{\beta}$ that is very positive, you might reject a true null hypothesis, H_0: $\beta \leq 0$, and conclude incorrectly that the alternative hypothesis H_A: $\beta > 0$ is true.

The second possibility is that the true β_1 is greater than 0, as stated in the alternative hypothesis. This is the same as saying that the impact of income on new car sales is indeed positive. Depending on the specific value of the population β_1 (and other factors) it's possible to obtain an estimate of β_1 that is close enough to zero (or negative) to be considered "not significantly positive." This occurs because the sampling distribution of $\hat{\beta}_1$, even if unbiased, has a portion of its area in the region of $\beta_1 \leq 0$. Such a result may lead the researcher to "accept" the hypothesis that $\beta_1 \leq 0$ when in truth $\beta_1 > 0$. This is a Type II Error; we have failed to reject a false null hypothesis! In the new car sales model this would mean "accepting" the null hypothesis of a nonpositive impact of income on sales even though it's false. A Type II Error is graphed in Figure 5.2. (The specific value of $\beta_1 = 1$ was selected as the true value in that figure purely for illustrative purposes.)

The following chart summarizes this example when the null hypothesis being tested is that $\beta_1 \leq 0$ and the alternative hypothesis is that $\beta_1 > 0$.

Inference from the test	The truth about the coefficient β_1: $\beta_1 \leq 0$ ("Strawman")	$\beta_1 > 0$ ("Pet theory")
$\beta_1 \leq 0$	no error	Type II Error
$\beta_1 > 0$	Type I Error	no error

5.2.3 Decision Rules of Hypothesis Testing

In testing a hypothesis, a sample statistic must be calculated that allows the null hypothesis to be "accepted" or rejected depending on the magnitude of that sample statistic compared with a preselected *critical* value found in tables such as those at the end of this text; this procedure is referred to as a **decision rule.**

A decision rule is formulated before regression estimates are obtained. The range of possible values of $\hat{\beta}$ is divided into two regions, an *"acceptance" region* and a *rejection region,* where the terms are expressed relative to the null hypothesis. To define these regions, we must determine a *critical value* (or, for a two-tailed test, two critical values) of $\hat{\beta}$. A **critical value** is thus a $\hat{\beta}$ value that divides the "acceptance" region from the rejection region when testing a null hypothesis.

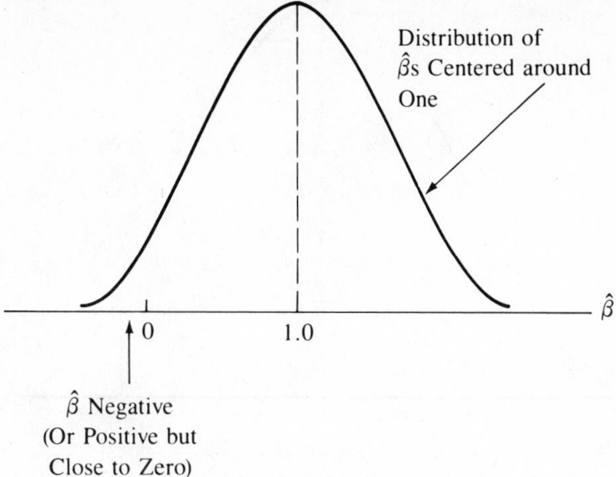

Figure 5.2 FAILURE TO REJECT A FALSE NULL HYPOTHESIS IS A TYPE II ERROR

If $\beta = 1$, but you observe a $\hat{\beta}$ that is negative or close to zero, you might fail to reject a false null hypothesis, H_0: $\beta \leq 0$, and incorrectly ignore that the alternative hypothesis H_A: $\beta > 0$ is true.

Look again at the example of the new car demand model in Equation 5.1, H_0: $\beta_1 \leq 0$ and H_A: $\beta_1 > 0$. Once we've chosen a critical value, let's call it β_c, then the decision rules becomes:

$$\begin{array}{ll} \text{Do not reject} & H_0 \text{ if } \hat{\beta}_1 \leq \beta_c \\ \text{Reject} & H_0 \text{ if } \hat{\beta}_1 > \beta_c \end{array}$$

A graph of these "acceptance" and rejection regions is presented in Figure 5.3. For a two-tailed test that H_0: $\beta_1 = 0$ and H_A: $\beta_1 \neq 0$, there would be two critical values and the graph would have two different rejection regions as shown in Figure 5.4, but the principle would be identical.

To use a decision rule, we need to select the critical value β_c. Let's suppose that we arbitrarily select a $\beta_c = 0.8$. If the observed $\hat{\beta}_1$ is greater than 0.8 but the true $\beta_1 \leq 0$, then we would reject H_0 when in fact it is true, thus making a Type I Error. Suppose we now increased β_c from 0.8 to 1.6, what would happen to the probability of a Type I Error? Since we make a Type I Error only when a $\hat{\beta}_1$ coming from a distribution centered around zero (or a negative number) exceeds β_c, then increasing β_c would *decrease* the probability of a Type I Error. As can be seen in Figure 5.5, the probability of a Type I Error decreases

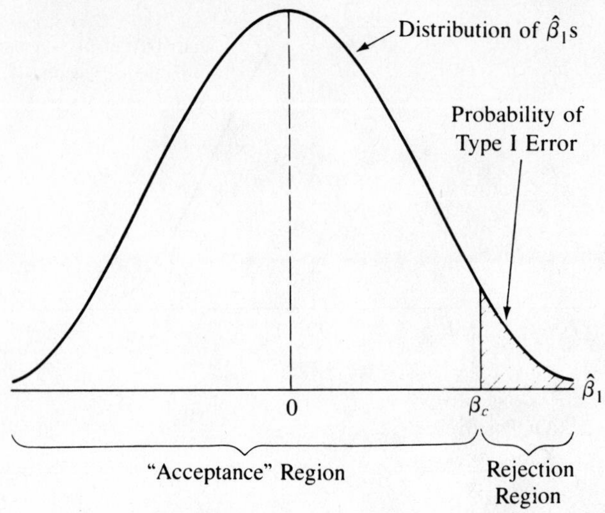

Figure 5.3 "ACCEPTANCE" AND REJECTION REGIONS FOR A ONE-SIDED TEST OF β_1

For a one-sided test of H_0: $\beta_1 \leq 0$ vs. H_A: $\beta_1 > 0$, the critical value β_c divides the distribution of $\hat{\beta}_1$ (centered around zero on the assumption that H_0 is true) into "acceptance" and rejection regions.

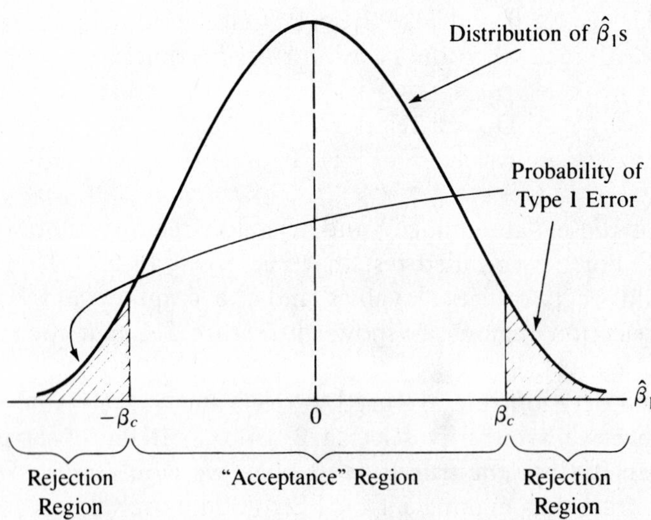

Figure 5.4 "ACCEPTANCE" AND REJECTION REGIONS FOR A TWO-SIDED TEST OF β_1

For a two-sided test of H_0: $\beta_1 = 0$ vs. H_A: $\beta_1 \neq 0$, the critical values $+\beta_c$ and $-\beta_c$ divide the distribution of $\hat{\beta}_1$ into an "acceptance" region and *two* rejection regions.

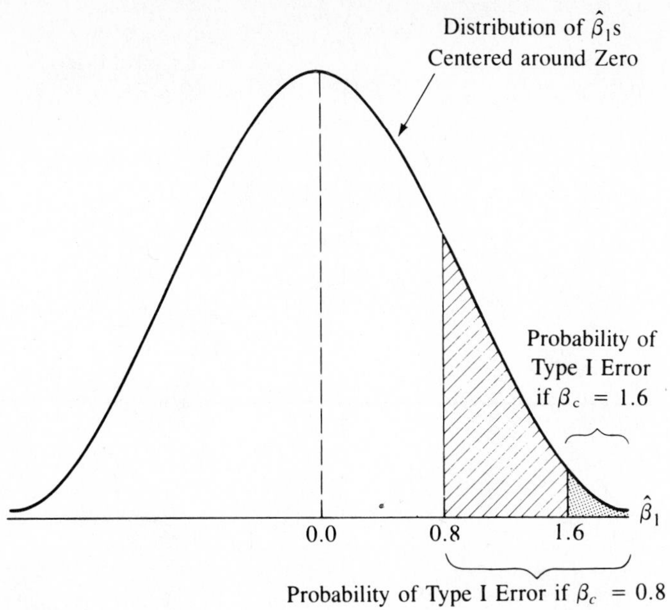

Figure 5.5 TYPE I ERROR DECREASES AS THE CRITICAL VALUE β_c INCREASES

For a distribution of $\hat{\beta}_1$s centered around zero for the one-sided test of H_0: $\beta_1 \leq 0$ vs. H_A: $\beta_1 > 0$, the probability of a Type I Error (rejection of a true H_0) decreases as the critical value is increased from 0.8 to 1.6. For this particular null hypothesis, the probability of a Type I Error is represented by the area under the curve to the right of the critical value.

by more than tenfold. Thus, if Type I Errors were the only possible kind (or the only kind that mattered to the researcher) then it would make sense to set β_c quite high.

What would such a decision do to the probability of making a Type II Error? Well, when the observed $\hat{\beta}_1$ is less than 0.8 but the true $\beta_1 > 0$, then we "accept" H_0 when in fact it is false, thus making a Type II Error. If we increase β_c to 1.6, this would increase the likelihood of making a Type II Error because more $\hat{\beta}_1$s centered around a positive value would fall below 1.6 than would fall below 0.8. In Figure 5.6, the probability of a Type II Error more than doubles if we increase β_c from 0.8 to 1.6 when the true β_1 equals one instead of zero.

Thus, to decrease the probability of a Type II Error, we should keep β_c low, but this will increase the probability of a Type I Error, as shown in Figure 5.5. What should we do? We should never consider trying to eliminate Type I Error completely because to do so would increase Type II Error considerably. Similarly, it is foolhardy to try to

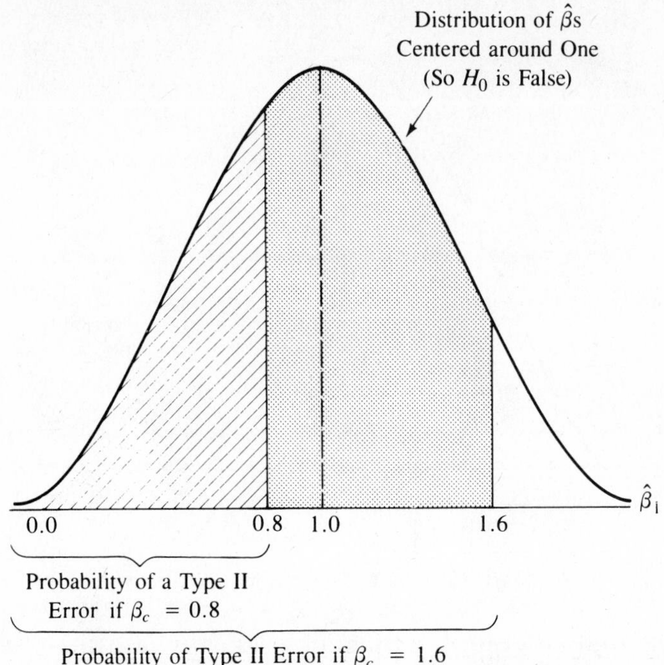

Figure 5.6 TYPE II ERROR INCREASES AS THE CRITICAL VALUE β_c INCREASES
For a distribution of $\hat{\beta}_1$s centered around one for the one-sided test of H_0: $\beta_1 \leq 0$ vs. H_A: $\beta_1 > 0$, the probability of a Type II Error ("acceptance" of a false H_0) increases as the critical value is increased from 0.8 to 1.6. For this particular null hypothesis, the probability of a Type II Error is represented by the area under the curve to the left of the critical value.

eliminate Type II Error totally (by decreasing β_c) because such a decision rule would mean rejecting virtually every null hypothesis that comes along.

Since it seems clear that we must make a trade-off between Type I and Type II Errors, how would we go about choosing β_c? Is there some easy rule to always follow? Unfortunately, the answer is no; the choice of a critical value depends on the details of the equation and variable in question. One important factor is the costs to you of making each kind of error. In some circumstances, you might want to be extremely confident that you have not "accepted" a false null hypothesis (which, in the example of the FDA, would permit the marketing of a potentially harmful product). In such a case, you would want to be quite conservative because even a hint of harmful side effects would be enough to reject the product. On the other hand, there might be situations in which you would be more concerned with not rejecting a true null hypothesis (for instance, when dealing with a new approach to the consumption function). In such circumstances, you would want to be fairly liberal for fear of rejecting a potentially valuable new theory before giving it a fair test.

As an example of choosing between levels of Type I and Type II Errors, let's suppose that you're on a jury in a murder case.[7] In such a situation, the presumption of "innocent until proven guilty" implies that:

H_0: The defendant is innocent.
H_A: The defendant is guilty.

What would a Type I Error be? Rejecting the null hypothesis would mean sending the defendant to jail, so a Type I Error, rejecting a true null hypothesis, would mean:

Type I Error = sending an innocent defendant to jail.

Similarly:

Type II Error = freeing a guilty defendant.

Most reasonable jury members would want both levels of error to be quite small, but such certainty is almost impossible. After all, couldn't there be a mistaken identification or a lying witness? In the real world, decreasing the probability of a Type I Error (sending an innocent defendant to jail) means increasing the probability of a Type II Error (freeing a guilty defendant). If we never sent an innocent defendant to jail, we'd be freeing quite a few murderers!

Given that, how would you choose between Type I and Type II Errors? Is it worse to send an innocent defendant to jail or to free a murderer? Typical econometric practice is to choose decision rules and critical values that control the probability of Type I Error, thus implying that it's more important to avoid rejecting a true null hypothesis. Exceptions to this general rule aren't unusual, however, especially when the strawman principle has been violated. In essence, it's up to the researcher to decide which kind of error is more important to avoid and to set up hypothesis testing procedures accordingly. Typically, when we are concerned with avoiding Type II Error, we let Type I Error be larger, and we interpret the statistical results with this in mind.

7. This example comes from and is discussed in much more detail in Edward E. Leamer, *Specification Searches* (New York: John Wiley and Sons, 1978), pp. 93–98.

5.3 The *t*-Test

Rather than compare $\hat{\beta}$s with critical β-values, econometricians usually use the *t*-test to test hypotheses about individual regression slope coefficients. Tests of more than one coefficient at a time (joint hypotheses) are typically done with the *F*-test, presented in Sections 5.6 and 5.8.

The *t*-test is easy to use because it accounts for differences in the units of measurement of the variables and in the standard deviations of the estimated coefficients (both of which would affect the shape of the distribution of $\hat{\beta}$ and the location of the critical value β_c). More important, the t-statistic is the appropriate test to use when the stochastic error terms are normally distributed and when the variance of that distribution must be estimated. Since these usually are the case, the use of the *t*-test for hypothesis testing has become standard practice in econometrics.

5.3.1 The t-Statistic

For a typical multiple regression equation:

$$Y_i = \beta_0 + \beta_1 X_{1i} + \beta_2 X_{2i} + \epsilon_i \tag{5.2}$$

we can calculate t-values for each of the estimated coefficients in the equation. For reasons that will be explained in Section 7.4.2, *t*-tests are usually done only on the slope coefficients; for these, the relevant general form of the **t-statistic** for the *k*th coefficient is:

$$t_k = \frac{(\hat{\beta}_k - \beta_{H_0})}{SE(\hat{\beta}_k)} \qquad (k = 1,2,\ldots,K) \tag{5.3}$$

where: $\hat{\beta}_k$ = the estimated regression coefficient of the *k*th variable

β_{H_0} = the border value (usually zero) implied by the null hypothesis for β_k

$SE(\hat{\beta}_k)$ = the estimated standard error of $\hat{\beta}_k$ (that is, the square root of the estimated variance of the distribution of the $\hat{\beta}_k$; note that there is no "hat" attached to SE because SE is already defined as an estimate).

How do you decide what *border* is implied by the null hypothesis? Some null hypotheses specify a particular value. For these, β_{H_0} is simply that value; if H_0: $\beta = S$, then $\beta_{H_0} = S$. Other null hypotheses involve ranges, but we are concerned only with the value in the null hypothesis that is closest to the border between the "acceptance" region and the rejection region. This border value then becomes the β_{H_0}; for example, if H_0: $\beta \geq 0$ and H_A: $\beta < 0$, then the value in the null hypothesis closest to the border is zero, and $\beta_{H_0} = 0$.

Since most regression hypotheses test whether a particular regression coefficient is significantly different from zero, β_{H_0} is typically zero, and the most-used form of the t-statistic becomes

$$t_k = \frac{(\hat{\beta}_k - 0)}{SE(\hat{\beta}_k)} \qquad (k = 1,2,\ldots,K)$$

which simplifies to

$$t_k = \frac{\hat{\beta}_k}{SE(\hat{\beta}_k)} \qquad (k = 1,2,\ldots,K) \tag{5.4}$$

or the estimated coefficient divided by the estimate of its standard error. This is the t-statistic formula used by most computer programs.

For an example of this calculation, let's reconsider the equation for the check volume at Woody's restaurants from Section 3.4:

$$\begin{aligned} \hat{Y}_i &= 102{,}192 - 9075C_i + 0.3547P_i + 1.288I_i \\ &\qquad\qquad\quad (2053) \qquad (0.0727) \qquad (0.543) \\ t &= \qquad\quad\; -4.42 \qquad\quad 4.88 \qquad\quad 2.37 \\ n &= 33 \qquad\qquad \bar{R}^2 = .579 \end{aligned} \tag{5.5}$$

In Equation 5.5, the numbers in parentheses underneath the estimated regression coefficients are the estimated standard errors of the estimated $\hat{\beta}$s, and the numbers below them are t-values calculated according to Equation 5.4. The format used to document Equation 5.5 above is the one we'll use whenever possible throughout this text. Note that the sign of the t-value is always the same as that of the estimated regression coefficient, while the standard error is always positive.

Using the regression results in Equation 5.5, let's calculate the t-value for the estimated coefficient of P, the population variable. Given the values in Equation 5.5 of 0.3547 for $\hat{\beta}_p$ and 0.0727 for $SE(\hat{\beta}_p)$, and given H_0: $\beta \leq 0$, the relevant t-value is indeed 4.88 as specified in Equation 5.5:

$$t_P = \frac{\hat{\beta}_P}{SE(\hat{\beta}_P)} = \frac{0.3547}{0.0727} = 4.88$$

The larger in absolute value this t-value is, the greater the likelihood that the estimated regression coefficient is significantly different from zero.

5.3.2 The Critical t-Value and the *t*-Test Decision Rule

To reject a null hypothesis based on a calculated t-value, we use a critical t-value (which parallels the critical β-value discussed in Section 5.2.3). A *critical t-value* is the value that distinguishes the "acceptance" region from the rejection region. The critical t-value, t_c, is selected from a t-table (see Statistical Table B-1 in the back of the book) depending on whether the test is one-sided or two-sided, on the level of Type I Error you specify, and on the degrees of freedom, which we have defined as the number of observations minus the number of coefficients estimated (including the constant) or $n - K - 1$. The level of Type I Error in a hypothesis test is also called the *level of significance* of that test and will be discussed in more detail later in this section. The t-table was created to save time during research; it consists of critical t-values given specified areas underneath curves such as those in Figure 5.5 for Type I Errors. A critical t-value is thus a function of the probability of Type I Error that the researcher wants to specify.

Once you have obtained a calculated t-value and a critical t-value, you reject the null hypothesis if the calculated t-value is greater in absolute value than the critical t-value and if the calculated t-value has the sign implied by H_A.

Thus, the rule to apply when testing a single regression coefficient is that you should:

> Reject H_0 if $|t_k| > t_c$ and if t_k also has the sign implied by H_A.
> Do Not Reject H_0 otherwise.

This decision rule works for calculated t-values and critical t-values for one-sided hypotheses around zero:

$$H_0\colon \beta_k \leq 0$$
$$H_A\colon \beta_k > 0$$

$$H_0\colon \beta_k \geq 0$$
$$H_A\colon \beta_k < 0$$

for two-sided hypotheses around zero:

$$H_0\colon \beta_k = 0$$
$$H_A\colon \beta_k \neq 0$$

for one-sided hypotheses based on hypothesized values other than zero:

$$H_0\colon \beta_k \leq S$$
$$H_A\colon \beta_k > S$$

$$H_0\colon \beta_k \geq S$$
$$H_A\colon \beta_k < S$$

and for two-sided hypotheses based on hypothesized values other than zero:

$$H_0\colon \beta_k = S$$
$$H_A\colon \beta_k \neq S$$

The decision rule is the same: Reject the null hypothesis if the appropriately calculated t-value, t_k, is greater in absolute value than the critical t-value, t_c, as long as the sign of t_k is the same as the sign of the coefficient implied in H_A. Otherwise, "accept" H_0. Always use Equation 5.3 whenever the hypothesized value is not zero.

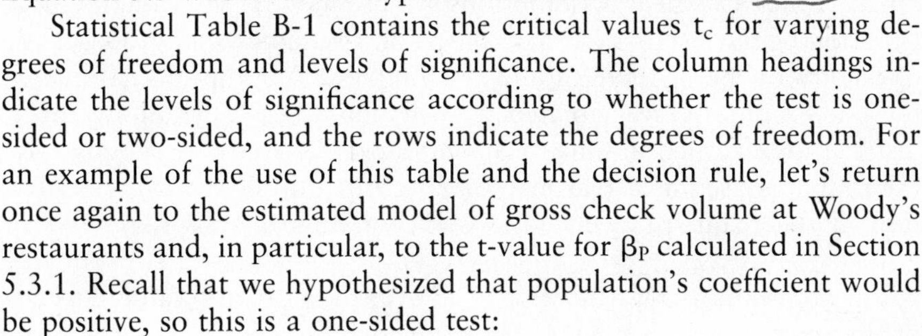

Statistical Table B-1 contains the critical values t_c for varying degrees of freedom and levels of significance. The column headings indicate the levels of significance according to whether the test is one-sided or two-sided, and the rows indicate the degrees of freedom. For an example of the use of this table and the decision rule, let's return once again to the estimated model of gross check volume at Woody's restaurants and, in particular, to the t-value for β_P calculated in Section 5.3.1. Recall that we hypothesized that population's coefficient would be positive, so this is a one-sided test:

$$H_0\colon \beta_P \leq 0$$
$$H_A\colon \beta_P > 0$$

There are 29 degrees of freedom (equal to $n-K-1$, or $33-3-1$) in this regression, so the appropriate critical t-value with which to test the calculated t-value is a one-tailed critical t-value with 29 degrees of freedom. To find this value, pick a level of significance, say five percent, and turn to Statistical Table B-1. The number there is 1.699; should you reject the null hypothesis?

The decision rule is to reject H_0 if $|t_k| > t_c$ and if t_k has the sign implied by H_A. Since the 5 percent, one-sided, 29 degrees of freedom critical t-value is 1.699, and since the sign implied by H_A is positive, decision rule (for this specific case) becomes:

$$\text{Reject } H_0 \text{ if } |t_P| > 1.699 \text{ and if } t_P \text{ is positive.}$$

or, combining the two conditions:

$$\text{Reject } H_0 \text{ if } t_P > 1.699.$$

What was t_P? In the previous section, we found that t_P was +4.88, so we would reject the null hypothesis and conclude that population does indeed tend to have a positive relationship with Woody's check volume (holding constant the other variables in the equation).

This decision rule is based on the fact that, since both $\hat{\beta}$ and $SE(\hat{\beta})$ have known sampling distributions, so does their ratio, the t-statistic. The sampling distribution of $\hat{\beta}$ was shown in Chapter 4 and is based on the assumption of the normality of the error term ϵ_i and on the other Classical Assumptions. Consequently, the sampling distribution of the t-statistic is also based on the same assumption of the normality of the error term and the Classical Assumptions. If any of these assumptions are violated, t_c will not necessarily follow the t-distribution detailed in Statistical Table B-1. In many cases, however, the t-table is used as a reasonable approximation of the true distribution of the t-statistic even when some of these assumptions do not hold.

In addition, as was mentioned above, the critical t-value depends on the number of degrees of freedom, on the level of Type I Error (referred to as the level of statistical significance), and on wheher the hypothesis is a one-tailed or two-tailed one. Figure 5.7 illustrates the dependence of the critical value t_c on two of these factors. For the simple regression model with 30 observations and two coefficients to estimate (the slope and the intercept), there are 28 degrees of freedom. The "acceptance" and rejection regions are stated in terms of the decision rule for several levels of statistical significance and for one-sided (H_A: $\beta > 0$) and two-sided (H_A: $\beta \neq 0$) alternatives.

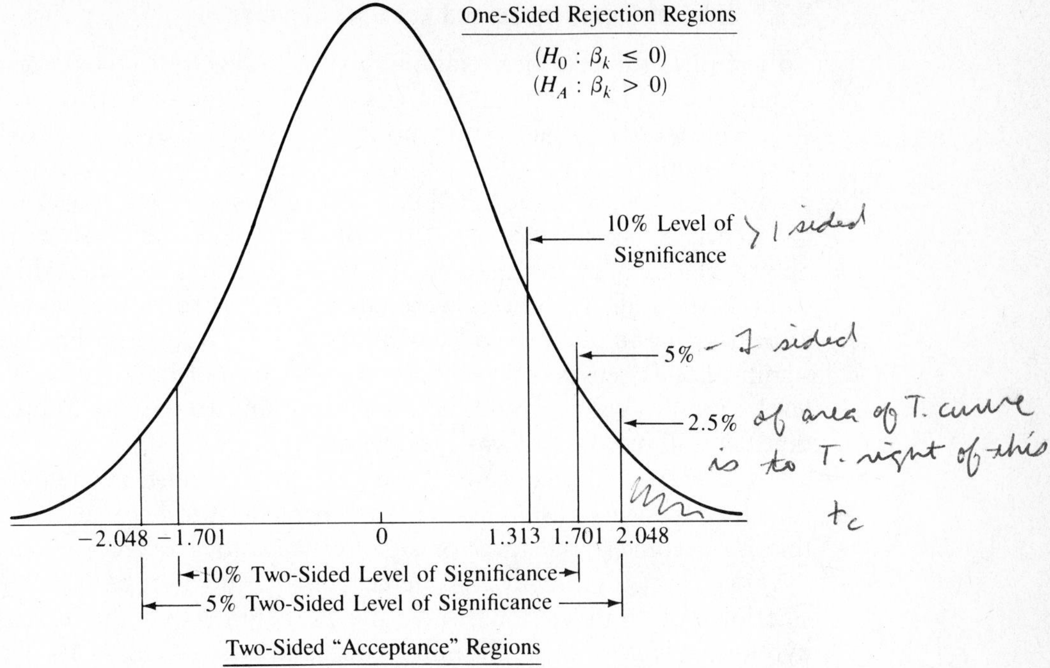

Figure 5.7 ONE-SIDED AND TWO-SIDED *t*-TESTS FOR VARIOUS LEVELS OF SIGNIFICANCE AND 28 DEGREES OF FREEDOM

The critical t-value depends on whether the *t*-test is two-sided or one-sided and on the chosen level of significance. In particular, the t_c for a one-sided test at a given level of significance is equal exactly to the t_c for a two-sided test with twice the level of significance of the one-sided test. For example, $t_c = 1.701$ for a 10 percent two-sided and a 5 percent one-sided test.

Note from Statistical Table B-1 that the critical t-value for a one-tailed test at a given level of significance is exactly equal to the critical t-value for a two-tailed test at twice the level of significance as the one-tailed test. This property arises because the t-statistic is symmetrical. For example, if five percent of the area under the curve is to the right of t_c, then five percent will also be to the left of $-t_c$, and the two tails will sum to ten percent. This relationship between one-sided and two-sided tests is illustrated in Figure 5.7. The critical value $t_c = 1.701$ is for a one-sided, five percent level of significance, but it also represents a two-sided, ten percent level of significance because if one tail represents five percent, then both tails added together represent ten percent.

5.3.3 Level of Significance and Level of Confidence

To complete the previous example, it was necessary to pick a level of significance before a critical t-value could be found in Statistical Table B-1. The words "significantly positive" usually carry the statistical interpretation that H_0 ($\beta \leq 0$) was rejected in favor of H_A ($\beta > 0$) according to the preestablished decision rule, which was set up with a given level of significance. The **level of significance** indicates the probability of observing an estimated t-value greater than the critical t-value if the null hypothesis were correct. It measures the amount of Type I Error implied by a particular critical t-value. If the level of significance is ten percent and we reject the null hypothesis at that level, then this result would have occurred only 10 percent of the time that the null hypothesis was indeed correct.

If a ten percent chance of Type I Error is specified in the decision rule, then the level of significance is ten percent. Another way of stating this is to convert the level of significance into a degree or *level of confidence* that is exactly 100 percent minus the level of significance. The **level of confidence** indicates the probability that the alternative hypothesis is correct if the null hypothesis is rejected. A 10 percent level of significance can also be stated as a 90 percent level of confidence, and results can be summarized by saying that a coefficient has been shown to be "statistically significantly positive," or just "statistically significant" at the 10 percent level of significance or the 90 percent level of confidence. Thus, if the alternative hypothesis is accepted, we have 90 percent confidence that the alternative hypothesis is correct.

Many economists use an arbitrary value of two for the critical value of the t-statistic. For a two-tailed test, a value of $t_c = 2$ provides approximately a five percent chance of a Type I Error for ten or more degrees of freedom, and for a one-tailed test, a value of $t_c = 2$ produces a probability of Type I Error equal to about 2.5 percent for 20 or more degrees of freedom. Although the choice of the permissible level of Type I Error and therefore the choice of t_c is essentially arbitrary, beginners are advised to use the more "scientific" approach outlined here rather than the rule of thumb $t_c = 2$; though it often saves quite a bit of time to use $t_c = 2$ as a preliminary benchmark when judging the significance of a particular estimated regression coefficient. Some researchers do not even choose a level of significance or a critical t-value before running the regression. Instead, they simply state the highest degree of confidence possible for any given estimated regression coefficient. Such a use of the critical t-value should be regarded as a descriptive, rather than hypothesis testing, use of statistics.

5.4 Examples of *t*-Tests

5.4.1 Examples of One-Sided *t*-Tests

Researchers usually want to know whether the regression results indicate that β is of the particular sign they have hypothesized; this is a one-sided test. After all, given a positive expected sign, a negative $\hat{\beta}$ of any size would cause a researcher to question whether the results are consistent with theory even without the use of the *t*-test. If the observed $\hat{\beta}$ is positive but fairly close to zero, however, then a one-sided *t*-test should be used to determine whether the $\hat{\beta}$ is different enough from zero to allow the rejection of the null hypothesis. Recall that in order to be able to control the amount of Type I Error we make, such a theory implies an alternative hypothesis of H_A: $\beta > 0$ (the expected sign) and a null hypothesis of H_0: $\beta \leq 0$. Let's look at some complete examples of these kinds of one-sided *t*-tests.

Return one last time to the model of aggregate retail sales of new cars first discussed in Section 5.2. The original model hypothesized was that sales of new cars (Y) are a function of real disposable income (X_1) and the average retail price of a new car adjusted by the consumer price index (X_2). Suppose you spend some time reviewing the literature on the automobile industry and are inspired to test a new theory. You decide to add a third independent variable, the number of four-wheel drive trucks sold (X_3), to take account of the fact that some potential new car buyers now buy car-like trucks instead. You hypothesize the following model:

$$Y = f(\overset{+}{X_1}, \overset{-}{X_2}, \overset{-}{X_3}) \tag{5.6}$$

β_1 is expected to be positive and β_2 and β_3, negative. This makes sense, since you'd expect higher incomes, lower prices, or lower numbers of four-wheel trucks sold to increase new car sales, holding the other variables in the equation constant. Although in theory a single test for all three slope coefficients could be applied here, nearly every researcher examines each coefficient separately with the *t*-test. The four steps to use when working with the *t*-test are:

1. Set up the null and alternative hypotheses.
2. Choose a level of significance and therefore a critical t-value.
3. Run the regression and obtain an estimated t-value (or t-score).

4. Apply the decision rule by comparing the calculated t-value with the critical t-value in order to reject or "accept" the null hypothesis.

1. *Set up the null and alternative hypotheses.*[8] From Equation 5.6, the one-sided hypotheses are set up as:

$$1.\ H_0: \beta_1 \leq 0 \qquad H_A: \beta_1 > 0$$

$$2.\ H_0: \beta_2 \geq 0 \qquad H_A: \beta_2 < 0$$

$$3.\ H_0: \beta_3 \geq 0 \qquad H_A: \beta_3 < 0$$

Remember that a *t*-test typically is not run on the estimate of the constant term β_0, as already mentioned.

2. *Choose a level of significance and therefore a critical t-value.* Assume that you have considered the various costs involved in making Type I and Type II Errors and have chosen five percent as the level of significance with which you want to test. There are ten observations in the data-set that is going to be used to test these hypotheses, and so there are $10 - 3 - 1 = 6$ degrees of freedom. At a 5 percent level of significance (or a 95 percent level of confidence) the critical t-value, t_c, can be found in Statistical Table B-1 to be 1.943. Note that the level of significance does not have to be the same for all the coefficients in the same regression equation. It could well be that the costs involved in an incorrectly rejected null hypothesis for one coefficient are much higher than for another, and so lower levels of significance would be used. In this equation, though, for all three variables:

$$t_c = 1.943$$

3. *Run the regression and obtain an estimated t-value.* You now use the data (annual from 1981 to 1990) to run the regression on your computer's OLS package, getting:

8. Recall from footnote 4 that a one-sided hypothesis can be stated either as H_0: $\beta \leq 0$ or H_0: $\beta = 0$ because the value used to test H_0: $\beta \leq 0$ is the value in the null hypothesis closest to the border between the acceptance and the rejection regions. When the amount of Type I Error is calculated, this border value of β is the one that is used because, over the whole range of $\beta \leq 0$, the value $\beta = 0$ gives the maximum amount of Type I Error. The classical approach limits this maximum amount to a preassigned level, the chosen level of significance.

$$\begin{array}{rlll} \hat{Y}_t = 1.30 + & 4.91X_{1t} + & 0.00123X_{2t} - & 7.14X_{3t} \\ & (2.38) & (0.00022) & (71.38) \\ t = & 2.1 & 5.6 & -0.1 \end{array} \tag{5.7}$$

where: Y = new car sales (in hundreds of thousands of units) in year t

X_1 = real U.S. disposable income (in hundreds of billions of dollars)

X_2 = the average retail real price of a new car in year t (in dollars)

X_3 = the number of four-wheel drive trucks sold in year t (in millions)

Once again, we use our standard documentation notation, so the figures in parentheses are the estimated standard errors of the $\hat{\beta}$s. The t-values to be used in these hypothesis tests are printed out by most standard OLS programs, because the programs are written to test the null hypothesis that $\beta = 0$ (or, equivalently, $\beta \leq$ or ≥ 0). If the program does not calculate the t-scores automatically, one may plug the $\hat{\beta}$s and their estimated standard errors into Equation 5.4, repeated here:

$$t_k = \frac{\hat{\beta}_k}{SE(\hat{\beta}_k)} \qquad (k = 1,2,\ldots,K) \tag{5.4}$$

For example, the estimated coefficient of X_3 divided by its estimated standard error is $-7.14/71.38 = -0.1$. Note that since standard errors are always positive, a negative estimated coefficient implies a negative t-value.

4. *Apply the decision rule by comparing the calculated t-value with the critical t-value in order to reject or "accept" the null hypothesis.* As stated in Section 5.3, the decision rule for the *t*-test is to

Reject H_0 if $|t_k| > t_c$ and if t_k also has the sign implied by H_A.
Do Not Reject H_0 otherwise.

What would these decision rules would be for the three hypotheses, given the relevant critical t-value (1.943), and the calculated t-values?

For β_1: Reject H_0 if $|2.1| > 1.943$ and if 2.1 is positive.

In the case of disposable income, you reject the null hypothesis that

$\beta_1 \leq 0$ with 95 percent confidence since 2.1 is indeed greater than 1.943. This result (that is, H_A: $\beta_1 > 0$) is as you expected on the basis of theory since the more income in the country, the more new car sales you'd expect.

For β_2: Reject H_0: if $|5.6| > 1.943$ and if 5.6 is negative.

For prices, the t-statistic is large in absolute value (being greater than 1.943) but has a sign that is contrary to our expectations, since the alternative hypothesis implies a negative sign. Since both conditions in the decision rule must be met before we can reject H_0, you cannot reject the null hypothesis that $\beta_2 \geq 0$. That is, you cannot reject the hypothesis that prices have a zero or positive effect on new car sales! This is an extremely small data-set that covers a time period of dramatic economic swings, but even so, you're surprised by this result. Despite your surprise, you stick with your contention that prices belong in the equation and that their expected impact should be negative.

Notice that the coefficient of X_2 is quite small, 0.00123, but that this size has no effect on the t-calculation other than its relationship to the standard error of the estimated coefficient. In other words, the absolute magnitude of any $\hat{\beta}$ is of no particular importance in determining statistical significance because a change in the units of measurement of X_2 will change both $\hat{\beta}_2$ and $SE(\hat{\beta}_2)$ in exactly the same way, so the calculated t-value (the ratio of the two) is unchanged.

For β_3: Reject H_0 if $|-0.1| > 1.943$ and if -0.1 is negative.

For sales of four-wheel drive trucks, the coefficient $\hat{\beta}_3$ is not statistically different from zero since $|-0.1| < 1.943$, and you cannot reject the null hypothesis that $\beta \geq 0$ even though the estimated coefficient has the sign implied by the alternative hypothesis. After thinking this model over again, you come to the conclusion that you were hasty in adding the four-wheel drive variable to the equation.

Figure 5.8 illustrates all three of these outcomes by plotting the critical t-value and the calculated t-values for all three null hypotheses on a t-distribution that is centered around zero (the value in the null hypothesis closest to the border between the acceptance and rejection regions). Students are urged to analyze the results of tests on the estimated coefficients of Equation 5.7 assuming different numbers of observations and different levels of significance. Exercise 4 has a number of such specific combinations, with answers in the back of the book.

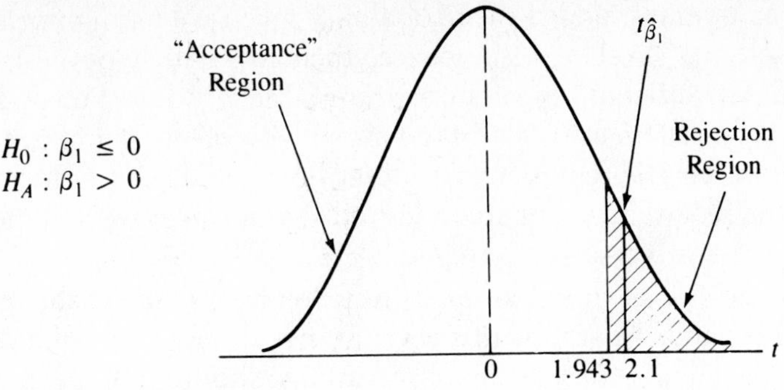

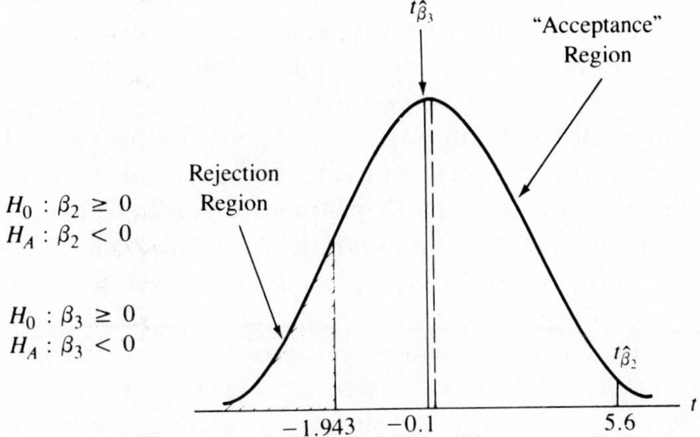

Figure 5.8 ONE-SIDED *t*-TESTS OF THE COEFFICIENTS OF THE NEW CAR SALES MODEL

Given the estimates in Equation 5.7 and the critical t-value of 1.943 for a 5 percent level of significance, one-sided, 6 degree of freedom *t*-test, we can reject the null hypothesis for $\hat{\beta}_1$, but not for $\hat{\beta}_2$ or $\hat{\beta}_3$.

Researchers sometimes note in their results the maximum level of confidence achieved by an estimated coefficient. For example, in Figure 5.8, the area under the t-statistic curve to the right of 2.1 for β_1 is the level of significance (about four percent in this case). Since the level of significance chosen is subjective, such an approach allows readers to form their own conclusions about the acceptance or rejection of hypotheses. It also conveys the information that the null hypothesis can be rejected with more confidence for some estimated coefficients than for others. Computer programs often give such probabilities of significance for t-values, and if the probability given is less than or equal to the preselected level of significance, then the null hypothesis can be rejected. The availability of such probabilities should not deceive beginning researchers into waiting to state the levels of significance to be used until after the regressions are run, however, because the researchers run the risk of adapting their desired significance levels to the results rather than vice versa.

The purpose of this example is to provide practice in the testing of hypotheses, and the results of such a poorly thought-through equation for such a small number of observations should not be taken too seriously. Given all that, however, it's still instructive to note that you did not react the same way to your inability to reject the null hypotheses for the price and four-wheel drive truck variables. That is, the failure of the four-wheel drive variable's coefficient to be significantly negative caused you to realize that perhaps the addition of this variable was ill-advised. The failure of the price variable's coefficient to be significantly negative did not cause you to consider the possibility that price has no effect on new car sales. Put differently, estimation results should never be allowed to cause you to want to adjust theoretically sound variables or hypotheses, but if they make you realize you have made a serious mistake, then it would be foolhardy to ignore that mistake. What to do about the positive coefficient of price, on the other hand, is what the "art" of econometrics is all about. Surely a positive coefficient is unsatisfactory, but throwing the price variable out of the equation seems even more so. Possible answers to such issues are addressed more than once in the chapters that follow.

5.4.2 Examples of Two-Sided *t*-Tests

As mentioned above, most hypotheses in regression analysis can be tested with one-sided *t*-tests, but in particular situations, two-sided tests on a single coefficient are appropriate. Researchers sometimes encounter hypotheses that should be rejected if estimated coefficients are significantly different from β in either direction. This situation requires a two-sided *t*-test. The kinds of circumstances that call for a two-sided test fall into two categories:

1. Two-sided tests of whether an estimated coefficient is "significantly different from zero," and
2. Two-sided tests of whether an estimated coefficient is significantly different from a specific nonzero value.

1. Testing whether a $\hat{\beta}$ is statistically different from zero. The first case for a two-sided test of $\hat{\beta}$ arises when there are two or more conflicting hypotheses about the expected sign of a coefficient. For example, in the Woody's restaurant equation of Section 3.4, the impact of the average income of an area on the expected number of Woody's customers in that area is ambiguous. A high-income neighborhood might have more total customers going out to dinner, but those customers might decide to eat at a more formal restaurant than Woody's. As a result, you could run a two-sided *t*-test around zero to determine whether or not the estimated coefficient of income is significantly different from zero in *either* direction. In other words, since there are reasonable cases to be made for either a positive or a negative coefficient, it is appropriate to test the $\hat{\beta}$ for income with a two-sided *t*-test:

$$H_0: \beta_I = 0$$
$$H_A: \beta_I \neq 0$$

As Figure 5.9 illustrates, a two-sided test implies two different rejection regions (one positive and one negative) surrounding the acceptance region. A critical t-value, t_c, must be increased in order to achieve the same level of significance with a two-sided test as can be achieved with a one-sided test.[9] As a result, there is an advantage to testing hypotheses with a one-sided test if the underlying theory allows because, for the same t-values, the possibility of Type I Error is half as much for a one-sided test as for a two-sided test. In cases where there are powerful theoretical arguments on both sides, however, the researcher has no alternative to using a two-sided *t*-test around zero. To see how this works, let's follow through the Woody's income variable example in more detail.

1. *Set up the null and alternative hypotheses.*

$$H_0: \beta_I = 0$$
$$H_A: \beta_I \neq 0$$

2. *Choose a level of significance and therefore a critical t-value.* You decide to keep the level of significance at 5 percent, but now this

9. See Figure 5.7 in Section 5.3. In that figure, the same critical t-value has double the level of significance for a two-sided test as for a one-sided test.

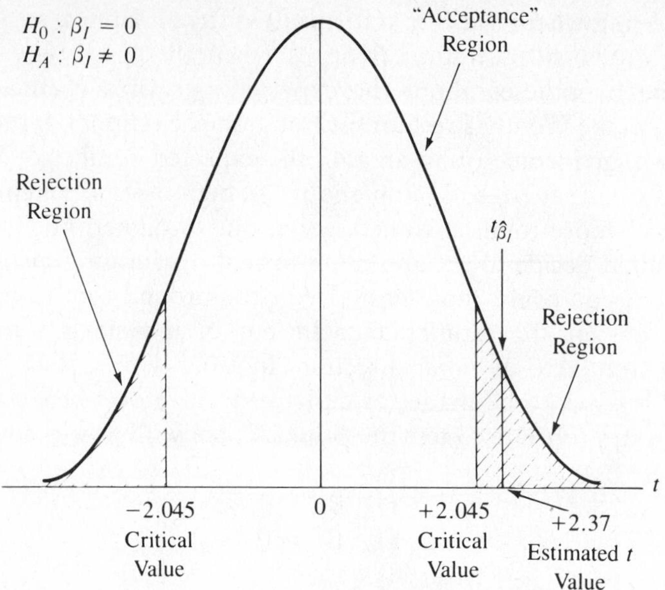

Figure 5.9 A TWO-SIDED *t*-TEST OF THE COEFFICIENT OF INCOME IN THE WOODY'S MODEL
Given the estimates of Equation 5.5 and the critical t-values of ±2.045 for a 5 percent level significance, two-sided, 29 degree of freedom *t*-test, we can reject the null hypothesis that $\beta_I = 0$.

amount must be distributed between two rejection regions for 29 degrees of freedom. Hence, the correct critical t-value is 2.045 (found in Statistical Table B-1 for 29 degrees of freedom and a 5 percent, two-sided test). Note that technically there now are two critical t-values, +2.045 and −2.045.

3. *Run the regression and obtain an estimated t-value.* Since the value implied by the null hypothesis is still zero, the estimated t-value of +2.37 given in Equation 5.5 is applicable.

4. *Apply the decision rule by comparing the calculated t-value with the critical t-value in order to reject or "accept" the null hypothesis.* We once again use the decision rule stated in Section 5.3, but since the alternative hypothesis specifies either sign, the decision rule simplifies to:

$$\text{For } \beta_I\text{: Reject } H_0 \text{ if } |2.37| > 2.045$$

In this case, you reject the null hypothesis that β_I equals zero because 2.37 is greater than 2.045 (see Figure 5.9). Note that the positive sign implies that, at least for Woody's restaurants, income increases customer volume (holding constant population and competition). Given this result, we might well choose to run a one-sided *t*-test on the next year's Woody's data-set. For more practice with two-sided *t*-tests, see Exercise 6.

2. Two-sided *t*-tests of a specific nonzero coefficient value. The second case for a two-sided *t*-test arises when there is reason to expect a specific nonzero value for an estimated coefficient. For example, if a previous researcher has stated that the true value of some coefficient almost surely equals a particular number, β_{H_0}, then that number would be the one to test by creating a two-sided *t*-test around the hypothesized value, β_{H_0}. To the extent that you feel that the hypothesized value is theoretically correct, you also violate the normal practice of using the null hypothesis to state the hypothesis you expect to reject, or the "strawman" approach.[10]

In such a case, the null and alternative hypotheses become:

$$H_0\colon \beta_k = \beta_{H_0}$$
$$H_A\colon \beta_k \neq \beta_{H_0}$$

where β_{H_0} is the specific nonzero value hypothesized.

Since the hypothesized β value is no longer zero, the formula with which to calculate the estimated t-value is Equation 5.3, repeated here:

$$t_k = \frac{(\hat{\beta}_k - \beta_{H_0})}{SE(\hat{\beta}_k)} \qquad (k = 1,2,\ldots,K) \tag{5.3}$$

This t-statistic is still distributed around zero if the null hypothesis is correct, because we have subtracted β_{H_0} from the estimated regression coefficient whose expected value is supposed to be β_{H_0} when H_0 is true. Since the t-statistic is still centered around zero, the decision

10. Instead of being able to reject an incorrect theory based on the evidence, the researcher who violates the strawman approach is reduced to "not rejecting" the β value expected to be true. This makes a big difference because to "accept" H_0 is merely to say that H_0 is not rejected by the data. However, there are many theories that are not rejected by the data, and the researcher is left with a regrettably weak conclusion. One way to accommodate violations of the strawman approach is to increase the level of significance, thereby increasing the likelihood of a Type I Error.

rules developed earlier are still applicable. In other words, the techniques used above are precisely the same as for a two-sided *t*-test of a specific nonzero coefficient. For practice with this kind of *t*-test, see Exercise 6.

5.4.3 Other Uses of the *t*-Test

From the previous sections, it'd be easy to get the impression that the *t*-test is used only for tests of regression coefficients, but that's hardly the case. It turns out that there is a variety of applications for the *t*-test that don't involve $\hat{\beta}$s.

The most immediately useful of these applications is a *t*-test of the simple correlation coefficient, r. Recall that r is a measure of the strength and direction of the linear relationship between two variables. We presented the formula for the simple correlation coefficient between X_1 and X_2 in Equation 2.25:

$$r_{12} = \frac{\Sigma[(X_{1i} - \overline{X}_1)(X_{2i} - \overline{X}_2)]}{\sqrt{\Sigma(X_{1i} - \overline{X}_1)^2 \Sigma(X_{2i} - \overline{X}_2)^2}} \tag{2.25}$$

One of the major uses of the simple correlation coefficient r is to test the hypothesis that two explanatory variables are correlated in a less than perfect but still significant (multicollinear) way. For imperfect multicollinearity to occur in this two-variable case, the simple correlation coefficient must be fairly large in the direction indicated by theory. In order to test this hypothesis, r can be converted into a t-statistic using:

$$t = \frac{r\sqrt{n - 2}}{\sqrt{(1 - r^2)}} \tag{5.8}$$

where n is the size of the sample. The statistic defined in Equation 5.8 follows the t-distribution with n − 2 degrees of freedom. Since t is directly related to r, a large positive r will convert into a large positive t, and so on.

Tests of hypotheses about t (and therefore about r) can be undertaken using the critical t-values and decision rules outlined in Section 5.3 and Table B-1. For example, suppose you encounter a simple correlation coefficient of 0.946 between two variables you expect to be positively correlated in a data-set with 28 observations. In this case,

$$H_0: r \leq 0$$
$$H_A: r > 0$$

and we can reject the null hypothesis of no positive collinearity if the calculated t-score is larger in absolute value than the critical t-value of 1.706 (for 26 degrees of freedom at the 5 percent, one-sided level of significance) and if t has the positive sign implied by the alternative hypothesis. If we substitute r = 0.946 and n = 28 into Equation 5.8, we obtain 14.880, and so the null hypothesis of no positive collinearity can be rejected. (If theory provides no expected direction, a two-sided test should be used.) For practice in hypothesis tests of simple correlation coefficients, see Exercise 10.

5.5 Limitations of the *t*-Test

A problem with the *t*-test is that it is easy to misuse; t-scores are so frequently printed out by computer regression packages and the *t*-test seems so easy to work with that beginning researchers sometimes attempt to use the *t*-test to "prove" things that it was never intended to even test. For that reason, it's probably just as important to know the limitations of the *t*-test as it is to know the applications of that test. Perhaps the most important of these limitations, that the usefulness of the *t*-test diminishes rapidly as more and more specifications are estimated and tested, is the subject of Section 6.4. The purpose of the present section is to give additional examples of how the *t*-test should *not* be used.

5.5.1 The *t*-Test Does Not Test Theoretical Validity

Recall that the purpose of the *t*-test is to help the researcher make inferences about a particular population coefficient based on an estimate obtained from a sample of that population. Some beginning researchers conclude that any *statistically* significant result is also a *theoretically* correct one. This is dangerous because such a conclusion confuses statistical significance with theoretical validity.

Consider for instance, the following estimated regression[11] that explains the consumer price index in the United Kingdom:

$$\begin{aligned} \hat{P} = 10.9 \; - \; & 3.2C \; + \; 0.39C^2 \\ & (0.23) \quad\; (0.02) \\ t = \; & -13.9 \quad\;\; 19.5 \\ \bar{R}^2 = .982 \quad & n = 21 \end{aligned} \tag{5.9}$$

11. These results, and others similar to them, can be found in David F. Hendry, "Econometrics—Alchemy or Science?" *Economica*, Nov. 1980, pp. 383–406.

Apply the *t*-test to these estimates. Do you agree that the two slope coefficients are statistically significant? As a quick check of Statistical Table B-1 shows, the critical t-value for 18 degrees of freedom and a 5 percent two-tailed level of significance is 2.101, so we can reject the null hypothesis of no effect in these cases and conclude that C and C^2 are indeed statistically significant variables in explaining P.

The catch is that P is the consumer price index and C is the cumulative amount of rainfall in the United Kingdom! We have just shown that rain is statistically significant in explaining consumer prices; does that also show that the underlying theory is valid? Of course not. Why is the statistical result so significant? The answer is that at a 5 percent level of significance, there is a 1-in-20 chance of rejecting a true null hypothesis. If we try 20 or more different tests, the odds are good that eventually we will be able to reject a correct null hypothesis. This almost always inappropriate technique (called data-mining) was used to obtain the unrealistic results above. The moral should be clear: Never conclude that statistical significance, as shown by the *t*-test, is the same as theoretical validity.

Occasionally, estimated coefficients will be significant in the direction opposite from that hypothesized, and some beginning researchers may be tempted to change their hypotheses. For example, a student might run a regression in which the hypothesized sign is positive, get a "statistically significant" negative sign, and be tempted to change the theoretical expectations to "expect" a negative sign. Naturally, changed hypotheses would be accompanied by detailed explanations of how the particular result makes sense. Although it is admirable to be willing to reexamine incorrect theories on the basis of new evidence, that evidence should be, for the most part, theoretical in nature. In the case cited above, the students should have been concerned that the evidence did not support the theory, but that lack of support should not have caused the theory itself to change completely. If the evidence causes a researcher to go back to the theoretical underpinnings of a model and find a mistake, then the null hypothesis should be changed, but then this new hypothesis should be tested using a completely different dataset. After all, we already know what the result will be if the hypothesis is tested on the old one.

5.5.2 The *t*-Test Does Not Test "Importance"

One possible use of a regression equation is to help determine which independent variable has the largest relative effect (importance) on the dependent variable. Some beginning researchers draw the unwarranted conclusion that the most statistically significant variable in their estimated regression is also the most important in terms of explaining the largest portion of the movement of the dependent variable. Statistical significance indicates the likelihood that a particular sample result could have been obtained by chance, but it says little if anything about which variables determine the major portion of the variation in the dependent variable. To determine importance, a measure such as the size of the coefficient multiplied by the average size of the independent variable would make much more sense.[12] Consider the following hypothetical equation:

$$\begin{aligned} \hat{Y} &= 30.0 + \underset{(1.0)}{10.0X_1} + \underset{(25.0)}{200.0X_2} \\ t &= \qquad 10.0 \qquad\quad 8.0 \\ \bar{R}^2 &= .90 \qquad n = 30 \end{aligned} \tag{5.10}$$

where all three variables are measured in dollars:

Y = mail order sales of "O'Henry's Oyster Recipes"
X_1 = advertising expenditures for "The Arlington Gourmet"
X_2 = advertising expenditures for "How to Lose 20 Pounds in a Week"
(assume prices do not change during the estimation period)

Where should O'Henry be spending its advertising money? That is, which independent variable has the biggest impact per dollar on Y? Given that X_2's coefficient is 20 times X_1's coefficient, you'd have to agree that X_2 was more important as defined above, and yet which one is more statistically significantly different from zero? With a t-statistic of 10.0, X_1 is more statistically significant than X_2 and its 8.0, but all that means is that we have more confidence that the coefficient is positive, not that the variable itself is necessarily more important in determining Y. The theoretical underpinnings of a result and the actual result itself are at least as important as the statistical significance of that result.

12. Some useful statistical measures of "importance" have been developed, but none is fully satisfactory because of the presence of multicollinearity (to be discussed in Chapter 8). See J. M. Shanks, "The Importance of Importance" (Berkeley: Survey Research Center, University of California, 1982).

5.5.3 The *t*-Test Is Not Intended for Tests of the Entire Population

The *t*-test helps make inferences about the true value of a parameter from an estimate calculated from a sample of the *population* (the group from which the sample is being drawn). As the size of the sample approaches the size of the population, an unbiased estimated coefficient approaches the true population value. If a coefficient is calculated from the entire population, then an unbiased estimate already measures the population value and a significant *t*-test adds nothing to this knowledge. One might forget this property and attach too much importance to t-scores that have been obtained from samples that approximate the population in size. All the *t*-test does is help decide how likely it is that a particular small sample will cause a researcher to make a mistake in rejecting hypotheses about the true population parameters.

This point can perhaps best be seen by remembering that the t-statistic is the estimated regression coefficient divided by the standard error of that estimated regression coefficient. If the sample size is large enough to approach the population, then the standard error will fall to about zero because the distribution of estimates becomes more and more narrowly distributed around the true parameter (if this is an unbiased estimate). The standard error will approach zero as the sample size approaches infinity. Thus the t-statistic will eventually become:

$$t = \frac{\hat{\beta}}{0} = \infty$$

The mere existence of a large t-score for a huge sample has no real substantive significance because if the sample size is large enough, you can reject almost any null hypothesis! It is true that sample sizes in econometrics can never approach infinity, but many are quite large; and others, even though fairly small, are not really samples of a population but contain the entire population in one data-set.[13]

5.6 The *F*-Test of Overall Significance

While the *t*-test is invaluable for hypotheses about individual regression coefficients, it can't be used to test hypotheses about more than one coefficient at a time. Such a limitation is unfortunate, since it's possible

13. Donald N. McCloskey, "The Loss Function Has Been Mislaid: The Rhetoric of Significance Tests," *American Economic Review*, May 1985, p. 204.

to imagine quite a few interesting hypotheses that involve more than one coefficient. For example, suppose you wanted to test the hypothesis that two regression coefficients were equal to each other? In such a situation, most researchers would use a different statistical test, the *F*-test.

The *F*-test is a method of testing a null hypothesis that includes more than one coefficient; it works by determining whether the overall fit of an equation is significantly reduced by constraining the equation to conform to the null hypothesis. If the fit is significantly reduced, then we can reject the null hypothesis. If the fit is not reduced significantly, then we can't reject the null hypothesis. The *F*-test is used most frequently in econometrics to test the overall significance of a regression equation, the topic of this section. We'll investigate other uses of the *F*-test in Section 5.8.

While R^2 and $\bar{R}^2$ measure the overall degree of fit of an equation, they don't provide a formal hypothesis test of the level of significance of that overall fit. Such a test is provided by the *F*-test. The null hypothesis in an *F*-test of overall significance is that all the slope coefficients in the equation equal zero simultaneously. For an equation with K independent variables, this means that the null[14] and alternative hypotheses would be:

$$H_0: \beta_1 = \beta_2 = \cdots = \beta_K = 0$$
$$H_A: H_0 \text{ is not true}$$

To show that the overall fit of the estimated equation is statistically significant, we must be able to reject this null hypothesis using the *F*-test.

The equation for the *F*-test of overall significance is:

$$F = \frac{ESS/K}{RSS/(n - K - 1)} = \frac{\Sigma(\hat{Y}_i - \bar{Y})^2/K}{\Sigma e_i^2/(n - K - 1)} \tag{5.11}$$

This is the ratio of the explained sum of squares (ESS) to the residual sum of squares (RSS), adjusted for the number of independent variables (K) and the number of observations in the sample (n). In this case, the "constrained equation" to which we're comparing the overall fit is:

$$Y_i = \beta_0 + \epsilon_i \tag{5.12}$$

14. Note that we don't hypothesize that $\beta_0 = 0$. This would imply that $\bar{Y} = 0$.

which is nothing more than saying $Y_i = \overline{Y}$ (plus an error term). Thus the *F*-test of overall significance is really testing the null hypothesis that the fit of the equation isn't significantly better than that provided by using the mean alone.

The decision rule to use in the *F*-test is to reject the null hypothesis if the calculated F-ratio (F) from Equation 5.11 is greater than the appropriate critical F-value (F_c):

$$\begin{aligned} \text{Reject } H_0 \text{ if } \quad & F \geq F_c \\ \text{Do Not Reject } H_0 \text{ if } \quad & F < F_c \end{aligned}$$

The critical F-value, F_c, is determined from Statistical Tables B-2 or B-3, depending on a level of significance chosen by the researcher and on the degrees of freedom. The F-statistic has two types of degrees of freedom: the degrees of freedom for the numerator of Equation 5.11 (K, the number of constraints implied by the null hypothesis) and the degrees of freedom for the denominator of Equation 5.11 ($n - K - 1$, the degrees of freedom in the regression equation). The underlying principle here is that if the F-ratio is greater than the critical value, then the estimated equation's fit is significantly better than the constrained equation's fit (in this case, just using $\overline{Y}$), and we can reject the null hypothesis of no effect.

As an example of the use of the *F*-test, let's test the overall significance of the Woody's restaurant model of Equation 3.14. Since there are three independent variables, the null and alternative hypotheses are:

$$\begin{aligned} &H_0\colon \beta_c = \beta_P = \beta_I = 0 \\ &H_A\colon H_0 \text{ is not true} \end{aligned}$$

To decide whether to reject or not reject this null hypothesis, we need to calculate Equation 5.11 for the Woody's example. There are three constraints in the null hypothesis, so $K = 3$. If we check the ECSTAT computer output for the Woody's equation on page 86, we can see that $n = 33$ and $RSS = 6{,}133{,}280{,}000$. In addition, it can be calculated that ESS equals 9,929,450,000. Thus the appropriate F-ratio is:

$$F = \frac{ESS/K}{RSS/(n - K - 1)} = \frac{9{,}929{,}450{,}000/3}{6{,}133{,}280{,}000/29} = 15.65 \qquad (5.13)$$

In practice, this calculation is never necessary, since virtually every computer regression package routinely provides the computed F-ratio for a test of overall significance as a matter of course. On the Woody's computer output, the value of the F-statistic (15.64894) can be found four lines below the R^2.

Our decision rule tells us to reject the null hypothesis if the calculated F-value is greater than the critical F-value. To determine that critical F-value, we need to know the level of significance and the degrees of freedom. If we assume a five percent level of significance, the appropriate table to use is Statistical Table B-2. The numerator degrees of freedom equal 3 (K), and the denominator degrees of freedom equal 29 $(n - K - 1)$, so we need to look in Table B-2 for the critical F-value for 3 and 29 degrees of freedom. As the reader can verify,[15] $F_c = 2.93$ is well below the calculated F-value of 15.65, so we can reject the null hypothesis and conclude that the Woody's equation does indeed have a significant overall fit.

Two final comments about the *F*-test of overall significance are in order. First, if there is only one independent variable, then an *F*-test and a *t*-test of whether the slope coefficient equals zero will always produce the same answer. (Indeed, it can be shown mathematically that the two tests are identical.) This property does not hold if there are two or more independent variables. In such a situation, an *F*-test could determine that the coefficients *jointly* are not significantly different from zero even though a *t*-test on one of the coefficients might show that *individually* it is significantly different from zero (or vice versa).

Second, as we will see in Exercise 8, the F-statistic can be shown to be a direct function of R^2. The larger R^2 is, the larger the F-ratio. Thus the *F*-test of overall significance is a test of the significance of R^2 itself.

5.7 Summary

1. Hypothesis testing makes inferences about the validity of specific economic (or other) theories from a sample of the population for

15. Note that this critical F-value must be interpolated. The critical value for 30 denominator degrees of freedom is 2.92, and the critical value for 25 denominator degrees of freedom is 2.97. Since both numbers are well below the calculated F-value of 15.65, however, the interpolation isn't necessary to reject the null hypothesis. As a result, many researchers don't bother with such interpolations unless the calculated F-value is inside the range of the interpolation.

which the theories are supposed to be true. The four basic steps of hypothesis testing (using a *t*-test as an example) are:

a. Set up the null and alternative hypotheses.
b. Choose a level of significance and therefore a critical t-value.
c. Run the regression and obtain an estimated t-value.
d. Apply the decision rule by comparing the calculated t-value with the critical t-value in order to reject or "accept" the null hypothesis.

2. The null hypothesis states the range of values that the regression coefficient is expected to take on if the researcher's theory is not correct. The alternative hypothesis is a statement of the range of values that the regression coefficient is expected to take if the researcher's theory is correct.

3. The two kinds of errors we can make in such hypothesis testing are:

Type I : We reject a null hypothesis that is true.
Type II: We do not reject a null hypothesis that is false.

4. A decision rule states critical t-values above or below which observed sample t-values will lead to the rejection or "acceptance" of hypotheses concerning population parameters. Critical values are selected from a t-distribution table depending on the chosen level of significance, the degrees of freedom involved, and the specifics of the particular hypothesis.

5. The *t*-test tests hypotheses about individual coefficients from regression equations. The general form for the t-statistic is

$$t_k = \frac{(\hat{\beta}_k - \beta_{H_0})}{SE(\hat{\beta}_k)} \qquad (k = 1,2,\ldots,K)$$

In many regression applications, B_{H_0} is zero. Once you have calculated a t-value and chosen a critical t-value, you reject the null hypothesis if the t-value is greater in absolute value than the critical t-value and if (for a one-sided test) the t-value has the sign implied by the alternative hypothesis.

6. The *t*-test is easy to use for a number of reasons, but care should be taken when using the *t*-test to avoid confusing statistical significance with theoretical validity or empirical importance.

7. The *F*-test is a method of testing a null hypothesis that includes more than one coefficient. The *F*-test is used most frequently in econometrics to test the overall significance of a regression equation with the following equation:

$$F = \frac{ESS/K}{RSS/(n - K - 1)} = \frac{\Sigma(\hat{Y}_i - \overline{Y})^2/K}{\Sigma e_i^2/(n - K - 1)}$$

Once you've calculated an F and chosen a critical F-value, F_c, then you can reject the null hypothesis that the overall fit of an equation is not significant if $F \geq F_c$.

Exercises

(Answers to even-numbered exercises are in Appendix A.)

1. Write the meaning of each of the following terms without reference to the book (or your notes), and compare your definition with the version in the text for each.
   a. null hypothesis
   b. alternative hypothesis
   c. Type I Error
   d. level of significance
   e. two-sided test
   f. decision rule
   g. critical value
   h. t-statistic
   i. *t*-test of the simple correlation coefficient
   j. *F*-test
   k. strawman principle

2. Create null and alternative hypotheses for the following coefficients:
   a. the impact of height on weight (Section 1.3)
   b. all the coefficients in Equation A in Exercise 7, Chapter 2
   c. all the coefficients in Y = f(X_1, X_2, and X_3) where Y is total gasoline used on a particular trip, X_1 is miles traveled, X_2 is the weight of the car, and X_3 is the average speed traveled
   d. the impact of the decibel level of the grunt of a shot-putter on the length of the throw involved (shot-putters are known to make loud noises when they throw, but there is little theory about the impact of this yelling on the length of the put). Assume all relevant "nongrunt" variables are included in the equation.

3. Think of examples other than the ones in this chapter in which:

a. It would be more important to keep Type I Error low than to keep Type II Error low.
b. It would be more important to keep Type II Error low than to keep Type I Error low.

4. Return to Section 5.4 and test the hypotheses in Equation 5.6 with the results in Equation 5.7 for all three coefficients under the following circumstances:
   a. 10 percent significance and 15 observations
   b. 90 percent confidence and 28 observations
   c. 99 percent confidence and 10 observations

5. Return to Section 5.3 and test the appropriate hypotheses with the results in Equation 5.5 for all three coefficients under the following circumstances:
   a. 5 percent significance and 6 degrees of freedom
   b. 90 percent confidence and 29 degrees of freedom
   c. 99 percent confidence and 2 degrees of freedom

6. Using the techniques of Section 5.4, test the following two-sided hypotheses:
   a. For Equation 5.10, test the hypothesis that:

$$H_0\colon \beta_2 = 160.0$$
$$H_A\colon \beta_2 \neq 160.0$$

   at the 5 percent level of significance.
   b. For Equation 5.5, test the hypothesis that:

$$H_0\colon \beta_3 = 0$$
$$H_A\colon \beta_3 \neq 0$$

   at 99 percent level of confidence.
   c. For Equation 5.7, test the hypothesis that:

$$H_0\colon \beta_2 = 0$$
$$H_A\colon \beta_2 \neq 0$$

   at the 5 percent level of significance.

7. For all three tests in Exercise 6, under what circumstances would you worry about possible violations of the strawman principle? In particular, what would your theoretical expectations have to be in order to avoid violating the strawman principle on Exercise 6a?

8. It turns out that the F-ratio can be expressed as a function of R^2.
   a. As an exercise, substitute Equation 2.21 into Equation 5.11 to derive the exact relationship between F and R^2.
   b. If one can be expressed as a function of the other, why do we need both? What reason is there for computer regression packages to typically print out both R^2 and the F-ratio?

9. Test the overall significance of equations that have the following F-values (Using Statistical Table B-2):
   a. F = 5.63 with 4 degrees of freedom in the numerator and 30 degrees of freedom in the denominator
   b. F = 1.53 with 3 degrees of freedom in the numerator and 24 degrees of freedom in the denominator
   c. F = 57.84 with 5 degrees of freedom in the numerator and 60 degrees of freedom in the denominator

10. Given the following simple correlation coefficients between two explanatory variables, use the *t*-test (and Equation 5.8) to test the possibility of significant collinearity in the specified circumstances:
   a. r = .905, n = 18, 5 percent level, positive expected relationship
   b. r = .958, n = 27, 2.5 percent level, positive expected relationship
   c. r = .821, n = 7, 1 percent level, positive expected relationship
   d. r = −.753, n = 42, 10 percent level, negative expected relationship
   e. r = .519, n = 30, 5 percent level, ambiguous expected relationship

11. Consider the following hypothetical equation for a sample of divorced men who failed to make at least one child support payment in the last four years (standard errors in parentheses):

$$\hat{P}_i = \underset{}{2.0} + \underset{(0.10)}{0.50M_i} + \underset{(20.0)}{25.0Y_i} + \underset{(1.00)}{0.80A_i} + \underset{(3.0)}{3.0B_i} - \underset{(0.05)}{0.15C_i}$$

where: P_i = the number of monthly child support payments that the *i*th man missed in the last four years
M_i = the number of months the *i*th man was unemployed in the last four years
Y_i = the ratio of the dollar value of the average child support payment to average monthly disposable income for the *i*th man

A_i = the age in years of the ith man
B_i = the religious beliefs of the ith man (a scale of 1 to 4, with 4 being the most religious)
C_i = the number of children the ith man has fathered

a. Your friend expects the coefficients of M and Y to be positive. Test these hypotheses. (Use the 95 percent level and n = 20.)
b. Test the hypothesis that the coefficient of A is different from zero. (Use the 1 percent level and n = 25.)
c. Develop and test hypotheses for the coefficients of B and C. (Use the 90 percent level and n = 17.)

12. Test the overall significance of the Magic Flag Farm weight/height estimated equation in Section 1.3 by using the F-ratio and Statistical Table B-2 in the back of the book. (Hint: The first step is to calculate the F-ratio from the information given in Table 2.1.)

13. Bruggink and Rose[16] estimated a regression for the annual team revenue for Major League Baseball franchises:

$$\hat{R}_i = -1522.5 + \underset{(9.1)}{53.1P_i} + \underset{(233.6)}{1469.4M_i} + \underset{(1363.6)}{1322.7S_i} - \underset{(2255.7)}{7376.3T_i}$$
$$t = \quad 5.8 \qquad 6.3 \qquad 1.0 \qquad -3.3$$
$$\bar{R}^2 = .682 \qquad n = 78 \text{ (1984-1986)} \qquad F = 42.2$$

where: R_i = team revenue from attendance, broadcasting, and concessions (in thousands of dollars)
P_i = the percentage of the ith team's games that the team won (in thousands, so 1000 = high)
M_i = the population of the ith team's metropolitan area (in millions)
S_i = a dummy equal to 1 if the ith team's stadium was built before 1940, 0 otherwise
T_i = a dummy equal to 1 if the ith team's city has two Major League Baseball teams, 0 otherwise

a. Develop and test appropriate hypotheses about the individual coefficients and the overall fit at the 5 percent level. (Hint: You do not have to be a sports fan to do this question correctly.)

16. Thomas H. Bruggink and David R. Rose, Jr., "Financial Restraint in the Free Agent Labor Market for Major League Baseball: Players Look at Strike Three," *Southern Economic Journal*, April 1990, pp. 1029–1043.

b. The authors originally expected a negative coefficient for S. Their explanation for the unexpected positive sign was that teams in older stadiums have greater revenue because they're better known and have more faithful fans. Since this $\hat{\beta}$ is just one observation from the sampling distribution of $\hat{\beta}$s, do you think they should have changed their expected sign based on one $\hat{\beta}$?

c. On the other hand, Keynes reportedly said, "When I'm wrong, I change my mind, what do you do?" If one $\hat{\beta}$ lets you realize an error, shouldn't you be allowed to change your expectation? How would you go about resolving this difficulty?

d. Assume that your team is in last place with P = 350. According to this regression equation, would it be profitable to pay $4 million a year to a free agent superstar (like Darryl Strawberry) who would raise the team's winning percentage (P) to 500? Be specific.

14. Develop appropriate hypotheses for each slope coefficient in each of the following equations, and then calculate t-scores and test each null hypothesis at the 5 percent level:
   a. Exercise 11 in Chapter 4
   b. Exercise 11 in Chapter 3 (Hint: Assume 28 degrees of freedom. Would your answer change if there were only 5 degrees of freedom? How?)
   c. Exercise 12 in Chapter 3

15. Consider the following equation estimated by Fred McChesney[17] to determine whether the *Washington Post*'s Pulitzer Prize winning coverage of the Watergate political crisis of the 1970s had an effect on the newspaper's circulation (t-scores in parentheses):

$$\hat{C}_t = 290.10 + \underset{(14.27)}{0.761J_t} + \underset{(6.07)}{0.325S_t} + \underset{(1.31)}{0.058W_t}$$

$$\bar{R}^2 = .97 \qquad n = 26 \text{ (annual)} \qquad F = 168.05$$

where: C_t = circulation of the *Post* in year t
J_t = circulation of the *Wall Street Journal* in year t
S_t = the number of months during year t that the *Washington Star*, the *Post*'s main local competitor at the time, did not publish

17. Fred S. McChesney, "Sensationalism, Newspaper Profits, and the Marginal Value of Watergate," *Economic Inquiry*, Jan. 1987, pp. 135–144. (n is hypothetical.)

W_t = a dummy variable equal to 1 during years of Watergate coverage and 0 otherwise

a. Develop and test (at the 5 percent level) appropriate hypotheses about the slope coefficients using the *t*-test. (Hint: Note that t-scores, not standard errors, are given in parentheses. As mentioned in the chapter, not all published regression results follow our documentation format.)
b. Test the overall significance of the equation using the *F*-test (at the 1 percent level).
c. The variable J_t was chosen because the *Wall Street Journal* did relatively little reporting on Watergate. Why not use an industry circulation figure or circulation from the *New York Times*?
d. What economic conclusion can you draw about the effect of Watergate on the *Post*'s circulation?

5.8 Appendix: Other Uses of the *F*-Test

While the *F*-test is used most commonly to test the overall significance of a regression, it has many other uses. In fact, the *F*-test of overall significance is nothing more than a specialized application of the general *F*-test. This general *F*-test can be used to conduct tests of any (linear) hypothesis that involves more than one coefficient at a time. Such tests should be used whenever the underlying economic theory implies a hypothesis that simultaneously specifies values for more than one coefficient ("joint" or "compound" hypotheses).

The way in which the *F*-test evaluates hypotheses about more than one coefficient at a time is fairly ingenious. The first step is to translate the particular null hypothesis in question into constraints that will be placed on the equation. The resulting constrained equation can be thought of as what the equation would look like if the null hypothesis were correct; you substitute the hypothesized values into the regression equation in order to see what would happen if the equation was constrained to agree with the null hypothesis. As a result, in the *F*-test the null hypothesis always leads to a constrained equation, whether or not the constrained equation is the one which corresponds to the researcher's strawman hypothesis.

The second step in an *F*-test is to estimate this constrained equation with OLS and compare the fit of the constrained equation with the fit of the unconstrained equation. If the fit of the constrained equation and the unconstrained equation are not significantly different, the null hypothesis should not be rejected. If the fit of the unconstrained equation is significantly better than that of the constrained equation, then we reject the null hypothesis. The fit of the constrained equation is never superior to the fit of the unconstrained equation, as we'll explain below.

The fits of the equations are compared with a special F-statistic:

$$F = \frac{(RSS_M - RSS)/M}{RSS/(n - K - 1)} \tag{5.14}$$

where: RSS is the residual sum of squares from the unconstrained equation

RSS_M is the residual sum of squares from the constrained equation

M is the number of constraints placed on the equation (usually equal to the number of βs eliminated from the unconstrained equation)

(n − K − 1) is the degrees of freedom in the unconstrained equation

RSS_M is always greater than or equal to RSS; imposing constraints on the coefficients instead of allowing OLS to select their values can never decrease the summed squared residuals. (Recall that OLS selects that combination of values of the coefficients that minimizes RSS.) At the extreme, if the unconstrained regression yields exactly the same estimated coefficients as does the constrained regression, then the RSS are equal, and the F-statistic is zero. In this case, H_0 is not rejected because the data indicate that the constraints appear to be correct. As the difference between the constrained coefficients and the unconstrained coefficients increases, the data indicate that the null hypothesis is less likely to be true. Thus when F gets larger than the critical F-value, the hypothesized restrictions specified in the null hypothesis are rejected by the test.

The decision rule for the *F*-test is:

$$\text{Reject } H_0 \text{ if } F \geq F_c$$
$$\text{Do Not Reject } H_0 \text{ if } F < F_c$$

where F_c is the critical F-value found in the appropriate F-table.

As an example, let's look at a linearized annual Cobb-Douglas production function for the U.S.:

$$Q_t = \beta_0 + \beta_1 L_t + \beta_2 K_t + \epsilon_t \tag{5.15}$$

where: Q_t = the log of total output in the U.S. in year t
L_t = the log of labor input in the U.S. in year t
K_t = the log of capital input in the U.S. in year t
ϵ_t = a well-behaved stochastic error term

We'll discuss this particular functional form (called a "double-log") in more detail in Chapter 7, but for now all that's important is to know that one of the properties of this equation is that the coefficients of Equation 5.15 can be used to test for constant returns to scale. (Constant returns to scale refers to a situation in which a given percentage increase in inputs translates to exactly that percentage increase in output.) It can be shown that a Cobb-Douglas production function with constant returns to scale is one where β_1 and β_2 add up to exactly one, so the null hypothesis to be tested is:

$$H_0: \beta_1 + \beta_2 = 1$$
$$H_A: \text{otherwise}$$

To test this null hypothesis with the *F*-test, we must run regressions on the unconstrained Equation 5.15 and an equation that is constrained to conform to the null hypothesis. To create such a constrained equation, we solve the null hypothesis for β_2 and substitute it into Equation 5.15, obtaining:

$$\begin{aligned} Q_t &= \beta_0 + \beta_1 L_t + (1 - \beta_1)K_t + \epsilon_t \\ &= \beta_0 + \beta_1(L_t - K_t) + K_t + \epsilon_t \end{aligned} \tag{5.16}$$

If we move K_t to the left-hand side of the equation, we obtain our constrained equation:

$$(Q_t - K_t) = \beta_0 + \beta_1(L_t - K_t) + \epsilon_t \tag{5.17}$$

Equation 5.17 is the equation that would hold if our null hypothesis were correct.

To run an *F*-test on our null hypothesis of constant returns to scale, we need to run regressions on the constrained Equation 5.17 and the unconstrained Equation 5.15 and compare the fits of the two equations with the F-ratio from Equation 5.14. It turns out that if we use annual data from 1950 through 1970, we obtain an unconstrained equation of:

$$\begin{aligned} \hat{Q}_t &= -2.28 + 1.48L_t + 0.42K_t \qquad (5.18) \\ & \qquad\qquad (0.14) \qquad (0.04) \\ t &= \qquad\quad 10.3 \qquad\quad 11.8 \\ n &= 21 \qquad \bar{R}^2 = .996 \end{aligned}$$

If we run the constrained equation and substitute the appropriate RSS into Equation 5.14, we obtain F = 15.30. When this F is compared to a 5 percent critical F-value of only 4.41 (for 1 and 18 degrees of freedom) we must reject the null hypothesis that constant returns to scale characterized the U.S. economy in the 1950s and 1960s. (Note that the degrees of freedom in the numerator equal one, since only one coefficient has been eliminated from the equation by the constraint.)

Interestingly, the point estimate of $\hat{\beta}_1 + \hat{\beta}_2 = 1.48 + 0.42 = 1.90$ indicates dramatically increasing returns to scale. However, since $\hat{\beta}_1 = 1.48$, and since economic theory suggests that the slope coefficient of a Cobb–Douglas production function should be between zero and one, we should be extremely cautious. There are problems in the equation that need to be resolved before we can feel comfortable with this conclusion.

The *F*-test can be used with null hypotheses and constrained equations that apply to various subsets of the coefficients in the equation. For example, if

$$Y_i = \beta_0 + \beta_1 X_{1i} + \beta_2 X_{2i} + \beta_3 X_{3i} + \epsilon_i$$

then the only way to test a null hypothesis involving two of the slope coefficients (for example $H_0: \beta_1 = \beta_2$) would be to estimate constrained and unconstrained equations and to compare their fits with the *F*-test.

An illustration of the use of the *F*-test to test null hypotheses that involve only a subset of the slope coefficients can be obtained by looking at the problem of testing the significance of *seasonal dummies*. When a researcher is using quarterly (or monthly) data, it is common, as mentioned in Chapter 3, to include dummy variables that are equal to one in a given quarter (or month) and equal to zero otherwise. Since only three dummies are included in a quarterly model (to avoid perfect multicollinearity), the coefficients of the dummies are the expected difference between the effect of the specified quarter and the base quarter on the dependent variable, holding constant all the variables in the equation. As a result, to test the hypothesis of significant seasonality in the data, one must test the hypothesis that all the dummies equal zero simultaneously rather than test the dummies one at a time. In other words, the appropriate test of seasonality in a regression model using seasonal dummies involves the use of the *F*-test instead of the *t*-test.

Suppose the original equation was

$$Y = \beta_0 + \beta_1 D_1 + \beta_2 D_2 + \beta_3 D_3 + \beta_4 X + \epsilon \qquad (5.19)$$

where: Y and X are typical dependent and independent variables, respectively, and ϵ is a classical error term
D_1, D_2, and D_3 are seasonal dummies equal to one in the first, second, and third quarters of the year, respectively, and equal to zero otherwise (note that the fourth quarter is the omitted condition in this model)

In this case, the null hypothesis is that there is *no* seasonality:

$$H_0: \beta_1 = \beta_2 = \beta_3 = 0$$
$$H_A: H_0 \text{ is not true}$$

The constrained equation would then be $Y = \beta_0 + \beta_4 X + \epsilon$. To determine whether the whole set of seasonal dummies should be included, the fit of the estimated constrained equation would be compared to the fit of the estimated unconstrained equation by using the *F*-test. Note that this example does indeed use the *F*-test to test null hypotheses that include only a subset of the slope coefficients.

The exclusion of some seasonal dummies because their estimated coefficients have low t-scores for statistics is not recommended. Instead, testing seasonal dummy coefficients should be done with the *F*-test instead of with the *t*-test because seasonality is usually a single compound hypothesis rather than 3 (or 11 with monthly data) individual hypotheses having to do with each quarter (or month). To the extent that a hypothesis is a joint one, it should be tested with the *F*-test. If the hypothesis of seasonal variation can be summarized into a single dummy variable, then the use of the *t*-test will cause no problems. Often, where seasonal dummies are unambiguously called for, no hypothesis testing at all is undertaken.

Another common use of the *F*-test is to test the equivalence of regression coefficients between two sets of data, that is, whether two sets of data contain significantly different regression coefficients for the same theoretical equation. This can be helpful when deciding if it is appropriate to combine two data-sets. For example, the null hypothesis may be that the slope coefficients are the same in two samples, such as before and after a major war. The conern is whether there has been a major structural shift in the economy from one set of data to the other. This application of the *F*-test is often referred to as a *Chow test,*[18] and it can be set up by using dummy variables that distinguish between data-sets.

A Chow test has four steps:

1. Run identically specified regressions on the two samples of data being tested and note the RSS from the two (RSS_1 and RSS_2).
2. Pool the data from the two samples, run an identically specified regression on the combined sample, and note this equation's RSS (RSS_T).
3. Calculate the following F-statistic:

$$F = \frac{(RSS_T - RSS_1 - RSS_2)/(K + 1)}{(RSS_1 + RSS_2)/(N_1 + N_2 - 2K - 2)} \tag{5.20}$$

where: K = the number of independent variables
N_1 = the number of observations in sample #1
N_2 = the number of observations in sample #2

4. Reject the null hypothesis that the two sets of regression coefficients are equivalent if $F \geq F_c$, where F_c is the critical F value for $(K + 1)$ numerator and $(N_1 + N_2 - 2K - 2)$ denominator degrees of freedom.

18. See Gregory C. Chow, "Tests of Equality Between Sets of Coefficients in Two Linear Regressions," *Econometrica,* July 1960, pp. 591–605, or any advanced econometrics textbook for the details of this test.

6

Specification: Choosing the Independent Variables

Before any equation can be estimated, it must be completely *specified.* Specifying an econometric equation consists of three parts: choosing the correct independent variables, the correct functional form, and the correct form of the stochastic error term.

A **specification error** results when any one of these choices is made incorrectly. This chapter is concerned with only the first of these, choosing the variables; the second and third will be taken up in subsequent chapters.

That researchers can decide which independent variables to include in regression equations is a source of both strength and weakness in econometrics. The strength is that the equations can be formulated to fit individual needs, but the weakness comes from researchers being able to estimate many different specifications until they find the one that "proves" their point, even if many other results disprove it. A major goal of this chapter is to help you understand how to choose variables for your regressions without falling prey to the various errors that result from misusing the choice.

The primary consideration in deciding if an independent variable belongs in an equation is whether the variable is essential to the regression on the basis of theory. If the answer is an unambiguous yes, then the variable definitely should be included in the equation, even if it seems to be lacking in statistical significance. If theory is ambivalent or less emphatic, a dilemma arises. Leaving a relevant variable out of an equation is likely to bias the remaining estimates, but including an irrelevant variable leads to higher variances of the estimated coefficients. Although we'll develop statistical tools to help us deal with this decision, it's difficult in practice to be sure that a variable is relevant, and so the problem often remains unresolved.

We devote the fourth section of the chapter to specification searches and the pros and cons of various approaches to such searches. For example, techniques like stepwise regression procedures or sequential specification searches often cause bias or make the usual tests of significance inapplicable, and we do not recommend them. Instead, we suggest trying to minimize the number of regressions estimated and relying as much as possible on theory rather than statistical fit when choosing variables. There are no pat answers, however, and so the final decisions must be left to each individual researcher.

6.1 Omitted Variables

Suppose that you forget to include all the relevant independent variables when you first specify an equation (after all, no one's perfect!). Or suppose that you can't get data (or a good proxy) for one of the variables that you *do* think of. The result in both of these situations is an **omitted variable,** defined as an important explanatory variable that has been left out of a regression equation.

Whenever you have an omitted (or *left-out*) variable, the interpretation and use of your estimated equation becomes suspect. Leaving out a relevant variable, like price from a demand equation, not only prevents you from getting an estimate of the coefficient of price but also usually causes bias in the estimated coefficients of the variables that are in the equation.

The bias caused by leaving a variable out of an equation is called

specification bias (or, more casually, *omitted variable bias*). In an equation with more than one independent variable, the coefficient β_k represents the change in the dependent variable Y caused by a one-unit change in the independent variable X_k, holding the values of all other independent variables in the equation constant. If a variable is omitted, then it is not included as an independent variable, and it is not held constant for the calculation and interpretation of $\hat{\beta}_k$. This omission can cause bias: It can change the expected value of the estimated coefficient away from the true value of the population coefficient.

The estimated value of a regression coefficient can change, depending on the other variables in the equation. Thus, omitting a relevant variable is usually evidence that the entire estimated equation is suspect because of the likely bias in the coefficients of the variables that remain in the equation. Let's look at this issue in more detail.

6.1.1 The Consequences of an Omitted Variable

Suppose the true regression model is

$$Y_i = \beta_0 + \beta_1 X_{1i} + \beta_2 X_{2i} + \epsilon_i \tag{6.1}$$

where ϵ_i is a classical error term. If a researcher inadvertently omits an important independent variable (or can't get data on that variable), then the equation becomes:

$$Y_i = \beta_0 + \beta_1 X_{1i} + \epsilon_i^* \tag{6.2}$$

where the error term of the mispecified equation can be seen to be:

$$\epsilon_i^* = \beta_2 X_{2i} + \epsilon_i \tag{6.3}$$

Take another look at Equations 6.2 and 6.3. The error term ϵ_i^* is not independent of the explanatory variable X_{1i}, as long as X_{1i} and X_{2i} are correlated because if X_{2i} changes, both X_{1i} and ϵ_i^* will change. In other words, if we leave an important variable out of an equation, we violate Classical Assumption III (that the explanatory variables are independent of the error term), unless the omitted variable is totally uncorrelated with all the included independent variables (which is extremely unlikely). Recall that the correlation between X_1 and X_2 can be measured by the simple correlation coefficient between the two variables (r_{12}) using Equation 2.25.

In general, when there is a violation of one of the Classical Assumptions, the Gauss-Markov Theorem does not hold, and the OLS estimates are not BLUE. Given linear estimators, this means that the estimated coefficients are no longer unbiased or are no longer minimum variance (for all linear unbiased estimators), or both. In such a circumstance, econometricians first determine the exact property (unbiasedness or minimum variance) that no longer holds and then suggest an alternative estimation technique that might, in some sense, be better than OLS.

An omitted variable causes the Classical Assumptions to be violated in a way that causes bias. The estimation of Equation 6.2 when Equation 6.1 is the truth will cause bias in the estimates of Equation 6.2. This means that:

$$E(\hat{\beta}_1) \neq \beta_1$$

Instead of having an expected value equal to the true β_1, the estimate $\hat{\beta}_1$ will compensate for the fact that X_2 is missing from the equation. If X_1 and X_2 are correlated and X_2 is omitted from the equation, then the OLS program will attribute to X_1 variations in Y actually caused by X_2, and a biased $\hat{\beta}_1$ will result.

To see how a left-out variable might cause bias, picture a production function that states that output (Y) depends on the amount of labor (X_1) and capital (X_2) used. What would happen if data on capital were unavailable for some reason and X_2 was omitted from the equation? In this case, we would be leaving out the impact of capital on output in our model. This omission would almost surely bias the estimate of the coefficient of labor because it is likely that capital and labor are positively correlated (an increase in capital usually requires at least some labor to utilize it and vice versa). As a result, the OLS program would attribute to labor the increase in output actually caused by capital to the extent that labor and capital were correlated. Thus the bias would be a function of the impact of capital on output (β_2) and the correlation between capital and labor (r_{12}).

To generalize for a model with two independent variables, the expected value of the coefficient of an included variable (β_1) when a relevant variable (X_2) is omitted from the equation equals:

$$E(\hat{\beta}_1) = \beta_1 + \beta_2 \cdot f(r_{12}) \tag{6.4}$$

This states that the expected value of the included variable's coefficient is equal to its true value plus the excluded variable's true coefficient times a function of the simple correlation coefficient between the included and excluded variables.[1] Thus bias exists unless:

(a) the true B_2 is zero (that is, X_2 is not a relevant variable in the true model), or

(b) r_{12}, the simple correlation coefficient between X_1 and X_2, is zero (the excluded and included variables are perfectly uncorrelated).

The term $\beta_2 \cdot f(r_{12})$ is the amount of specification bias introduced into the estimate of β_1 by leaving out X_2. For the production function example above, this term would equal the coefficient of capital (β_2) times a function of the simple correlation coefficient between labor and capital (r_{12}). If the included and excluded variables are uncorrelated, there will be no bias, but there is almost always some correlation between any two variables in the real world (even if it's just random), and so bias is almost always caused by the omission of a relevant variable.[2]

6.1.2 An Example of Specification Bias

Consider the following equation for the annual consumption of chicken in the U.S.:[3]

$$\begin{aligned} \hat{Y}_t = & -60.5 - 0.45PC_t + 0.12PB_t + 12.2LYD_t \qquad (6.5) \\ & \qquad\quad (0.07) \qquad\quad (0.05) \qquad\quad (1.2) \\ & t = -6.4 \qquad\quad 2.5 \qquad\quad 10.6 \\ & \bar{R}^2 = .984 \qquad n = 35 \text{ (annual: 1950–1984)} \end{aligned}$$

where: Y_t = per capita chicken consumption (in pounds) in year t
PC_t = the price of chicken (in cents per pound) in year t
PB_t = the price of beef (in cents per pound) in year t

1. This function, $f(r_{12})$, is: $f(r_{12}) = r_{12}\sqrt{\Sigma x_2^2/\Sigma x_1^2}$ where $x_1 = (X_{1i} - \bar{X}_1)$ and $x_2 = (X_{2i} - \bar{X}_2)$. This turns out to equal the slope coefficient of the linear regression that relates X_2 to X_1. Note that Equation 6.4 only holds when there are exactly two independent variables, but the more general equation is quite similar.

2. While the omission of a relevant variable almost always produces bias in the estimators of the coefficients of the included variables, the variances of these estimators are generally lower than they otherwise would be. One method of deciding whether this decreased variance in the distribution of the $\hat{\beta}$s is valuable enough to offset the bias is to compare different estimation techniques with a measure called Mean Square Error (MSE). MSE is equal to the variance plus the square of the bias. The lower the MSE, the better. For more on the MSE, see Section 6.7.2.

3. The data for this example are included in Exercise 5; t-scores differ due to rounding.

LYD_t = the natural log of U.S. per capita disposable income (in dollars) in year t

This equation is a simple demand for chicken equation that includes the prices of chicken and a close substitute (beef) and a logged income variable. (The log allows the impact of a one-unit increase in income to be less at higher levels of income that at lower ones. We will discuss this functional form, called a semi-log, in more detail in Chapter 7.) Note that the signs of the estimated coefficients agree with the signs you would have hypothesized before seeing any regression results.

If we estimate this equation without the price of the substitute, we obtain:

$$\begin{aligned} \hat{Y}_t = -\ 80.7 &- \underset{(0.06)}{0.34PC_t} + \underset{(0.42)}{15.0LYD_t} \\ t = &\quad -5.6 \qquad\quad 36.0 \\ n = 35 &\quad \bar{R}^2 = .981 \end{aligned} \tag{6.6}$$

Let's compare Equations 6.5 and 6.6 to see if dropping the price-of-beef variable had an impact on the estimated equations. If you compare the overall fit, for example, you can see that $\bar{R}^2$ fell slightly from .984 to .981 when PB was dropped, exactly what we'd expect to occur when a relevant variable is omitted.

More important, from the point of view of showing that an omitted variable causes bias, let's see if the coefficient estimates of the remaining variables changed. Sure enough, dropping PB causing $\hat{\beta}_{PC}$ to go from -0.45 to -0.34 and caused $\hat{\beta}_{LYD}$ to go from 12.2 all the way to 15.0. The direction of this bias, by the way, is considered positive because the biased coefficient of PC (-0.34) is more positive (less negative) than the suspected unbiased one (-0.45), and the biased coefficient of LYD (15.0) is more positive than the suspected unbiased one of (12.2).

The fact that the bias is positive could have been guessed before any regressions were run if Equation 6.4 had been used. The specification bias caused by omitting the price of beef is expected[4] to be

4. It is important to note the distinction between expected bias and any actual observed differences between coefficient estimates. Because of the random nature of the error term (and hence the $\hat{\beta}$s), the change in an estimated coefficient brought about by dropping a relevant variable from the equation will not necessarily be in the expected direction. Biasedness refers to the central tendency of the sampling distribution of the $\hat{\beta}$s, not to every single drawing from that distribution. However, we usually (and justifiably) rely on these general tendencies.

positive because the expected sign of the coefficient of PB is positive and because the expected correlation between the price of beef and the price of chicken itself is positive:

$$\text{expected bias in } \hat{\beta}_{PC} = \beta_{PB} \cdot f(r_{PC,PB}) = (+) \cdot (+) = (+)$$

Similarly for LYD:

$$\text{expected bias in } \hat{\beta}_{LYD} = \beta_{PB} \cdot f(r_{LYD,PB}) = (+) \cdot (+) = (+)$$

Note that both correlation coefficients are anticipated to be (and actually are) positive. (To see this, think of the impact of an increase in the price of chicken on the price of beef and then follow through the impact of an increase in income on the price of beef.)

To sum, if a relevant variable is left out of a regression equation,

1. there is no longer an estimate of the coefficient of that variable in the equation, and
2. the coefficients of the remaining variables are likely to be biased.

While the amount of the bias might not be very large in some cases (when, for instance, there is little correlation between the included and excluded variables), it is extremely likely that at least a small amount of specification bias will be present in all such situations.

6.1.3 Correcting for an Omitted Variable

In theory, the solution to a problem of specification bias seems easy: Simply add the omitted variable to the equation. Unfortunately, that's more easily said than done, for a couple of reasons.

First, omitted variable bias is hard to detect. As mentioned above, the amount of bias introduced can be small and not immediately detectable. This is especially true when there is no reason to believe that you have misspecified the model. While some indications of specification bias are obvious (such as an estimated coefficient that is significant in the direction opposite from that expected), others are not so clear. Could you tell from Equation 6.6 alone that a variable was missing? The best indicators of an omitted relevant variable are the theoretical underpinnings of the model itself. What variables *must* be included? What signs do you expect? Do you have any notions about the range into which the coefficient values should fall? Have you accidentally left out a variable that most researchers would agree is important? The best way to avoid omitting an important variable is to invest the time to think carefully through the equation before the data are entered into the computer.

A second source of complexity is the problem of choosing which variable to add to an equation once you decide that it is suffering from omitted variable bias. That is, a researcher faced with a clear case of specification bias (like an estimated $\hat{\beta}$ that is significantly different from zero in the unexpected direction) will often have no clue as to what variable could be causing the problem. Some beginning researchers, when faced with this dilemma, will add all the possible relevant variables to the equation at once, but this process leads to less precise estimates, as will be discussed in the next section. Other beginning researchers will test a number of different variables and keep the one in the equation that does the best statistical job of appearing to reduce the bias (by giving plausible signs and satisfactory t-values). This technique, adding a "left-out" variable to "fix" a strange-looking regression result, is invalid because the variable that best corrects a case of specification bias might do so only by chance rather than by being the true solution to the problem. In such an instance, the fixed equation may apparently give superb statistical results for the sample at hand but then do terribly when applied to other samples, because it does not describe the characteristics of the true population.

Dropping a variable will not help cure omitted variable bias. If the sign of an estimated coefficient is different from expected, it cannot be changed to the expected direction by dropping a variable that has a lower t-score (in absolute value) than the t-score of the coefficient estimate that has the unexpected sign. Furthermore, the sign in general will not likely change even if the variable to be deleted has a large t-score.[5]

If the estimated coefficient is significantly different from our expectations (either in sign or magnitude), then it is likely that some sort of specification bias exists in our model. Although it is true that a poor sample of data or a poorly theorized expectation may also yield statistically significant unexpected signs or magnitudes, these possibilities sometimes can be eliminated.

A legitimate technique for reducing the number of theoretically sound candidates to be the omitted variable is the investigation of the direction of the bias caused by the omission of a variable from an

5. Ignazio Visco, "On Obtaining the Right Sign of a Coefficient Estimate by Omitting a Variable from the Regression," *Journal of Econometrics*, Feb. 1978, pp. 115–117.

equation. If the sign of the expected bias can be shown to be in a direction opposite the observed, then that variable can be eliminated from consideration. The direction of the expected bias can be determined from the second term in Equation 6.4:

$$\text{expected bias in } \hat{\beta}_1 = \beta_2 \cdot f(r_{12})$$

In the example of omitting the price of beef from the chicken demand equation, the expected direction of the bias was positive since both the expected coefficient and the expected correlation between PB and PC were positive:

$$\text{expected bias in } \hat{\beta}_{PC} = \beta_{PB} \cdot f(r_{PC,PB}) = (+) \cdot (+) = (+)$$

Hence the price of beef was a reasonable candidate to be the omitted variable in Equation 6.6.

A bad choice for the omitted variable in the same equation would be the price of a good that is a complement to the consumption of chicken (such as dumplings) because the expected bias is negative. To see this, calculate the sign of the expected bias in the coefficient of the price of chicken due to the omission of the price of dumplings (PD) from the chicken demand equation:

$$\text{expected bias in } \hat{\beta}_{PC} = \beta_{PD} \cdot f(r_{PC,PD}) = (-) \cdot (+) = (-)$$

The expected sign of the price of dumplings in the chicken demand equation is negative because a high price of dumplings (the complement) would make it more expensive to consume mass quantities of chicken, shifting the demand curve for chicken downward. In addition, the prices of complements generally move together, because a change in the underlying demand for one would change the underlying demand for the other in the same direction.

The possible combinations of the expected signs of the coefficients and simple correlation coefficients (due to the omission of a single independent variable) are matched with the expected sign of the resultant specification bias in Table 6.1. To use this table, find the row that contains the expected sign of the coefficient of the candidate omitted variable (for price of dumplings, this is negative) and match it with the column that contains the expected sign of the simple correlation coefficient between the included and excluded variables (for price of dumplings, this is positive). The sign at the intersection of the appropriate row and column is the sign of the expected bias in the coefficient of

TABLE 6.1 THE EXPECTED SIGN OF THE BIAS CAUSED BY AN OMITTED VARIABLE

		Expected Sign of the Simple Correlation Coefficient between the included (in) and omitted (om) variables, $r_{in,om}$	
		$r_{in,om}$	
		+	−
Expected Sign of the Coefficient of the Omitted Variable, β_{om} (Based on Theory)	β_{om} +	+	−
	β_{om} −	−	+

the particular included variable due to leaving out the particular candidate excluded variable (for price of dumplings, this is negative).

While you can never actually observe bias (since you don't know the true β), the use of this technique to screen potential causes of specification bias should reduce the number of regressions run and therefore increase the statistical validity of the results. This technique will work best when only one (or one kind) of variable is omitted from the equation in question. With a number of different kinds of variables omitted simultaneously, the impact on the equation's coefficients is quite hard to specify.

A brief warning: It may be tempting to conduct what might be called "residual analysis" by examining a plot of the residuals in an attempt to find patterns that suggest variables that have been accidentally omitted. A major problem with this approach is that the coefficients of the estimated equation will possibly have some of the effects of the left-out variable already altering their estimated values. Thus, residuals from this equation may show a pattern that only vaguely resembles the pattern of the actual omitted variable. The chances are high that the pattern shown in the residuals may lead to the selection of an incorrect variable. In addition, care should be taken to use residual analysis only to choose between theoretically sound candidate variables rather than to generate those candidates.

6.2 Irrelevant Variables

What happens if you include a variable in an equation that doesn't belong there? This case, **irrelevant variables,** is the converse of omitted variables and can be analyzed using the model we developed in Section

6.1. Whereas the omitted variable model has more independent variables in the true model than in the estimated equation, the irrelevant variable model has more independent variables in the estimated equation than in the true one.

The addition of a variable to an equation where it doesn't belong does not cause bias, but it does increase the variances of the included variables' estimated coefficients.

6.2.1 Impact of Irrelevant Variables

If the true regression specification is

$$Y_i = \beta_0 + \beta_1 X_{1i} + \epsilon_i \tag{6.7}$$

but the researcher for some reason includes an extra variable,

$$Y_i + \beta_0 + \beta_1 X_{1i} + \beta_2 X_{2i} + \epsilon_i^{**} \tag{6.8}$$

the misspecified equation's error term can be seen to be:

$$\epsilon_i^{**} = \epsilon_i - \beta_2 X_{2i} \tag{6.9}$$

Such a mistake will not cause bias if the true coefficient of the extra (or irrelevant) variable is zero. In that case, $\epsilon_i = \epsilon_i^{**}$. That is, $\hat{\beta}_1$ in Equation 6.8 is unbiased when $\beta_2 = 0$.

The inclusion of an irrelevant variable will increase the variance of the estimated coefficients, and this increased variance will tend to decrease the absolute magnitude of their t-scores. Also, an irrelevant variable usually will decrease the $\bar{R}^2$ (but not the R^2). In a model of Y on X_1 and X_2, the variance of the OLS estimator of β_1 is:

$$\text{VAR}(\hat{\beta}_1) = \frac{\sigma^2}{(1 - r_{12}^2) \cdot \Sigma(X_1 - \bar{X}_1)^2} \tag{6.10}$$

But when $r_{12} = 0$ (or in the single independent variable model), then:

$$\text{VAR}(\hat{\beta}_1) = \frac{\sigma^2}{\Sigma(X_1 - \bar{X}_1)^2} \tag{6.11}$$

Thus, while the irrelevant variable causes no bias, it causes problems for the regression, because it reduces the precision of the regression.

To see why this is so, try plugging a nonzero value (between +1.0 and −1.0) for r_{12} into Equation 6.10 and note that VAR($\hat{\beta}_1$) has increased when compared to Equation 6.11. The equation with an included variable that does not belong in the equation usually has lower t-scores and a lower $\bar{R}^2$ than it otherwise would. This property holds, by the way, only when $r_{12} \neq 0$, but since this is the case in virtually every sample, the conclusion of increased variance due to irrelevant variables is a valid one. Table 6.2 summarizes the consequences of the omitted variable and the included irrelevant variable cases:

TABLE 6.2 SUMMARY OF THE IMPACTS OF AN OMITTED VARIABLE OR AN INCLUDED IRRELEVANT VARIABLE ON THE REMAINING COEFFICIENTS

Effect on Remaining Coefficient Estimates	Omitted Variable	Included Irrelevant Variable
Bias?	Yes*	No
Increases or Decreases Variance?	Decreases*	Increases*

*unless $r_{12} = 0$

6.2.2 An Example of an Irrelevant Variable

Let's return to the equation from Section 6.1 for the annual consumption of chicken and see what happens when we add an irrelevant variable to the equation. The original equation is:

$$\hat{Y}_t = -60.5 - \underset{(0.07)}{0.45}PC_t + \underset{(0.05)}{0.12}PB_t + \underset{(1.2)}{12.2}LYD_t \quad (6.12)$$

$$t = \quad -6.4 \qquad 2.5 \qquad 10.6$$

$$\bar{R}^2 = .984 \qquad n = 35$$

Suppose you hypothesize that the demand for chicken also depends on R, the interest rate (which perhaps confuses the demand for a nondurable good with an equation you saw for a consumer durable). If you now estimate the equation with the interest rate included, you obtain:

$$\hat{Y}_t = -61.4 - \underset{(0.08)}{0.47}PC_t + \underset{(0.06)}{0.14}PB_t + \underset{(1.2)}{12.4}LYD_t - \underset{(0.21)}{0.15}R_t \quad (6.13)$$

$$t = -6.0 \qquad 2.5 \qquad 10.4 \qquad -0.70$$

$$\bar{R}^2 = .983 \qquad n = 35$$

A comparison of Equations 6.12 and 6.13 will make the theory in Section 6.2.1 come to life. First of all, $\bar{R}^2$ has fallen slightly, indicating the reduction in fit adjusted for degrees of freedom. Second, none of the regression coefficients from the original equation changed significantly; compare these results with the larger differences between Equations 6.5 and 6.6. Further, slight increases in the standard errors of the estimated coefficients can be observed. Finally, the t-score for the potential variable (the interest rate) is very small, indicating that it is not significantly different from zero. Given the theoretical shakiness of the new variable, these results indicate that it is irrelevant and never should have been included in the regression.

6.2.3 Making Correct Specification Choices

We have now discussed at least four valid criteria to help decide whether a given variable belongs in the equation:

1. *Theory:* Is the variable's place in the equation unambiguous and theoretically sound?
2. *t-test:* Is the variable's estimated coefficient significantly different from zero?
3. $\bar{R}^2$: Does the overall fit of the equation (adjusted for degrees of freedom) improve when the variable is added to the equation?
4. *Bias:* Do other variables' coefficients change significantly when the variable is added to the equation?

If all these conditions hold, the variable belongs in the equation; if none of them do, the variable is irrelevant and can be safely excluded from the equation. When a typical omitted relevant variable is included in the equation, its inclusion probably would increase $\bar{R}^2$ and change other coefficients while having a significant t-score. If an irrelevant variable, on the other hand, is included, it would reduce $\bar{R}^2$, have an insignificant t-score, and have little impact on the other variables' coefficients.

In many cases, all four criteria do not agree. It is possible for a variable to have an insignificant t-score that is greater than one, for example. In such a case, it can be shown that $\bar{R}^2$ would go up and yet the t-score would still be insignificant. In another case, the variable might be comparatively uncorrelated with the included variables and thus have little effect on their estimated coefficients. What do you do in such circumstances?

Whenever the four criteria of whether or not a variable should be included in an equation disagree, the econometrician must use careful judgment. Researchers should not misuse this freedom by testing various combinations of variables until they find the results that appear to statistically support the point they want to make. All such decisions are a bit easier when you realize that the single most important determinant of a variable's relevance is its theoretical justification. No amount of statistical evidence should make a theoretical necessity into an "irrelevant" variable. Once in a while, a researcher is forced to leave a theoretically important variable out of an equation for lack of a better alternative; in such cases, the usefulness of the equation is limited.

6.3 An Illustration of the Misuse of Specification Criteria

At times, the criteria outlined in the previous section will lead the researcher to an incorrect conclusion if those criteria are applied blindly to a problem without the proper concern for common sense or economic principles. In particular, a t-score can often be insignificant for reasons other than the presence of an irrelevant variable. Since economic theory is the most important test for including a variable, an example of why a variable should not be dropped from an equation simply because it has an insignificant t-score is in order.

Suppose you believe that the demand for Brazilian coffee in the U.S. is a function of the real price of Brazilian coffee (P_{bc}), the real price of tea (P_t), and the real disposable income in the U.S. (Y_d).[6] Suppose further that you obtain the data, run the implied regression, and observe the following results:

$$\widehat{COFFEE} = 9.1 + 7.8P_{bc} + 2.4P_t + 0.0035Y_d \qquad (6.14)$$
$$\qquad\qquad (15.6) \qquad (1.2) \qquad (0.0010)$$
$$t = \quad 0.5 \qquad 2.0 \qquad 3.5$$
$$\bar{R}^2 = .60 \qquad n = 25$$

The coefficients of the second and third variables, P_t and Y_d, appear to be fairly significant in the direction you hypothesized, but the first variable, P_{bc}, appears to have an insignificant coefficient with an unexpected sign. If you think there is a possibility that the demand for

6. This example was inspired by a similar one concerning Ceylonese tea published in Potluri Rao and Roger LeRoy Miller, *Applied Econometrics* (Belmont, California: Wadsworth, 1971), pp. 38–40. This book is now out of print.

Brazilian coffee is perfectly price-inelastic (that is, its coefficient is zero), you might decide to run the same equation without the price variable, obtaining:

$$\widehat{COFFEE} = \underset{(1.0)}{9.3 + 2.6P_t} + \underset{(0.0009)}{0.0036Y_d} \tag{6.15}$$
$$t = \quad 2.6 \qquad 4.0$$
$$\bar{R}^2 = .61$$

By comparing Equations 6.14 and 6.15, we can apply our four criteria for the inclusion of a variable in an equation that were outlined in the previous section:

1. *Theory:* Since the demand for coffee could possibly be perfectly price-inelastic, the theory behind dropping the variable seems plausible.
2. *t-test:* The t-score of the possibly irrelevant variable is 0.5, insignificant at any level.
3. $\bar{R}^2$*:* $\bar{R}^2$ increases when the variable is dropped, indicating that the variable is irrelevant. (Since the t-score is less than one, this is to be expected.)
4. *Bias:* The remaining coefficients change only a small amount when P_{bc} is dropped, suggesting that there is little if any bias caused by excluding the variable.

Based upon this analysis, you might conclude that the demand for Brazilian coffee is perfectly price-inelastic and that the variable is therefore irrelevant and should be dropped from the model. As it turns out, this conclusion would be unwarranted. While the demand for coffee in general might be price-inelastic (actually, the evidence suggests that it is inelastic only over a particular range of prices), it is hard to believe that Brazilian coffee is immune to price competition from other kinds of coffee. Indeed, one would expect quite a bit of sensitivity in the demand for Brazilian coffee with respect to the price of, for example, Colombian coffee. To test this hypothesis, the price of Colombian coffee (P_{cc}) should be added to the original Equation 6.14:

$$\widehat{COFFEE} = 10.0 + \underset{(4.0)}{8.0P_{cc}} - \underset{(2.0)}{5.6P_{bc}} + \underset{(1.3)}{2.6P_t} + \underset{(0.0010)}{0.0030Y_d} \tag{6.16}$$
$$t = \quad 2.0 \qquad -2.8 \qquad 2.0 \qquad 3.0$$
$$\bar{R}^2 = .65$$

By comparing Equations 6.14 and 6.16, we can once again apply the four criteria:

1. *Theory:* Both prices should always have been included in the model; their logical justification is quite strong.
2. t-*test:* The t-score of the new variable, the price of Colombian coffee, is 2.0, significant at most levels.
3. $\bar{R}^2$: $\bar{R}^2$ increases with the addition of the variable, indicating that the variable was an omitted variable.
4. *Bias:* While two of the coefficients remain virtually unchanged, indicating that the correlations between these variables and the price of Colombian coffee variable are low, the coefficient for the price of Brazilian coffee did change significantly, indicating bias in the original result.

An examination of the bias question will also help us understand Equation 6.4, the equation for bias. Since the expected sign of the coefficient of the omitted variable (P_{cc}) is positive and since the simple correlation coefficient between the two competitive prices ($r_{P_{cc},P_{bc}}$) is also positive, the expected direction of the bias in $\hat{\beta}_{P_{bc}}$ in the estimation of Equation 6.14 is positive. If you compare Equations 6.14 and 6.16, that positive bias can be seen because the coefficient of P_{bc} is $+7.8$ instead of -5.6. The increase from -5.6 to $+7.8$ may be due to the positive bias that results from leaving out P_{cc}.

The moral to be drawn from this example is that theoretical considerations should never be discarded even in the face of statistical insignificance. If a variable known to be extremely important from a theoretical point of view turns out to be statistically insignificant in a particular sample, that variable should be left in the equation despite the fact that it makes the results look bad.

Don't conclude that the particular path outlined in this example is the correct way to specify an equation. Trying a long string of possible variables until you get the particular one that makes P_{bc} turn negative and significant is not the way to obtain a result that will stand up well to other samples or alternative hypotheses. The original equation should never have been run without the Columbian coffee variable. Instead, the problem should have been analyzed enough so that such errors of omission were unlikely before any regressions were attempted at all. The more thinking that's done before the first regression is run and the fewer alternative specifications that are estimated, the better the regression results are likely to be.

6.4 Specification Searches

One of the weaknesses of econometrics is that a researcher can potentially manipulate a data-set to produce almost *any* results by specifying different regressions until estimates with the desired properties are obtained. Thus, the integrity of all empirical work is potentially open to question.

Although the problem is a difficult one, it makes sense to attempt to minimize the number of equations estimated and to rely on theory rather than statistical fit as much as possible when choosing variables. We'll try to illustrate this by discussing three of the most commonly used *incorrect* techniques for specifying a regression equation. These techniques produce the best specification only by chance and at worst are possibly unethical in that they misrepresent the methods used to obtain the regression results and the significance of those results.

6.4.1 Data Mining

Almost surely the worst way to choose a specification is to simultaneously try a whole series of possible regression formulations and to then choose the equation that conforms the most to what the researcher wants the results to look like. In such a situation, the researcher would estimate virtually every possible combination of the various alternative independent variables, and the choice between them would be made on the basis of the results. This practice of simultaneously estimating a number of combinations of independent variables and selecting the best from them does not account for the fact that a number of specifications have been examined before the final one. To oversimplify, if you are 95% confident that a regression result didn't occur by chance and you run more than 20 regressions, how much confidence can you have in your result? Since you'll tend to keep regressions with high t-scores and discard ones with low t-scores, the reported t-scores overstate the degree of statistical significance of the estimated coefficients.

Furthermore, such "data mining" and "fishing expeditions" to obtain desired statistics for the final regression equation are potentially unethical methods of empirical research. These procedures include using not only many alternative combinations of independent variables but also many functional forms, lag structures, and what are offered as "sophisticated" or "advanced" estimating techniques. "If you just torture the data long enough, they will confess."[7] In other words, if

7. Thomas Mayer, "Economics as a Hard Science: Realistic Goal or Wishful Thinking?" *Economic Inquiry,* April 1980, p. 175.

enough alternatives are tried, the chances of obtaining the results desired by the researcher are increased tremendously, but the final result is essentially worthless. The researcher hasn't found any scientific evidence to support the original hypothesis; rather, prior expectations were imposed on the data in a way that is essentially misleading.

6.4.2 Stepwise Regression Procedures

A **stepwise regression** involves the use of a computer program to choose the independent variables to be used in the estimation of a particular equation. The computer program is given a "shopping list" of possible independent variables, and then it builds the equation in steps. It chooses as the first explanatory variable the one that by itself explains the largest amount of the variation of the dependent variable around its mean. It chooses as the second variable the one that adds the most to R^2, given that the first variable is already in the equation. The stepwise procedure continues until the next variable to be added fails to achieve some researcher-specified increase in R^2 (or all the variables are added). The measure of the supposed contribution of each independent variable is the increase in R^2 (which is sometimes called the "R^2 delete") caused by the addition of the variable.

Unfortunately, any correlation among the independent variables (called multicollinearity, which we will take up in more detail in Chapter 8) causes this procedure to be deficient. To the extent that the variables are related, it becomes difficult to tell the impact of one variable from another. As a result, in the presence of multicollinearity, it's impossible to determine unambiguously the individual contribution of each variable enough to say which one is more important and thus should be included first.[8] Even worse, there is no necessity that the particular combination of variables chosen has any theoretical justification.

Because of these problems, most researchers avoid stepwise procedures. The major pitfalls are that the coefficients may be biased, the calculated t-values no longer follow the t-distribution, relevant variables may be excluded because of the arbitrary order in which the

8. Some programs compute standardized beta coefficients, which are the estimated coefficients for an equation in which all variables have been standardized by subtracting their means from them and by dividing them by their own standard deviations. The higher the beta of an independent variable is in absolute value, the more important it is thought to be in explaining the movements in the dependent variable. Unfortunately, beta coefficients are deficient in the presence of multicollinearity, as are partial correlation coefficients, which measure the correlation between the dependent variable and a given independent variable holding all other independent variables constant.

selection takes place, and the signs of the estimated coefficients at intermediate or final stages of the routine may be different from the expected signs. Using a stepwise procedure is an admission of ignorance concerning which variables should be entered.

6.4.3 Sequential Specification Searches

To their credit, most econometricians avoid data mining and stepwise regressions. Instead, they tend to specify equations by estimating an initial equation and then sequentially dropping or adding variables (or changing functional forms) until a plausible equation is found with "good statistics." Faced with a situation of perhaps knowing that a few variables are relevant (on the basis of theory) but not knowing whether other additional variables are relevant, recourse to inspecting $\bar{R}^2$ and t-tests for all variables (both before and after selection or exclusion of some independent variables) appears to be the generally accepted practice. Indeed, it would be easy to draw from a casual reading of the previous sections the impression that such a sequential specification search is the best way to go about finding the "truth." Instead, as we shall see, there is a vast difference in approach between a sequential specification search and our recommended approach.

The **sequential specification search** technique allows a researcher to estimate an undisclosed number of regressions and then present a final choice (which is based upon an unspecified set of expectations about the signs and significance of the coefficients) as if it were the only specification estimated. Such a method misstates the statistical validity of the regression results for two reasons:

1. The statistical significance of the results is overestimated because the estimations of the previous regressions are ignored.
2. The set of expectations used by the researcher to choose between various regression results is rarely if ever disclosed.[9] Thus the reader has no way of knowing whether or not all the other regression results had opposite signs or insignificant coefficients for the important variables.

Unfortunately, there is no universally accepted way of conducting sequential searches, primarily because the appropriate test at one stage in the procedure depends on which tests were previously conducted,

9. As mentioned in Chapter 5, Bayesian regression is a technique for dealing systematically with these prior expectations. For more on this issue, see Edward E. Leamer, *Specification Searches* (New York: Wiley), 1978.

and also because the tests have been very difficult to invent. One possibility is to reduce the degrees of freedom in the "final" equation by one for each alternative specification attempted. This procedure is far from exact, but it does impose an explicit penalty for specification searches.

More generally, we recommend trying to keep the number of regressions estimated as low as possible; to focus on theoretical considerations when choosing variables, functional forms, and the like; and to reveal all the various specifications investigated. That is, we recommend combining parsimony (using theory and analysis to limit the number of specifications estimated) with disclosure (reporting all the equations estimated).

There is another side to this story, however. Some researchers feel that the true model will show through if given the chance and that the best statistical results (including signs of coefficients, etc.) are most likely to have come from the true specification. The problem with this philosophy is that the element of chance is ordinarily quite strong in any given application. In addition, reasonable people often disagree as to what the "true" model should look like. As a result, different researchers can look at the same data-set and come up with very different "best" equations. Because this can happen, the distinction between good and bad econometrics is not always as clear-cut as is implied by the previous paragraphs. As long as researchers have a healthy respect for the dangers inherent in specification searches, they are very likely to proceed in a reasonable way.

The lesson to be learned from this section should be quite clear. Most of the work of specifying an equation should be done before even attempting to estimate the equation on the computer. Since it is unreasonable to expect researchers to be perfect, there will be times when additional specifications must be estimated; however, these new estimates should be thoroughly grounded in theory and explicitly taken into account when testing for significance or summarizing results. In this way, the danger of misleading the reader about the statistical properties of estimates is reduced.

6.4.4 The Impact of Sequential Specification Searches

In the previous section, we stated that sequential specification searches are likely to mislead researchers about the statistical properties of the results. This section presents an example of a problem that can be encountered with a particular kind of sequential specification search.

The example will illustrate the fact that dropping variables from a model on the basis of *t*-tests alone will introduce systematic bias into the estimated equation.[10]

Say the hypothesized model for a particular dependent variable is:

$$Y_i = \beta_0 + \beta_1 X_{1i} + \beta_2 X_{2i} + \epsilon_i \tag{6.17}$$

Assume further that, on the basis of theory, we are certain that X_1 belongs in the equation but that we are not as certain that X_2 belongs. Even though we have stressed four criteria to determine whether X_2 should be included, many inexperienced researchers just use the *t*-test on $\hat{\beta}_2$ to determine whether X_2 should be included. If this preliminary *t*-test indicates that $\hat{\beta}_2$ is significantly different from zero, then these researchers leave X_2 in the equation, and they choose Equation 6.17 as their final model. If, however, the *t*-test does *not* indicate that $\hat{\beta}_2$ is significantly different from zero, then such researchers drop X_2 from the equation and consider Y as a function of X_1.

Two kinds of mistakes can be made using such a system. First, X_2 can sometimes be left in the equation when it does not belong there, but such a mistake does not change the expected value of $\hat{\beta}_1$. Second, X_2 can sometimes be dropped from the equation when it belongs, and then the estimated coefficient of X_1 will be biased by the value of the true β_2 to the extent that X_1 and X_2 are correlated. In other words, $\hat{\beta}_1$ will be biased every time X_2 belongs in the equation and is left out, and X_2 will be left out every time that its estimated coefficient is not significantly different from zero. That is, the expected value of $\hat{\beta}_1$ will not equal the true β_1, and we will have systematic bias in our equation:

$$E(\hat{\beta}_1) = \beta_1 + \beta_2 \cdot f(r_{x_1,x_2}) \cdot P \neq \beta_1$$

Where P indicates the probability of an insignificant t-score. It is also the case that the t-score of $\hat{\beta}_1$ no longer follows the t-distribution. In other words, the *t*-test is biased by sequential specification searches.

Since most researchers consider a number of different variables before settling on the final model, someone who relies on the *t*-test alone is likely to encounter this problem systematically. That is, the

10. For a number of better techniques, including sequential or "pretest" estimators and "Stein-rule" estimators, see George G. Judge, W. E. Griffiths, R. Carter-Hill, Helmut Lutkepohl, and Tsoung-Chao Lea, *The Theory and Practice of Econometrics* (New York: Wiley 1985).

practice of dropping a potential independent variable simply because its t-score indicates that its estimated coefficient is insignificantly different from zero will cause systematic bias in the estimated coefficients (and their t-scores) of the remaining variables.

6.5 An Example of Choosing Independent Variables

It's time to get some experience choosing independent variables. After all, every equation so far in the text has come with the specification already determined, but once you've finished this course you'll have to make all such specification decisions on your own. In future chapters, we'll use a technique called "interactive regression learning exercises" to allow you to make your own actual specification choices and get feedback on your choices, but to start with let's work through a specification together.

To keep things as simple as possible, we'll begin with a topic near and dear to your heart, your GPA! Suppose a friend surveys all 25 members of your econometrics class and obtains data on the variables listed below:

GPA_i = the cumulative college grade point average of the *i*th student on a four-point scale

$HGPA_i$ = the cumulative high school grade point average of the *i*th student on a four-point scale

$MSAT_i$ = the highest score earned by the *i*th student on the math section of the SAT test (800 maximum)

$VSAT_i$ = the highest score earned by the *i*th student on the verbal section of the SAT test (800 maximum)

$SAT_i = MSAT_i + VSAT_i$

$GREK_i$ = a dummy variable equal to 1 if the *i*th student is a member of a fraternity or sorority, 0 otherwise

HRS_i = the *i*th student's estimate of the average number of hours spent studying per course per week in college

$PRIV_i$ = a dummy variable equal to 1 if the *i*th student graduated from a private high school, 0 otherwise

$JOCK_i$ = a dummy variable equal to 1 if the *i*th student is or was a member of a varsity intercollegiate athletic team for at least one season, 0 otherwise

$lnEX_i$ = the natural log of the number of full courses that the *i*th student has completed in college

Assuming that GPA_i is the dependent variable, which independent variables would you choose? Before you answer, think through the possibilities carefully. What are the expected signs of each of the coefficients? How strong is the theory behind each variable? Which variables seem obviously important? Which variables seem potentially irrelevant or redundant? Are there any other variables that you wish your friend had collected?

To get the most out of this example, you should take the time to *write down* the exact specification that you would run:

$$GPA_i = f(?,?,?,?,?)$$

It's hard for most beginning econometricians to avoid the temptation of including *all* the above variables in a GPA equation and then dropping any variables that have insignificant t-scores. Even though we mentioned in the previous section that such a specification search procedure will result in biased coefficient estimates, most beginners don't trust their own judgment and tend to include too many variables. With this warning in mind, do you want to make any changes in your proposed specification?

No? OK, let's compare notes. We believe that grades are a function of a student's ability, how hard the student works, and the student's experience taking college courses. Consequently, our specification would be:

$$GPA_i = f(\overset{+}{HGPA_i}, \overset{+}{HRS_i}, \overset{+}{\ln EX_i})$$

We can already hear you complaining! What about SATs, you say? Everyone knows they're important. How about jocks and Greeks? Don't they have lower GPAs? Don't prep schools grade harder and prepare students better than public high schools?

Before we answer, it's important to note that we think of specification choice as choosing which variables to *include,* not which variables to *exclude*. That is, we don't assume automatically that a given variable should be included in an equation simply because we can't think of a good reason for dropping it.

Given that, however, why did we choose the variables we did? First, we think that the best predictor of a student's college GPA is his or her high school GPA, and we have a hunch that once you know HGPA, SATs are redundant. In addition, we're concerned that possible racial and gender bias in the SAT test makes it a questionable measure of academic potential, but we recognize that we could be wrong on this issue.

As for the other variables, we're more confident. For example, we feel that once we know how many hours a week a student spends studying, we couldn't care less what that student does with the rest of his or her time, so JOCK and GREK are superfluous once HRS is included. Finally, while we recognize that some private schools are superb and that some public schools are not, we'd guess that PRIV is irrelevant; it probably has only a minor effect.

If we estimate this specification on the 25 students, we obtain:

$$\begin{aligned} \widehat{GPA}_i = -\ 0.26 &+ 0.49HGPA_i &+ 0.06HRS_i &+ 0.42lnEX_i \qquad (6.18)\\ &\quad (0.21) &\quad (0.02) &\quad (0.14)\\ t = &\quad 2.33 &\quad 3.00 &\quad 3.00\\ n = 25 &\quad \bar{R}^2 = .585 &\quad F = 12.3 & \end{aligned}$$

Since we prefer this specification on theoretical grounds, since the overall fit seems reasonable, and since each coefficient meets our expectations in terms of sign, size, and significance, we consider this an acceptable equation. The only circumstance under which we'd consider estimating a second specification would be if we had theoretical reasons to believe that we had omitted a relevant variable. The only variable that might meet this description is SAT_i (which we prefer to the individual MSAT and VSAT):

$$\begin{aligned} \widehat{GPA}_i = -\ 0.92 &+ 0.47HGPA_i &+ 0.05HRS_i\\ &\quad (0.22) &\quad (0.02)\\ t = &\quad 2.12 &\quad 2.50\\ &+ 0.44lnEX_i &+ 0.00060SAT_i \qquad (6.19)\\ &\quad (0.14) &\quad (0.00064)\\ t = &\quad 3.12 &\quad 0.93\\ n = 25 &\quad \bar{R}^2 = .583 &\quad F = 9.4 \end{aligned}$$

Let's use our four specification criteria to compare Equations 6.18, and 6.19:

1. *Theory:* As discussed above, the theoretical validity of SAT tests is a matter of some academic controversy, but they still are one of the most-cited measures of academic potential in this country.
2. t-*test:* The coefficient of SAT is positive, as we'd expect, but it's not significantly different from zero.

3. $\bar{R}^2$: As you'd expect (since SAT's t-score is under one), $\bar{R}^2$ falls slightly when SAT is added.

4. *Bias:* None of the estimated slope coefficients changes significantly when SAT is added, though some of the t-scores do change because of the increase in the $SE(\hat{\beta})$s caused by the addition of SAT.

Thus the statistical criteria support our theoretical contention that SAT is irrelevant.

Finally, it's important to recognize that different researchers could come up with different final equations on this topic. A researcher whose prior expectation was that SAT unambiguously belonged in the equation would have estimated Equation 6.19 and accepted that equation without bothering to estimate Equation 6.18.

6.6 Summary

1. The omission of a variable from an equation will cause bias in the estimates of the remaining coefficients to the extent that the omitted variable is correlated with included variables.

2. The bias to be expected from leaving a variable out of an equation equals the coefficient of the excluded variable times a function of the simple correlation coefficient between the excluded variable and the particular included variable in question.

3. Including a variable in an equation in which it is actually irrelevant does not cause bias, but it will usually increase the variances of the included variables' estimated coefficients, thus lowering their t-values and lowering $\bar{R}^2$.

4. Four useful criteria for the inclusion of a variable in an equation are:
   a. Theory
   b. *t*-test
   c. $\bar{R}^2$
   d. Bias

5. Theory, not statistical fit, should be the most important criterion for the inclusion of a variable in a regression equation. To do otherwise runs the risk of producing incorrect and/or disbelieved results. For example, stepwise regression routines will generally give biased estimates and will almost always have test statistics

that will not follow the distribution necessary to use standard t-tables.

Exercises

(Answers to even-numbered questions are in Appendix A.)

1. Write the meaning of each of the following terms without reference to the book (or your notes), and compare your definition with the version in the text for each:
   a. omitted variable
   b. irrelevant variable
   c. specification bias
   d. stepwise regression
   e. sequential specification search
   f. specification error

2. For each of the following situations, determine the *sign* (and if possible comment on the likely size) of the bias introduced by omitting a variable:
   a. In an equation for the demand for peanut butter, the impact on the coefficient of disposable income of omitting the price-of-peanut butter variable. (Hint: Start by hypothesizing signs.)
   b. In an earnings equation for workers, the impact on the coefficient of experience of omitting the variable for age.
   c. In a production function for airplanes, the impact on the coefficient of labor of omitting the capital variable.
   d. In an equation for daily attendance at outdoor concerts, the impact on the coefficient of the weekend dummy variable (1 = weekend) of omitting a variable that measures the probability of precipitation at concert time (as estimated by the weather bureau).

3. Consider the following annual model of the death rate (per million population) due to coronary heart disease in the U.S. (Y_t):

$$
\begin{array}{rlll}
\hat{Y}_t = 140 + & 10.0C_t + & 4.0E_t - & 1.0M_t \\
 & (2.5) & (1.0) & (0.5) \\
t = & 4.0 & 4.0 & -2.0 \\
n = 31 & (1950\text{–}1980) & \bar{R}^2 = .678 &
\end{array}
$$

where: C_t = per capita cigarette consumption (pounds of tobacco) in year t

E_t = per capita consumption of edible saturated fats (pounds of butter, margarine, and lard) in year t
M_t = per capita consumption of meat (pounds) in year t

a. Create and test appropriate null hypotheses at the 10 percent level. What, if anything, seems to be wrong with the estimated coefficient of M?
b. The most likely cause of a coefficient that is significant in the unexpected direction is omitted variable bias. Which of the following variables could possibly be an omitted variable that is causing $\hat{\beta}_M$'s unexpected sign? Explain.
B_t = per capita consumption of hard liquor (gallons) in year t
F_t = average fat content (percentage) of the meat that was consumed in year t
W_t = per capita consumption of wine and beer (gallons) in year t
R_t = per capita number of miles run in year t
H_t = per capita open-heart surgeries in year t
O_t = per capita amount of oat bran eaten in year t
c. If you had to choose one variable to add to the equation, what would it be? Explain your answer. (Hint: You're not limited to the variables listed in part b above.)

4. The "term structure of interest rates" describes the effect that maturity (the length of time before the principal of the bond is to be repaid) has on the yield of debt instruments. The yield curve, which plots the yield of bonds against their terms to maturity (the bonds on any given curve differ only in their terms to maturity), graphically illustrates the term structure for any given point in time. Suppose you were given a cross-sectional set of data that included all available information on 25 different bonds. Unfortunately, the bonds differ with respect to more than just their terms to maturity; they have different amounts of risk, taxability, and so on. Is there a way you could still estimate the slope of the yield curve that existed at the point in time the data were collected? How?

5. The data-set in Table 6.3 is the one that was used to estimate the chicken demand examples of Sections 6.1.2 and 6.2.2.
a. Enter these data into your computer and attempt to reproduce the specifications in the chapter.

TABLE 6.3 DATA FOR THE CHICKEN DEMAND EQUATION

Y	PC	YD	PB	R	LYD	YEAR
20.60	22.20	1362	23.30	1.200	7.216	1950
21.70	25.00	1465	28.70	1.520	7.289	1951
22.10	22.10	1515	24.30	1.720	7.323	1952
21.90	22.10	1581	16.30	1.900	7.365	1953
22.80	16.80	1583	16.00	0.940	7.367	1954
21.40	18.60	1664	15.60	1.730	7.416	1955
24.40	16.00	1741	14.90	2.620	7.462	1956
25.50	13.70	1802	17.20	3.230	7.496	1957
28.10	14.00	1832	21.90	1.780	7.513	1958
28.90	11.00	1903	22.60	3.370	7.551	1959
28.10	12.20	1947	20.40	2.870	7.574	1960
30.20	10.10	1991	20.20	2.360	7.596	1961
30.00	10.20	2073	21.30	2.770	7.636	1962
30.80	10.00	2144	19.90	3.160	7.670	1963
31.20	9.20	2296	18.00	3.540	7.738	1964
33.30	8.90	2448	19.80	3.950	7.803	1965
35.60	9.70	2613	22.29	4.860	7.868	1966
36.50	7.90	2757	22.30	4.290	7.921	1967
36.70	8.20	2956	23.40	5.340	7.991	1968
38.40	9.70	3152	26.20	6.670	8.055	1969
40.50	8.80	3393	27.10	6.390	8.129	1970
40.30	7.70	3630	29.00	4.330	8.196	1971
41.80	9.00	3880	33.50	4.070	8.263	1972
40.40	15.10	4346	42.80	7.030	8.377	1973
40.70	9.70	4710	35.60	7.840	8.457	1974
40.10	9.90	5132	32.30	5.800	8.543	1975
42.70	12.90	5550	33.70	4.980	8.621	1976
44.10	12.00	6046	34.50	5.270	8.707	1977
46.70	12.40	6688	48.50	7.190	8.808	1978
50.60	13.90	7682	66.10	10.070	8.946	1979
50.10	11.00	8421	62.40	11.430	9.038	1980
51.60	11.10	9243	58.60	14.030	9.131	1981
53.00	10.30	9724	56.70	10.610	9.182	1982
53.80	12.70	10340	55.50	8.610	9.243	1983
55.60	15.90	11257	57.30	9.520	9.328	1984

Sources: U.S. Department of Agriculture, *Agricultural Statistics*
U.S. Bureau of the Census, *Historical Statistics of the United States*
U.S. Bureau of the Census, *Statistical Abstract of the United States*

b. Find data for the price of another substitute for chicken and add that variable to your version of Equation 6.5. Analyze your results. In particular, apply the four criteria for the inclusion of a variable to determine whether the price of the substitute is an irrelevant variable or previously was an omitted variable.

6. You have been retained by the "Expressive Expresso" company to help them decide where to build their next "Expressive Expresso" store. You decide to run a regression on the sales of the 30 existing "Expressive Expresso" stores as a function of the characteristics of the locations they are in and then use the equation to predict the sales at the various locations you are considering for the newest store. You end up estimating (standard errors in parentheses):

$$\hat{Y}_i = 30 + \underset{(0.02)}{0.1X_{1i}} + \underset{(0.01)}{0.01X_{2i}} + \underset{(1.0)}{10.0X_{3i}} + \underset{(1.0)}{3.0X_{4i}}$$

where: Y_i = average daily sales (in hundreds of dollars) of the *i*th store
X_{1i} = the number of cars that pass the *i*th location per hour
X_{2i} = average income in the area of the *i*th store
X_{3i} = number of tables in the *i*th store
X_{4i} = number of competing shops in the area of the *i*th store

a. Hypothesize expected signs, calculate the correct t-scores, and test the significance at the one percent level for each of the coefficients.
b. What problems appear to exist in the equation? What evidence of these problems do you have?
c. What suggestions would you make for a possible second run of this admittedly hypothetical equation? (Hint: Before recommending the inclusion of a potentially left-out variable, consider whether the exclusion of the variable could possibly have caused any observed bias.)

7. Discuss the topic of specification searches with various members of your econometrics class. What is so wrong with not mentioning previous (probably incorrect) estimates? Why should readers be suspicious when researchers attempt to find results that support their hypotheses; who would try to do the opposite? Do these concerns have any meaning in the world of business? In particular, if you're not trying to publish a paper, couldn't you use any specification search techniques you want to find the best equation?

8. Suppose you run a regression explaining the number of hamburgers that the campus fast-food store (let's call it "The Cooler") sells per day as a function of the price of their hamburgers (in dollars), the

weather (in degrees F), the price of hamburgers at a national chain nearby (also in dollars), and the number of students (in thousands) on campus that day. Assume that The Cooler stays open whether or not school is in session (for staff, etc.). Unfortunately, a lightning bolt strikes the computer and wipes out all the memory and you cannot tell which independent variable is which! Given the following regression results (standard errors in parentheses):

$$\hat{Y}_i = 10.6 + \underset{(2.6)}{28.4X_{1i}} + \underset{(6.3)}{12.7X_{2i}} + \underset{(0.61)}{0.61X_{3i}} - \underset{(5.9)}{5.9X_{4i}}$$
$$\bar{R}^2 = .63 \qquad n = 35$$

a. Attempt to identify which result corresponds to which variable.
b. Explain your reasoning for part a above.
c. Develop and test hypotheses about the coefficients assuming that your answer to Part a is correct. What suggestions would you have for changes in the equation for a re-run when the computer is back up again?

9. Many of the examples in the text so far have been demand-side equations or production functions, but economists often also have to quantify supply-side equations that are not true production functions. These equations attempt to explain the production of a product (for example, Brazilian coffee) as a function of the price of the product and various other attributes of the market that might have an impact on the total output of growers.
   a. What sign would you expect the coefficient of price to have in a supply-side equation? Why?
   b. What other variables can you think of that might be important in a supply-side equation?
   c. Many agricultural decisions are made months (if not a full year) before the results of those decisions appear in the market. How would you adjust your hypothesized equation to take account of these lags?
   d. Given all the above, carefully specify the exact equation you would use to attempt to explain Brazilian coffee production. Be sure to hypothesize the expected signs, be specific with respect to lags, and try to make sure you have not omitted an important independent variable.

10. If you think about the previous question, you'll realize that the same dependent variable (quantity of Brazilian coffee) can have different expected signs for the coefficient of the *same* independent

variable (the price of Brazilian coffee) depending on what other variables are in the regression.

a. How is this possible? That is, how is it possible to expect different signs in demand-side equations from what you would expect in supply-side ones?
b. Given that we will not discuss how to estimate simultaneous equations until Chapter 14, what can be done to avoid the "simultaneity bias" of getting the price coefficient from the demand equation in the supply equation and vice versa?
c. What can you do to systematically ensure that you do not have supply-side variables in your demand equation or demand-side variables in your supply equation?

11. You've been hired by "Indo," the new Indonesian automobile manufacturer, to build a model of U.S. car prices in order to help the company undercut our prices. Allowing Friedmaniac zeal to overwhelm any patriotic urges, you build the following model of the price of 35 different American-made 1991 U.S. sedans (standard errors in parentheses):

$$\text{Model A: } \hat{P}_i = \underset{}{3.0} + \underset{(0.07)}{0.28W_i} + \underset{(0.4)}{1.2T_i} + \underset{(2.9)}{5.8C_i} + \underset{(0.20)}{0.20L_i}$$
$$\bar{R}^2 = .92$$

where: P_i = list price of the ith car (thousands of dollars)
W_i = weight of the ith car (hundreds of pounds)
T_i = a dummy equal to 1 if the ith car has an automatic transmission, 0 otherwise
C_i = a dummy equal to 1 if the ith car has cruise control, 0 otherwise
L_i = the size of the engine of the ith car (in liters)

a. Your firm's pricing expert hypothesizes positive signs for all the slope coefficients in Model A. Test her expectations at the 95 percent level of confidence.
b. What econometric problems appear to exist in Model A? In particular, does the size of the coefficient of C cause any concern? Why? What could be the problem?
c. You decide to test the possibility that L is an irrelevant variable by dropping it and re-running the equation, obtaining Model T below. Which model do you prefer? Why?

$$\text{Model T: } \hat{P}_i = \underset{}{18} + \underset{(0.07)}{0.29W_i} + \underset{(0.3)}{1.2T_i} + \underset{(2.9)}{5.9C_i}$$
$$\bar{R}^2 = .92$$

12. Determine the sign (and if possible comment on the likely size) of the bias introduced by leaving a variable out of an equation in each of the following cases:
   a. In an annual equation for corn yields per acre (in year t), the impact on the coefficient of rainfall in year t of omitting average temperature that year. (Hint: Drought and cold weather both hurt corn yields.)
   b. In an equation for daily attendance at Los Angeles Lakers' home basketball games, the impact on the coefficient of the winning percentage of the opponent (as of the game in question) of omitting a dummy variable that equals one if the opponent's team includes a superstar (such as Michael Jordan).
   c. In an equation for annual consumption of apples in the U.S., the impact on the coefficient of the price of bananas of omitting the price of oranges.
   d. In an equation for student grades on the first midterm in this class, the impact on the coefficient of total hours studied (for the test) of omitting hours slept the night before the test.

13. Suppose that you run a regression to determine whether gender or race has any significant impact on scores on a test of the economic understanding of children.[11] You model the score of the *i*th student on the test of elementary economics (S_i) as a function of the composite score on the Iowa Tests of Basic Skills of the *i*th student, a dummy variable equal to 1 if the *i*th student is female (0 otherwise), the average number of years of education of the parents of the *i*th student, and a dummy variable equal to 1 if the *i*th student is nonwhite (0 otherwise). Unfortunately, a rainstorm floods the computer center and makes it impossible to read the part of the computer output that identifies which variable is which. All you know is that the regression results are (standard errors in parentheses):

11. These results have been jiggled to meet the needs of this question, but this research actually was done. See Stephen Buckles and Vera Freeman, "Male-Female Differences in the Stock and Flow of Economic Knowledge," *Review of Economics and Statistics*, May 1983, pp. 355–357.

$$\hat{S}_i = 5.7 - 0.63X_{1i} - 0.22X_{2i} + 0.16X_{3i} + 0.12X_{4i}$$
$$(0.63) \quad (0.88) \quad (0.08) \quad (0.01)$$
$$n = 24 \quad \bar{R}^2 = .54$$

a. Attempt to identify which result corresponds to which variable. Be specific.
b. Explain the reasoning behind your answer to part a above.
c. Assuming that your answer is correct, create and test appropriate hypotheses (at the five percent level) and come to conclusions about the effects of gender and race on the test scores of this particular sample.
d. Did you use a one-tail or two-tailed test in part c above? Why?

14. William Sander[12] estimated a 50-state cross-sectional model of the 1970 farm divorce rate as part of an effort to determine whether the national trend towards more divorces could be attributed in part to increases in the earning ability of women. His equation was (t-scores in parentheses):

$$\hat{Y}_i = -4.1 + 0.003P_i + 0.06L_i - 0.002A_i + 0.76N_i$$
$$(3.3) \quad (1.5) \quad (-0.6) \quad (13.5)$$

where: Y_i = the farm divorce rate in the *i*th state
P_i = the population density of the *i*th state
L_i = the labor force participation of farm women in the *i*th state
A_i = farm assets held by women in the *i*th state
N_i = the rural nonfarm divorce rate in that state

a. Develop and test hypotheses about the slope coefficients of Snider's equation at the five percent level.
b. What (if any) econometric problems (out of omitted variables and irrelevant variables) appear to exist in this equation? Justify your answer.
c. What one specification change in this equation would you suggest? Be specific.

12. William Sander, "Women, Work, and Divorce," *The American Economic Review*, June 1985, pp. 519–523.

d. Use our four specification criteria to decide whether you believe L is an irrelevant variable. The equation without L (t-scores again in parentheses) was:

$$\hat{Y}_i = -2.5 + \underset{(4.3)}{0.004P_i} - \underset{(-1.3)}{0.004A_i} + \underset{(14.8)}{0.79N_i}$$

6.7 Appendix: Additional Specification Criteria

So far in this chapter, we've suggested four criteria for choosing the independent variables (economic theory, $\bar{R}^2$, the *t*-test, and possible bias in the coefficients). Sometimes, however, these criteria don't provide enough information for a researcher to feel confident that a given specification is best. For instance, there can be two different specifications that both have excellent theoretical underpinnings. In such a situation, many econometricians use additional, often more formal, specification criteria to provide comparisons of the properties of the alternative estimated equations.

The use of formal specification criteria is not without problems, however. First, no test, no matter how sophisticated, can "prove" that a particular specification is the true one. The use of specification criteria, therefore, must be tempered with a healthy dose of economic theory and common sense. A second problem is that more than 20 such criteria have been proposed; how do we decide which one(s) to use? Because many of these criteria overlap with one another or have varying levels of complexity, a choice between the alternatives is a matter of personal preference.

In this section, we'll describe the use of two of the most popular specification criteria, Ramsey's RESET test and Amemiya's PC. Our inclusion of just two techniques does not imply that other tests and criteria are not appropriate or useful. Indeed, the reader will find that most other formal specification criteria have quite a bit in common with at least one of the two that we include. We think that you'll be more able to use and understand other formal specification criteria[13] once you've mastered these two.

13. In particular, the Likelihood Ratio test, versions of which will be covered in Sections 7.6 and 12.2, can be used as a specification test. For an introductory level summary of eight other specification criteria, see Ramu Ramanathan, *Introductory Econometrics* (San Diego: Harcourt Brace Jovanovich, 1989), pp. 165–167.

6.7.1 Ramsey's Regression Specification Error Test (RESET)

One of the most-used formal specification tests other than $\bar{R}^2$ is the Ramsey Regression Specification Error Test (RESET).[14] The **Ramsey RESET test** is a general test that determines the likelihood of an omitted variable or some other specification error by measuring whether the fit of a given equation can be significantly improved by the addition of $\hat{Y}^2$, $\hat{Y}^3$, and $\hat{Y}^4$ terms.

What's the intuition behind RESET? The additional terms act as proxies for any possible (unknown) omitted variables or incorrect functional forms. If the proxies can be shown by the *F*-test to have improved the overall fit of the original equation, then we have evidence that there is some sort of specification error in our equation. As we'll learn in Chapter 7, the $\hat{Y}^2$, $\hat{Y}^3$, and $\hat{Y}^4$ terms form a *polynomial* functional form. Such a polynomial is a powerful curve-fitting device that has a good chance of acting as a proxy for a specification error if one exists. If there is no specification error, then we'd expect the coefficients of the added terms to be insignificantly different from zero because there is nothing for them to act as a proxy for.

The Ramsey RESET test involves three steps:

1. Estimate the equation to be tested using OLS:

$$\hat{Y}_i = \hat{\beta}_0 + \hat{\beta}_1 X_{1i} + \hat{\beta}_2 X_{2i} \tag{6.20}$$

2. Take the $\hat{Y}_i$ values from Equation 6.20 and create $\hat{Y}_i^2$, $\hat{Y}_i^3$, and $\hat{Y}_i^4$ terms. Then add these terms to Equation 6.20 as additional explanatory variables and estimate the new equation with OLS:

$$Y_i = \beta_0 + \beta_1 X_{1i} + \beta_2 X_{2i} + \beta_3 \hat{Y}_i^2 + \beta_4 \hat{Y}_i^3 + \beta_5 \hat{Y}_i^4 + \epsilon_i \tag{6.21}$$

3. Compare the fits of Equation 6.20 and 6.21 using the *F*-test. If the two equations are significantly different in overall fit, we can conclude that it's likely that Equation 6.20 is misspecified.

While the Ramsey RESET test is fairly easy to use, it does little more than signal *when* a major specification error might exist. If you encounter a significant Ramsey RESET test, then you face the daunting task of figuring out exactly *what* the error is! Thus, the test often ends up being more useful in "supporting" (technically, not refuting) a

14. J. B. Ramsey, "Tests for Specification Errors in Classical Linear Least Squares Regression Analysis," *Journal of the Royal Statistical Society*, 1969, pp. 350–371.

researcher's contention that a given specification has no major specification errors than it is in helping find an otherwise undiscovered flaw.[15]

As an example of the Ramsey RESET test, let's return to the chicken demand model of this chapter to see if RESET can detect the known specification error (omitting the price of beef) in Equation 6.6. Step one involves running the original equation:

$$\begin{aligned} \hat{Y}_t = &- 80.7 - \underset{(0.06)}{0.34}PC_t + \underset{(0.42)}{15.0}LYD_t \\ t = &\quad\ \ -5.6 \qquad\qquad 36.0 \\ \bar{R}^2 = &.981 \quad n = 35 \quad RSS = 68.82 \end{aligned} \tag{6.6}$$

For step two, we take $\hat{Y}_t$ from Equation 6.6, calculate $\hat{Y}_t^2$, $\hat{Y}_t^3$, and $\hat{Y}_t^4$, and then reestimate Equation 6.6 with the three new terms added in:

$$\begin{aligned} Y_t = &\ 1627 + \underset{(1.80)}{6.38}PC_t - \underset{(80)}{275}LYD_t + \underset{(0.23)}{0.89}\hat{Y}_t^2 \\ t = &\qquad\quad 3.5 \qquad\quad -3.4 \qquad\quad 3.8 \\ &- \underset{(0.004)}{0.017}\hat{Y}_t^3 + \underset{(0.00003)}{0.00012}\hat{Y}_t^4 + e_t \\ t = &\quad -3.9 \qquad\quad 4.0 \\ \bar{R}^2 = &\ .986 \quad n = 35 \quad RSS = 44.23 \end{aligned} \tag{6.22}$$

In step three, we compare the fits of the two equations by using the *F*-test. Specifically, we test the hypothesis that the coefficients of all three of the added terms are equal to zero:

$$\begin{aligned} &H_0: \hat{\beta}_3 = \hat{\beta}_4 = \hat{\beta}_5 = 0 \\ &H_A: \text{otherwise} \end{aligned}$$

The appropriate F-statistic to use is Equation 5.14, repeated here:

$$F = \frac{(RSS_M - RSS)/M}{RSS/(n - K - 1)} \tag{5.14}$$

15. The particular version of the Ramsey RESET test we describe in this section is only one of a number of possible formulations of the test. For example, some researchers delete the $\hat{Y}^4$ term from Equation 6.21. In addition, versions of the Ramsey RESET test are useful in testing for functional form errors (to be described in Chapter 7) and serial correlation (to be described in Chapter 9).

where RSS_M is the residual sum of squares from the restricted equation (Equation 6.6), RSS is the residual sum of squares from the unrestricted equation (Equation 6.22), M is the number of restrictions (3), and $(n - K - 1)$ is the number of degrees of freedom in the unrestricted equation (29):

$$F = \frac{(68.62 - 44.23)/3}{44.23/29} = \frac{8.130}{1.525} = 5.33 \tag{6.23}$$

The critical F-value to use, 2.93, is found in Statistical Table B-2 at the 5 percent level of significance with 3 numerator and 29 denominator[16] degrees of freedom. Since 5.33 is greater than 2.93, we can reject the null hypothesis that the coefficients of the added variables are jointly zero, allowing us to conclude that there is indeed a specification error in Equation 6.6. Such a conclusion is no surprise, since we know that the price of beef was left out of the equation. Note, however, that the Ramsey RESET test tells us only that a specification error is likely to exist in Equation 6.6; it does not specify the details of that error.

6.7.2 Amemiya's Prediction Criterion (PC)

A second category of formal specification criterion involves adjusting the RSS by one factor or another to create another index of the fit of an equation. The most popular criterion of this type is $\bar{R}^2$, but a number of interesting alternatives have been proposed, the easiest to use of which is Amemiya's PC.[17] **Amemiya's Prediction Criterion (PC)** is a method of comparing alternative specifications by adjusting RSS for the sample size and the number of explanatory variables; the lower the PC, the better the specification. The equation for Amemiya's PC is:

$$PC = \frac{RSS \cdot (n + K)}{(n - K)} \tag{6.24}$$

16. Table B-2 does not list 29 numerator degrees of freedom, so, as mentioned in Chapter 5, you must interpolate between 30 and 25 numerator degrees of freedom to get the answer. In this case, some researchers would note that the calculated F-value exceeds both critical F-values and wouldn't bother with the interpolation.

17. T. Amemiya, "Selection of Regressors," *International Economic Review,* 1980, pp. 331–354. A similar criterion is Akaike's FPE; see H. Akaike, "Statistical Predictor Identification," *Annals of the Institute of Statistical Mathematics,* 1970, pp. 203–217. The criteria are so similar that many econometricians refer to Equation 6.24 as Akaike's FPE.

To use Amemiya's PC, calculate Equation 6.24 for each of the specifications being compared. If all other factors (like theoretical relevance) are equal, then the specification with the lowest value of PC should be chosen.

One pragmatic advantage of Amemiya's PC is that it offsets a problem inherent in the use of $\bar{R}^2$ as a method of comparing the statistical fit of different specifications. This problem arises because $\bar{R}^2$ increases any time a variable with a t-score greater than one (in absolute value) is added to an equation. Thus, there are many times when the $\bar{R}^2$ criterion and the *t*-test criterion will yield different answers. Amemiya's PC is quite similar to $\bar{R}^2$, but it penalizes the addition of another explanatory variable more than $\bar{R}^2$ does. As a result, PC will quite often[18] be minimized by an equation with fewer explanatory variables than the one that maximizes $\bar{R}^2$.

The theoretical basis for Amemiya's PC is that under certain circumstances PC minimizes the Mean Square Error. The **Mean Square Error (MSE)** is a specification selection criterion that allows a tradeoff between bias and variance:

$$\text{MSE} = \text{variance}(\hat{\beta}) + (\text{bias in } \hat{\beta})^2 \qquad (6.25)$$

This bias-variance tradeoff is important because including an additional explanatory variable may reduce bias, but at the cost of increasing the variance of the estimated coefficients. While we'd normally prefer an unbiased model, there can be situations in which adding a variable decreases bias by a tiny amount but introduces a considerable amount of variance. In such circumstances, we might choose the equation that has the smaller MSE instead of automatically choosing the one that minimizes bias. Unfortunately, MSE cannot be measured directly (because we never know the true βs), so Amemiya's PC is a very useful proxy for MSE. Minimizing PC tends to minimize MSE.

Let's apply Amemiya's PC to the same chicken demand example we used for Ramsey's RESET. To see if Amemiya's PC can detect the specification error we already know exists in Equation 6.6 (omitting the price of beef), we need to calculate Amemiya's PC for Equation 6.6 and compare it with the PC for Equation 6.5. The equation with

18. Using a Monte Carlo study, Judge et. al. showed that (given specific simplifying assumptions) a specification chosen by maximizing $\bar{R}^2$ is over 50 percent more likely to include an extraneous variable than is one chosen by minimizing PC. See George G. Judge et al., *Introduction to the Theory and Practice of Econometrics* (New York: Wiley, 1988), pp. 849–850. While Judge didn't show this, minimizing PC will omit a relevant explanatory variable more frequently than will maximizing $\bar{R}^2$.

the lower PC value will, other things being equal, be our preferred specification.

The original chicken demand model, Equation 6.5, was:

$$
\begin{aligned}
\hat{Y}_t = & -60.5 - \underset{(0.07)}{0.45PC_t} + \underset{(0.05)}{0.12PB_t} + \underset{(1.2)}{12.2LYD_t} \qquad (6.5) \\
t = & \qquad\quad -6.4 \qquad\quad 2.5 \qquad\quad 10.6 \\
\bar{R}^2 = & .984 \qquad n = 35 \qquad RSS = 56.72
\end{aligned}
$$

Amemiya's PC for Equation 6.5 is:

$$PC_{6.5} = \frac{56.72 \cdot 38}{32} = 67.36 \tag{6.26}$$

Equation 6.6 (repeated in the previous example), which omits the price of beef, has an RSS of 68.62 with $K = 2$. Thus,

$$PC_{6.6} = \frac{68.62 \cdot 37}{33} = 76.94 \tag{6.27}$$

Since $67.36 < 76.94$, Amemiya's PC criterion provides evidence that Equation 6.5 is preferable to Equation 6.6. That is, the price of beef appears to belong in the equation. As it turns out, then, both Ramsey's RESET and Amemiya's PC indicate the existence of a specification error when we leave the price of beef out of the equation. This result is not surprising, since we purposely left out a theoretically justified variable, but it provides an example of how useful these criteria could be when we're less sure about the underlying theory.

Note that the use of Amemiya's PC requires the researcher to come up with a particular alternative specification, while Ramsey's RESET does not. Such a distinction makes RESET easier to use, but it makes PC more informative if a specification error is found. Thus our two additional specification criteria serve different purposes. RESET is most useful as a general test for the existence of a specification error, while PC is most useful as a means of comparing two or more alternative specifications.

7

Specification: Choosing a Functional Form

Even after you've chosen your independent variables, the job of specifying the equation is not over. The next step is to choose the functional form of the relationship between each independent variable and the dependent variable. Should the equation go through the origin? Do you expect a curve instead of a straight line? Does the effect of a variable peak at some point and then start to decline? An affirmative answer to any of these questions suggests that an equation other than the standard "linear in the variables" model of the previous chapters might be appropriate. Such alternative specifications are important for two reasons; a correct explanatory variable may well appear to be insignificant or to have an unexpected sign if an inappropriate functional form is used, and the consequences for interpretation and forecasting can be severe.

Theoretical considerations usually dictate the form of a regression model. The basic technique involved in deciding on a functional form is to choose the shape that best exemplifies the expected underlying economic or business principles and then to use the mathematical form that produces that shape. To help with that choice, this chapter contains plots of the most commonly used functional forms along with the mathematical equations that correspond to each.

One may use dummy variables to allow the coefficients of independent variables to differ for qualitative conditions (slope dummies) or to actually change as the independent variable changes (piecewise regression). These techniques are fairly innovative ways to use dummy variables to create nonlinear-in-the-variables functional forms that are still linear in the coefficients. This chapter also includes a brief discussion of the constant term. In particular, we suggest that the constant term should be retained in equations even if theory suggests otherwise and that estimates of the constant term should not be relied on for inference or analysis. The chapter concludes with a short appendix on nonlinear in the coefficients regression.

7.1 Alternative Functional Forms

The choice of a functional form for an equation is a vital part of the specification of that equation. The use of OLS requires that the equation be linear in the coefficients, but there is a wide variety of functional forms that are linear in the coefficients while being nonlinear in the variables. Indeed, in previous chapters we've already used several equations that are linear in the coefficients and nonlinear in the variables, but we've said little about when to use such nonlinear equations. The purpose of the current section is to present the details of the most frequently used functional forms to help the reader develop the ability to choose the correct one when specifying an equation.

The choice of a functional form almost always should be based on the underlying economic or business theory and only rarely on which form provides the best fit. The logical form of the relationship between the dependent variable and the independent variable in question should be compared with the properties of various functional forms, and the one that comes closest to that underlying theory should be chosen. To allow such a comparison, the paragraphs that follow characterize the most frequently used functional forms in terms of graphs, equations, and examples. In some cases, more than one functional form will be applicable, but usually a choice between alternative functional forms can be made on the basis of the information we'll present.

7.1.1 Linear Form

The linear regression model, used almost exclusively in this text thus far, is based on the assumption that the slope of the relationship between the independent variable and the dependent variable is constant:[1]

$$\frac{\Delta Y}{\Delta X_k} = \beta_k \qquad k = 1,2,\ldots,K \tag{7.1}$$

The slope is constant, so the **elasticity** of Y with respect to X (the percentage change in the dependent variable caused by a one percent change in the independent variable, holding the other variables in the equation constant) is not constant:

$$\eta_{Y,X_k} = \frac{\Delta Y/Y}{\Delta X_k/X_k} = \frac{\Delta Y}{\Delta X_k} \cdot \frac{X_k}{Y} = \beta_k \frac{X_k}{Y} \tag{7.2}$$

If the hypothesized relationship between Y and X is such that the slope of the relationship can be expected to be constant, then the linear functional form should be used.

Unfortunately, theory frequently predicts only the sign of a relationship and not its functional form. When there is little theory on which to base an expected functional form, the linear form should be used until strong evidence that it is inappropriate is found. Unless theory, common sense, or experience justifies using some other functional form, you should use the linear model. Because it's in effect being used by default, this model is sometimes referred to as the *default* functional form. (Some researchers use the double-log model as the default functional form.)

7.1.2 Double-Log or Exponential Form

The most common functional form that is nonlinear in the variables (but still linear in the coefficients) is the double-log form. A double-log form is often used because a researcher has specified that, contrary

1. Throughout this section, the "delta" notation (Δ) will be used instead of the proper calculus notation to make for easier reading. The specific definition of Δ is "change," and it implies a small change in the variable it is attached to. For example, the term ΔX should be read as "change in X." Since a regression coefficient represents the change in the expected value of Y brought about by a one-unit change in X (holding constant all other variables in the equation), then $\beta_k = \Delta Y/\Delta X_k$. Those comfortable with calculus should substitute partial derivative signs for Δs.

to the linear model, the elasticities rather than the slopes are constant. If an elasticity is assumed to be constant, that means:

$$\eta_{Y,X_k} = \beta_k = \text{a constant} \tag{7.3}$$

Given the assumption of constant elasticity, the proper form is the **exponential functional form:**

$$Y = e^{\beta_0} X_1^{\beta_1} X_2^{\beta_2} e^{\epsilon} \tag{7.4}$$

where e is the base of the natural logarithm. A logarithmic transformation can be applied to Equation 7.4 by taking the log of both sides of the equation to make it linear in the coefficients. This transformation converts Equation 7.4 into Equation 7.5, the **double-log functional form:**

$$\ln Y = \beta_0 + \beta_1 \ln X_1 + \beta_2 \ln X_2 + \epsilon \tag{7.5}$$

Where "$\ln Y_i$" refers to the natural log of Y_i, etc. In a double-log equation, an individual regression coefficient, for example β_k, can be interpreted as an elasticity because:

$$\beta_k = \frac{\Delta(\ln Y)}{\Delta(\ln X_k)} \approx \frac{\Delta Y/Y}{\Delta X_k/X_k} = \eta_{Y,X_k} \tag{7.6}$$

Since regression coefficients are constant, the condition that the model have a constant elasticity is met by the double-log equation.[2]

The way to interpret β_k in a double-log equation is that if X_k changes by one percent while the other Xs are held constant, then Y will change by β_k percent. Since elasticities are constant, the slopes are now no longer constant.

Figure 7.1 is a graph of the double-log or exponential function (ignoring the error term). The panel on the left shows the economic concept of a production function (or an indifference curve). Isoquants from production functions show the different combinations of factors X_1 and X_2, probably capital and labor, that can be used to produce a given level of output Y. This kind of double-log production function is called a Cobb-Douglas production function; for an example of the estimation of such a function, see Exercise 7. The panel on the right of Figure 7.1 shows the relationship between Y and X_1 that would

2. The ≈ means "approximately equal to," and the equivalence denoted by the sign is justified by the fact that the derivative of lnY with respect to Y equals ΔY/Y.

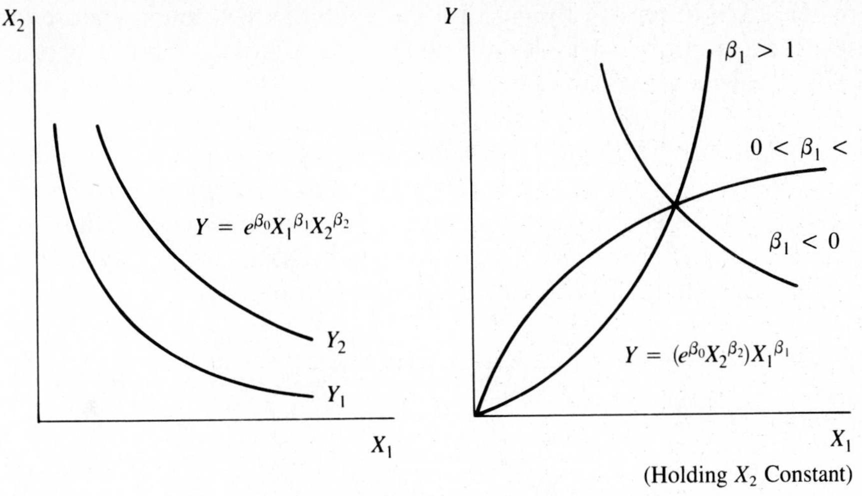

Figure 7.1 DOUBLE-LOG FUNCTIONS
Depending on the values of the regression coefficients, the double-log functional form can take on a number of shapes. The left panel shows the use of a double-log function to depict a shape useful in describing the economic concept of a production function (or an indifference curve). The right panel shows various shapes that can be achieved with a double-log function if X_2 is held constant or is not included in the equation.

exist if X_2 were held constant or were not included in the model. Note that the shape of the curve depends on the sign and magnitude of coefficient β_1.

Before using a double-log model, make sure that there are no negative or zero observations in the data-set. Since the log of a non-positive number is undefined, a regression cannot be run. Double-log models should be run only when all the variables take on positive values. Dummy variables, which can take on the value of zero, should not be logged even if they are in a double-log equation.[3]

7.1.3 Semi-Log Form

The **semi-log functional form** is a variant of the double-log equation in which some but not all of the variables (dependent and independent)

3. If it is necessary to take the log of a dummy variable, that variable needs to be transformed to avoid the possibility of taking the log of zero. The best way is to redefine the entire dummy variable so that, instead of taking on the values of zero and one, it takes on the values of one and e (the base of the natural logarithm). The log of this newly defined dummy then takes on the values of zero and one, and the interpretation of β remains the same as in a linear equation. Such a transformation changes the coefficient value but not the usefulness or theoretical validity of the dummy variable.

are expressed in terms of their logs. For example, you might choose to use as explanatory variables the logarithms of one or more of the original independent variables as in:

$$Y_i = \beta_0 + \beta_1 \ln X_{1i} + \beta_2 X_{2i} + \epsilon_i \tag{7.7}$$

In this case, the economic meanings of the two slope coefficients are different, since X_2 is linearly related to Y while X_1 is nonlinearly related to Y. In particular, calculus can be used to show that:

$$\Delta Y/\Delta X_1 = \beta_1/X_1 \tag{7.8}$$

or, solving for β_1:

$$\beta_1 = \Delta Y/(\Delta X_1/X_1) \tag{7.9}$$

In words, if X_1 changes by one percent, then Y will change by $\beta_1/100$ units (to see this, substitute a one percent change in X_1 into Equation 7.9; recall that values of X_1 must be positive in order to take a log). The elasticity of Y with respect to X_1 is thus:

$$\eta_{Y,X_1} = \frac{\Delta Y}{\Delta X_1} \cdot \frac{X_1}{Y} = \frac{\beta_1}{Y} \tag{7.10}$$

which decreases as Y increases.

Figure 7.2 shows the relationship between Y and X_1 when X_2 is held constant. Note that if β_1 is greater than zero, the impact of changes in X_1 on Y decreases as X_1 gets bigger. Thus the semi-log functional form should be used when the relationship between X_1 and Y is hypothesized to have this "tailing off" form.

Applications of the semi-log form are quite frequent in economics and business. For example, most consumption functions tend to increase at a decreasing rate past some level of income. These *Engel curves* tend to flatten out because as incomes get higher, a smaller percentage of income goes to consumption and a greater percentage goes to saving. Consumption thus increases at a decreasing rate. If Y is the consumption of an item and X_1 is disposable income (with X_2 standing for all the other independent variables), then the use of the semi-log functional form is justified whenever the item's consumption can be expected to tail off as income increases.

For example, recall the chicken consumption Equation 6.5 from the previous chapter:

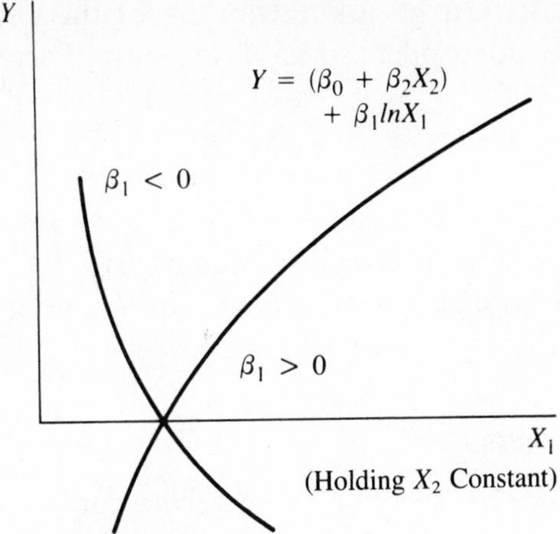

Figure 7.2 SEMI-LOG FUNCTIONS
The semi-log functional form (in the case of taking the log of one of the independent variables) can be used to depict a situation in which the impact of X_1 on Y is expected to tail off as X_1 gets bigger as long as β_1 is greater than zero (holding X_2 constant). An application of this type of semi-log model is a curve that describes quantity consumed as a function of income.

$$\hat{Y}_t = -60.5 - 0.45PC_t + 0.12PB_t + 12.2LYD_t \qquad (6.5)$$

In this equation, the independent variables include the two price variables (PC and PB) and the *log* of disposable income, because it was hypothesized that as income rose, consumption would increase at a decreasing rate. For other products, perhaps like yachts or summer homes, no such decreasing rate could be hypothesized, and the semi-log function would not be appropriate.

Note from Equations 6.5 and 7.7 that various combinations of the functional forms are possible. Thus, the form taken by X_1 may be different from the form taken by X_2. In addition, Y may assume yet another different functional form.[4]

An example of a situation in which the functional form of Y is different from that of the rest of the equation is a kind of semi-log

4. One example of such a combination functional form is called the *translog function*. The translog function combines three different functional forms to come up with an equation perfectly general for estimating various kinds of cost functions. For more on the translog function, see Laurits R. Christensen and William H. Greene, "Economies of Scale in U.S. Electrical Power Generation," *Journal of Political Economy*, Aug. 1976, pp. 655–676.

function that is derived by taking the log of the dependent variable while leaving the independent variables in linear form:

$$\ln Y_i = \beta_0 + \beta_1 X_{1i} + \beta_2 X_{2i} + \epsilon_i \tag{7.11}$$

This model has neither a constant slope nor a constant elasticity. If X_1 changes by one *unit*, then Y will change by $\beta_1 \cdot 100$ percent, holding X_2 constant. For an example of this version of the semi-log function, see Exercise 4.

7.1.4 Polynomial Form

In most cost functions, the slope of the cost curve changes as output changes. If the slopes of a relationship are expected to depend on the level of the variable itself (for example, get steeper as output increases) then a polynomial model should be considered. **Polynomial functional forms** express Y as a function of independent variables, some of which are raised to powers other than one. For example, in a second-degree polynomial (also called a quadratic) equation, at least one independent variable is squared:

$$Y_i = \beta_0 + \beta_1 X_{1i} + \beta_2 (X_{1i})^2 + \beta_3 X_{2i} + \epsilon_i \tag{7.12}$$

Such a model can indeed produce slopes that change as the independent variables change. The slopes of Y with respect to the Xs in Equation 7.12 are:

$$\frac{\Delta Y}{\Delta X_1} = \beta_1 + 2\beta_2 X_1 \quad \text{and} \quad \frac{\Delta Y}{\Delta X_2} = \beta_3 \tag{7.13}$$

Note that the first slope depends on the level of X_1 and the second slope is constant. If this were a cost function, with Y being the average cost of production and X_1 being the level of output of the firm, then we would expect β_1 to be negative and would expect β_2 to be positive if the firm has the typical U-shaped cost curve depicted in the left half of Figure 7.3.

For another example, consider a model of annual employee earnings as a function of the age of each employee and a number of other measures of productivity such as education. What is the expected impact of age on earnings? As a young worker gets older, his or her earnings will typically increase. Beyond some point, however, an increase in age will not increase earnings by very much at all, and around

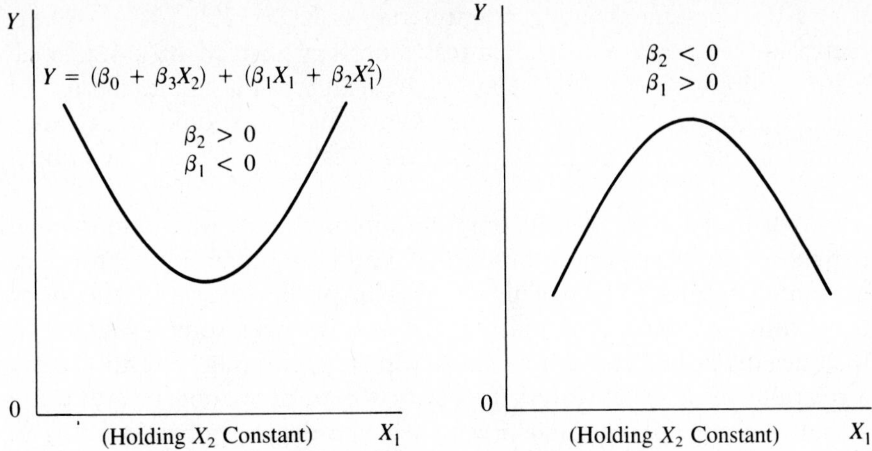

Figure 7.3 QUADRATIC (POLYNOMIAL) FUNCTIONS
Quadratic functional forms (polynomials with squared terms) take on U or inverted U shapes depending on the values of the coefficients (holding X_2 constant). The left panel shows the shape of a quadratic function that could be used to show a typical cost curve, while the right panel allows the description of an impact that rises and then falls (like the impact of age on earnings).

retirement we expect earnings to start to decrease with age. As a result, a logical relationship between earnings and age might look something like the right half of Figure 7.3; earnings would rise, level out, and then fall as age increased. Such a theoretical relationship could be modeled with a quadratic equation:

$$\text{Earnings}_i = \beta_0 + \beta_1 \text{Age}_i + \beta_2 \text{Age}_i^2 + \cdots + \epsilon_i \qquad (7.14)$$

What would the expected signs of $\hat{\beta}_1$ and $\hat{\beta}_2$ be? As a worker got older, the difference between "Age" and "Age^2" would increase dramatically, because "Age^2" would become quite large. As a result, the coefficient of "Age" would be more important at lower ages than it would be at higher ages. Conversely, the coefficient of "Age^2" would be more important at higher ages. Since you expect the impact of age to rise and fall, you'd thus expect $\hat{\beta}_1$ to be positive and $\hat{\beta}_2$ to be negative (all else being equal). In fact, this is exactly what many researchers in labor economics have observed.

Unfortunately, all polynomials can be used as a curve-fitting device. In fact, any n observations can be fitted exactly (that is, all residuals would be zero) to a regression curve that is a polynomial of degree n − 1 (i.e., having as independent variables X, X^2, X^3,...,X^{n-1}). Here, regression becomes a mathematical tautology instead of a statistical relationship and gives false pictures of reality. As a result, the use of higher degree polynomials in regression analysis should be avoided unless the underlying theory specifically calls for such a functional form.

With polynomial regressions, the interpretation of the individual regression coefficients becomes difficult, and the equation may produce unwanted results for particular ranges of X. For example, the slope for a third-degree polynomial can be positive over some range of X, then negative over the next range, and then positive again. Unless such a relationship is called for by theory, it would be inappropriate to use a higher-degree polynomial. Even a second-degree polynomial, as in Equation 7.12, imposes a particular symmetric shape (a U-shape or its inverse) that might be unreasonable in some cases. For example, review the rain equation in Section 5.5, where it seems obvious that the squared term was added solely to provide a better fit to this admittedly cooked-up equation. To avoid such curve-fitting, some researchers use just the square of the independent variable and exclude its linear form from the equation. In any event, great care must be taken when using a polynomial regression equation to ensure that the functional form will achieve what is intended by the researcher and no more.

7.1.5 Inverse Form

The **inverse functional form** expresses Y as a function of the reciprocal (or inverse) of one or more of the independent variables (in this case, X_1):

$$Y_i = \beta_0 + \beta_1(1/X_{1i}) + \beta_2 X_{2i} + \epsilon_i \tag{7.15}$$

The inverse (or reciprocal) functional form should be used when the impact of a particular independent variable is expected to approach zero as that independent variable increases and eventually approaches infinity. To see this, note that as X_1 gets larger, its impact on Y decreases.

In Equation 7.15, X_1 cannot equal zero, since if X_1 equaled zero, dividing it into anything would result in infinite or undefined values. The slopes are:

$$\frac{\Delta Y}{\Delta X_1} = \frac{-\beta_1}{X_1^2} \quad \text{and} \quad \frac{\Delta Y}{\Delta X_2} = \beta_2 \tag{7.16}$$

The slopes for X_1 fall into two categories, both of which are depicted in Figure 7.4:

1. When β_1 is positive, the slope with respect to X_1 is negative and decreases in absolute value as X_1 increases. As a result, the relationship between Y and X_1 holding X_2 constant approaches $\beta_0 + \beta_2X_2$ as X_1 increases (ignoring the error term).
2. When β_1 is negative, the relationship intersects the X_1 axis at $-\beta_1/(\beta_0 + \beta_2X_2)$ and slopes upward toward the same horizontal line (called an asymptote) that it approaches when β_1 is positive.

Applications of reciprocals or inverses exist in a number of areas in economic theory and the real world. For example, one way to think of the once-popular Phillips curve, a nonlinear relationship between the rate of unemployment and the percentage change in wages, is to posit that the percentage change in wages (W) is negatively related to the rate of unemployment (U), but that past some level of unemployment, further increases in the unemployment rate do reduce the level of wage increases any further because of institutional or other reasons. Such a hypothesis could be tested with an inverse functional form:

$$W_t = \beta_0 + \beta_1(1/U_t) + \epsilon_t \qquad (7.17)$$

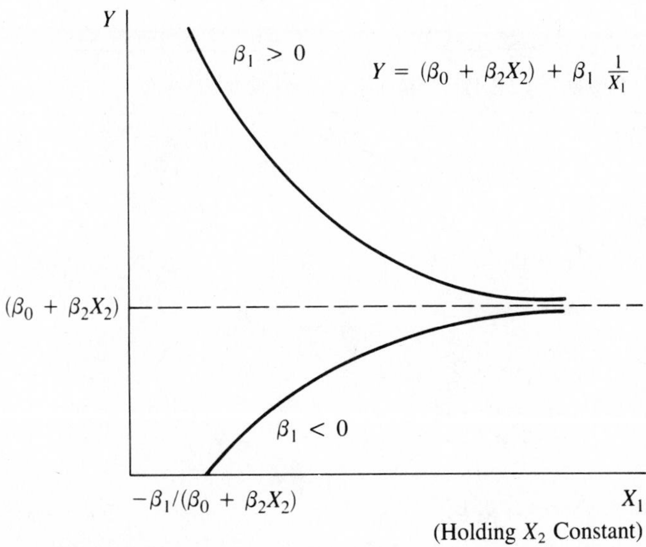

Figure 7.4 INVERSE FUNCTIONS
Inverse (or reciprocal) functional forms allow the impact of an X_1 on Y to approach zero as X_1 increases in size. The inverse functions approach the same value (the asymptote) from the top or bottom depending on the sign of β_1.

Estimating this equation using OLS gives the following:

$$\hat{W}_t = 0.00679 + 0.1842(1/U_t) \qquad R^2 = .397 \qquad (7.18)$$
$$(0.0590)$$
$$t = 3.20$$

This indicates that W and U are related in a way similar to that hypothesized (as shown in Figure 7.4 when β_1 is positive), but it doesn't provide any evidence that the inverse functional form is the best way to depict this particular theory. For more on this example, see Exercise 5.

7.2 Problems with Incorrect Functional Forms

The best way to choose a functional form for a regression model is to choose a specification that matches the underlying theory of the equation. In a majority of cases, the linear form will be adequate, and for most of the rest, common sense will point out a fairly easy choice from among the alternatives outlined above. Table 7.1 contains a summary of the properties of the various alternative functional forms.

TABLE 7.1 SUMMARY OF ALTERNATIVE FUNCTIONAL FORMS

Functional Form	Equation (one X only)	Slope $= \left(\frac{\Delta Y}{\Delta X}\right)$	Elasticity $= \left(\frac{\Delta Y}{\Delta X} \cdot \frac{X}{Y}\right)$
Linear	$Y_i = \beta_0 + \beta_1 X_i + \epsilon_i$	β_1	$\beta_1\left(\frac{X_i}{Y_i}\right)$
Double-Log	$\ln Y_i = \beta_0 + \beta_1 \ln X_i + \epsilon_i$	$\beta_1\left(\frac{Y_i}{X_i}\right)$	β_1
Semi-Log (lnX)	$Y_i = \beta_0 + \beta_1 \ln X_i + \epsilon_i$	$\beta_1\left(\frac{1}{X_i}\right)$	$\beta_1\left(\frac{1}{Y_i}\right)$
Semi-Log (lnY)	$\ln Y_i = \beta_0 + \beta_1 X_i + \epsilon_i$	$\beta_1 Y_i$	$\beta_1 X_i$
Polynomial	$Y_i = \beta_0 + \beta_1 X_i + \beta_2 X_i^2 + \epsilon_i$	$\beta_1 + 2\beta_2 X_i$	$\beta_1\left(\frac{X_i}{Y_i}\right) + 2\beta_2\left(\frac{X_i^2}{Y_i}\right)$
Inverse	$Y_i = \beta_0 + \beta_1\left(\frac{1}{X_i}\right) + \epsilon_i$	$-\beta_1\left(\frac{1}{X_i^2}\right)$	$-\beta_1\left(\frac{1}{X_i Y_i}\right)$

Note: Slopes and elasticities that include X_i or Y_i are not constant; they vary from point to point depending on the value of X_i or Y_i. If general slopes or elastiticies are desired, $\bar{X}$ and $\bar{Y}$ can be substituted into the equations.

Once in a while, however, a circumstance will arise in which the model is logically nonlinear in the variables, but the exact form of this nonlinearity is hard to specify. In such a case, the linear form is not correct, and yet a choice between the various nonlinear forms cannot be made on the basis of economic theory. Even in these cases, however, it still pays (in terms of understanding the true relationships) to avoid choosing a functional form on the basis of fit alone.

For example, recall the estimated Phillips curve in Equation 7.18. While the negative relationship between unemployment and inflation (using the percentage increase in wages as a proxy) implied by the Phillips curve suggests a downward-sloping nonlinear curve, there are a number of other functional forms that could produce such a curve. In addition to the inverse relationship that was actually used, the double-log form and various semi-log and exponential forms could also give shapes that would fit the hypothesis fairly well. If all the functional forms are so similar and if theory does not specify exactly which form to use, why should we try to avoid using goodness of fit over the sample to determine which equation to use? This section will highlight two answers to this question:

1. R^2s are difficult to compare if the dependent variable is transformed.
2. An incorrect functional form may provide a reasonable fit within the sample but have the potential to make large forecasting errors when used outside the range of the sample.

7.2.1 R^2s are Difficult to Compare When Y is Transformed

When the dependent variable is transformed from its linear version, the overall measure of fit, the R^2, cannot be used for comparing the fit of the nonlinear equation with the original linear one. This problem is not especially important in most cases because the emphasis in applied regression analysis is usually on the coefficient estimates. However, if R^2s (or $\bar{R}^2$s) are ever used to compare the fit of two different functional forms, then it becomes crucial that this lack of comparability be remembered. For example, suppose you were trying to compare a linear equation

$$Y = \beta_0 + \beta_1 X_1 + \beta_2 X_2 + \epsilon \tag{7.19}$$

with a semi-log version of the same equation (using the version of a semi-log function that takes the log of the dependent variable):

$$\ln Y = \beta_0 + \beta_1 X_1 + \beta_2 X_2 + \epsilon \qquad (7.20)$$

Notice that the only difference between Equations 7.19 and 7.20 is the functional form of the dependent variable. The reason that the R^2s of the respective equations cannot be used to compare overall fits of the two equations is that the total sum of squares (TSS) of the dependent variable around its mean is different in the two formulations. That is, the R^2s are not comparable because the dependent variables are different. There is no reason to expect that different dependent variables will have the identical (or easily comparable) degrees of dispersion around their means. Since the TSS are different, the R^2s (or $\bar{R}^2$s) will not be comparable.

The way to get around this problem is to create a "quasi-R^2" by transforming the predicted values of the nonlinear dependent variable into a form that is directly comparable to the original dependent variable. This transformed dependent variable is then used to calculate the quasi-R^2. In essence, then, a **quasi-R^2** is an R^2 that allows the comparison of the overall fits of equations with different functional forms by transforming the predicted values of one of the dependent variables into the functional form of the other dependent variable.

For the example of the previous paragraph, this would mean taking the following steps:

1. Estimate Equation 7.20 and create a set of $\widehat{\ln Y}$s for the sample.
2. Transform the $\widehat{\ln Y}$s by taking their anti-logs (an anti-log reverses the log function: anti-log[lnY] = Y).
3. Calculate quasi-R^2 (or quasi-$\bar{R}^2$) by using the newly calculated anti-logs as $\hat{Y}$s to get the residuals needed in the R^2 equation:

$$\text{quasi-}R^2 = 1 - \frac{\Sigma[Y_i - \text{anti-log}(\widehat{\ln Y_i})^2]}{\Sigma[Y_i - \bar{Y}]^2} \qquad (7.21)$$

This quasi-R^2 for Equation 7.20 is directly comparable to the conventional R^2 for Equation 7.19. Do not merely apply Equation 7.21 automatically, however. Each different functional form (of the dependent variable) requires a different transformation to calculate the appropriate quasi-R^2. Whenever the dependent variable is logged, though, Equation 7.21 should be used.[5]

Let's try an example of the comparison of the overall fit of two different functional forms for the Woody's restaurant model originally estimated on data presented in Chapter 3 and analyzed more fully in Equation 5.5:

$$\begin{array}{llll} \hat{Y}_i = 102{,}192 & -\ 9075C_i & +\ 0.3547P_i & +\ 1.288I_i \qquad (5.5) \\ & \quad(2053) & \quad(0.0727) & \quad(0.543) \\ \quad t = & \ -4.42 & \quad 4.88 & \quad 2.37 \\ \quad n = 33 & \bar{R}^2 = .579 & R^2 = .618 & \end{array}$$

If, for instance, we estimated a double-log version of this equation (perhaps because of an idea that the impacts of C, P, and I were nonlinear, with constant elasticities) then:

$$\begin{array}{llll} \widehat{\ln Y_i} = 6.66 & -\ 0.378\ln C_i & +\ 0.352\ln P_i & +\ 0.159\ln I_i \qquad (7.22) \\ & \quad(0.065) & \quad(0.056) & \quad(0.085) \\ \quad t = & \ -5.82 & \quad 6.29 & \quad 1.88 \\ \quad n = 33 & \bar{R}^2 = .674 & R^2 = .705 & \end{array}$$

Note that the R^2 in Equation 7.22 is quite a bit higher than that in 5.5. This result does not mean that the double-log equation fits substantially better; all a higher R^2 means is that the double-log equation explains a higher portion of the variation of the movement of the *log* of Y around its mean than the portion of the variation of the movement of Y around its mean that is explained by the linear equation.

To compare the fits, we need to calculate a quasi-R^2 for Equation 7.22. We take the 33 $\widehat{\ln Y}$s obtained by plugging the values for C, P,

5. One other problem with the quasi-R^2 should be mentioned. It is possible to find examples where the use of the quasi-R^2 to compare the overall fits of equations with different functional forms can give different answers depending on which functional form is considered the "original" functional form and which is considered the "transformed" functional form. In terms of Equation 7.21, this would mean that the R^2 from the linear equation could be greater than the quasi-R^2 obtained by using the logged equation and taking anti-logs, but that the R^2 from the log equation could be greater than the quasi-R^2 obtained by using the linear equation and converting to logs. Happily, such a circumstance is not frequent. In such a case, we advise choosing whichever functional form is easier to work with (i.e., linear, if it is one of the two involved functions). For more, see H. Theil, *Introduction to Econometrics* (Englewood Cliffs, N.J., Prentice-Hall, 1978), pp. 271–277.

and I into Equation 7.22 and then use those $\widehat{\ln Y}$s to calculate the quasi-R^2 as in Equation 7.21. The quasi-R^2 for Equation 7.22 is .688, implying that the double-log equation actually provides a better overall fit than the linear one. In deciding which equation to use, of course, this overall fit would not be the most important factor. For more practice with this procedure, see Exercise 6. Another method of choosing the functional form of the dependent variable is the *Box-Cox test,* described in Exercise 14.

7.2.2 Incorrect Functional Forms Outside the Range of the Sample

If an incorrect functional form is used, then the probability of mistaken inferences about the true population parameters will increase. Using an incorrect functional form is a kind of specification error that is similar to the omitted variable bias discussed in Section 6.1. Although the characteristics of any specification errors depend on the exact details of the particular situation, there is no reason to expect that coefficient estimates obtained from an incorrect functional form will necessarily be unbiased and minimum variance. Even if an incorrect functional form provides good statistics within a sample, though, large residuals almost surely will arise when the misspecified equation is used on data that were not part of the sample used to estimate the coefficients.

In general, the extrapolation of a regression equation to data that are outside the range over which the equation was estimated runs increased risks of large forecasting errors and incorrect conclusions about population values. This risk is heightened if the regression uses a functional form that is inappropriate for the particular variables being studied; nonlinear functional forms should be used with extreme caution for data outside the range of the sample because nonlinear functional forms by definition change their slopes. It is entirely possible that the slope of a particular nonlinear function could change to an unrealistic value outside the range of the sample even if the form produced reasonable slopes within the sample. Of course, even a linear function could be inappropriate in this way. If the true relationship changed slope outside the sample range, the linear functional form's constant slope would be quite likely to lead to large forecasting errors outside the sample range.

As a result, two functional forms that behave similarly over the range of the sample may behave quite differently outside that range. If the functional form is chosen on the basis of theory, then the researcher can take into account how the equation would act over any range of values, even if some of those values are outside the range of the sample. If functional forms are chosen on the basis of fit, then extrapolating outside the sample becomes tenuous.

Figure 7.5 contains a number of hypothetical examples. As can be seen, some functional forms have the potential to fit quite poorly

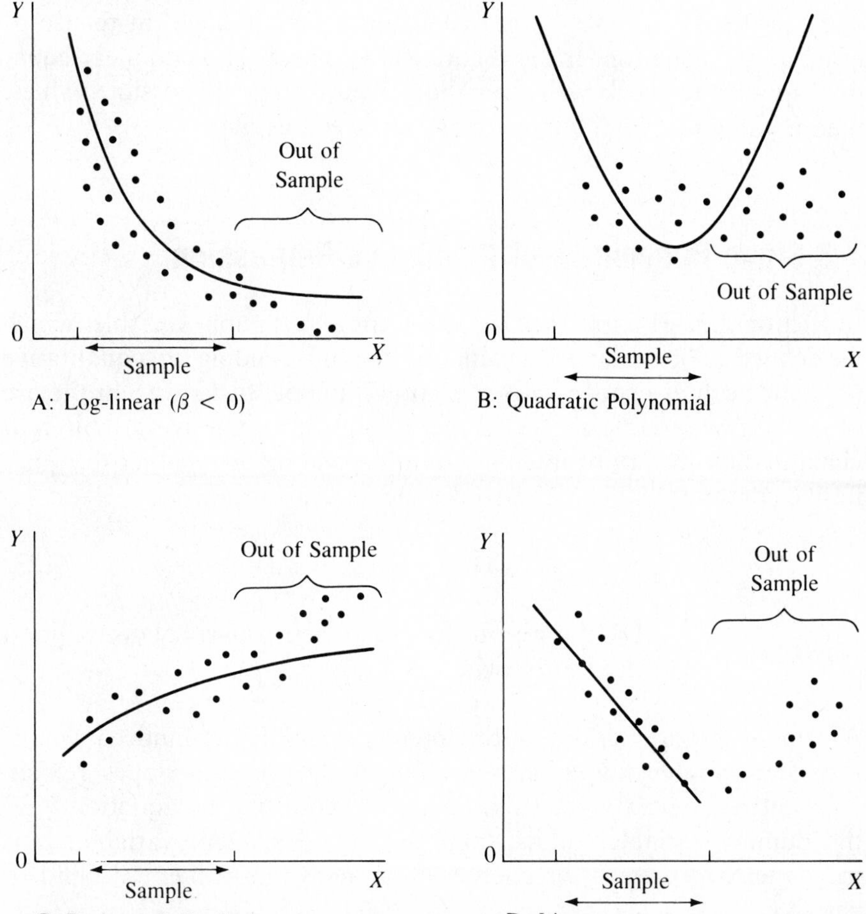

Figure 7.5 INCORRECT FUNCTIONAL FORMS OUTSIDE THE SAMPLE
If an incorrect functional form is applied to data outside the range of the sample on which it was estimated, the probability of large mistakes increases. In particular, note how the polynomial functional form can change slope rapidly outside the sample range (Panel B) and that even a linear form can cause large mistakes if the true functional form is nonlinear (Panel D).

outside the sample range. Others seem less likely to encounter this problem. Such graphs are meant as examples of what could happen, not as statements of what necessarily will happen, when incorrect functional forms are pushed outside the range of the sample over which they were estimated. Do not conclude from these diagrams that nonlinear functions should be avoided completely. If the true relationship is nonlinear, then the linear functional form also will make large forecasting errors outside the sample. Instead, the researcher must take the time to think through how the equation will act for values both inside and outside the sample before choosing a functional form to use to estimate the equation. If the theoretically appropriate nonlinear equation appears to work well over the relevant range of possible values, then it should be used without concern over this issue.

7.3 Slope Dummies and Piecewise Regression

In Section 3.2, we introduced the concept of a dummy variable, which we defined as one taking the values of 0 or 1 depending on a qualitative attribute such as gender. In that section, our sole focus was on the use of a dummy variable as an **intercept dummy,** a dummy variable that changes the constant or intercept term depending on whether the qualitative condition is met. These take the general form:

$$Y_i = \beta_0 + \beta_1 X_{1i} + \beta_2 X_{2i} + \beta_3 D_i + \epsilon_i \qquad (7.23)$$

where: $D_i = \begin{cases} 1 \text{ if the } i\text{th observation meets a particular condition} \\ \text{and 0 otherwise} \end{cases}$

As can be seen in Figure 7.6, the intercept dummy does indeed change the intercept depending on the value of D, but the slopes remain constant no matter what value D takes. Note that in Equation 7.23 the dummy variable stands alone as an independent variable; it is multiplied by its regression coefficient but not by any other independent variable.

The purpose of this section is to introduce two extremely useful alternate functional forms that involve dummy variables that are not intercept dummies:

1. **Slope dummies:** dummy variables that allow slopes to be different depending on whether the condition specified by the dummy variable is met.

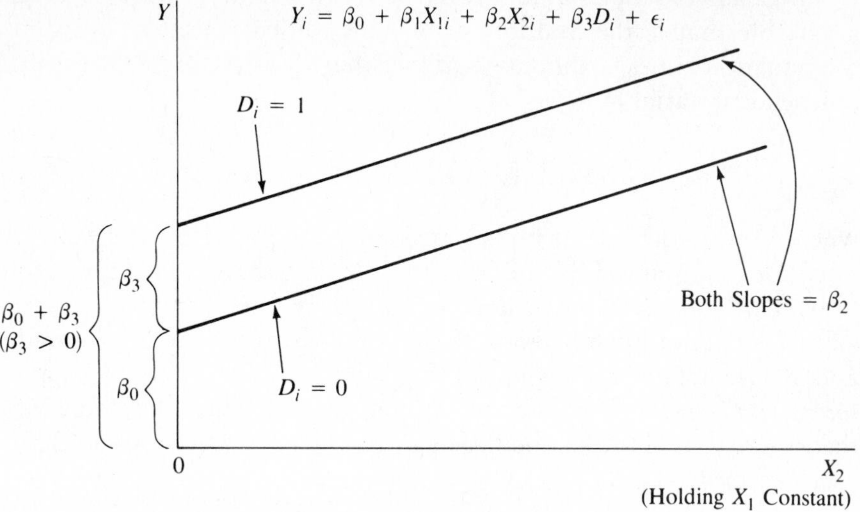

Figure 7.6 AN INTERCEPT DUMMY
If an intercept dummy ($\beta_3 D_i$) is added to an equation, a graph of the equation will have different intercepts for the two qualitative conditions specified by the dummy variable. The difference between the two intercepts is β_3. The slopes are constant with respect to the qualitative condition.

2. **Piecewise regression:** the use of dummy variables to allow slopes to be different depending on whether a particular nondummy independent variable exceeds a specified value.

Another possibility, the use of dummy variables as dependent variables, is presented in Chapter 13.

7.3.1 Slope Dummies

Slope dummies can be employed whenever the slope that measures an independent variable's impact on the dependent variable is hypothesized to change if some qualitative condition is met. For example, in the weight-height model of Chapter 1, if being a football player changed the expected increase in weight caused by a one-inch increase in height, then one way to model that relationship would be with a slope dummy variable. Similarly, if being at war decreased a country's marginal propensity to consume, this change in the coefficient of income in the consumption function could be modeled with a slope dummy that was equal to one during war years and equal to zero otherwise (or vice versa).

In general a slope dummy is introduced by adding to the equation a variable that is the multiple of an independent variable already in the equation times a dummy variable that does not depend on that independent variable:

$$Y_i = \beta_0 + \beta_1 X_{1i} + \beta_2 X_{2i} + \beta_3 D_i + \beta_4 X_{2i} D_i + \epsilon_i \tag{7.24}$$

where D is as defined in Equation 7.23. Note the difference between Equations 7.23 and 7.24. In Equation 7.24, we have added a variable in which the dummy variable is multiplied by one of the independent variables but not by the others. This multiple between the slope dummy and X_2 is a specific example of a general group of variables called *interaction terms*.[6] Such a model is appropriate when the underlying theory suggests that the coefficient of X_2 *changes* when the condition specified by D is met:

$$\text{When } D = 0, \frac{\Delta Y}{\Delta X_2} = \beta_2$$

$$\text{When } D = 1, \frac{\Delta Y}{\Delta X_2} = \beta_2 + \beta_4$$

$$\text{In both cases, } \frac{\Delta Y}{\Delta X_1} = \beta_1$$

To see this, substitute D = 1 and D = 0, respectively, into Equation 7.24. Note that X_1's relationship with Y is not hypothesized to depend on D.

Such a dummy term ($X_{2i}D_i$) is called a slope dummy because, as can be seen in Figure 7.7, the slope of X_2 changes depending on the value of D. Note that when a slope dummy is used, the researcher almost always will hypothesize that both a slope dummy and an intercept dummy are appropriate. Such a specification should be used in all but highly unusual and forced conditions. Using both dummies allows both the intercept and the slope with respect to X_2 to change with D.

6. Interaction terms involve multiples of independent variables as in Equation 7.24. They are used when the change in Y with respect to one independent variable (in this case, X_2) depends on the level of the other independent variable (in this case, D). For an example of interaction terms involving wages as a function of unionism and concentration, see F. M. Scherer, *Industrial Market Structure and Economic Performance* (Chicago: Rand McNally, 1977), pp. 298–302.

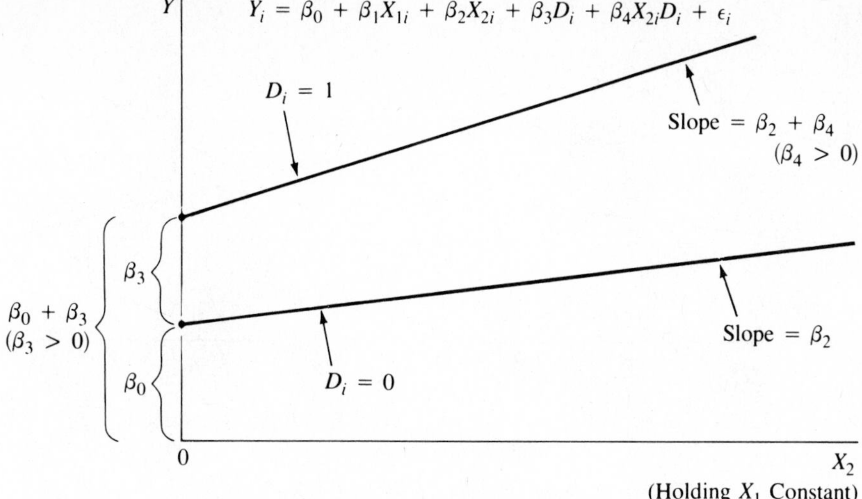

Figure 7.7 SLOPE AND INTERCEPT DUMMIES
If slope dummy ($\beta_4 X_{2i} D_i$) and intercept dummy ($\beta_3 D_i$) terms are added to an equation, a graph of the equation will have different intercepts and different slopes depending on the value of the qualitative condition specified by the dummy variable. The difference between the two intercepts is β_3, while the difference between the two slopes is β_4.

Using just the slope dummy constrains its coefficient because it has to have the same intercept as the rest of the equation.

In Figure 7.7, with both a slope dummy and an intercept dummy, the intercept will be β_0 when D = 0 and $\beta_0 + \beta_3$ when D = 1. In addition, the slope of Y with respect to X_2 (holding X_1 constant) will be β_2 when D = 0 and $\beta_2 + \beta_4$ when D = 1. As a result, there really are two equations that have only the $\beta_1 X_{1i}$ term in common:

$$Y_i = \beta_0 + \beta_1 X_{1i} + \beta_2 X_{2i} + \epsilon_i \quad \text{[when D = 0]}$$

$$Y_i = (\beta_0 + \beta_3) + \beta_1 X_{1i} + (\beta_2 + \beta_4) X_{2i} + \epsilon_i \quad \text{[when D = 1]}$$

As can be seen in Figure 7.8, an equation with both a slope and an intercept dummy can take on a number of different shapes depending on the signs and absolute values of the coefficients. As a result, slope dummies can be used to model a wide variety of relationships, but it's necessary to be fairly specific when hypothesizing values of the coefficients of the various dummy terms.

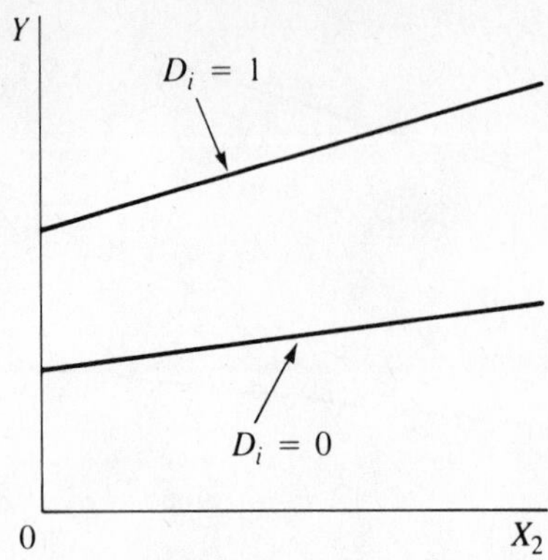

Panel A. All βs > 0

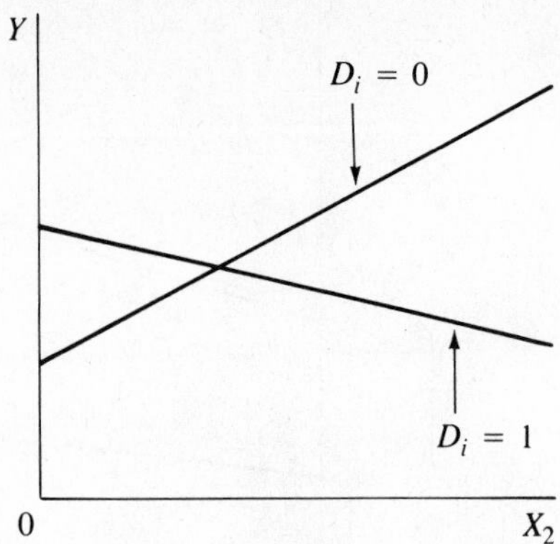

Panel B. All βs >0 except β_4, $|\beta_4| > |\beta_2|$

$$Y_i = \beta_0 + \beta_1 X_{1i} + \beta_2 X_{2i} + \beta_3 D_i + \beta_4 X_{2i} D_i + \epsilon_i$$

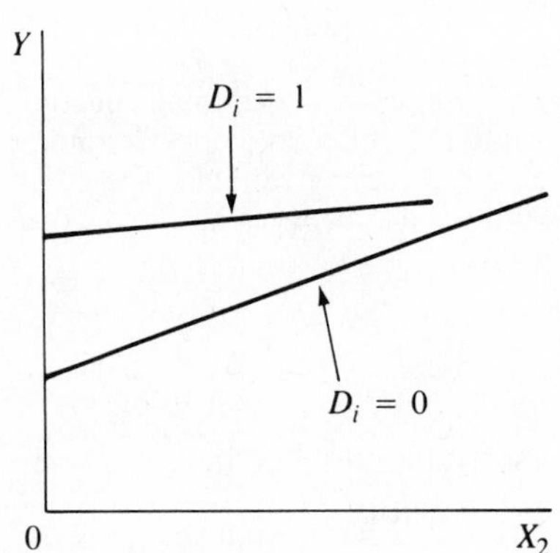

Panel C. $\beta_4 < 0$, $|\beta_4| < |\beta_2|$, $\beta_3 > 0$

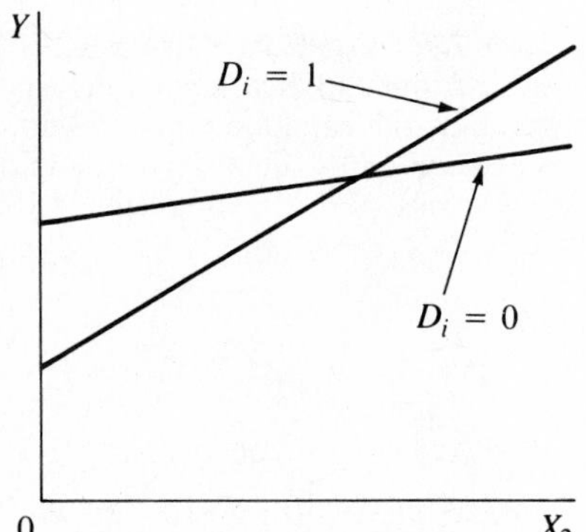

Panel D. All βs > 0 except β_3, $|\beta_3| < |\beta_0|$

Figure 7.8 VARIOUS SHAPES OF EQUATIONS WITH SLOPE AND INTERCEPT DUMMIES
Depending on the values of the coefficients of the slope (β_4) and intercept (β_3) dummies, equations with both slope and intercept dummies in them can take on a number of different shapes (holding X_1 constant). When using such equations, it is therefore necessary to be fairly specific when hypothesizing values of the coefficients of the various dummy terms.

For example, consider the question of earnings differentials between men and women. While there is little argument that these differentials exist, there is quite a bit of controversy over the extent to which these differentials are caused by sexual discrimination (as opposed to other factors). Suppose you decide to build a model of earnings to get a better view of this controversy. If you hypothesized that men earn more than women on average, then you would want to use an intercept dummy variable for gender in an earnings equation that included measures of experience, special skills, education, etc. as independent variables:

$$\text{Earnings}_i = \beta_0 + \beta_1 D_i + \beta_2 \text{EXP}_i + \cdots + \epsilon_i \qquad (7.25)$$

where: D_i = 1 if the *i*th worker is female and 0 otherwise
EXP_i = the years experience of the *i*th worker
ϵ_i = a classical error term

In Equation 7.25, $\hat{\beta}_1$ would be an estimate of the average difference between males and females holding constant their experience and the other factors in the equation. Equation 7.25 also forces the impact of increases in experience (and the other factors in the equation) to have the same effect for females as for males because the slopes are the same for both genders.

If you hypothesized that men also increase their earnings more per year of experience than women, then you would include a slope dummy as well as an intercept dummy in such a model:

$$\text{Earnings}_i = \beta_0 + \beta_1 D_i + \beta_2 \text{EXP}_i + \beta_3 D_i \text{EXP}_i + \cdots + \epsilon_i \qquad (7.26)$$

In Equation 7.26, $\hat{\beta}_3$ would be an estimate of the differential impact of an extra year of experience on earnings between men and women. We could test the possibility of a negative true β_3 by running a one-tailed *t*-test on $\hat{\beta}_3$. If $\hat{\beta}_3$ were significantly different from zero in a negative direction, then we could reject the null hypothesis of no difference due to gender in the impact of experience on earnings holding constant the other variables in the equation.[7]

7. Another approach to this problem is to use the Chow test suggested at the very end of Section 5.8 (the appendix on the *F*-test). To apply the Chow test to the question of earnings differentials between genders, use Equation 7.26 as the unconstrained equation (with all independent variables also having slope dummy formulations) and

$$\text{Earnings}_i = \beta_0 + \beta_1 \text{EXP}_i + \cdots + \epsilon_i \qquad (7.26a)$$

as the constrained equation. If the *F*-test shows that the fit of Equation 7.26 is significantly better than the fit of Equation 7.26a, then we would reject the null hypothesis of equivalence between the male and female slope coefficients in the earnings equation.

7.3.2 Piecewise Regression

A second use of dummy variables that goes beyond the concept of an intercept dummy to create a new functional form is piecewise regression. **Piecewise regression** is the use of dummy variables to allow slopes to be different depending on whether a particular nondummy independent variable exceeds a specified threshold value. This formulation allows the impact of X on Y to change as X itself changes. These are sometimes called "jack-knifed" equations because this use of dummy variables allows "bent" regression lines that can sometimes take on the shape of a partially opened jack-knife, as in Figure 7.9. Thus, such a piecewise regression could be used to model a kinked demand curve.[8]

If X is larger than the threshold value, then one slope is expected, but if X is less than the threshold value, then a different slope is expected. The definition of the dummy variable to be used in a piecewise equation is thus fairly unusual:

$$D_i = \begin{cases} 1 \text{ if } X_{2i} > X_2^* \\ 0 \text{ if } X_{2i} \leq X_2^* \end{cases}$$

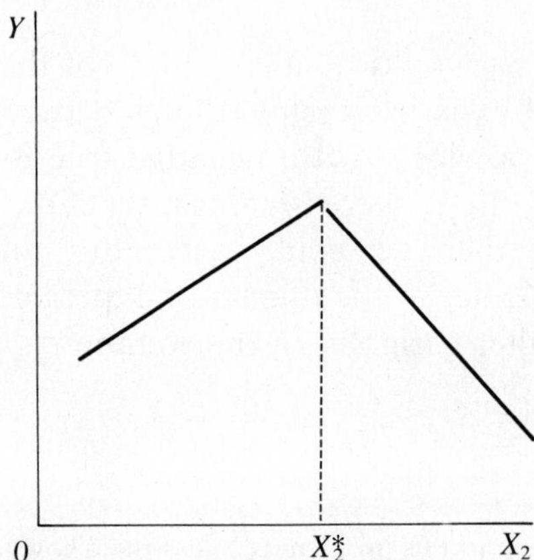

Figure 7.9 A PIECEWISE REGRESSION EQUATION
The use of the piecewise regression functional form allows the equation to actually change slope for different values of the quantitative independent variable (holding X_1 constant).

8. Or kinked budget constraints. See Robert Moffitt, "The Econometrics of Kinked Budget Constraints," *Journal of Economic Perspectives*, Spring 1990, pp. 119–139.

where X_2 is the independent variable in the equation whose slope will change and X_2^* is the threshold value at which the slope change takes place. Researchers should resist the urge to estimate a large number of piecewise equations in order to choose X^* on the basis of fit, because a good fit in one sample does not assure a good fit in other samples, unless sample-specific inference is their goal. In other words, theory or prior expectations should be the basis for the choice of X^*.

Except for the change in the definition of the dummy variable, the equation to be used in piecewise regression is identical to the slope dummy equation, Equation 7.24:

$$Y_i = \beta_0 + \beta_1 X_{1i} + \beta_2 X_{2i} + \beta_3 D_i + \beta_4 X_{2i} D_i + \epsilon_i \qquad (7.24)$$

It is vital, as with slope dummy equations, that piecewise equations include both the intercept dummy term and the slope dummy term.

If the particulars of the equation are carefully investigated, a number of fairly detailed hypotheses about the coefficients of piecewise equations can be developed. For example, if the slope below X_2^* is expected to be positive but the slope above X_2^* is expected to be negative, then it is not enough to hypothesize that $\beta_4 < 0$, because it must be negative and greater in absolute value than $\hat{\beta}_2$ for their sum (the new slope) to also be negative.

7.4 The Use and Interpretation of the Constant Term

In the linear regression model, β_0 is the intercept or constant term. It is the expected value of Y when all the explanatory variables (and the error term) equal zero. At times, β_0 is of theoretical importance. Consider, for example, the following cost equation:

$$C_i = \beta_0 + \beta_1 Q_i + \epsilon_i$$

where C_i is the total operator cost of a production process that is producing the level of output Q_i. The term $\beta_1 Q_i$ represents the total variable cost associated with output level Q_i, and β_0 represents the total fixed cost, defined as the cost when output $Q_i = 0$. Thus, a regression equation might seem useful to a researcher who wanted to determine the relative magnitudes of fixed and variable costs. This would be an example of relying on the constant term for inference.

On the other hand, the product involved might be one for which it is known that there are few if any fixed costs. In such a case, a researcher might want to set the constant term to zero; to do so would conform to the notion of zero fixed costs and would conserve a degree of freedom (which would presumably make the estimate of β_1 more precise). This would be an example of suppressing the constant term.

Neither suppressing the constant term nor relying on it for inference is advisable, however, and the reasons for these conclusions are explained in the following sections.

7.4.1 Do Not Suppress the Constant Term

Chapter 4 explained that one of the rationales behind the assumption of the normality of the error term is that the error term absorbs the effects of a number of variables, each of which is not important enough to be included as an independent variable in the equation. Chapter 4 stressed that Assumption II (the error term has an expected value of zero) requires that the constant term absorb the mean effect of all these variables. Thus, suppressing the constant term can lead to a violation of this classical assumption. The only time that this assumption would not be violated by leaving out the intercept term is when the mean effect of the unobserved error term (without a constant term) is zero over all the observations.

The consequence of suppressing the constant term is that the slope coefficient estimates are potentially biased and their t-scores are inflated. This is demonstrated in Figure 7.10. Given the pattern of the X and Y observations, estimating a regression equation with a constant term would likely produce an estimated regression line very similar to the true regression line, which has a constant term (β_0) quite different from zero. The slope of this estimated line is very low, and the t-score of the estimated slope coefficient may be very close to zero, implying that the slope coefficient is statistically insignificant, that is, it does not differ much from zero.

These results should be accepted by the researcher because in this case the true relationship has this appearance. However, if the researcher were to suppress the constant term, which implies that the estimated regression line must pass through the origin, then the estimated regression line shown in Figure 7.10 would result. The slope coefficient is now large. That is, it is biased upward compared with the true slope coefficient. Thus, the t-score is biased upward, and it may very well be large enough to indicate that the estimated slope

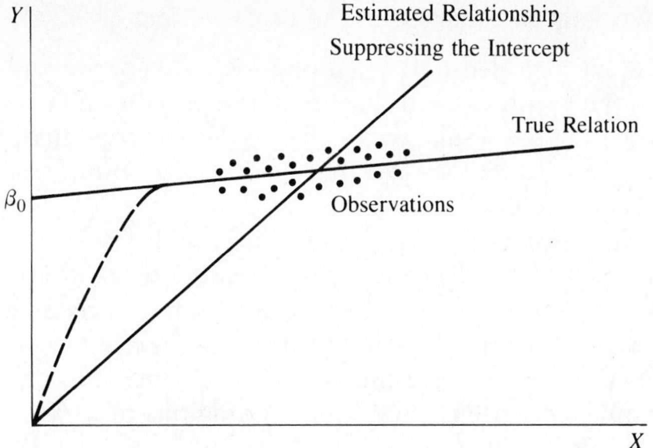

Figure 7.10 THE HARMFUL EFFECT OF SUPPRESSING THE CONSTANT TERM
If the constant or intercept term is suppressed, the estimated regression will go through the origin. Such an effect potentially biases the $\hat{\beta}$s and inflates their t-scores. In this particular example, the true β is close to zero in the range of the sample, but forcing the regression through the origin makes $\hat{\beta}$ appear to be significantly positive. In other examples, though, the difference might not be so pronounced.

coefficient is statistically significantly positive. Such a conclusion would be incorrect.

On the other hand, it's possible that the true relationship is nonlinear and passes through the origin. In Figure 7.10, such a relationship is shown by the appended dashed line that deviates from the straight portion of the true line and passes through the origin. If this nonlinear relationship were to be approximated by a linear regression line, it would be important not to suppress the constant term. Over the relevant range of the observations (that is, the sample range), the estimated regression line with the constant suppressed doesn't provide an adequate approximation of the true regression line, compared with an estimated regression equation that includes the constant term. It is a legitimate exercise in applied econometrics to use linear approximations of nonlinear functional forms; suppressing the constant term doesn't permit an accurate approximation over the sample range of observations.

Thus, even though some regression packages allow the constant term to be suppressed (set to zero), the general rule is: *Don't* even if theory specifically calls for it.

7.4.2 Do Not Rely on Estimates of the Constant Term

It would seem logical that if it's a bad idea to suppress the constant term, then the constant term must be an important analytical tool to use in evaluating the results of the regression. Unfortunately, there are at least two reasons that suggest that the intercept should *not* be relied on for purposes of analysis or inference.

First, the error term is generated, in part, by the omission of a number of marginal independent variables, the mean effect of which is placed in the constant term. The constant term acts as a garbage collector, with an unknown amount of this mean effect being dumped into it. The constant term's estimated coefficient may be different from what it would have been without performing this task, which is done for the sake of the equation as a whole.

Second, while the constant term is the value of the dependent variable when all the independent variables and the error term are zero, the values of variables used for economic analysis are usually positive. Thus, the origin often lies *outside* the range of sample observations (as can be seen in Figure 7.10). Since the constant term is an estimate of Y when the Xs are outside the range of the sample observations, estimates of it are tenuous. Estimating the constant term is like forecasting beyond the range of the sample data, a procedure that inherently contains greater error than within-sample forecasts. For more on this, see Figure 7.5 and Chapter 15.

7.5 Summary

1. The choice of a functional form should be based on the underlying economic theory to the extent to which theory suggests a shape similar to that provided by a particular functional form. A form that is linear in the variables should be used unless a specific hypothesis suggests otherwise.

2. Functional forms that are nonlinear in the variables include the double-log form, the semi-log form, the polynomial form, and the inverse form. The double-log form is especially useful if the elasticities involved are expected to be constant. The semi-log and inverse forms have the advantage of allowing the effect of an independent variable to tail off as that variable increases. The polynomial form is useful if the slopes are expected to depend on the level of an independent variable, but polynomials of degree higher than two should be avoided unless the underlying theory specifically calls for them.

3. The use of nonlinear functional forms has a number of potential problems. In particular, the R^2s are difficult to compare if Y has been transformed, and the residuals are potentially large if an incorrect functional form is used for forecasting outside the range of the sample.

4. A slope dummy is a dummy variable that is multiplied by an independent variable to allow the slope of the relationship between the dependent variable and the particular independent variable to change, depending on whether or not a particular condition is met. A similar technique, piecewise regression, bases the slope dummy's condition on the value of the independent variable in question and allows the slope to "kink" if X exceeds a particular value.

5. Do not suppress the constant term even if it appears to be theoretically likely to equal zero. On the other hand, don't rely on estimates of the constant term for inference even if it appears to be statistically significant.

Exercises

(Answers to even-numbered exercises in Appendix A.)

1. Write out the meaning of each of the following terms without reference to the book (or your notes), and compare your definition with the version in the text for each:
   a. elasticity
   b. double-log or exponential functional form
   c. semi-log functional form
   d. polynomial functional form
   e. reciprocal or inverse functional form
   f. slope dummy
   g. piecewise regression
   h. quasi-R^2

2. For each of the following pairs of dependent (Y) and independent (X) variables, pick the functional form that you think is likely to be appropriate, and then explain your reasoning (assume that all other relevant independent variables are included in the equation):
   a. Y = sales of shoes; X = disposable income
   b. Y = the attendance at the Hollywood Bowl outdoor symphony concerts on a given night;
      X = whether the orchestra's most famous conductor was scheduled to conduct that night

c. Y = aggregate consumption of goods and services in the U.S.;
   X = aggregate disposable income in the U.S.
d. Y = the money supply in the U.S.;
   X = the interest rate on Treasury Bills (in a demand function)
e. Y = average cost of production of a box of pasta;
   X = number of boxes of pasta produced
f. How would your answer to part e change if you knew there was a significant outlier due to a capacity constraint and a rush order one year?

3. Can either (or both) of the following equations be estimated with OLS? Why?
   a. $Y_i = \beta_0 X_{1i}^{\beta_1} X_{2i}^{\beta_2} + e^{u_i}$
   b. $Y_i = e^{\beta_0} X_{1i}^{\beta_1} X_{2i}^{\beta_2} + u_i$

   where u_i is a typical classical error term and e is the base of the natural logarithm

4. Consider the following estimated semi-log equation:

$$\widehat{\ln SAL_i} = 8.10 + \underset{(0.025)}{0.100} ED_i + \underset{(0.050)}{0.110} EXP_i$$
$$\bar{R}^2 = .48 \qquad n = 28$$

where: $\ln SAL_i$ = the log of the salary of the *i*th worker
ED_i = the years of education of the *i*th worker
EXP_i = the years of experience of the *i*th worker

   a. Make appropriate hypotheses for signs, calculate t-scores (standard errors are in parentheses), and test your hypotheses.
   b. What is the economic meaning of the constant in this equation?
   c. Why do you think this particular semi-log functional form is used in this model? More specifically, what are the elasticities of salary with respect to education and experience? What happens to the slopes of the independent variables as they increase?
   d. Suppose you ran the linear version of this equation and obtained an $\bar{R}^2$ of .46. What can you conclude from this result?

5. The Phillips curve discussed in Section 7.1.5 is a good example of the use of econometrics to develop, refine, and test theory. The curve was originally "discovered" in an empirical study and once

was firmly believed to be true. Today, the Phillips curve is not as highly regarded as it used to be in part because of empirical results. Since data for estimating a Phillips curve are readily available, you can test the validity of the Phillips curve yourself.

a. Search the literature (starting with the *Journal of Economic Literature*) and follow the controversy surrounding the topic of the Phillips curve and its estimation.[9]
b. Review the possible functional forms summarized in Section 7.1. What else besides an inverse function could have been used to estimate the model? If possible, collect data and compare alternative functional forms for the Phillips curve.
c. From the middle 1970s to the early 1980s, a Phillips curve estimated from data on the U.S. economy might have shown a positive slope. What inference should you draw from such an unexpected sign? Why?

6. Given the following data on personal consumption and personal income of various students at "Tiger College" last year:

Student Name	Consumption per week	Income per week
Manuel	$180	$200
Cindy	100	120
Mike	60	80
Betty	50	80
Mary	40	80
Pete	30	50
Bill	20	40
Terri	10	30

a. Run linear and double-log regressions to explain weekly consumption as a function of weekly income.
b. Compare the R^2s from the two equations. Which fits better?
c. Calculate a "quasi-R^2" for the double-log equation and compare it to the R^2 for the linear form of the same equation.

7. In their 1957 study, Murti and Sastri[10] investigated the production characteristics of seven Indian industries, including cotton and

9. For example, see Nancy Wulwick, "Phillips' Approximate Regression," in Neil de Marchi and Christopher Gilbert, *History and Methodology of Econometrics* (Oxford: Clarendon Press, 1989), pp. 170–188. This book contains many other interesting articles on econometrics.

10. V. N. Murti and V. K. Sastri, "Production Functions for Indian Industry," *Econometrica*, April 1957, pp. 205–221.

sugar. They specified Cobb-Douglas production functions for output (Q) as a double-log function of labor (L) and capital (K)

$$\ln Q_i = \beta_0 + \beta_1 \ln L_i + \beta_2 \ln K_i + \epsilon_i$$

and obtained the following estimates (standard errors in parentheses):

Industry	$\hat{\beta}_0$	$\hat{\beta}_1$	$\hat{\beta}_2$	R^2
Cotton	0.97	0.92	0.12	.98
		(0.03)	(0.04)	
Sugar	2.70	0.59	0.33	.80
		(0.14)	(0.17)	

a. Hypothesize and test appropriate null hypotheses at the five percent level of significance. (Hint: This is harder than it looks!)
b. Graph the involved relationships. Does a double-log function seem theoretically justified in this case?
c. What are the elasticities of output with respect to labor and capital for each industry?
d. What economic significance does the sum $(\hat{\beta}_1 + \hat{\beta}_2)$ have?

8. Suppose you are studying the rate of growth of income in a country as a function of the rate of growth of capital in that country and of the per capita income of that country. You're using a cross-sectional data-set that includes both developed and developing countries. Suppose further that the underlying theory suggests that income growth rates will increase as per capita income increases and then start decreasing past a particular point. Describe how you would model this relationship with each of the following functional forms:
   a. Piecewise regression (assume that the threshold occurs at $2000)
   b. a quadratic function
   c. a semi-log function

9. A study of hotel investments in Waikiki between 1965 and 1973 estimated this revenue production function:

$$R = AL^{\alpha}K^{\beta}e^{\epsilon}$$

where: A = a constant term
R = annual net revenue of the hotel (in thousands of dollars)

L = land input (site area in square feet)
K = capital input (construction cost in thousands of dollars)
ϵ = the classical error term
e = the base of the natural log

a. What are your general expectations for the population values of α and β; what are their theoretical interpretations?
b. Create specific null and alternative hypotheses for this equation.
c. Create specific decision rules for the two sets of hypotheses above (5 percent level with 25 degrees of freedom).
d. Calculate the appropriate *t*-values and run *t*-tests given the following regression result (standard errors in parentheses):

$$\widehat{\ln R} = -0.91750 + \underset{(0.135)}{0.273 \ln L} + \underset{(0.125)}{0.733 \ln K}$$

Did you reject or "accept" your null hypotheses?
e. Is this result surprising? If you were going to build a Waikiki hotel, what input would you most want to use for investment? Is there an additional piece of information you would need to know before you could answer?

10. Comanor and Wilson[11] specified the following regression in their study of advertising's effect on the profit rates of 41 consumer goods firms:

$$PR_i = \beta_0 + \beta_1 ADV_i/SALES_i + \beta_2 \ln CAP_i + \beta \ln ES_i + \beta_4 \ln DG_i + \epsilon_i$$

where: PR_i = a variable measuring the profit rate of the *i*th firm
ADV_i = the advertising expenditures in the *i*th firm (in dollars)
$SALES_i$ = the total gross sales of the *i*th firm (in dollars)
CAP_i = a variable measuring the capital needed to enter the *i*th firm's market at an efficient size
ES_i = a variable measuring the degree to which economies of scale exist in the *i*th firm's industry

11. William S. Comanor and Thomas A. Wilson, "Advertising, Market Structure and Performance," *Review of Economics and Statistics*, Nov. 1967, p. 432.

DG_i = the percent of growth in sales (demand) of the *i*th firm over the last ten years

ln = natural logarithm

ϵ_i = a classical error term

a. Hypothesize expected signs for each of the slope coefficients.
b. Note that there are two different kinds of nonlinear (in the variables) relationships in this equation. For each independent variable, determine the shape that the chosen functional form implies, and state whether you agree or disagree with this shape. Explain your reasoning in each case.
c. Comanor and Wilson state that the simple correlation coefficient between $ADV_i/SALES_i$ and each of the other independent variables is positive. If one of these remaining variables were omitted, in which direction would $\hat{\beta}_1$ likely be biased?

11. Suggest the appropriate functional forms for the relationships between the following variables. Be sure to explain your reasoning:
    a. The age of the *i*th house in a cross-sectional equation for the sales price of houses in Cooperstown, New York. (Hint: Cooperstown is known as a lovely town with a number of elegant historic homes.)
    b. The price of natural gas in year t in a time-series equation for the consumption of natural gas in the U.S.
    c. The income of the *i*th individual in a cross-sectional equation for the number of suits owned by individuals.
    d. A dummy variable for being a student (1 = yes) in the equation specified in part c.
    e. The number of long-distance telephone calls handled per year in a cross-sectional equation for the marginal cost of a telephone call faced by various competing long-distance telephone carriers.

12. Suppose you've been hired by a union that wants to convince workers in local dry-cleaning establishments that joining the union will improve their well-being. As your first assignment, your boss asks you to build a model of wages for dry-cleaning workers that measures the impact of union membership on those wages. Your first equation (standard errors in parentheses) is:

$$\hat{W}_i = -11.40 + \underset{(0.10)}{0.30A_i} - \underset{(0.002)}{0.003A_i^2} + \underset{(0.20)}{1.00S_i} + \underset{(1.00)}{1.20U_i}$$

$$n = 34 \quad \bar{R}^2 = .14$$

where: W_i = the hourly wage (in dollars) of the *i*th worker
A_i = the age of the *i*th worker
S_i = the number of years of education of the *i*th worker
U_i = a dummy variable = 1 if the *i*th worker is a union member, 0 otherwise

a. Evaluate the equation. How do $\bar{R}^2$ and the signs and significance of the coefficients compare with your expectations?
b. What is the meaning of the A^2 term? What relationship between A and W does it imply? Why doesn't the inclusion of A and A^2 violate the classical assumption of no perfect collinearity between two independent variables?
c. Do you think you should have used the log of W as your dependent variable? Why or why not? (Hint: Compare this equation to the one in Exercise 4.)
d. Even though we've been told not to analyze the value of the intercept, isn't −\$11.40 too low to ignore? What should be done to correct this problem?
e. On the basis of your regression, should the workers be convinced that joining the union will improve their well-being? Why or why not?

13. Your boss manages to use the regression results in Exercise 12 to convince the dry-cleaning workers to join your union. About a year later, they go on strike, a strike that turns violent. Now your union is being sued by all the local dry-cleaning establishments for some of the revenues lost during the strike. Their claim is that the violence has intimidated replacement workers, thus decreasing production. Your boss doesn't believe that the violence has had a significant impact on production efficiency and asks you to test his hypothesis with a regression. Your results (standard errors in parentheses) are:

$$\widehat{LE}_t = 3.08 + \underset{(0.04)}{0.16LQ_t} - \underset{(0.010)}{0.020A_t} - \underset{(0.0008)}{0.0001V_t}$$
$$n = 24 \quad \bar{R}^2 = .855$$

where: LE_t = the natural log of the efficiency rate (defined as the ratio of actual total output to the goal output in week t)
LQ_t = the natural log of actual total output in week t

A_t = the absentee rate (%) during week t
V_t = the number of incidents of violence during week t

a. Hypothesize signs and develop and test the appropriate hypotheses for the individual estimated coefficients (95 percent level).
b. If the functional form is correct, what does its use suggest about the theoretical elasticity of E with respect to Q compared with the elasticities of E with respect to A and V?
c. On the basis of this result, do you think the court will conclude that the violence had a significant impact on the efficiency rate? Why or why not?
d. What problems appear to exist in this equation? (Hint: The problems may be theoretical as well as econometric.) If you could make one change in the specification of this equation, what would it be?

14. The **Box-Cox test** is a method of choosing the functional form of an equation based on the relative fit of regressions that have been standardized by dividing each dependent variable by its geometric mean. Most typically, the Box-Cox test[12] is used if we're theoretically undecided between a linear equation:

$$Y_i = \beta_0 + \beta_1 X_i + \epsilon_i$$

and a semi-log equation (with lnY as the dependent variable):

$$\ln Y_i = \beta_0 + \beta_1 X_i + \epsilon_i$$

To use the Box-Cox test in this situation, first calculate the geometric mean of each dependent variable (the geometric mean of n items is the nth root of the product of all n items). Then divide each observation of each dependent variable by its geometric mean and estimate both equations with OLS. The functional form with the lower RSS can be judged to have the better relative fit. To get some practice with the Box-Cox test, let's go back to the personal consumption example of Exercise 6:

12. This is just one specific application of the Box-Cox test. For more on the general test and the associated Box-Cox transformation, see G. E. P. Box and D. R. Cox, "An Analysis of Transformations," *Journal of the Royal Econometric Society,* 1962B, pp. 211–243, and G. S. Maddala, *Econometrics* (New York: McGraw-Hill, 1977), pp. 316–317.

a. Calculate the geometric means of consumption and of the log of consumption.
b. Divide each dependent variable by its geometric mean and run linear and semi-log (lnY) versions of the personal consumption function.
c. Which functional form has the lower RSS?
d. Does this result mean that you should definitely use that particular functional form? Why or why not?

15. Walter Primeaux[13] used slope dummies to help test his hypothesis that publicly owned monopolies tend to advertise less intensively than do duopolies in the electric utility industry. His estimated equation (which also included a number of geographic dummies and a time variable) was (t-scores in parentheses):

$$\hat{Y}_i = 0.15 + \underset{(4.5)}{5.0S_i} + \underset{(0.4)}{0.015G_i} + \underset{(2.9)}{0.35D_i}$$
$$- \underset{(-5.0)}{20.0S_i \cdot D_i} + \underset{(2.3)}{0.49G_i \cdot D_i}$$
$$\bar{R}^2 = .456 \qquad n = 350$$

where: Y_i = advertising and promotional expense (in dollars) per 1000 residential kilowatt hours (KWH) of the *i*th electric utility
S_i = number of residential customers of the *i*th utility (hundreds of thousands)
G_i = annual percentage growth in residental KWH of the *i*th utility
D_i = a dummy variable equal to 1 if the *i*th utility is a duopoly, 0 if a monopoly

a. Hypothesize and test the relevant null hypotheses with the *t*-test at the 5 percent level of significance. (Hint: Note that *both* independent variables have slope dummies.)
b. Assuming that Primeaux's equation is correct, graph the relationship between advertising (Y_i) and size (S_i) for monopolies and for duopolies.
c. Assuming that Primeaux's equation is correct, graph the rela-

13. Walter J. Primeaux, Jr., "An Assessment of the Effect of Competition on Advertising Intensity," *Economic Inquiry,* Oct. 1981, pp. 613–625.

tionship between advertising (Y_i) and growth (G_i) for monopolies and for duopolies.

d. For those who read Appendix 5.8, compare this approach (slope dummies for all variables) to the Chow test. Could you apply the Chow test to this equation? If not, what modifications would you need?

7.6 Appendix: Nonlinear (in the Coefficients) Regression

Unfortunately, economic theory sometimes leads to equations that are inherently nonlinear in the coefficients, such as:

$$Y_i = \beta_0 + \beta_1 X_{1i}^{\beta_2} + \beta_3 X_{2i}^{\beta_4} + \epsilon_i \qquad (7.27)$$

or

$$Y_i = \alpha(1 - 1/[1 + X_i]^{\beta}) + \epsilon_i \qquad (7.28)$$

Neither of these equations can be estimated by OLS because they cannot be easily transformed into equations that are linear in the coefficients.

7.6.1 Estimating Nonlinear (in the Coefficients) Equations

Given an equation that is nonlinear in the coefficients such as Equation 7.27 or 7.28, estimates can be obtained through either nonlinear least squares or maximum likelihood techniques.

Nonlinear least squares estimates coefficients by iteratively minimizing the summed squared residuals. This technique allows researchers to fit a nonlinear-in-the-coefficients model directly to the data. Thus coefficients can be estimated even for the most complex theoretically nonlinear models.

To use this technique, derivatives[14] of Σe^2 with respect to each coefficient are calculated. The derivatives are functions of the coefficients as well as of the data. Values of the coefficients that make all

14. Derivatives are calculus notations similar to the delta (Δ) notation used in this chapter. If these derivatives are set equal to zero and solved simultaneously, the resulting coefficient values will be those that minimize Σe^2. Thus, to use nonlinear least squares, the derivatives, which are functions of the coefficients as well as of the data, must be known.

the derivatives zero (and therefore minimize the summed squared residuals) are found by iterating through a number of different sets of $\hat{\beta}$ estimates before arriving at the best $\hat{\beta}$. The iterative estimation procedure starts with an initial set of researcher-specified $\hat{\beta}$ coefficients, calculates Σe^2 for those $\hat{\beta}$s, uses the derivatives to find $\hat{\beta}$s that will decrease Σe^2, and then starts over again with the new set of $\hat{\beta}$s. When further changes in the $\hat{\beta}$s will no longer improve things, the program stops and prints out the last set of $\hat{\beta}$s.[15]

An alternative approach to estimating equations that are nonlinear in the coefficients is to apply **maximum likelihood** (ML) estimation. This approach is inherently different from least squares in that it chooses coefficient estimates that *maximize* the *likelihood* of the sample data-set being observed.

The first step in ML estimation is to explicitly assume a particular probability distribution (typically the normal distribution) for the error terms of the equation to be estimated. Once this has been completed, a *likelihood function* for the equation is created. The **likelihood function** measures the probability of observing the particular set of dependent variable values ($Y_1, Y_2, \ldots, Y_n$) that occur in the sample. It is written as the probability of the product of the observed Y_is:

$$\text{Likelihood Function:} \quad L = \Pr(Y_1 \cdot Y_2 \cdot Y_3 \cdot \cdots Y_n) \qquad (7.29)$$

where $\Pr(Y_1 \cdot Y_2)$ refers to the joint probability that both Y_1 and Y_2 will occur.[16] The higher L is, the higher the probability of observing the set of Ys found in the sample.

The final step in ML estimation is to find the set of coefficient values ($\hat{\beta}$s) that makes L, the likelihood function, as high as possible given the sample observations. Maximizing L means maximizing the probability of observing the particular set of Ys that was in the sample

15. Another least squares approach is to "linearize" the continuous nonlinear curve into a series of linked (and extremely short) straight lines. This is usually done with a "Taylor series expansion," which uses first and second derivatives of the function in question with respect to the coefficients. In such a case, OLS can be applied to the individual equation estimations within the iteration. For more on this approach and a number of other useful techniques, see S. M. Goldfeld and R. E. Quandt, *Nonlinear Methods in Econometrics* (Amsterdam: North-Holland, 1972), or A. R. Gallant, "Nonlinear Regression," *The American Statistician*, May 1975, pp. 73–81.

16. Under certain circumstances, such likelihood functions can be equivalently stated in terms of error terms, as in $L = \Pr(\epsilon_1 \cdot \epsilon_2 \cdot \epsilon_3 \cdots \epsilon_n)$. This is because Y is a function of the Xs, the βs, and the ϵs. Note that if the Ys (or ϵs) are independent of each other, as they should be, then L can be restated as the product of the probabilities.

(given the error term distribution chosen). Computational reasons often incline researchers to maximize the logarithm of the likelihood function, called a log-likelihood function, rather than the likelihood function itself, but the underlying principles are identical in either case.

One of the reasons that maximum likelihood is frequently used is that, in general, it has desirable large sample properties. In particular, ML has better large sample properties[17] than does least squares for a number of advanced estimation problems. Interestingly, the OLS estimation procedure and the ML estimation procedure do not necessarily produce different $\hat{\beta}$s; for a linear equation that meets the Classical Assumptions (including the normality assumption), the OLS estimates are identical to the ML estimates. Indeed, one of the reasons that OLS estimation is attractive is because it is equivalent to maximum likelihood in this case.

The estimation of ML equations is fairly complicated and usually takes a considerable amount of computer time. Luckily, continuing developments in computer programming and the recent decreases in the dollar cost of computer time have now made ML estimation procedures reasonably inexpensive. Some fairly general computer packages have been developed to estimate equations using the ML approach, and some specific computer packages have been developed to calculate ML estimates. Interestingly, if the error term ϵ of the inherently nonlinear equation is normally distributed, the ML estimates and the nonlinear least squares estimates are identical, leading most researchers to use whichever method is computationally easier given their computer system.

Indeed, many software packages for ML or nonlinear regression now are available. Given the data, derivatives, and starting values above, these programs iterate and attempt to converge on a set of coefficient estimates. Since these programs also usually provide standard evaluation statistics like R^2 and t-scores, coefficient estimates produced by nonlinear (in the coefficients) regression packages can be analyzed using the tools presented in previous chapters without regard

17. Maximum likelihood estimators have desirable large sample properties: consistency, asymptotic efficiency, and asymptotic normality. Consistency means that the probability that the $\hat{\beta}$s equal the true βs approaches one as the sample size gets larger. Asymptotic efficiency means that the estimator is consistent and has the smallest variance of all the consistent estimators. Asymptotic normality means that the distribution of the estimates approaches the normal distribution as the sample size approaches infinity. For more on maximum likelihood, see G. S. Maddala, *Econometrics* (New York: McGraw-Hill, 1977).

to the estimation procedure, with the exception of the likelihood ratio test, to be discussed in the example that follows.

7.6.2 An Example of Nonlinear (in the Coefficients) Estimation

Let's look at an example of the application of an ML estimation technique to an inherently nonlinear equation. It can be shown that Equation 7.30 is the theoretically appropriate functional form to use in a cross-sectional study of aggregate capital requirements by country:[18]

$$Y_i = \alpha(1 - 1/[1 + X_i]^{\beta}) + \epsilon_i \qquad (7.30)$$

where: Y_i = the gross incremental output/capital ratio for country i
X_i = the rate of growth of output for country i
ϵ_i = a classical normal error term

The gross incremental output/capital ratio (the change in output in a country during a particular time period divided by the gross investment during that time period) is useful in determining the capital needs of a country. To quantify the movement of Y with respect to X, data for a sample of 62 countries were collected, and Equation 7.30 was estimated using ML estimation with the following results:

$$\hat{Y}_i = 0.56(1 - 1/[1 + X_i]^{17.0}) \qquad (7.31)$$
$$(0.14) \qquad\qquad (6.8)$$
$$t = \ 4.0 \qquad\qquad 2.5$$
$$LL = 60.28 \qquad n = 62 \qquad \text{iterations} = 13 \qquad \bar{R}^2 = .37$$

where LL refers to −2 times the log of the likelihood ratio (to be discussed shortly) that tests the overall significance of the equation.

The ML estimates in Equation 7.31 can be used and analyzed in much the same way that we have treated other estimated equations throughout this text. For example, both estimated coefficients are significantly different from zero in the direction hypothesized in the ar-

18. J. Vanek, et al., "Towards a Better Understanding of the Incremental Capital-Output Ratio," *Quarterly Journal of Economics,* Oct. 1968, pp. 452–464. This article also contains the data-set used for the example in this section. While we do not derive Equation 7.30 here, we can interpret α as the long-run (or "golden-age") net incremental output/capital ratio and β as the average length of time that a typical capital asset lasts before it is scrapped.

ticle, as shown by t-scores greater than the 5 percent one-tailed critical t-value of approximately 1.645.

Note, however, that tests of the overall significance of an equation estimated with ML are done with the **likelihood ratio test,** which consists of comparing the value of a likelihood function for an unrestricted equation with one that has been restricted by the joint hypothesis in question. If the two likelihood functions are significantly different, then the joint hypothesis can be rejected because the imposition of the hypothesis has signicantly reduced the fit. If the two likelihood functions are not significantly different, then the null hypothesis cannot be rejected. This likelihood ratio turns out to be distributed according to the Chi-Square distribution (Statistical Table B-8) with degrees of freedom equal to the number of restrictions in the hypothesis. In Equation 7.31, the joint null hypothesis that $\alpha = 0$ and $\beta = 0$ simultaneously (versus the alternative hypothsis that this is not the case) can be rejected at the 5 percent level of significance, because the LL of 60.28 is greater than the critical Chi-Square value of 5.99 for two degrees of freedom as found in Table B-8.

8

Multicollinearity

The next three chapters deal with violations of the Classical Assumptions and remedies for those violations. This chapter addresses multicollinearity; the next two chapters are on serial correlation and heteroskedasticity. For each of the three problems, we will attempt to answer the following questions:

1. What is the nature of the problem?
2. What are the consequences of the problem?
3. How is the problem diagnosed?
4. What remedies for the problem are available?

Strictly speaking, **multicollinearity** is the violation of the assumption that no independent variable is a perfect linear function of one or more other independent variables (Classical Assumption VI). Perfect multicollinearity is rare, but severe imperfect multicollinearity (where two or more independent variables are highly correlated in the particular data-set being studied), while not violating Classical Assumption VI, still causes substantial problems.

Recall that the coefficient β_k can be thought of as the impact on the dependent variable of a one-unit change in the independent variable X_k, holding constant the other independent variables in the equation. But if two explanatory variables are significantly related in a particular sample, whenever one changes, the other will tend to change too, and the OLS computer program will find it difficult to distinguish the effects of one variable from the effects of the other. Since the Xs can move together more in one sample than they do in another, the severity of multicollinearity can vary tremendously.

In essence, the more highly correlated two (or more) independent variables are, the more difficult it becomes to accurately estimate the coefficients of the true model. We are usually less concerned about the existence of multicollinearity in a sample (it almost always exists to some degree) than with how severe the multicollinearity is. If two variables move identically, then there is no hope of distinguishing between the impacts of the two, but if the variables are only roughly correlated, then we still might be able to estimate the two impacts accurately enough for most purposes.

8.1 Perfect vs. Imperfect Multicollinearity

8.1.1. Perfect Multicollinearity

Perfect multicollinearity[1] violates Classical Assumption VI, which specifies that no explanatory variable is a perfect linear function of any other explanatory variables. The word "perfect" in this context implies that the variation in one explanatory variable can be *completely* explained by movements in another explanatory variable. Such a perfect linear function between two independent variables would be:

$$X_{1i} = \alpha_0 + \alpha_1 X_{2i} \tag{8.1}$$

1. The word *collinearity* describes a linear correlation between two independent variables, and *multicollinearity* indicates that more than two independent variables are involved. In common usage, multicollinearity is used to apply to both cases, and so we'll typically use that term in this text even though many of the examples and techniques discussed relate, strictly speaking, to collinearity.

where the αs are constants; the Xs are independent variables in:

$$Y_i = \beta_0 + \beta_1 X_{1i} + \beta_2 X_{2i} + \epsilon_i \tag{8.2}$$

Notice that there is no error term in Equation 8.1. This implies that X_1 can be calculated exactly given X_2 and the equation. Examples of such perfect linear relationships would be:

$$X_{1i} = 3X_{2i} \quad \text{or} \quad X_{1i} = 6 + X_{2i} \quad \text{or} \quad X_{1i} = 2 + 4X_{2i} \tag{8.3}$$

Figure 8.1 shows a graph of explanatory variables that are perfectly correlated. As can be seen in Figure 8.1, a perfect linear function has all data points on the same straight line. There is none of the variation that accompanies the data-set from a typical regression.

Some examples of perfect multicollinearity were briefly mentioned in Section 4.1. Recall what happens when nominal and real interest rates are both included as explanatory variables in an equation. Usually, the relationship between nominal and real interest rates continually changes because the difference between the two, the rate of inflation, is always changing. If the rate of inflation somehow was constant (during extremely strict price controls, for instance), then the difference

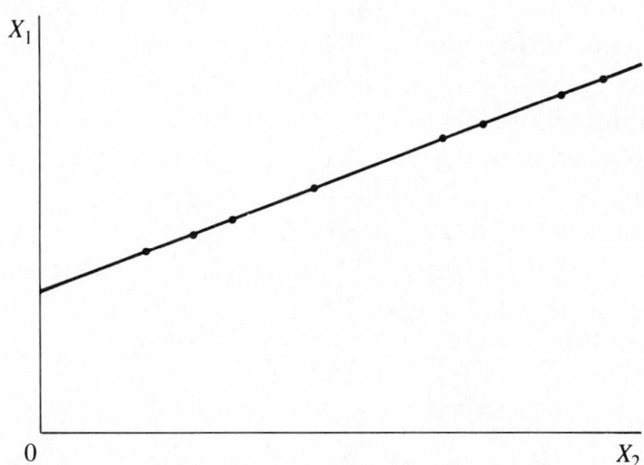

Figure 8.1 PERFECT MULTICOLLINEARITY
With perfect multicollinearity, an independent variable can be completely explained by the movements of one or more other independent variables. Perfect multicollinearity can usually be avoided by careful screening of the independent variables before a regression is run.

between the two would be constant, the two would be perfectly linearly related, and perfect multicollinearity would result:

$$in_t = ir_t + inf_t = ir_t + \alpha \qquad (8.4)$$

where: in_t = the nominal (or money) interest rate in time t
ir_t = the real interest rate in time t
inf_t = the rate of inflation in time t
α = the constant rate of inflation

What happens to the estimation of an econometric equation where there is perfect multicollinearity? OLS is incapable of generating estimates of the regression coefficients. Using Equation 8.2 as an example, we theoretically would obtain the following estimated coefficients and standard errors:

$$\hat{\beta}_1 = \text{indeterminate} \qquad SE(\hat{\beta}_1) = \infty \qquad (8.5)$$
$$\hat{\beta}_2 = \text{indeterminate} \qquad SE(\hat{\beta}_2) = \infty$$

Perfect multicollinearity ruins our ability to estimate the coefficients because the two variables cannot be distinguished. You cannot "hold all the other independent variables in the equation constant" if every time one variable changes, another changes in an identical manner.[2]

Fortunately, instances in which one explanatory variable is a perfect linear function of another are rare. More important, perfect multicollinearity should be fairly easy to discover before a regression is run. You can detect perfect multicollinearity by asking whether one variable equals a multiple of another or if one variable can be derived by adding a constant to another. If so, then one of the variables should be dropped because there is no essential difference between the two.

A special case related to perfect multicollinearity occurs when a variable that is definitionally related to the dependent variable is included as an independent variable in a regression equation. Such a **dominant variable** is so highly correlated with the dependent variable

2. Most OLS estimation programs will print out an error message when faced with perfect multicollinearity rather than fruitlessly attempt to calculate something indeterminate. A few computer programs contain rounding errors that will produce estimates (admittedly highly unreliable estimates) in the face of perfect multicollinearity, but these programs are in the minority. In such circumstances, the standard errors would not be infinite but merely very large, and the resulting t-scores would still be quite low.

that it completely masks the effects of all other independent variables in the equation. In a sense, this is a case of perfect collinearity between the dependent and an independent variable.

For example, if you include a variable measuring the amount of raw materials used by the shoe industry in a production function for that industry, the raw materials variable would have an extremely high t-score, but otherwise important variables like labor and capital would have quite insignificant t-scores. Why? In essence, if you knew how much leather was used by a shoe factory, you could predict the number of pairs of shoes produced without knowing *anything* about labor or capital. The relationship is definitional, and the dominant variable should be dropped from the equation to get reasonable estimates of the coefficients of the other variables.

A dominant variable involves a tautology; it is defined in such a way that you can calculate the dependent variable from it without any knowledge of the underlying theory. Be careful, though; dominant variables shouldn't be confused with highly significant or important explanatory variables. Instead, they should be recognized as being virtually identical to the dependent variable. While the fit between the two is superb, knowledge of that fit could have been obtained from the definitions of the variables without any econometric estimation.

8.1.2 Imperfect Multicollinearity

Since perfect multicollinearity is fairly easy to avoid, econometricians almost never talk about perfect multicollinearity. Instead, when we use the word multicollinearity, we are really talking about severe imperfect multicollinearity. **Imperfect multicollinearity** can be defined as a linear functional relationship between two or more independent variables that is so strong that it can significantly affect the estimation of the coefficients of the variables.

In other words, imperfect multicollinearity occurs when two (or more) explanatory variables are imperfectly linearly related as in:

$$X_{1i} = \alpha_0 + \alpha_1 X_{2i} + u_i \qquad (8.6)$$

Compare Equation 8.6 to Equation 8.1; notice that Equation 8.6 includes u_t, a stochastic error term. This implies that while the relationship between X_1 and X_2 might be fairly strong, it is not strong enough to allow X_1 to be completely explained by X_2; some unexplained variation still remains. Figure 8.2 shows the graph of two explanatory variables that might be considered multicollinear. Notice that while all

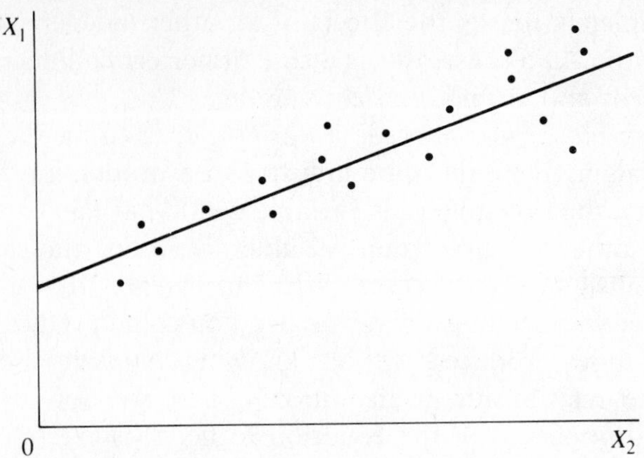

Figure 8.2 IMPERFECT MULTICOLLINEARITY
With imperfect multicollinearity, an independent variable is a strong but not perfectly linear function of one or more other independent variables. Imperfect multicollinearity varies in degree from sample to sample.

the observations in the sample are fairly close to the straight line, there is still some variation in X_1 that has not been explained by X_2.

Imperfect multicollinearity is a strong linear relationship between the explanatory variables. The stronger the relationship between the two (or more) explanatory variables, the more likely it is that they'll be considered significantly multicollinear. Whether explanatory variables are multicollinear in a given equation depends on the theoretical relationship between the variables and on the particular sample chosen. Two variables that might be only slightly related in one sample might be so strongly related in another that they could be considered to be imperfectly multicollinear. In this sense, it is fair to say that multicollinearity is a sample phenomenon as well as a theoretical one. Whether the data are correlated enough to have a significant effect on the estimation of the equation depends on the particular sample drawn, and each sample must be investigated (using, for example, the simple correlation coefficient to measure collinearity) before multicollinearity can be diagnosed. This contrasts with perfect multicollinearity because two variables that are perfectly related probably can be detected on a logical basis. The detection of multicollinearity will be discussed in more detail in Section 8.3.

Some research projects are inherently more likely to suffer from multicollinearity than others. For example, almost all macroeconomic time-series data-sets have potential multicollinearity in them because aggregates have tended to rise unambiguously over the past half century. The labor force, national income, consumption, taxes, and almost every other measure of economic activity have increased as the years have gone by. In addition, increases in productivity and the general rise in prices have augmented this tendency towards co-movement.

Consider a model of the impact of taxes on savings over the last 30 years:

$$S_t = f(\overset{+}{Yd_t}, \overset{+}{i_t}, \overset{-}{T_t}, \overset{-}{SS_t}) \tag{8.7}$$

where: S_t = savings in year t (nominal, excluding Social Security)
Yd_t = disposable income in year t (nominal)
i_t = the nominal interest rate on savings in year t
T_t = the average tax rate in year t
SS_t = contributions to Social Security in year t (nominal)

In this model, savings are hypothesized to rise as disposable income rises. In addition, the incentive to save is hypothesized to increase as the interest rate paid on savings increases but decrease as the tax rate on income from savings increases. To the extent that Social Security is considered savings by the average individual, then contributions to Social Security would tend to reduce the perceived need for saving.

Given the growth in the economy over the years, almost all the variables in this savings model have probably increased over time; the increase in population, in income, and in the rate of Social Security deductions make it seem quite likely that many of the independent variables in Equation 8.7 will be highly correlated. In such a model, multicollinearity is quite likely to be fairly severe, and the consequences of such severe imperfect multicollinearity need to be investigated.

8.2 The Consequences of Multicollinearity

If the multicollinearity in a particular sample is severe, what will happen to estimates calculated from that sample? Since perfect multicollinearity means that the estimation of an equation is impossible, what consequences does significant imperfect multicollinearity imply? The purpose of this section is to explain the consequences of multicollinearity and then to explore some examples of such consequences.

Recall the properties of OLS estimators that might be affected by this or some other econometric problem. In Chapter 4, we stated that the OLS estimators are BLUE (or MvLUE) if the Classical Assumptions hold. This means that OLS estimates can be thought of as being unbiased and having the minimum variance possible for unbiased linear estimators.

8.2.1 An Overview of the Consequences of Multicollinearity

The general consequences of multicollinearity are:

1. *Estimates will remain unbiased.* Even if an equation has significant multicollinearity, the estimates of the βs will still be centered around the true population βs if all the Classical Assumptions are met for a correctly specified equation.
2. *The variances of the estimates will increase.* This is the major consequence of multicollinearity. Since two or more of the explanatory variables are significantly related, it becomes difficult to precisely identify the separate effects of the multicollinear variables. In essence, we are asking the regression to tell us something about which we have little information. When it becomes hard to distinguish the effect of one variable from the effect of another, then we're much more likely to make large errors in estimating the βs than we were before we encountered multicollinearity. As a result, the estimated coefficients, while still unbiased, now come from distributions with much larger variances.[3]

   To see this, recall the equation for the standard error (the square root of the variance) of an estimated slope coefficient in a multivariate regression model with exactly two independent variables. That equation was Equation 2.19:

$$SE(\hat{\beta}_1) = \sqrt{\frac{\sum e_i^2/(n-3)}{\sum(X_{1i} - \overline{X}_1)^2(1 - r^2_{12})}} \tag{2.19}$$

   What happens to $SE(\hat{\beta}_1)$, and therefore to the variance, in the face of severe multicollinearity? With multicollinearity, the simple correlation coefficient between X_1 and X_2, r_{12}, will be high. If r_{12} is high, then $(1 - r^2_{12})$ will be low, causing $SE(\hat{\beta}_1)$ to be high. Thus multicollinearity causes $SE(\hat{\beta}_1)$ and the variance of the estimated

3. Even though the variances are larger with multicollinearity than they are without it, OLS is still BLUE when multicollinearity exists. That is, no other linear unbiased estimation technique can get lower variances than OLS even in the presence of multicollinearity. Thus, while the effect of multicollinearity is to increase the variance of the estimated coefficients, OLS still has the property of minimum variance (these "minimum variances" are just fairly large).

coefficients to be higher than they would be without such correlation.

Figure 8.3 compares a distribution of $\hat{\beta}$s from a sample with severe multicollinearity to one with virtually no correlation between any of the independent variables. Notice that the two distributions have the same mean, indicating that multicollinearity does not cause bias. Also note how much wider the distribution of $\hat{\beta}$ becomes when multicollinearity is severe; this is the result of the increase in the variance of $\hat{\beta}$ that is caused by multicollinearity.

In particular, with multicollinearity, there is a higher probability of obtaining a $\hat{\beta}$ that is dramatically different from the true β. For example, it turns out that multicollinearity increases the likelihood of obtaining an unexpected sign[4] for a coefficient even though, as mentioned above, multicollinearity causes no bias. For more on this see Exercise 6.

3. *The computed t-scores will fall.* Multicollinearity tends to decrease the t-scores of the estimated coefficients mainly because of the formula for the t-statistic:

$$t_k = \frac{(\hat{\beta}_k - \hat{\beta}_{H_0})}{SE(\hat{\beta}_k)} \tag{8.8}$$

Notice that this equation is divided by the standard error of the estimated coefficient. Multicollinearity increases the variance, estimated variance, and therefore the standard error of the estimated coefficient. If the standard error increases, then the t-score must fall, as can be seen from Equation 8.8. Not surprisingly, it's quite common to observe low t-scores in equations with severe multicollinearity.

A second factor also tends to cause more insignificant t-scores to be observed with severe multicollinearity than without. Because multicollinearity increases the variance of the $\hat{\beta}$s, the coefficient estimates are likely to be farther from the true parameter value

4. These unexpected signs generally occur because the distribution of the $\hat{\beta}$s with multicollinearity is wider than without it, increasing the chance that a particular observed $\hat{\beta}$ will be on the other side of zero from the true β (have an unexpected sign). More specifically, particular combinations of multicollinear variables can make such unexpected signs occur quite frequently. For instance, if two independent variables both have positive true coefficients and positive simple correlation coefficients with Y in the observed sample, and if the simple correlation coefficient between the independent variables in the sample is higher than either of the two simple correlation coefficients between Y and the Xs, then one of the two slope coefficients is virtually assured of having an unexpected sign.

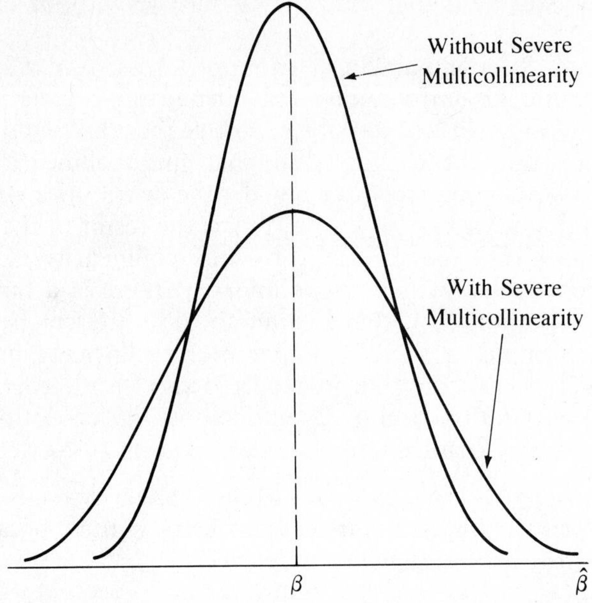

Figure 8.3 SEVERE MULTICOLLINEARITY INCREASES THE VARIANCES OF THE $\hat{\beta}$s

Severe multicollinearity produces a distribution of the $\hat{\beta}$s that is centered around the true β but which has a much larger variance. Thus the distribution of $\hat{\beta}$s with multicollinearity is much wider than otherwise.

than they would have been with less multicollinearity. This "pushes" a portion of the distribution of the $\hat{\beta}$s towards zero, making it more likely that a t-score will be insignificantly different from zero (or will have an unexpected sign). This "pushing" goes in both directions, so multicollinearity is also likely to cause some higher-than-expected $\hat{\beta}$s (and therefore t-scores). However, this positive pushing is often overshadowed by the larger standard errors mentioned in the previous paragraph.

4. *Estimates will become very sensitive to changes in specification.* The addition or deletion of an explanatory variable or of a few observations will often cause major changes in the values of the $\hat{\beta}$s when significant multicollinearity exists. If you drop a variable, even one that appears to be statistically insignificant, the coefficients of the remaining variables in the equation will sometimes change dramatically.

These large changes occur because OLS estimation is sometimes forced to emphasize small differences between variables in order to distinguish the effect of one multicollinear variable from another. Thus, even a minor specification change can cause a major change in the attribute that the computer program had "focused on," and the estimated coefficients can change significantly. If two variables are virtually identical throughout most of the sample, the estimation procedure relies on the observations in which the variables move differently in order to distinguish between them. As a result, a specification change that drops a variable that had an unusual value for one of these crucial observations can cause the estimated coefficients of the multicollinear variables to change dramatically.

5. *The overall fit of the equation will be largely unaffected.* Even though the individual t-scores are often quite low in a multicollinear equation, the overall fit of the equation, as measured by R^2 or the F-test, will not fall much, if at all, in the face of significant multicollinearity. As a result, it's not uncommon to encounter multicollinear equations that have quite high R^2s and yet have no individual independent variable's coefficient even close to being statistically significantly different from zero.

   Because multicollinearity has little effect on the overall fit of the equation, it will also have little effect on the use of that equation for prediction or forecasting, as long as the independent variables maintain the same pattern of multicollinearity in the forecast period that they demonstrated in the sample.

6. *The estimation of nonmulticollinear (orthogonal) variables will be unaffected.* If an explanatory variable in an equation is not multicollinear (also called orthogonal) with the other variables, then the estimation of its coefficient and standard error usually will not be affected. It's unusual to find an explanatory variable that's totally uncorrelated with any other explanatory variable. If this were to occur, though, then the multicollinearity in the rest of the equation would not change the estimated coefficient or the t-score of the nonmulticollinear variable.

7. *The severity of multicollinearity worsens its consequences.* It is intuitively logical that the more severe the multicollinearity, the more severe the impact on the estimates. After all, if perfect multicollinearity makes the estimation of the equation impossible, then almost perfect multicollinearity should cause much more damage to estimates than virtually nonexistent multicollinearity.

   Indeed, the higher the simple correlation between the multicollinear variables (in the two-variable case), the higher the estimated variances and the lower the calculated t-values; the variances of the $\hat{\beta}$s calculated with OLS increase as the simple correlation coefficient between the two independent variables increases. When $r = 0$ (no multicollinearity), the variance of $\hat{\beta}$ equals its minimum (that is, nonmulticollinear) value. As the absolute value of r increases from 0 to 1, holding other things constant, the variance slowly increases to infinity, and the t-score goes to zero. For example, if a t-score of 4.00 were observed when $r = 0$, it may fall to 1.75 if $r = 0.90$ and to 1.25 if $r = 0.95$. The same tendency also holds (only more so) when there are three or more multicollinear explanatory variables.

8.2.2 Two Examples of the Consequences of Multicollinearity

To see what severe multicollinearity does to an estimated equation, let's look at a hypothetical example. Suppose you decide to estimate a "student consumption function" that relates better to student problems than some of the aggregate macroeconomic consumption functions you've read about. After the appropriate preliminary work, you come up with the following hypothesized equation:

$$C_i = f(\overset{+}{Yd_i}, \overset{+}{LA_i}) = \beta_0 + \beta_1 Yd_i + \beta_2 LA_i + \epsilon_i \tag{8.9}$$

where: C_i = the annual consumption expenditures of the *i*th student

Yd_i = the annual disposable income (including gifts) of that student

LA_i = the liquid assets (savings, etc.) of the *i*th student's family

ϵ_i = a stochastic error term

You then collect a small amount of data from people who are sitting near you in class:

Student	C_i	Yd_i	LA_i
Courtney	\$2000	\$2500	\$25000
Ruth	2300	3000	31000
May	2800	3500	33000
Morgan	3800	4000	39000
Duncan	3500	4500	48000
Allen	5000	5000	54000
Reed	4500	5500	55000

If you run an OLS regression on your data-set for Equation 8.9, you obtain:

$$\hat{C}_i = -367.83 + 0.5113Yd_i + 0.0427LA_i \qquad (8.10)$$
$$(1.0307) \qquad (0.0942)$$
$$t = 0.496 \qquad 0.453 \qquad \bar{R}^2 = .835$$

On the other hand, if you had consumption as a function of disposable income alone, then you would have obtained:

$$\hat{C}_i = -471.43 + 0.9714Yd_i \qquad (8.11)$$
$$(0.157)$$
$$t = 6.187 \qquad \bar{R}^2 = .861$$

Notice from Equations 8.10 and 8.11 that the t-score for disposable income increases more than tenfold when the liquid assets variable is dropped from the equation. Why does this happen? First of all, the simple correlation coefficient between Yd and LA is quite high: $r_{Yd,LA} = .986$. This high degree of correlation causes the standard errors of the estimated coefficients to increase dramatically. In the case of $\hat{\beta}_{Yd}$, the standard error goes from 0.157 to 1.03! In addition, the coefficient estimate itself changes somewhat. Further, note that the $\bar{R}^2$s of the two equations are quite similar despite the large differences in the significance of the explanatory variables in the two equations. It's quite common for $\bar{R}^2$ to stay virtually unchanged when multicollinear variables are dropped. All of these results are typical of equations with multicollinearity.

Which equation is better? If the liquid assets variable theoretically belongs in the equation, then to drop it will run the risk of left-out variable bias, but to include the variable will mean certain multicollinearity. There is no automatic answer when dealing with multicollinearity. We'll discuss this issue in more detail in Sections 8.4 and 8.5.

A second example of the consequences of multicollinearity is based on actual, rather than hypothetical, data. Suppose you've decided to build a cross-sectional model of the demand for gasoline by state:

$$PCON_i = f(\overset{+}{UHM_i}, \overset{-}{TAX_i}, \overset{+}{REG_i}) \qquad (8.12)$$

where: $PCON_i$ = petroleum consumption in the *i*th state (trillions of BTUs)

UHM_i = urban highway miles within the *i*th state
TAX_i = the gasoline tax rate in the *i*th state (cents per gallon)
REG_i = motor vehicle registrations in the *i*th state (thousands)

A complete listing of the data for this model is contained in Section 10.5, so let's move on to the estimation of Equation 8.12 in a linear functional form (assuming a stochastic error term):

$$\widehat{PCON_i} = 389.6 + 60.8UHM_i - 36.5TAX_i - 0.061REG_i \tag{8.13}$$
$$\qquad\qquad (10.3) \qquad\qquad (13.2) \qquad\qquad (0.043)$$
$$t = \quad 5.92 \qquad\qquad -2.77 \qquad\qquad -1.43$$
$$n = 50 \qquad \bar{R}^2 = .919$$

What's wrong with this equation? Motor vehicle registrations has an insignificant coefficient with an unexpected sign, but it's hard to believe that the variable is irrelevant. Is a left-out variable causing bias? It's possible, but adding a variable is unlikely to fix things. Does it help to know that the simple correlation coefficient between REG and UHM is 0.98? Given that, it seems fair to say that one of the two variables is redundant; both variables are really measuring the *size* of the state, so we have multicollinearity.

Notice the impact of the multicollinearity on the equation. The coefficient of a variable such as motor vehicle registrations, which has a very strong theoretical relationship to petroleum consumption, is insignificant and has a sign contrary to our expectations. This is mainly because the multicollinearity has increased the variance of the distribution of the estimated $\hat{\beta}$s.

What would happen if we were to drop one of the multicollinear variables?

$$\widehat{PCON_i} = 551.7 - 53.6TAX_i + 0.186REG_i \tag{8.14}$$
$$\qquad\qquad (16.9) \qquad\qquad (0.012)$$
$$t = \quad -3.18 \qquad\qquad 15.88$$
$$n = 50 \qquad \bar{R}^2 = .861$$

Dropping UHM has made REG extremely significant. Why did this occur? The answer is that the standard error of the coefficient of REG has fallen substantially (from 0.043 to 0.012) now that the multicollinearity has been removed from the equation. Also note that the sign of the estimated coefficient has now become positive as hypothesized.

The reason is that REG and UHM are virtually indistinguishable from an empirical point of view, and so the OLS program latched onto minor differences between the variables to explain the movements of PCON. Once the multicollinearity was removed, the direct positive relationship between REG and PCON was obvious. Note, however, that the coefficient of the REG variable now measures the effect of both REG and UHM on PCON. Since we've dropped a variable, the remaining coefficient soaks up the effect of the left-out variable.

Either UHM or REG could have been dropped with similar results because the two variables are, in a quantitative sense, virtually identical as indicated by the high simple correlation coefficient between them. In this case, REG was judged to be theoretically superior to UHM. Note also that $\bar{R}^2$ fell when UHM was dropped, and yet Equation 8.14 should be considered superior to Equation 8.13. This is an example of the point, originally made in Chapter 3, that the fit of an equation is not the most important criterion to be used in determining its overall quality.

8.3 The Detection of Multicollinearity

How do we decide whether an equation has a severe multicollinearity problem? A first step is to recognize that some multicollinearity exists in every equation. It's virtually impossible in a real-world example to find a set of explanatory variables that are totally uncorrelated with each other. Our main purpose in this section will be to learn to determine *how much* multicollinearity exists in an equation, not *whether* any multicollinearity exists.

A second key point is that multicollinearity is a sample phenomenon as well as a theoretical one. That is, the severity of multicollinearity in a given equation can change from sample to sample depending on the characteristics of that sample. As a result, the theoretical underpinnings of the equation are not quite as important in the detection of multicollinearity as they are in the detection of an omitted variable or an incorrect functional form. Instead, we tend to rely more on data-oriented techniques to determine the severity of the multicollinearity in a given sample. Of course, we can never ignore the theory behind an equation. The trick is to find variables that are theoretically relevant (for meaningful interpretation) and that are also statistically non-multicollinear (for meaningful inference).

Because multicollinearity is a sample phenomenon whose damaging impact is a matter of degree, many of the methods used to detect it are informal tests without critical values or levels of significance. Indeed, there are no generally accepted true statistical tests for multicollinearity. Most researchers develop a general feeling for the severity of multicollinearity in an estimated equation by looking at a number of the characteristics of that equation. Let's examine three of the most-used of those characteristics.

8.3.1 High $\bar{R}^2$ with All Low t-Scores

One of the unique consequences of multicollinearity is that the overall level of significance of an equation is affected far less by multicollinearity than are the levels of significance of the individual regression coefficients. Multicollinearity severe enough to lower t-scores substantially does little to decrease $\bar{R}^2$ and the F-statistic. Given this, one of the first indications of severe multicollinearity is the combination of a high $\bar{R}^2$ with low t-values for *all* the individual regression coefficients.

For example, return to Equation 8.10 and note that the $\bar{R}^2$ is .835 even though no individual t-score is higher than 0.5. Such a combination is a sure sign of severe multicollinearity. It is almost impossible to get a significant overall fit with none of the slope coefficients significantly different from zero unless you have a severe multicollinearity problem.

This approach has a problem, however. Equations with high levels of multicollinearity can still have one or more regression coefficients significantly different from zero. This is possible for two reasons. First, nonmulticollinear explanatory variables can have significant coefficients even if there is multicollinearity between two or more other explanatory variables. Second, multicollinearity often causes some, but not all, of the coefficients of the multicollinear variables to be insignificant. Thus "high $\bar{R}^2$ with all low t-scores" must be considered a sufficient but not necessary test for severe multicollinearity. While every equation with a high $\bar{R}^2$ and all low t-scores will have multicollinearity of some sort, the lack of these characteristics is not proof of the absence of multicollinearity.

At the other extreme, if all the estimated coefficients *are* significantly different from zero in the expected direction, then we can conclude that severe multicollinearity is *not* likely to be a problem. This is because the two major effects of multicollinearity are to increase the estimated variances of the estimated coefficients and to lower their computed t-scores. An equation with all of its t-scores significant in the expected direction may have multicollinearity between some of its explanatory variables, but that multicollinearity is not severe enough to cause consequences worth worrying about.

In summary, we can gain important inferences about multicollinearity by studying $\bar{R}^2$, the F-test of overall significance, and the *t*-tests of individual estimated coefficient significance. If $\bar{R}^2$ is high (if the F-test is significant), then the likelihood of severe multicollinearity is directly related to the proportion of t-scores that are insignificant:

Proportion of t-Scores that are Insignificant	Likelihood* of Severe Multicollinearity
all	highly probable
some	possible
none	not a problem

* assuming $\bar{R}^2$ is high

8.3.2 High Simple Correlation Coefficients

Another way to detect severe multicollinearity is to examine the simple correlation coefficients between the explanatory variables. If an r is high in absolute value, then we know that the two particular Xs are quite correlated and that multicollinearity is a potential problem. For example, in Equation 8.10, the simple correlation coefficient between disposable income and liquid assets is 0.986. A simple correlation coefficient this high, especially in an equation with only two independent variables, is a certain indication of severe multicollinearity.

How high is high? Some researchers pick an arbitrary number, such as 0.80, and become concerned about multicollinearity any time the absolute value of a simple correlation coefficient exceeds 0.80.

A more systematic method is to test the significance of individual simple correlation coefficients using the *t*-test as described in Equation 5.8 in Section 5.4. (For practice in using this test, see Exercise 10 of Chapter 5.) Unfortunately, the *t*-test on r rejects the null hypothesis that r = 0 for simple correlation coefficients with absolute values well below 0.80. Some researchers avoid this problem by adjusting the null hypothesis to test whether r is significantly different from a specific value (like 0.30). To do this, subtract the specific value from the absolute value of r and substitute the resulting difference for r in Equation 5.8.

Be careful; all tests of simple correlation coefficients as an indication of the extent of multicollinearity share a major limitation if there are more than two explanatory variables. It is quite possible for groups of independent variables, acting together, to cause multicollinearity without any single simple correlation coefficient being high enough to prove that multicollinearity is indeed severe. As a result, tests of simple correlation coefficients must also be considered to be sufficient but not necessary tests for multicollinearity. While a high r does indeed indicate the probability of severe multicollinearity, a low r by no means proves otherwise.[5]

8.3.3 High Variance Inflation Factors (VIFs)

The use of tests to give an indication of the severity of multicollinearity in a particular sample is controversial. Some econometricians reject even the simple indicators described above, mainly because of the limitations cited. Others tend to use a variety of more formal tests.[6]

One measure of the severity of multicollinearity that is easy to use and that is gaining in popularity is the Variance Inflation Factor. The **Variance Inflation Factor** (VIF) is a method of detecting the severity of multicollinearity by looking at the extent to which a given explanatory variable can be explained by all the other explanatory variables in the equation. The VIF is an estimate of how much multicollinearity has increased the variance of an estimated coefficient; thus there is a VIF for each explanatory variable in an equation. A high VIF indicates that multicollinearity has increased the estimated variance of the estimated coefficient, yielding a decreased t-score.

Suppose you want to use the VIF to attempt to detect multicollinearity in an original equation with k independent variables:

$$Y = \beta_0 + \beta_1 X_1 + \beta_2 X_2 + \cdots + \beta_k X_k + \epsilon$$

5. Most authors criticize the use of simple correlation coefficients to detect multicollinearity in equations with large numbers of explanatory variables, but many researchers continue to do so because a scan of the simple correlation coefficients is a "quick and dirty" way to get a feel for the degree of multicollinearity in an equation.

6. Perhaps the two most used of these are the Farrar-Glauber test and the Condition number. For more on the Farrar-Glauber test, which uses partial correlation coefficients, see D. E. Farrar and R. R. Glauber, "Multicollinearity in Regression Analysis: The Problem Revisited," *Review of Economics and Statistics,* 1967, pp. 92–107. For more on the Condition number, which is a single index of the degree of multicollinearity in the overall equation, see D. A. Belsley, E. Kuh, and R. E. Welsch, *Regression Diagnostics, Identifying Influential Data and Sources of Collinearity* (New York: Wiley, 1980), Chapter 3.

Doing so requires calculating k different VIFs, one for each X_i. Calculating the VIF for a given X_i involves three steps:

1. *Run an OLS regression that has X_i as a function of all the other explanatory variables in the equation.* For example, if i = 1, then this equation would be:

$$X_1 = \alpha_1 + \alpha_2 X_2 + \alpha_3 X_3 + \cdots + \alpha_k X_k + v \qquad (8.15)$$

where v is a typical stochastic error term. Note that X_1 is not included on the right-hand side of Equation 8.15, which is referred to as an auxiliary regression. Thus there are k auxiliary regressions, one for each independent variable in the original equation.

2. *Calculate the Variance Inflation Factor for $\hat{\beta}_i$:*

$$\text{VIF}(\hat{\beta}_i) = \frac{1}{(1 - R_i^2)} \qquad (8.16)$$

where R_i^2 is the coefficient of determination (the unadjusted R^2) of the auxiliary regression in step one. Since there is a separate auxiliary regression for each independent variable in the original equation, there also is an R_i^2 and a $\text{VIF}(\hat{\beta}_i)$ for each X_i.

3. *Analyze the degree of multicollinearity by evaluating the size of the $\text{VIF}(\hat{\beta}_i)$.* The higher a given variable's VIF, the higher the variance of that variable's estimated coefficient (holding constant the variance of the error term.) Hence, the higher the VIF, the more severe the effects of multicollinearity.

Why will a high VIF indicate multicollinearity? The $\text{VIF}(\hat{\beta}_i)$ can be thought of as the ratio of the estimated variance of $\hat{\beta}_i$ to what the variance would be with no correlation between X_i and the other Xs in the equation. How high is high? An R_i^2 of one, indicating perfect multicollinearity, produces a VIF of infinity, while an R_i^2 of zero, indicating no multicollinearity at all, produces a VIF of one. Where there is no table of formal critical VIF values, a common rule of thumb is that if $\text{VIF}(\hat{\beta}_i) > 5$, the multicollinearity is severe.

For example, let's return to Equation 8.10 and calculate the VIFs for both independent variables. Both VIFs equal 36.0, confirming the quite severe multicollinearity we already know exists. It's no coincidence that the VIFs for the two variables are equal. In an equation with exactly two independent variables, the auxiliary equations are $X_1 = f(X_2)$ and $X_2 = f(X_1)$, so the two auxiliary equations will have identical R_i^2s, leading to equal VIFs.

Thus the VIF is a method of detecting multicollinearity that takes into account all the explanatory variables at once. Some authors and statistical software programs replace the VIF with its reciprocal, $(1 - R_i^2)$, called *Tolerance,* or TOL. Whether we calculate VIF or TOL is a matter of personal preference, but either way, the general approach is the most comprehensive multicollinearity detection technique we've discussed in this text.

Unfortunately, there are a few problems with using VIFs. First of all, the approach involves a lot of "busy work" because we need to calculate a VIF for each estimated slope coefficient in every equation, and to do so we need to run an auxiliary regression for each VIF. Second, as mentioned, there is no hard-and-fast VIF decision rule; many authors suggest using VIF > 10 as a rule of thumb instead of VIF > 5, especially in equations with many explanatory variables. Finally, it's possible to have multicollinear effects in an equation that has no large VIFs. For instance, if the simple correlation coefficient between X_1 and X_2 is 0.88, multicollinear effects are quite likely, and yet the VIF for the equation (assuming no other Xs) is only 4.4.

In essence, then, the VIF is a sufficient but not necessary test for multicollinearity, just like all the other tests described in this section. Indeed, as is probably obvious to the reader by now, there is no test that allows a researcher to reject the possibility of multicollinearity with any real certainty.

8.4 Remedies for Multicollinearity

What can be done to minimize the consequences that severe multicollinearity might have on your estimated equation? There is no automatic answer to this question because multicollinearity is a phenomenon that could change from sample to sample even for the same specification of a regression equation. The purpose of this section is to outline a number of alternative remedies for multicollinearity that might be appropriate under certain circumstances.

8.4.1 Do Nothing

The first step to take once severe multicollinearity has been diagnosed is to decide whether anything should be done at all. As we'll see, it turns out that every remedy for multicollinearity has a drawback of some sort, and so it often happens that doing nothing is the correct course of action.

The major reason for seriously considering doing nothing is that multicollinearity in an equation will not always reduce the t-scores enough to make them insignificant or change the $\hat{\beta}$s enough to make them differ significantly from expectations. In other words, the mere existence of multicollinearity does not necessarily mean anything. A remedy for multicollinearity should only be considered if and when the consequences cause insignificant t-scores or unreliable estimated coefficients. For example, it's possible to observe a simple correlation coefficient of .97 between two explanatory variables, each of which has an individual t-score that is significant at the 95 percent level of confidence. It makes no sense to consider remedial action in such a case, because any remedy for multicollinearity would probably cause other problems for the equation. In a sense, multicollinearity is similar to a non-life-threatening human disease that requires general anesthesia to operate on: The risk of the operation should only be undertaken if the disease is causing a significant problem.

The easiest remedy for severe multicollinearity is to drop one or more of the multicollinear variables from the equation. Unfortunately, the deletion of a multicollinear variable that theoretically belongs in an equation is fairly dangerous because now the equation will be subject to specification bias. If we drop such a variable, then we are *purposely* creating bias. Given all the effort typically expended to avoid omitted variables, it seems foolhardy to consider running that risk on purpose. As a result, experienced econometricians often will leave multicollinear variables in equations despite potential decreases in t-scores.

The final reason for considering doing nothing to offset multicollinearity is a theoretical one that would apply to all equations. Every time a regression is re-run, we're taking the risk of encountering a specification that fits because it accidentally works for the particular data-set involved, not because it is the truth. The larger the number of experiments, the greater the chances of finding the accidental result. When there is significant multicollinearity in the sample, the odds of strange results increase rapidly because of the sensitivity of the coefficient estimates to slight specification changes. Thus the case against sequential specification searches outlined in Chapter 6 is even stronger in the face of severe multicollinearity.

To sum, it is often best to leave an equation unadjusted in the face of all but extreme multicollinearity. Such advice might be difficult for beginning researchers to take, however, if they think that it's embarrassing to report that their final regression is one with insignificant t-scores. Compared to the alternatives of possible omitted variable bias or accidentally significant regression results, the low t-scores seem like a minor problem. For an example of "doing nothing" in the face of severe multicollinearity, see Section 8.5.1.

8.4.2 Drop One or More of the Multicollinear Variables

Perhaps the surest way to rid an equation of significant multicollinearity is to drop all but one of the multicollinear variables. Multicollinearity is caused by correlation between the explanatory variables; without all the multicollinear variables in the equation, the correlation no longer exists, and any multicollinear consequences also cease to exist. The coefficient of the remaining included variable also now measures almost all of the joint impact on the dependent variable of the excluded multicollinear explanatory variables.

To see how this solution would work, let's return to the student consumption function example of Equation 8.10:

$$\begin{aligned}\hat{C}_i = -367.83 &+ 0.5113Yd_i + 0.0427LA_i \\ &\quad(1.0307) \qquad\quad (0.0942) \\ t = &\quad 0.496 \qquad\qquad 0.453 \qquad \bar{R}^2 = .835\end{aligned} \tag{8.10}$$

where C = consumption, Yd = disposable income, and LA = liquid assets. When we first discussed this example, we compared this result to the same equation without the liquid assets variable (also reproduced):

$$\begin{aligned}\hat{C}_i = -471.43 &+ 0.9714Yd_i \\ &\quad(0.157) \\ t = &\quad 6.187 \qquad \bar{R}^2 = .861\end{aligned} \tag{8.11}$$

If we had instead dropped the disposable income variable, we would have obtained:

$$\begin{aligned}\hat{C}_i = -199.44 &+ 0.08876LA_i \\ &\quad(0.01443) \\ t = &\quad 6.153 \qquad \bar{R}^2 = .860\end{aligned} \tag{8.17}$$

Note that dropping one of the multicollinear variables has eliminated both the multicollinearity between the two explanatory variables and also the low t-score of the coefficient of the remaining variable. By dropping Yd, we were able to increase t_{LA} from 0.453 to 6.153. Since dropping a variable changes the meaning of the remaining coefficient (because the dropped variable is no longer being held constant), such dramatic changes are not unusual.

Assuming you want to drop a variable, how do you decide which variable to drop? In cases of severe multicollinearity, it makes no statistical difference which variable is dropped. To see this, compare the $\bar{R}^2$ and the t-score from Equation 8.11 with those in 8.17. Note that they are virtually identical. This is hardly a surprise, since the variables themselves move in virtually identical patterns. As a result, it doesn't make sense to pick the variable to be dropped on the basis of which one gives superior fit or which one is more significant (or has the expected sign) in the original equation. Instead, the theoretical underpinnings of the model should be the basis for such a decision. In the example of the student consumption function, there is more theoretical support for the hypothesis that disposable income determines consumption than there is for the liquid assets hypothesis. Therefore, Equation 8.11 should be preferred to Equation 8.17.

On occasion, the simple solution of dropping one of the multicollinear variables is a good one. For example, some inexperienced researchers include too many variables in their regressions, not wanting to have to face left-out variable bias. As a result, they often have two or more variables in their equations that are measuring essentially the same thing. In such a case, the multicollinear variables are not irrelevant, since any one of them is quite probably theoretically and statistically sound. Instead, the variables might be called **redundant;** only one of them is needed to represent the effect on the dependent variable that all of them currently represent. For example, in an aggregate demand function, it would not make sense to include disposable income and GNP because both are measuring the same thing: income. A bit more subtle is the inference that population and disposable income should not both be included in the same aggregate demand function because, once again, they really are measuring the same thing: the size of the aggregate market. As population rises, so too will income. Dropping these kinds of redundant multicollinear variables is doing nothing more than making up for a specification error; the variables should never have been included in the first place.

8.4.3 Transform the Multicollinear Variables

Often, in equations where the consequences of multicollinearity are serious enough to warrant the consideration of remedial action, the variables are all extremely important on theoretical grounds. In these cases, neither inaction nor dropping a variable is especially helpful. However, it is sometimes possible to transform the variables in the equation to get rid of at least some of the multicollinearity. The two most common such transformations are to:

1. Form a linear combination of the multicollinear variables.
2. Transform the equation into first differences (or logs).

The technique of forming a **linear combination** of two or more of the multicollinear variables consists of:

a. creating a new variable that is a function of the multicollinear variables

b. using the new variable to replace the old ones in the regression equation

For example, if X_1 and X_2 are highly multicollinear, a new variable, $X_3 = X_1 + X_2$ (or more generally, any linear combination of the two variables like $k_1X_1 + k_2X_2$) might be substituted for both of the multicollinear variables in a reestimation of the model. This technique is especially useful if the equation is going to be applied to data outside the sample, since the multicollinearity outside the sample might not exist or might not follow the same pattern that it did inside the sample.

A major disadvantage of the technique is that both portions of the linear combination are forced to have the same coefficient in the reestimated equation. For example, if $X_{3i} = X_{1i} + X_{2i}$:

$$Y_i = \beta_0 + \beta_3 X_{3i} + \epsilon_i = \beta_0 + \beta_3(X_{1i} + X_{2i}) + \epsilon_i \quad (8.18)$$

Care must be taken not to include, in a linear combination, variables with different expected coefficients (such as different expected signs) or dramatically different mean values (such as different orders of magnitude) without adjusting for these differences by using appropriate constants (ks) in the more general equation $X_3 = k_1X_1 + k_2X_2$. For example, if the two multicollinear variables were GNP and the rate of inflation, then a simple sum might swamp the inflation variable completely (depending on the units of measurement of the variables):

$$X_{3i} = GNP_i + INF_i = 3{,}250 + 0.08 = 3{,}250.08 \quad (8.19)$$

Consider how X_3 changes as GNP or INF change. If GNP doubles, so too does X_3, but if INF doubles, X_3 hardly changes at all. In most linear combinations, then, careful account must be taken of the average size and expected coefficients of the variables used to form the combination. Otherwise, the variables might cancel each other out or swamp one another in magnitude.

To see an example of this, let's form a linear combination of disposable income and liquid assets in the student consumption function and then re-run Equation 8.10 with the linear combination as the explanatory variable. As can be seen from an examination of the original data in Section 8.2.2 or from the regression run in Exercise 5, liquid assets are about ten times the size of disposable income in the sample. To bring the two into balance, disposable income could be multiplied by ten, yielding:

$$X_{3i} = 10(Yd_i) + LA_i \tag{8.20}$$

The constants in such linear combinations end up being arbitrary, but they can work fairly well. If X_3 is used to replace both explanatory variables in Equation 8.10 and a regression is estimated, we obtain:

$$\begin{aligned} \hat{C}_i &= -355.43 + 0.0467X_{3i} \\ &\qquad\qquad\quad (0.0073) \\ t &= \qquad\qquad 6.362 \qquad\qquad \bar{R}^2 = .868 \end{aligned} \tag{8.21}$$

Compare this equation with Equations 8.11 and 8.17. Notice that once again the deletion of the multicollinearity has significantly raised the t-score of the explanatory variable while having little effect on the overall significance of the equation. Interestingly, the estimated coefficients of Equation 8.21 can be calculated from the previous estimates and the equation of the linear combination.

The second kind of transformation to consider as a possible remedy for severe multicollinearity is to change the functional form of the equation. Let's look at how the conversion of an equation to first differences might decrease the amount of multicollinearity in the sample. (We won't discuss conversions to double-log or other functional forms, but the principles involved are quite similar.) A **first difference** is nothing more than the change in a variable from the previous time-period to the current time period (which we've referred to as "delta" or Δ). That is, we shall define a first difference as:

$$\Delta X_t = X_t - X_{t-1}$$

If an equation (or some of the variables in an equation) is switched from its normal specification to a first difference specification, it's quite likely that the degree of multicollinearity will be significantly reduced for two reasons. First, any change in the definitions of the variables (except a simple linear change) will change the degree of multicollinearity. Second, multicollinearity takes place most frequently (although certainly not exclusively) in time-series data, in which first differences are far less likely to move steadily upward than are the aggregates from which they are calculated. For example, while GNP might grow only five or six percent from year to year, the *change in GNP* (or the first difference) could fluctuate severely. As a result, switching all or parts of an equation to a first difference specification is likely to decrease the possibility of multicollinearity in a time-series model.

While the severity of multicollinearity sometimes can be diminished by switching to a first difference (or other) format, changing the functional form of an equation simply to avoid multicollinearity often is not worth the possible theoretical complications. For example, modeling capital stock is not the same as modeling the change in capital stock, which is investment, even though one equation can be derived from the other. If the basic purpose of running the regression were to model first differences, then the model should have been specified that way. In addition, one observation will be used up to calculate the first differences, and so the degrees of freedom will fall by one.

8.4.4 Increase the Size of the Sample

Another way to deal with multicollinearity is to attempt to increase the size of the sample so as to reduce the degree of multicollinearity. While such increases may be impossible when limitations of some sort exist, they are useful alternatives to be considered when they are feasible.

The idea behind increasing the size of the sample is that a larger data-set (often requiring new data collection) will allow more accurate estimates than a small one, since the large sample normally will reduce somewhat the variance of the estimated coefficients, diminishing the impact of the multicollinearity even if the degree of multicollinearity remains the same.

For most economic and business applications, however, this solution isn't feasible. After all, samples are typically drawn by getting all the available data that seem comparable. As a result, new data are generally impossible or quite expensive to find. Going out and generating new data is much easier in an experimental situation than it is when the samples must be generated by the passage of time.

One way to increase the sample is to pool cross-sectional and time-series data. Such a combination of data sources usually consists of the addition of cross-sectional data (typically nonmulticollinear) to multicollinear time-series data, thus potentially reducing the multicollinearity in the total sample. The major problem with this pooling is in the interpretation and use of the estimates that are generated. Unless there is reason to believe that the underlying theoretical model is the same in both settings, the parameter estimates obtained will be some sort of joint functions of the true time-series model and the true cross-sectional model. In general, such combining of different kinds of data is not recommended as a means of avoiding multicollinearity. In most cases, the unknown interpretation difficulties are worse than the known consequences of the multicollinearity.

8.5 Choosing the Proper Remedy

Of all the possibilities listed, how do you go about making a choice? There is no automatic answer to this question; an adjustment for multicollinearity that might be useful in one equation could be inappropriate in another. As a result, all that this section can accomplish is to illustrate general guidelines to follow when attempting to rid an equation of severe multicollinearity.

8.5.1 An Example of Multicollinearity Left Unadjusted

Our first case provides an example of the idea that multicollinearity is often best left unadjusted. Suppose you work in the marketing department of a hypothetical soft drink company, Mr. T's, and you build a model of the impact on sales of your firm's advertising (which is centered around the slogan, "It Packs a Punch"):

$$\begin{aligned} \hat{S}_t = 3080 - &\ 75{,}000P_t + 4.23A_t - 1.04B_t \qquad (8.22)\\ &(25{,}000) \quad\ (1.06) \quad\ (0.51)\\ t = &\ -3.00 \qquad\ 3.99 \qquad -2.04\\ &\bar{R}^2 = .825 \qquad n = 28 \end{aligned}$$

where: S_t = sales of Mr. T's soft drink in year t
P_t = average relative price of Mr. T's in year t
A_t = advertising expenditures for Mr. T's in year t
B_t = advertising expenditures for Mr. T's main competitor in year t

(Assume that there are no left-out variables. All variables are measured in real dollars; that is, the nominal values are divided, or deflated, by a price index.)

On the face of it, this is a reasonable-looking result. Estimated coefficients are significant in the directions implied by the underlying theory, and both the overall fit and the size of the coefficients seem acceptable. Suppose you now were told that advertising in the soft drink industry is cutthroat in nature and that firms tend to match their main competitor's advertising expenditures. This would lead you to suspect that significant multicollinearity was possible. Further suppose that the simple correlation coefficient between the two advertising variables is over 0.97:

$$r_{A,B} = 0.974$$

Such a correlation coefficient is evidence that there is severe multicollinearity in the equation, but there is no reason even to consider doing anything about it, because the coefficients are so powerful that their t-scores remain significant even in the face of severe multicollinearity. Unless multicollinearity causes problems in the equation, it should not be adjusted for. To change the specification might give us better looking results, but the adjustment would be likely to decrease our chances of obtaining the best possible estimates of the true coefficients. While it's certainly lucky that there were no major problems due to multicollinearity in this example, that luck is no reason to try to fix something that isn't broken.

Similarly, remember that when a variable is dropped from an equation, its effect will be absorbed by the other explanatory variables (in the form of specification bias) to the extent that they are correlated with the newly omitted variable. In the case of multicollinear variables, it's likely that the remaining multicollinear variable(s) will absorb virtually all the bias, since the variables are highly correlated. This bias may destroy whatever usefulness the estimates had before the variable was dropped. For example, if a variable, say B, was dropped from the Mr. T's equation to fix the multicollinearity, then the following would occur:

$$\begin{aligned} \hat{S}_t &= 2586 - 78{,}000P_t + 0.52A_t \\ &\qquad\quad (24{,}000) \qquad (4.32) \\ t &= \qquad -3.25 \qquad\quad 0.12 \\ \bar{R}^2 &= .531 \qquad n = 28 \end{aligned} \tag{8.23}$$

What's going on here? The Mr. T's advertising coefficient has become less instead of more significant when the multicollinear variable is dropped. To see why, first note that the expected bias on $\hat{\beta}_A$ is negative since the product of the correlation between A and B (positive) and the expected sign of the coefficient of B (negative) is negative:

$$\text{bias} = \beta_B \cdot f(r_{A,B}) = (-) \cdot (+) = - \qquad (8.24)$$

Second, this negative bias is strong enough to decrease the estimated coefficient of A until it is insignificant. While this problem could have been avoided by using a relative advertising variable (A divided by B, for instance), that formulation would have forced identical absolute coefficients on the two advertising effects. Such identical coefficients will sometimes be theoretically expected or empirically reasonable, but in most cases these kinds of constraints will force bias onto an equation that previously had none.

This example is simplistic, but its results are typical of cases in which equations are adjusted for multicollinearity by dropping a variable without regard to the effect that the deletion is going to have. The point here is that it's quite often theoretically or operationally unwise to drop a variable from an equation and that multicollinearity in such cases is best left unadjusted.

8.5.2 A More Complete Example of Multicollinearity

Finally, let's work through a more complete example of dealing with significant multicollinearity, a model of the annual demand for fish in the U.S. from 1946 to 1970.[7] Suppose that you decide to try to confirm your idea that the Pope's 1966 decision to allow Catholics to eat meat on (non-Lent) Fridays caused a shift in the demand function for fish (instead of just changing the days of the week when fish was eaten without changing the total amount of fish consumed). Let's say your hypothesized equation was:

$$F_t = f(\overset{-}{PF_t}, \overset{+}{PB_t}, \overset{+}{Yd_t}, \overset{+}{C_t}, \overset{-}{D_t}) \qquad (8.25)$$

7. The data used in this study were obtained from *Historical Statistics of the U.S., Colonial Times to 1970 Part I* (Washington, D.C.: U.S. Bureau of the Census, 1975).

where: F_t = average pounds of fish consumed per capita in year t
PF_t = price index for fish in year t
PB_t = price index for beef in year t
Yd_t = real per capita disposable income in year t (in billions of dollars)
C_t = number of Catholics in the U.S. in year t (tens of thousands)
D_t = a dummy variable equal to zero before the Pope's 1966 decision and one afterwards

and that you chose the following functional form:

$$F_t = \beta_0 + \beta_1 PF_t + \beta_2 PB_t + \beta_3 \ln Yd_t + \beta_4 C_t + \beta_5 D_t + \epsilon_t \quad (8.26)$$

A few words about this specification are in order. First, note that the method you have chosen to test your hypothesis is an intercept dummy. Since you've stated that you expect this coefficient to be negative, the null hypothesis should be the "strawman" $H_0{:}\beta_5 \geq 0$. Second, you've chosen a semi-log function to relate disposable income to the quantity of fish consumed; this is consistent with the theory that as income rises, the portion of that extra income devoted to the consumption of fish will decrease. Third, notice that you make no mention of any aggregate supply function for fish; you have perhaps assumed that "fish" is traded in an internationally competitive market in which the U.S. price plays little role (thus there is no simultaneity problem). Leaving other valid criticisms of the model aside, let's investigate the model and the consequences of multicollinearity for it.

After collecting the data (which are in Table 8.1 at the end of this section), you obtain the following OLS estimates:

$$\begin{aligned}
\hat{F}_t = & -1.99 + \underset{(0.031)}{0.039PF_t} - \underset{(0.02020)}{0.00077PB_t} + \underset{(1.87)}{1.77\ln Yd_t} \\
t = & \qquad\quad 1.27 \qquad\quad -0.0384 \qquad\quad 0.945 \\
& - \underset{(0.0033)}{0.0031C_t} - \underset{(0.353)}{0.355D_t} \qquad (8.27) \\
t = & \quad -0.958 \qquad -1.01 \\
& \bar{R}^2 = .666 \qquad n = 25
\end{aligned}$$

This result is not encouraging, since you don't have to look at a t-table to know that none of your estimated cofficients is significantly different from zero with 19 degrees of freedom. In addition, three of your coefficients have unexpected signs. Your problems could have been caused, for example, by omitted variables (biasing the coefficients), irrelevant variables (not belonging in the equation), or multicollinearity (a good guess, since this is the topic of the current chapter).

Where do you start? If you have confidence in your literature review and the theoretical work you did before estimating the equation, a good place would be to see if there are any signs of multicollinearity. Sure enough, the $\bar{R}^2$ of .666 (an ominous number in a religious regression) seems fairly high for such unanimously low t-scores. Thus the first of our "sufficient but not necessary" measures of severe multicollinearity appears to exist.

The second aspect of severe multicollinearity we mentioned has to do with the simple correlation coefficients. Looking at the variables without knowing those statistics, which pairs (or sets) of variables look likely to be significantly correlated? It appears that per capita disposable income and the number of Catholics are quite likely to be highly correlated in virtually any time-series sample from the U.S., and both appear to have been included in the equation to measure buying power. Sure enough, the correlation coefficient between C_t and $\ln Yd_t$ is .946.

In addition, it's not unreasonable to think that food prices might move together. Since the prices that we observe are equilibrium prices, supply and demand shocks might affect beef and fish price indices in similar ways. For example, an oil spill that makes fish unmarketable will admittedly raise the price of fish, but that fish price rise will almost surely shift the demand for beef upward, thus increasing the price of beef. Thus it is quite possible for prices of substitutes to tend to move together. As it turns out, the simple correlation coefficient between the two price variables is .958. With multicollinearity this severe between two variables with opposite expected signs, it is no surprise that the two coefficients "switched signs." As multicollinearity increases, the distribution of the $\hat{\beta}$s widens, and the probability of observing an unexpected sign increases.

The third method of detecting multicollinearity, the size of the Variance Inflation Factor, also indicates severe problems. All the VIFs for Equation 8.27 except VIF_D are well above the $VIF > 10$ or $VIF > 5$ indicators of severe multicollinearity:

$$
\begin{aligned}
VIF_{PF} &= 42.9 \\
VIF_{\ln Yd} &= 23.5 \\
VIF_{PB} &= 18.8 \\
VIF_{C} &= 18.5 \\
VIF_{D} &= 4.4
\end{aligned}
$$

So there appears to be significant multicollinearity in the model. What, if anything, should you do about it? Before going on with this section, go back over Equation 8.27 and review not only the estimates but also the underlying theory.

The easiest multicollinearity to cope with is between income and the number of Catholics. Independently, either variable is quite likely to be significant because each represents the increase in the buying power of the market over time. Together, however, they ruin each other's chances because of multicollinearity. As a result, one should be dropped as a "redundant" multicollinear variable; they should never have been included together in the first place. Given that the logic behind including the number of Catholics in a per capita fish demand equation is fairly weak, you decide to drop C.

At this point, you would normally drop C and re-run the equation without any other changes, since simultaneously making more than one adjustment to your specification makes it difficult to distinguish the effects of one change from another. Unfortunately, while dropping C certainly eliminates a redundant variable from the equation, the new equation still has severe multicollinearity as measured by all three of our detection techniques. Because the new equation[8] is quite similar to Equation 8.27, let's save space by thinking about how we might solve the obvious multicollinearity involving the price variables before presenting any further estimates.

In the case of the prices, we don't have the option of dropping one of the multicollinear variables because both PB and PF are too theoretically important to the model. In such a situation, it's worth investigating another of our potential remedies, transforming the variables. For example, one alternative would be to create a transformation of the two price variables by dividing one by the other to form a relative price variable:

$$RP_t = PF_t/PB_t$$

Such a variable would make sense only if theory called for keeping both variables in the equation and if the two coefficients could be

8. The actual results are:

$$\begin{aligned} \hat{F}_i = 7.96 &+ \underset{(0.03)}{0.03}PF_i + \underset{(0.019)}{0.0047}PB_i + \underset{(1.15)}{0.36}\ln Yd_i - \underset{(0.26)}{0.12}D_i \\ t = &\qquad 0.98 \qquad\quad 0.24 \qquad\qquad 0.31 \qquad\quad -0.48 \\ &\bar{R}^2 = .667 \qquad n = 25 \end{aligned}$$

expected to be close in absolute value but of opposite signs.[9] Choosing to use a relative price variable in effect would be hypothesizing that while consumers might not be sophisticated enough always to consider real prices, they do compare prices of substitutes before making their purchases. Depending on your perception of the underlying theory, you could make a strong case for either approach (that is, dropping C and shifting to real prices or dropping C and replacing both prices with a relative price variable). For the purpose of discussion, suppose you decide to estimate the latter equation:

$$F = f(\overset{-}{RP}, \overset{+}{Yd}, \overset{-}{D}) \tag{8.28}$$

obtaining

$$\begin{aligned} \hat{F}_t = & -5.17 - 1.93RP_t + 2.71\ \ln Yd_t + 0.0052D_t \\ & \qquad\quad (1.43) \qquad\quad (0.66) \qquad\quad (0.2801) \\ t = & \qquad -1.35 \qquad\quad 4.13 \qquad\quad 0.019 \\ & \bar{R}^2 = .588 \qquad n = 25 \end{aligned} \tag{8.29}$$

Although these are all questions of judgment, the two changes appear to have worked reasonably well in terms of ridding the equation of much of its severe multicollinearity. More important, once we decide that this specification is good enough, we can now test the hypothesis that was the real reason for the research project. What was the result? If this specification is at all close to the best one, then the null hypothesis of no effect cannot be rejected. For all intents and purposes, it appears that the Pope's decision did not cut down on consumption of fish (the coefficient is quite insignificant).[10]

9. To see why opposite signs are required, note that an increase in PF will increase RP while an increase in PB will decrease it. Unless PF and PB are hypothesized to have opposite effects on the dependent variable, this relative price variable will not work at all. To test your understanding of this point, attempt to figure out the expected sign of the coefficient of RP in this equation before going on with this example. Note, by the way, that a relative price ratio such as RP is a real variable even if PF and PB are not.

10. This is in contrast with the findings of the original empirical work on the issue, Frederick Bell's "The Pope and the Price of Fish," *American Economic Review*, Dec. 1968, pp. 1346–1350. Bell built monthly models of the price of seven different species of fish and determined that the Pope's decision had a significant negative impact on the demand for fish in New England in the first nine months after the decision. Since our example was misspecified purposely to cause multicollinearity and then respecified in part to allow an example of the use of a relative price variable, Equation 8.27 should not be considered to refute Bell's result. It is interesting, however, that none of the specifications considered in constructing this example included a significantly negative coefficient of the dummy variable.

TABLE 8.1 DATA FOR THE FISH/POPE EXAMPLE

Year	F	PF	PB	Yd
1946	12.8	56.0	50.1	1606
1947	12.3	64.3	71.3	1513
1948	13.1	74.1	81.0	1567
1949	12.9	74.5	76.2	1547
1950	13.8	73.1	80.3	1646
1951	13.2	83.4	91.0	1657
1952	13.3	81.3	90.2	1678
1953	13.6	78.2	84.2	1726
1954	13.5	78.7	83.7	1714
1955	12.9	77.1	77.1	1795
1956	12.9	77.0	74.5	1839
1957	12.8	78.0	82.8	1844
1958	13.3	83.4	92.2	1831
1959	13.7	84.9	88.8	1881
1960	13.2	85.0	87.2	1883
1961	13.7	86.9	88.3	1909
1962	13.6	90.5	90.1	1969
1963	13.7	90.3	88.7	2015
1964	13.5	88.2	87.3	2126
1965	13.9	90.8	93.9	2239
1966	13.9	96.7	102.6	2335
1967	13.6	100.0	100.0	2403
1968	14.0	101.6	102.3	2486
1969	14.2	107.2	111.4	2534
1970	14.8	118.0	117.6	2610

Source: *Historical Statistics of the U.S., Colonial Times to 1970 Part 1*

Finally, notice that someone else might take a completely different approach to alleviating the severe multicollinearity in this sample. There is no obviously correct remedy. Indeed, if you want to be sure that your choice of a specification did not influence your inability to reject the null hypothesis about β_D, you might see how sensitive that conclusion is to an alternative approach towards fixing the multicollinearity. (In such a case, both results would have to be part of the research report.)

8.6 Summary

1. Perfect multicollinearity is the violation of the assumption that no explanatory variable is a perfect linear function of other explanatory variables. Perfect multicollinearity results in indeterminate es-

timates of the regression coefficients and infinite standard errors of those estimates.

2. Imperfect multicollinearity, which is what is typically meant when the word "multicollinearity" is used, is a linear relationship between two or more independent variables that is so strong that it can significantly affect the estimation of that equation. Multicollinearity is a sample phenomenon as well as a theoretical one. Different samples can exhibit different degrees of multicollinearity.

3. The major consequence of severe multicollinearity is to increase the variances of the estimated regression coefficients and therefore decrease the calculated t-scores of those coefficients. Multicollinearity causes no bias in the estimated coefficients, and it has little effect on the overall significance of the regression or on the estimates (or variances) of any nonmulticollinear explanatory variables.

4. Severe multicollinearity causes difficulty in the identification of the separate effects of the multicollinear variables in a regression equation. In addition, coefficient estimates will become very sensitive to changes in specification in the presence of multicollinearity.

5. The more severe the multicollinearity, the worse the consequences of that multicollinearity. Since multicollinearity exists, to one degree or another, in virtually every data-set, the question to be asked in detection is how severe the multicollinearity in a particular sample is.

6. A useful method for the detection of severe multicollinearity consists of three questions:
   a. Is $\bar{R}^2$ high with all low individual t-scores?
   b. Are the simple correlation coefficients between the explanatory variables high?
   c. Are the variance inflation factors high?

   If the answers are yes, then multicollinearity certainly exists, but multicollinearity can also exist even if the answers are no.

7. The four most common remedies for multicollinearity are:
   a. Do nothing (and thus avoid specification bias).
   b. Drop some multicollinear variables (especially "redundant" ones).
   c. Transform the multicollinear variables or the equation.
   d. Increase the sample.

8. Quite often, doing nothing is the best remedy for multicollinearity. If the multicollinearity has not decreased t-scores to the point of insignificance, then no remedy should even be considered. Even if the t-scores are insignificant, remedies should be undertaken cautiously, because all impose costs on the estimation that may be greater than the potential benefit of ridding the equation of multicollinearity.

Exercises

(Answers to even-numbered exercises are in Appendix A.)

1. Write out the meaning of each of the following terms without reference to the book (or your notes), and then compare your definition with the version in the text for each:
   a. perfect multicollinearity
   b. severe imperfect multicollinearity
   c. dominant variable
   d. linear combination
   e. first difference
   f. variance inflation factor

2. Beginning researchers quite often believe that they have multicollinearity when they've accidentally included in their equation two or more explanatory variables that basically serve the same purpose or are in essence measuring the same thing. Which of the following pairs of variables are likely to include such a "redundant" variable?
   a. GNP and NNP in a macroeconomic equation of some sort
   b. the price of refrigerators and the price of washing machines in a durable-goods demand function
   c. the number of acres harvested and the amount of seed used in an agricultural supply function
   d. long-term interest rates and the money supply in an investment function

3. A researcher once attempted to estimate an asset-demand equation that included the following three explanatory variables: current wealth W_t, wealth in the previous quarter, W_{t-1}, and the change in wealth $\Delta W_t = W_t - W_{t-1}$. What problem did this researcher encounter? What should have been done to solve this problem?

4. In each of the following situations, determine whether the variable involved is a "dominant variable":

a. "games lost in year t" in an equation for the number of games won in year t by a baseball team that plays the same number of games each year
b. "number of Woody's restaurants" in a model of the total sales of the entire Woody's chain of restaurants
c. "disposable income" in an equation for aggregate consumption expenditures
d. "number of tires purchased" in an annual model of the production of automobiles for a "Big Three" auto maker that does not make its own tires
e. "number of acres planted" in an agricultural supply function

5. The formation of linear combinations is an arbitrary process. The linear combination used between liquid assets and disposable income in Section 8.4.3 [$X_{3i} = 10(Yd_i) + LA_i$] could have been justified in two ways. First, the mean of the liquid assets variable is almost exactly ten times the mean of the disposable income variable. To ensure that one does not overwhelm the other, an adjustment by a factor of ten makes sense. Other researchers prefer to regress one of the explanatory variables on the other. In this case, we also obtain evidence that a multiple of ten makes sense:

$$\widehat{LA}_t = -2428.6 + 10.786Yd_t$$
$$(0.8125)$$
$$t = 13.274 \qquad \bar{R}^2 = .967$$

Use this same general technique to form linear combinations of the following variables:
a. height and weight in Table 1.1 (assume both are explanatory variables)
b. P and I from the Woody's data-set in Table 3.1
c. Y and Yd in Table 6.3 (assume both are explanatory Xs)

6. You've been hired by the Dean of Students' Office to help reduce damage done to dorms by rowdy students, and your first step is to build a cross-sectional model of last term's damage to each dorm as a function of the attributes of that dorm:

$$\hat{D}_i = 210 + 733F_i - 0.805S_i + 74.0A_i$$
$$(253) \qquad (0.752) \qquad (12.4)$$
$$n = 33 \qquad \bar{R}^2 = .84$$

where: D_i = the amount of damage (in dollars) done to the *i*th dorm last term

F_i = the percentage of the *i*th dorm residents who are frosh

S_i = the number of students who live in the *i*th dorm

A_i = the number of incidents involving alcohol that were reported to the Dean of Students' Office from the *i*th dorm last term (incidents involving alcohol may or may not involve damage to the dorm)

a. Hypothesize signs, calculate t-scores, and test hypotheses for this result (five percent level).
b. What problems (out of left-out variables, irrelevant variables, and multicollinearity) appear to exist in this equation? Why?
c. Suppose you were now told that the simple correlation coefficient between S_i and A_i was .94; would that change your answer? How?
d. Is it possible that the unexpected sign of $\hat{\beta}_s$ could have been caused by multicollinearity? Why?

7. Suppose your friend was modeling the impact that changes in income had on consumption in a quarterly model and discovered that increases in income do not complete their impact on consumption until at least a year has gone by. As a result, your friend estimated the following model:

$$C_t = \beta_0 + \beta_1 Yd_t + \beta_2 Yd_{t-1} + \beta_3 Yd_{t-2} + \beta_4 Yd_{t-3} + \epsilon_t$$

a. Would this equation be subject to perfect multicollinearity?
b. Would this equation be subject to imperfect multicollinearity?
c. What, if anything, could be done to rid this equation of any multicollinearity it might have? (One answer to this question, the autoregressive approach to distributed lags, will be covered in Chapter 12.)

8. You decide to see if the number of votes a baseball player receives in the Most Valuable Player election is more a function of batting average than it is of home runs or runs batted in, and you collect the following data-set from the 1983 National League:

Name	Votes	BA	HR	RBI
Murphy	371	.302	36	121
Dawson	249	.299	32	113
Schmidt	223	.255	40	109
Guererro	212	.298	32	103
Raines	97	.298	11	71
Cruz	87	.318	14	92
Thon	78	.286	20	79
Madlock	52	.323	12	68

Just as you are about to run the regression, your friend (trying to get back at you for your comments on Exercise 7) warns you that you probably have multicollinearity in the data-set.

a. What should you do about your friend's warning before running the regression?
b. Run the regression implied above: $V = f(\overset{+}{BA}, \overset{+}{HR}, \overset{+}{RBI})$ on the data-set above. What signs of multicollinearity are there?
c. What suggestions would you make for another run of this equation? (If you did not get a chance to run the equation yourself, refer to Appendix A before answering this part of the question.) In particular, what would you do about multicollinearity?

9. A full-scale regression model for the total annual gross sales in thousands of dollars of J. C. Quarter's durable goods for the years 1965–1990 produces the following result (all measurements are in real dollars—or billions of real dollars). Standard errors are in parentheses:

$$\widehat{SQ}_t = -7.2 + \underset{(250.1)}{200.3PC_t} - \underset{(125.6)}{150.6PQ_t} + \underset{(40.1)}{20.6Y_t} - \underset{(10.6)}{15.8C_t} + \underset{(103.8)}{201.1\ N_t}$$

where: SQ_t = sales of durable goods at J. C. Quarter's in year t
PC_t = average price of durables in year t at J. C. Quarter's main competition

PQ_t = average price of durables at J. C. Quarter's in year t
Y_t = U.S. gross national product in year t
C_t = U.S. aggregate consumption in year t
N_t = number of J. C. Quarter's stores open in year t

a. Hypothesize signs, calculate t-scores, and test hypotheses for this result (five percent level).
b. What problems (out of omitted variables, irrelevant variables, and multicollinearity) appear to exist in this equation? Explain.
c. Suppose you were now told that the $\bar{R}^2$ was .821, that $r_{Y,C}$ was .993, and that $r_{PC,PQ}$ was .813. Would this change your answer to the above question? How?
d. What recommendation would you make for a re-run of this equation with different explanatory variables? Why?

10. A cross-sectional regression was run on a sample of 44 states in an effort to understand Federal defense spending by state (standard errors in parentheses):

$$\hat{S}_i = -148.0 + \underset{(0.027)}{0.841C_i} - \underset{(0.1664)}{0.0115P_i} - \underset{(0.0092)}{0.0078E_i}$$

where: S_i = spending (millions of dollars) on defense in the *i*th state in 1982
C_i = contracts (millions of dollars) awarded in the *i*th state in 1982 (contracts are often for many years of service)
P_i = payroll (millions of dollars) for workers in defense-oriented industries in the *i*th state in 1982
E_i = number of civilians employed in defense-oriented industries in the *i*th state in 1982

a. Hypothesize signs, calculate t-scores, and test hypotheses for this result (five percent level).
b. The VIFs for this equation are all above 20, and those for P and C are above 30. What conclusion does this information allow you to draw?
c. What recommendation would you make for a re-run of this equation with a different specification? Explain your answer.

11. Consider the following regression result paraphrased from a study

conducted by the admission office at the Stanford Business School (standard errors in parentheses):

$$\hat{G}_i = 1.00 + \underset{(0.001)}{0.005M_i} - \underset{(0.10)}{0.10A_i} + \underset{(0.20)}{0.20B_i} + \underset{(0.10)}{0.25E_i}$$
$$\bar{R}^2 = 0.20 \qquad n = 1000$$

where: G_i = the Stanford Business School GPA of the *i*th student (4 = high)
M_i = the score on the graduate management admission test of the *i*th student (800 = high)
A_i = the age of the *i*th student
B_i = the number of years of business experience of the *i*th student
E_i = equal to 1 if the *i*th student was an economics major and 0 otherwise

a. Theorize the expected signs of all the coefficients (try not to look at the results) and test these expectations with appropriate hypotheses (including choosing a significance level).
b. Do any problems appear to exist in this equation? Explain your answer.
c. How would you react if someone suggested a polynomial functional form for A? Why?
d. What suggestions (if any) would you have for another run of this equation?

12. Calculating VIFs typically involves running sets of auxiliary regressions, one regression for each independent variable in an equation. To get practice with this procedure, calculate the following:
a. the VIFs for C, P, and I from the Woody's data in Table 3.1.
b. the VIFs for PB, PC, and LYD from the chicken demand data in Table 6.3 (using Equation 6.5).

13. Calculating a VIF doesn't always involve running an auxiliary regression. To see this, calculate the following:
a. the VIF for X_1 in an equation where X_1 and X_2 are the only independent variables, given that the VIF for X_2 is 3.8 and n is 28
b. the VIF for X_1 in an equation where X_1 and X_2 are the only independent variables, given that the simple correlation coefficient between X_1 and X_2 is 0.80 and n is 15

14. Let's assume that you were hired by the Department of Agriculture to do a cross-sectional study of weekly expenditures for food consumed at home by the *i*th household (F_i) and that you estimated the following equation (standard errors in parentheses):

$$\hat{F}_i = -10.50 + \underset{(0.7)}{2.1Y_i} - \underset{(.05)}{.04Y_i^2} + \underset{(2.0)}{13.0H_i} - \underset{(2.0)}{2.0A_i}$$
$$\bar{R}^2 = .46 \qquad n = 235$$

where: Y_i = the weekly disposable income of the *i*th household
H_i = the number of people in the *i*th household
A_i = the number of children (under 19) in the *i*th household

a. Create and test appropriate hypotheses at the 90 percent level.
b. Does the functional form of this equation appear reasonable? Isn't the estimated coefficient for Y impossible? (There's no way that people can spend twice their income on food.) Explain your answer.
c. Which econometric problems (out of omitted variables, irrelevant variables, and multicollinearity) appear to exist in this equation? Explain your answer.
d. Suppose that you were now told that the VIFs for A and H were both between 5 and 10. How does this change your answer to part c above?
e. Would you suggest changing this specification for one final run of this equation? Why? How? What are the possible econometric costs of estimating another specification?

15. Suppose you hear that because of the asymmetry of the human heart, the heartbeat of any individual is a function of the difference between the lengths of that individual's legs rather than of the length of either leg. You decide to collect data and build a regression model to test this hypothesis, but you can't decide which of the following two models to estimate:[11]

$$\text{Model A: } H_i = \alpha_0 + \alpha_1 R_i + \alpha_2 L_i + \epsilon_i$$
$$\text{Model B: } H_i = \beta_0 + \beta_1 R_i + \beta_2 (L_i - R_i) + \epsilon_i$$

11. Potluri Rao and Roger Miller, *Applied Econometrics* (Belmont, CA: Wadsworth, 1971), p. 48.

where: H_i = the heartbeat of the *i*th cardiac patient
R_i = the length of the *i*th patient's right leg
L_i = the length of the *i*th patient's left leg

a. Model A seems more likely to encounter multicollinearity than does Model B, at least as measured by the simple correlation coefficient. Why? What remedy for this multicollinearity would you recommend?
b. Suppose you estimate a set of coefficients for Model A. Can you calculate estimates of the coefficients of Model B from this information? If so, how? If not, why not?
c. What does your answer to part b tell you about which of the two models is more vulnerable to multicollinearity?
d. Suppose you had dropped L_i from Model A because of the high simple correlation coefficient between L_i and R_i. What would this deletion have done to your answers to parts b and c?

8.7 Appendix: The SAT Interactive Regression Learning Exercise

Econometrics is difficult to learn by reading examples, no matter how expertly done and explained they are. Most econometricians, the authors included, had trouble understanding how to use econometrics, particularly in the area of specification choice, until they ran their own regression projects. This is because there's an element of econometric understanding that is better learned by *doing* than by reading about what someone else is doing.

Unfortunately, mastering the art of econometrics by running your own regression projects without any feedback is also difficult because it takes quite a while to learn to avoid some fairly simple mistakes. Probably the best way to learn is to work on your own regression project, analyzing your own problems and making your own decisions, but with a more experienced econometrician nearby to give you one-on-one feedback on exactly which of your decisions were inspired and which were flawed (and why).

This section is an attempt to give you an opportunity to make independent specification decisions and to then get feedback on the advantages or disadvantages of those decisions. Using the interactive learning exercise of this section requires neither a computer nor a tutor, although either would certainly be useful. Instead, we have designed an exercise that can be used on its own to help to bridge the gap between the typical econometrics examples (which require no decision making) and the typical econometrics projects (which give little feedback). Three additional interactive learning exercises are presented in Chapter 11.

STOP!

To get the most out of the exercise, it's important to follow the instructions carefully. Reading the pages "in order" as with any other example will waste your time, because once you have seen even a few of the results, the benefits to you of making specification decisions will diminish. In addition, you shouldn't look at any of the regression results until you have specified your first equation. This same warning applies to the interactive learning exercises in Chapter 11.

8.7.1 Building a Model of Scholastic Aptitude Test Scores

The dependent variable for this interactive learning exercise is the combined SAT score, math plus verbal, earned by students in the class of 1989 at Arcadia High School. Arcadia is an upper-middle-class suburban community located near Los Angeles, California. Out of a graduating class of about 640, a total of 65 students who had taken the SATs were randomly selected for inclusion in the data-set. In cases where a student had taken the test more than once, the highest score was recorded.

A review of the literature on the SAT shows many more psychological studies and popular press articles than econometric regressions. Many articles have been authored by critics of the SAT, who maintain (among other things) that it is biased against women and minorities. In support of this argument, these critics have pointed to national average scores for women and some minorities, which in recent years have been significantly lower than the national averages for white males. Any reader interested in reviewing a portion of the applicable literature should do so now before continuing on with the section.[12]

12. See Jay Amberg, "The SAT," *American Scholar,* Autumn 1982, pp. 535–542, and James Fallows, "The Tests and the 'Brightest': How Fair are the College Boards?" *The Atlantic,* Feb. 1980, pp. 37–48. We are grateful to former Occidental student Bob Sego for his help in preparing this interactive exercise.

If you were going to build a single-equation linear model of SAT scores, what factors would you consider? First of all, you'd want to include some measures of a student's academic ability. Three such variables are cumulative high school grade point average (GPA) and participation in advanced placement math and English courses (APMATH and APENG). Advanced placement (AP) classes are academically rigorous courses that may help a student do well on the SAT. More important, students are invited to be in AP classes on the basis of academic potential, and students who choose to take AP classes are revealing their interest in academic subjects, both of which bode well for SAT scores. GPAs at Arcadia High School are weighted GPAs; each semester that a student takes an AP class adds one extra point to his or her total grade points. (For example, a semester grade of "A" in an AP math class counts for five grade points as opposed to the conventional four points.)

A second set of important considerations includes qualitative factors which may affect performance on the SAT. Available dummy variables in this category include measures of a student's gender (GEND), ethnicity (RACE), and native language (ESL). All of the students in the sample are either Asian or Caucasian, and RACE is assigned a value of one if a student is Asian. Asian students currently make up about 30 percent of the student body at Arcadia High. The ESL dummy is given a value of one if English is a student's second language. In addition, studying for the test may be relevant, so a dummy variable indicating whether or not a student has attended an SAT seminar or preparation class (PREP) is also included in the dataset.

To sum, the explanatory variables available for you to choose for your model are:

GPA_i = the weighted GPA of the *i*th student

$APMATH_i$ = a dummy variable equal to 1 if the *i*th student has taken AP math, 0 otherwise

$APENG_i$ = a dummy variable equal to 1 if the *i*th student has taken AP English, 0 otherwise

AP_i = a dummy variable equal to 1 if the *i*th student has taken AP math and/or AP English, 0 if the *i*th student has taken neither

ESL_i = a dummy variable equal to 1 if English is not the *i*th student's first language, 0 otherwise

$RACE_i$ = a dummy variable equal to 1 if the *i*th student is Asian, 0 if the student is Caucasian

$GEND_i$ = a dummy variable equal to 1 if the *i*th student is male, 0 if the student is female

$PREP_i$ = a dummy variable equal to 1 if the *i*th student has attended a SAT preparation course, 0 otherwise

The data for these variables are presented in Table 8.2.

TABLE 8.2 DATA FOR THE SAT INTERACTIVE LEARNING EXERCISE

SAT	GPA	APMATH	APENG	AP	ESL	GEND	PREP	RACE
1060	3.74	0	1	1	0	0	0	0
740	2.71	0	0	0	0	0	1	0
1070	3.92	0	1	1	0	0	1	0
1070	3.43	0	1	1	0	0	1	0
1330	4.35	1	1	1	0	0	1	0
1220	3.02	0	1	1	0	1	1	0
1130	3.98	1	1	1	1	0	1	0
770	2.94	0	0	0	0	0	1	0
1050	3.49	0	1	1	0	0	1	0
1250	3.87	1	1	1	0	1	1	0
1000	3.49	0	0	0	0	0	1	0
1010	3.24	0	1	1	0	0	1	0
1320	4.22	1	1	1	1	1	0	1
1230	3.61	1	1	1	1	1	1	1
840	2.48	1	0	1	1	1	0	1
940	2.26	1	0	1	1	0	0	1
910	2.32	0	0	0	1	1	1	1
1240	3.89	1	1	1	0	1	1	0
1020	3.67	0	0	0	0	1	0	0
630	2.54	0	0	0	0	0	1	0
850	3.16	0	0	0	0	0	1	0
1300	4.16	1	1	1	1	1	1	0
950	2.94	0	0	0	0	1	1	0
1350	3.79	1	1	1	0	1	1	0
1070	2.56	0	0	0	0	1	0	0
1000	3.00	0	0	0	0	1	1	0
770	2.79	0	0	0	0	0	1	0
1280	3.70	1	0	1	1	0	1	1
590	3.23	0	0	0	1	0	1	1
1060	3.98	1	1	1	1	1	0	1
1050	2.64	1	0	1	0	0	0	0
1220	4.15	1	1	1	1	1	1	1
930	2.73	0	0	0	0	1	1	0
940	3.10	1	1	1	1	0	0	1
980	2.70	0	0	0	1	1	1	1
1280	3.73	1	1	1	0	1	1	0
700	1.64	0	0	0	1	0	1	1
1040	4.03	1	1	1	1	0	1	1
1070	3.24	0	1	1	0	1	1	0
900	3.42	0	0	0	0	1	1	0

TABLE 8.2 DATA FOR THE SAT INTERACTIVE LEARNING EXERCISE

SAT	GPA	APMATH	APENG	AP	ESL	GEND	PREP	RACE
1430	4.29	1	1	1	0	1	0	0
1290	3.33	0	0	0	0	1	0	0
1070	3.61	1	0	1	1	0	1	1
1100	3.58	1	1	1	0	0	1	0
1030	3.52	0	1	1	0	0	1	0
1070	2.94	0	0	0	0	1	1	0
1170	3.98	1	1	1	1	1	1	0
1300	3.89	1	1	1	0	1	0	0
1410	4.34	1	1	1	1	0	1	1
1160	3.43	1	1	1	0	1	1	0
1170	3.56	1	1	1	0	0	0	0
1280	4.11	1	1	1	0	0	1	0
1060	3.58	1	1	1	1	0	1	0
1250	3.47	1	1	1	0	1	1	0
1020	2.92	1	0	1	1	1	1	1
1000	4.05	0	1	1	1	0	0	1
1090	3.24	1	1	1	1	1	1	1
1430	4.38	1	1	1	1	0	0	1
860	2.62	1	0	1	1	0	0	1
1050	2.37	0	0	0	0	1	0	0
920	2.77	0	0	0	0	0	1	0
1100	2.54	0	0	0	0	1	1	0
1160	3.55	1	0	1	1	1	1	1
1360	2.98	0	1	1	1	0	1	0
970	3.64	1	1	1	0	0	1	0

Now:

1. Hypothesize expected signs for each of these variables in an equation for the SAT score of the *i*th student. Examine each variable carefully; what is the theoretical content of your hypothesis?
2. Choose carefully the best set of explanatory variables. Start off by including GPA, APMATH, and APENG; what other variables do you think should be specified? Don't simply include all the variables, intending to drop the insignificant ones. Instead, think through the problem carefully and find the best possible equation.

Once you've specified your equation, you're ready to move to Section 8.7.2. Keep following the instructions in the exercise until you have specified your equation completely. You may take some time to think over the questions contained in Section 8.7.2 or take a break, but when you return to the interactive exercise make sure to go back to the exact point from which you left rather than starting all over again. To the extent you can do it, try to avoid looking at the hints until after you've completed the entire project. The hints are there to help you if you get stuck, not to allow you to check every decision you make.

One final bit of advice: Each regression result is accompanied by a series of questions. Take the time to answer all these questions, in writing if possible. Rushing through this interactive exercise will lessen its effectiveness.

8.7.2 The SAT Score Interactive Regression Exercise

To start, choose the specification you'd like to estimate, find the regression run number[13] of that specification in the following table, and then turn to that regression. Note that the simple correlation coefficient matrix for this data-set is in Table 8.3 just before the results begin.

All the equations include SAT as the dependent variable and GPA, APMATH, and APENG as explanatory variables. Find below the combination of explanatory variables (from ESL, GEND, PREP, and RACE) that you wish to include and go to the indicated regression:

None of them, go to regression #8.1
ESL only, go to regression #8.2
GEND only, go to regression #8.3
PREP only, go to regression #8.4
RACE only, go to regression #8.5
ESL and GEND, go to regression #8.6
ESL and PREP, go to regression #8.7
ESL and RACE, go to regression #8.8
GEND and PREP, go to regression #8.9
GEND and RACE, go to regression #8.10
PREP and RACE, go to regression #8.11
ESL, GEND and PREP, go to regression #8.12
ESL, GEND and RACE, go to regression #8.13
ESL, PREP and RACE, go to regression #8.14
GEND, PREP and RACE, go to regression #8.15
All four, go to regression #8.16

13. All the regression results in this book were obtained using the ECSTAT regression package. A student version of ECSTAT (on a 5¼ inch PC-compatible diskette) is attached to the inside front cover of this text. ECSTAT is an extremely user-friendly software package, but this book is designed to be compatible with any standard computer regression package.

TABLE 8.3 MEANS, VARIANCES, AND SIMPLE CORRELATION COEFFICIENTS FOR THE SAT INTERACTIVE REGRESSION LEARNING EXERCISE

Means, Variances, and Correlations

Sample Range: 1-65

Variable	Mean	Standard Dev	Variance
SAT	1075.538	189.8828	36055.48
GPA	3.362308	0.608008	0.369673
APMATH	0.523077	0.499467	0.249467
APENG	0.553846	0.497092	0.247101
AP	0.676923	0.467652	0.218698
ESL	0.400000	0.489898	0.240000
GEND	0.492308	0.499941	0.249941
PREP	0.738462	0.439473	0.193136
RACE	0.323077	0.467652	0.218698

	Correlation Coeff		Correlation Coeff
APMATH,GPA	0.497	GPA,SAT	0.678
APENG,SAT	0.608	APMATH,SAT	0.512
APENG,APMATH	0.444	APENG,GPA	0.709
AP,SAT	0.579	AP,GPA	0.585
AP,APMATH	0.723	AP,APENG	0.769
ESL,GPA	0.071	ESL,SAT	0.024
ESL,APENG	0.037	ESL,APMATH	0.402
GEND,GPA	−0.008	ESL,AP	0.295
GEND,APENG	−0.044	GEND,SAT	0.293
GEND,ESL	−0.050	GEND,APMATH	0.077
PREP,SAT	−0.100	GEND,AP	−0.109
PREP,APMATH	−0.147	PREP,GPA	0.001
PREP,AP	−0.111	PREP,APENG	0.029
PREP,GEND	−0.044	PREP,ESL	−0.085
RACE,SAT	−0.085	RACE,GPA	−0.025
RACE,APMATH	0.330	RACE,APENG	−0.107
RACE,AP	0.195	RACE,ESL	0.846
RACE,GEND	−0.022	RACE,PREP	−0.187

Regression Run 8.1

```
ORDINARY LEAST SQUARES                    DEPENDENT VARIABLE IS SAT
SAMPLE RANGE:              1-65

                        COEFFICIENT        STANDARD ERROR        T-SCORE
    CONST               545.2537             117.8141           4.628086
    GPA                 131.8512             40.86212           3.226735
    APMATH              78.60445             39.13018           2.008793
    APENG               82.77424             48.40687           1.709969

  R-squared               0.524341        Mean of depend var      1075.538
  Adjusted R-squared      0.500948        Std dev depend var      191.3605
  Std err of regress      135.1840        Residual sum            1.79625E-11
  Durbin Watson stat      1.998585        Sum squared resid       1114757.
  F Statistic             22.41440
```

Answer each of the following questions for the above regression run.

a. Evaluate this result with respect to its economic meaning, overall fit, and the signs and significance of the individual coefficients.

b. What econometric problems (out of omitted variables, irrelevant variables, or multicollinearity) does this regression have? Why? If you need feedback on your answer, see hint #2 in the material on this chapter in Appendix A.

c. Which of the following statements comes closest to your recommendation for further action to be taken in the estimation of this equation?
   i. No further specification changes are advisable (go to Section 8.7.3).
   ii. I would like to add ESL to the equation (go to run #8.2).
   iii. I would like to add GEND to the equation (go to run #8.3).
   iv. I would like to add PREP to the equation (go to run #8.4).
   v. I would like to add RACE to the equation (go to run #8.5).

If you need feedback on your answer, see hint #6 in the material on this chapter in Appendix A.

Regression Run 8.2

```
ORDINARY LEAST SQUARES                        DEPENDENT VARIABLE IS SAT
SAMPLE RANGE:               1-65

                         COEFFICIENT        STANDARD ERROR        T-SCORE
      CONST               566.7551            118.6016           4.778644
      GPA                 128.3402            40.78800           3.146519
      APMATH              101.5886            43.19023           2.352121
      APENG               77.30713            48.40462           1.597102
      ESL                -46.72721            37.88203          -1.233493

   R-squared              0.536104       Mean of depend var      1075.538
   Adjusted R-squared     0.505178       Std dev depend var      191.3605
   Std err of regress     134.6098       Residual sum            1.54046E-11
   Durbin Watson stat     2.027210       Sum squared resid       1087187.
   F Statistic            17.33489
```

Answer each of the following questions for the above regression run.

a. Evaluate this result with respect to its economic meaning, overall fit, and the signs and significance of the individual coefficients.

b. What econometric problems (out of omitted variables, irrelevant variables, or multicollinearity) does this regression have? Why? If you need feedback on your answer, see hint #3 in the material on this chapter in Appendix A.

c. Which of the following statements comes closest to your recommendation for further action to be taken in the estimation of this equation?
   i. No further specification changes are advisable (go to Section 8.7.3).
   ii. I would like to drop ESL to the equation (go to run #8.1).
   iii. I would like to add GEND to the equation (go to run #8.6).
   iv. I would like to add RACE to the equation (go to run #8.8).
   v. I would like to add PREP to the equation (go to run #8.7).

If you need feedback on your answer, see hint #6 in the material on this chapter in Appendix A.

Regression Run 8.3

```
ORDINARY LEAST SQUARES                    DEPENDENT VARIABLE IS SAT
SAMPLE RANGE:            1-65

                         COEFFICIENT       STANDARD ERROR        T-SCORE
      CONST              491.8225            108.5429            4.531135
      GPA                131.5798            37.29970            3.527638
      APMATH             65.04046            35.91313            1.811049
      APENG              94.10841            44.29652            2.124510
      GEND               112.0465            30.82961            3.634379

   R-squared               0.610162        Mean of depend var      1075.538
   Adjusted R-squared      0.584172        Std dev depend var      191.3605
   Std err of regress      123.3982        Residual sum            5.03064E-12
   Durbin Watson stat      2.104997        Sum squared resid       913626.4
   F Statistic             23.47754
```

Answer each of the following questions for the above regression run.

a. Evaluate this result with respect to its economic meaning, overall fit, and the signs and significance of the individual coefficients.

b. What econometric problems (out of omitted variables, irrelevant variables, or multicollinearity) does this regression have? Why? If you need feedback on your answer, see hint #5 in the material on this chapter in Appendix A.

c. Which of the following statements comes closest to your recommendation for further action to be taken in the estimation of this equation?
   i. No further specification changes are advisable (go to Section 8.7.3).
   ii. I would like to add ESL to the equation (go to run #8.6).
   iii. I would like to add PREP to the equation (go to run #8.9).
   iv. I would like to add RACE to the equation (go to run #8.10).

If you need feedback on your answer, see hint #19 in the material on this chapter in Appendix A.

Regression Run 8.4

```
ORDINARY LEAST SQUARES                         DEPENDENT VARIABLE IS SAT
SAMPLE RANGE:              1-65

                           COEFFICIENT      STANDARD ERROR        T-SCORE
      CONST                 569.2532           121.1058           4.700463
      GPA                   132.7666           40.94846           3.242287
      APMATH                72.29444           39.84456           1.814412
      APENG                 85.68562           48.60529           1.762887
      PREP                 -34.38129           38.88201          -0.884246

   R-squared                 0.530460        Mean of depend var     1075.538
   Adjusted R-squared        0.499157        Std dev depend var     191.3605
   Std err of regress        135.4263        Residual sum           1.76215E-11
   Durbin Watson stat        1.976378        Sum squared resid      1100417.
   F Statistic               16.94616
```

Answer each of the following questions for the above regression run.

a. Evaluate this result with respect to its economic meaning, overall fit, and the signs and significance of the individual coefficients.

b. What econometric problems (out of omitted variables, irrelevant variables, or multicollinearity) does this regression have? Why? If you need feedback on your answer, see hint #8 in the material on this chapter in Appendix A.

c. Which of the following statements comes closest to your recommendation for further action to be taken in the estimation of this equation?
   i. No further specification changes are advisable (go to Section 8.7.3).
   ii. I would like to drop PREP from the equation (go to run #8.1).
   iii. I would like to add ESL to the equation (go to run #8.7).
   iv. I would like to add GEND to the equation (go to run #8.9).
   v. I would like to replace APMATH and APENG with AP, a linear combination of the two variables (go to run #8.17).

If you need feedback on your answer, see hint #12 in the material on this chapter in Appendix A.

Regression Run 8.5

```
ORDINARY LEAST SQUARES                        DEPENDENT VARIABLE IS SAT
SAMPLE RANGE:              1-65

                       COEFFICIENT       STANDARD ERROR         T-SCORE
     CONST              570.8148           117.7382            4.848172
     GPA                128.2798           40.48924            3.168244
     APMATH             106.2137           42.71559            2.486533
     APENG              67.42362           48.92704            1.378044
     RACE              -60.33471           39.47330           -1.528494

  R-squared               0.542168      Mean of depend var      1075.538
  Adjusted R-squared      0.511646      Std dev depend var      191.3605
  Std err of regress      133.7271      Residual sum            8.52651E-12
  Durbin Watson stat      2.033014      Sum squared resid       1072977.
  F Statistic             17.76314
```

Answer each of the following questions for the above regression run.

a. Evaluate this result with respect to its economic meaning, overall fit, and the signs and significance of the individual coefficients.

b. What econometric problems (out of omitted variables, irrelevant variables, or multicollinearity) does this regression have? Why? If you need feedback on your answer, see hint #3 in the material on this chapter in Appendix A.

c. Which of the following statements comes closest to your recommendation for further action to be taken in the estimation of this equation?
   i. No further specification changes are advisable (go to Section 8.7.3).
   ii. I would like to drop RACE from the equation (go to run #8.1).
   iii. I would like to add ESL to the equation (go to run #8.8).
   iv. I would like to add GEND to the equation (go to run #8.10).
   v. I would like to add PREP to the equation (go to run #8.11).

If you need feedback on your answer, see hint #14 in the material on this chapter in Appendix A.

Regression Run 8.6

```
ORDINARY LEAST SQUARES                        DEPENDENT VARIABLE IS SAT
SAMPLE RANGE:              1-65

                       COEFFICIENT        STANDARD ERROR          T-SCORE
     CONST              508.8237            110.0355             4.624179
     GPA                129.0595            37.41416             3.449484
     APMATH             81.97538            40.00950             2.048898
     APENG              89.84960            44.54376             2.017109
     ESL               -33.64469            34.94751            -0.962720
     GEND               108.8598            31.02552             3.508717

   R-squared              0.616191     Mean of depend var       1075.538
   Adjusted R-squared     0.583665     Std dev depend var       191.3605
   Std err of regress     123.4735     Residual sum             9.94760E-12
   Durbin Watson stat     2.142956     Sum squared resid        899496.2
   F Statistic            18.94449
```

Answer each of the following questions for the above regression run.

a. Evaluate this result with respect to its economic meaning, overall fit, and the signs and significance of the individual coefficients.

b. What econometric problems (out of omitted variables, irrelevant variables, or multicollinearity) does this regression have? Why? If you need feedback on your answer, see hint #7 in the material on this chapter in Appendix A.

c. Which of the following statements comes closest to your recommendation for further action to be taken in the estimation of this equation?
   i. No further specification changes are advisable (go to Section 8.7.3).
   ii. I would like to drop ESL from the equation (go to run #8.3).
   iii. I would like to add PREP to the equation (go to run #8.12).
   iv. I would like to add RACE to the equation (go to run #8.13).

If you need feedback on your answer, see hint #4 in the material on this chapter in Appendix A.

Regression Run 8.7

```
ORDINARY LEAST SQUARES                         DEPENDENT VARIABLE IS SAT
SAMPLE RANGE:              1-65

                        COEFFICIENT        STANDARD ERROR        T-SCORE
      CONST              591.2047            121.8609           4.851472
      GPA                129.2439            40.86539           3.162673
      APMATH             95.35163            43.81128           2.176417
      APENG              80.21916            48.58978           1.650947
      ESL               -47.03944            37.94402          -1.239706
      PREP              -34.82031            38.71083          -0.899498

   R-squared              0.542380      Mean of depend var       1075.538
   Adjusted R-squared     0.503599      Std dev depend var       191.3605
   Std err of regress     134.8244      Residual sum             9.89075E-12
   Durbin Watson stat     2.008613      Sum squared resid        1072480.
   F Statistic            13.98561
```

Answer each of the following questions for the above regression run.

a. Evaluate this result with respect to its economic meaning, overall fit, and the signs and significance of the individual coefficients.

b. What econometric problems (out of omitted variables, irrelevant variables, or multicollinearity) does this regression have? Why? If you need feedback on your answer, see hint #8 in the material on this chapter in Appendix A.

c. Which of the following statements comes closest to your recommendation for further action to be taken in the estimation of this equation?
   i. No further specification changes are advisable (go to Section 8.7.3).
   ii. I would like to drop ESL from the equation (go to run #8.4).
   iii. I would like to drop PREP from the equation (go to run #8.2).
   iv. I would like to add GEND to the equation (go to run #8.12).
   v. I would like to add RACE to the equation (go to run #8.14).

If you need feedback on your answer, see hint #18 in the material on this chapter in Appendix A.

Regression Run 8.8

ORDINARY LEAST SQUARES DEPENDENT VARIABLE IS SAT
SAMPLE RANGE: 1-65

	COEFFICIENT	STANDARD ERROR	T-SCORE
CONST	570.6367	118.8985	4.799359
GPA	128.3251	40.86223	3.140434
APMATH	106.0310	43.55940	2.434170
APENG	67.23015	49.81328	1.349643
ESL	1.885689	66.79448	0.028231
RACE	-61.96231	70.05962	-0.884422

R-squared	0.542174	Mean of depend var	1075.538
Adjusted R-squared	0.503375	Std dev depend var	191.3605
Std err of regress	134.8548	Residual sum	8.12861E-12
Durbin Watson stat	2.032924	Sum squared resid	1072962.
F Statistic	13.97402		

Answer each of the following questions for the above regression run.

a. Evaluate this result with respect to its economic meaning, overall fit, and the signs and significance of the individual coefficients.

b. What econometric problems (out of omitted variables, irrelevant variables, or multicollinearity) does this regression have? Why? If you need feedback on your answer, see hint #9 in the material on this chapter in Appendix A.

c. Which of the following statements comes closest to your recommendation for further action to be taken in the estimation of this equation?
   i. No further specification changes are advisable (go to Section 8.7.3).
   ii. I would like to drop ESL from the equation (go to run #8.5).
   iii. I would like to drop RACE from the equation (go to run #8.2).
   iv. I would like to add GEND to the equation (go to run #8.13).
   v. I would like to add PREP to the equation (go to run #8.14).

If you need feedback on your answer, see hint #15 in the material on this chapter in Appendix A.

Regression Run 8.9

```
ORDINARY LEAST SQUARES                    DEPENDENT VARIABLE IS SAT
SAMPLE RANGE:            1-65

                     COEFFICIENT        STANDARD ERROR        T-SCORE
    CONST             513.9945            111.6115            4.605210
    GPA               132.4152            37.38088            3.542326
    APMATH            59.37168            36.54919            1.624432
    APENG             96.69438            44.47540            2.174109
    GEND              111.3943            30.89564            3.605501
    PREP             -31.31762            35.50451           -0.882074

R-squared               0.615236    Mean of depend var     1075.538
Adjusted R-squared      0.582629    Std dev depend var     191.3605
Std err of regress      123.6270    Residual sum           1.56035E-11
Durbin Watson stat      2.065021    Sum squared resid      901734.9
F Statistic             18.86816
```

Answer each of the following questions for the above regression run.

a. Evaluate this result with respect to its economic meaning, overall fit, and the signs and significance of the individual coefficients.

b. What econometric problems (out of omitted variables, irrelevant variables, or multicollinearity) does this regression have? Why? If you need feedback on your answer, see hint #8 in the material on this chapter in Appendix A.

c. Which of the following statements comes closest to your recommendation for further action to be taken in the estimation of this equation?

i. No further specification changes are advisable (go to Section 8.7.3).
ii. I would like to drop PREP from the equation (go to run #8.3).
iii. I would like to add ESL to the equation (go to run #8.12).
iv. I would like to add RACE to the equation (go to run #8.15).

If you need feedback on your answer, see hint #17 in the material on this chapter in Appendix A.

Regression Run 8.10

```
ORDINARY LEAST SQUARES                    DEPENDENT VARIABLE IS SAT
SAMPLE RANGE:               1-65

                          COEFFICIENT         STANDARD ERROR        T-SCORE
      CONST               514.5822              109.0157           4.720259
      GPA                 128.6381              37.08886           3.468376
      APMATH              88.26401              39.45591           2.237029
      APENG               81.07941              44.98391           1.802409
      GEND                108.5953              30.70716           3.536482
      RACE               -49.83756              36.27973          -1.373703

   R-squared               0.622244         Mean of depend var      1075.538
   Adjusted R-squared      0.590231         Std dev depend var      191.3605
   Std err of regress      122.4960         Residual sum            1.15392E-11
   Durbin Watson stat      2.148211         Sum squared resid       885310.6
   F Statistic             19.43712
```

Answer each of the following questions for the above regression run.

a. Evaluate this result with respect to its economic meaning, overall fit, and the signs and significance of the individual coefficients.

b. What econometric problems (out of omitted variables, irrelevant variables, or multicollinearity) does this regression have? Why? If you need feedback on your answer, see hint #10 in the material on this chapter in Appendix A.

c. Which of the following statements comes closest to your recommendation for further action to be taken in the estimation of this equation?
   i. No further specification changes are advisable (go to Section 8.7.3).
   ii. I would like to drop RACE from the equation (go to run #8.3).
   iii. I would like to add ESL to the equation (go to run #8.13).
   iv. I would like to add PREP to the equation (go to run #8.15).

If you need feedback on your answer, see hint #4 in the material on this chapter in Appendix A.

Regression Run 8.11

```
ORDINARY LEAST SQUARES                          DEPENDENT VARIABLE IS SAT
SAMPLE RANGE:              1-65

                      COEFFICIENT          STANDARD ERROR          T-SCORE
      CONST            602.4718              121.0769              4.975943
      GPA              129.0898              40.43172              3.192785
      APMATH           100.8919              42.92558              2.350391
      APENG            69.65070              48.89190              1.424586
      PREP            -42.14969              38.62038             -1.091385
      RACE            -65.60984              39.70586             -1.652397

   R-squared               0.551228      Mean of depend var       1075.538
   Adjusted R-squared      0.513196      Std dev depend var       191.3605
   Std err of regress      133.5147      Residual sum             1.71951E-11
   Durbin Watson stat      2.020544      Sum squared resid        1051744.
   F Statistic             14.49400
```

Answer each of the following questions for the above regression run.

a. Evaluate this result with respect to its economic meaning, overall fit, and the signs and significance of the individual coefficients.

b. What econometric problems (out of omitted variables, irrelevant variables, or multicollinearity) does this regression have? Why? If you need feedback on your answer, see hint #8 in the material on this chapter in Appendix A.

c. Which of the following statements comes closest to your recommendation for further action to be taken in the estimation of this equation?
   i. No further specification changes are advisable (go to Section 8.7.3).
   ii. I would like to drop PREP from the equation (go to run #8.5).
   iii. I would like to drop RACE from the equation (go to run #8.4).
   iv. I would like to add GEND to the equation (go to run #8.15).
   v. I would like to replace APMATH and APENG with AP, a linear combination of the two variables (go to run #8.18).

If you need feedback on your answer, see hint #18 in the material on this chapter in Appendix A.

Regression Run 8.12

```
ORDINARY LEAST SQUARES                       DEPENDENT VARIABLE IS SAT
SAMPLE RANGE:            1-65

                     COEFFICIENT        STANDARD ERROR        T-SCORE
      CONST           531.4692            113.1041            4.698939
      GPA             129.8782            37.48974            3.464368
      APMATH          76.41832            40.55854            1.884149
      APENG           92.42253            44.71331            2.067002
      ESL            -34.01275            35.01006           -0.971513
      GEND            108.1642            31.08865            3.479219
      PREP           -31.72391            35.52388           -0.893030

   R-squared                0.621397      Mean of depend var       1075.538
   Adjusted R-squared       0.582231      Std dev depend var       191.3605
   Std err of regress       123.6859      Residual sum             2.07478E-11
   Durbin Watson stat       2.106229      Sum squared resid        887295.9
   F Statistic              15.86581
```

Answer each of the following questions for the above regression run.

a. Evaluate this result with respect to its economic meaning, overall fit, and the signs and significance of the individual coefficients.

b. What econometric problems (out of omitted variables, irrelevant variables, or multicollinearity) does this regression have? Why? If you need feedback on your answer, see hint #8 in the material on this chapter in Appendix A.

c. Which of the following statements comes closest to your recommendation for further action to be taken in the estimation of this equation?

  i. No further specification changes are advisable (go to Section 8.7.3).

  ii. I would like to drop ESL from the equation (go to run #8.9).

  iii. I would like to drop PREP from the equation (go to run #8.6).

  iv. I would like to add RACE to the equation (go to run #8.16).

If you need feedback on your answer, see hint #17 in the material on this chapter in Appendix A.

Regression Run 8.13

```
ORDINARY LEAST SQUARES                      DEPENDENT VARIABLE IS SAT
SAMPLE RANGE:             1-65

                      COEFFICIENT       STANDARD ERROR        T-SCORE
     CONST             512.6796           110.0966           4.656635
     GPA               129.0460           37.41213           3.449311
     APMATH            86.52973           40.26408           2.149055
     APENG             79.42187           45.73811           1.736449
     ESL               16.88299           61.30223           0.275405
     GEND              109.1893           31.02557           3.519333
     RACE             -64.35243           64.14694          -1.003204

  R-squared              0.622737     Mean of depend var      1075.538
  Adjusted R-squared     0.583710     Std dev depend var      191.3605
  Std err of regress     123.4668     Residual sum            1.33582E-11
  Durbin Watson stat     2.143234     Sum squared resid       884154.4
  F Statistic            15.95653
```

Answer each of the following questions for the above regression run.

a. Evaluate this result with respect to its economic meaning, overall fit, and the signs and significance of the individual coefficients.

b. What econometric problems (out of omitted variables, irrelevant variables, or multicollinearity) does this regression have? Why? If you need feedback on your answer, see hint #9 in the material on this chapter in Appendix A.

c. Which of the following statements comes closest to your recommendation for further action to be taken in the estimation of this equation?

   i. No further specification changes are advisable (go to Section 8.7.3).

   ii. I would like to drop ESL from the equation (go to run #8.10).

   iii. I would like to drop RACE from the equation (go to run #8.6).

   iv. I would like to add PREP to the equation (go to run #8.16).

If you need feedback on your answer, see hint #15 in the material on this chapter in Appendix A.

Regression Run 8.14

```
ORDINARY LEAST SQUARES                      DEPENDENT VARIABLE IS SAT
SAMPLE RANGE:                1-65

                         COEFFICIENT         STANDARD ERROR          T-SCORE
      CONST              602.1427              122.0822             4.932274
      GPA                129.4491              40.80133             3.172669
      APMATH             99.37976              43.89816             2.263871
      APENG              68.29405              49.73286             1.373218
      ESL                13.89708              67.55991             0.205700
      PREP              -43.45964              39.45502            -1.101498
      RACE              -77.76882              71.39042            -1.089345

   R-squared                0.551555        Mean of depend var      1075.538
   Adjusted R-squared       0.505164        Std dev depend var      191.3605
   Std err of regress       134.6116        Residual sum            1.56035E-11
   Durbin Watson stat       2.020634        Sum squared resid       1050977.
   F Statistic              11.88933
```

Answer each of the following questions for the above regression run.

a. Evaluate this result with respect to its economic meaning, overall fit, and the signs and significance of the individual coefficients.

b. What econometric problems (out of omitted variables, irrelevant variables, or multicollinearity) does this regression have? Why? If you need feedback on your answer, see hint #9 in the material on this chapter in Appendix A.

c. Which of the following statements comes closest to your recommendation for further action to be taken in the estimation of this equation?

   i. No further specification changes are advisable (go to Section 8.7.3).
   ii. I would like to drop ESL from the equation (go to run #8.11).
   iii. I would like to drop PREP from the equation (go to run #8.8).
   iv. I would like to add GEND to the equation (go to run #8.16).
   v. I would like to replace APMATH and APENG with AP, a linear combination of the two variables (go to run #8.19).

If you need feedback on your answer, see hint #15 in the material on this chapter in Appendix A.

Regression Run 8.15

```
ORDINARY LEAST SQUARES                    DEPENDENT VARIABLE IS SAT
SAMPLE RANGE:            1-65

                         COEFFICIENT        STANDARD ERROR        T-SCORE
      CONST              543.6309             112.2128            4.844641
      GPA                129.3628             37.04936            3.491632
      APMATH             83.66463             39.64091            2.110563
      APENG              82.94048             44.96213            1.844674
      GEND               107.4700             30.68735            3.502094
      PREP               -37.90098            35.41026           -1.070339
      RACE               -54.68974            36.51752           -1.497630

   R-squared                0.629561        Mean of depend var      1075.538
   Adjusted R-squared       0.591240        Std dev depend var      191.3605
   Std err of regress       122.3451        Residual sum            6.96332E-12
   Durbin Watson stat       2.114836        Sum squared resid       868162.5
   F Statistic              16.42852
```

Answer each of the following questions for the above regression run.

a. Evaluate this result with respect to its economic meaning, overall fit, and the signs and significance of the individual coefficients.

b. What econometric problems (out of omitted variables, irrelevant variables, or multicollinearity) does this regression have? Why? If you need feedback on your answer, see hint #8 in the material on this chapter in Appendix A.

c. Which of the following statements comes closest to your recommendation for further action to be taken in the estimation of this equation?

   i. No further specification changes are advisable (go to Section 8.7.3).
   ii. I would like to drop PREP from the equation (go to run #8.10).
   iii. I would like to drop RACE from the equation (go to run #8.9).
   iv. I would like to add ESL to the equation (go to run #8.16).

If you need feedback on your answer, see hint #17 in the material on this chapter in Appendix A.

Regression Run 8.16

```
ORDINARY LEAST SQUARES                    DEPENDENT VARIABLE IS SAT
SAMPLE RANGE:             1-65

                      COEFFICIENT        STANDARD ERROR        T-SCORE
     CONST             542.4723             113.0203           4.799777
     GPA               130.0882             37.34094           3.483794
     APMATH            80.47642             40.53608           1.985303
     APENG             80.32262             45.64401           1.759762
     ESL               27.96510             61.95989           0.451341
     GEND              108.3766             30.96543           3.499924
     PREP             -40.50116             36.11828          -1.121348
     RACE             -79.06514             65.33603          -1.210131

  R-squared                0.630880      Mean of depend var       1075.538
  Adjusted R-squared       0.585550      Std dev depend var       191.3605
  Std err of regress       123.1937      Residual sum             2.03215E-11
  Durbin Watson stat       2.106524      Sum squared resid        865070.9
  F Statistic              13.91736
```

Answer each of the following questions for the above regression run.

a. Evaluate this result with respect to its economic meaning, overall fit, and the signs and significance of the individual coefficients.

b. What econometric problems (out of omitted variables, irrelevant variables, or multicollinearity) does this regression have? Why? If you need feedback on your answer, see hint #9 in the material on this chapter in Appendix A.

c. Which of the following statements comes closest to your recommendation for further action to be taken in the estimation of this equation?
   i. No further specification changes are advisable (go to Section 8.7.3).
   ii. I would like to drop ESL from the equation (go to run #8.15).
   iii. I would like to drop PREP from the equation (go to run #8.13).
   iv. I would like to drop RACE from the equation (go to run #8.12).

If you need feedback on your answer, see hint #15 in the material on this chapter in Appendix A.

Regression Run 8.17

```
ORDINARY LEAST SQUARES                    DEPENDENT VARIABLE IS SAT
SAMPLE RANGE:              1-65

                        COEFFICIENT      STANDARD ERROR        T-SCORE
      CONST              475.7963          104.7275           4.543185
      GPA                163.4716          34.41783           4.749619
      AP                 107.7460          45.02942           2.392790
      PREP              -30.92277          38.84976          -0.795957

   R-squared              0.516299      Mean of depend var     1075.538
   Adjusted R-squared     0.492510      Std dev depend var     191.3605
   Std err of regress     136.3219      Residual sum           9.89075E-12
   Durbin Watson stat     1.912398      Sum squared resid      1133604.
   F Statistic            21.70368
```

Answer each of the following questions for the above regression run.

a. Evaluate this result with respect to its economic meaning, overall fit, and the signs and significance of the individual coefficients.

b. What econometric problems (out of omitted variables, irrelevant variables, or multicollinearity) does this regression have? Why? If you need feedback on your answer, see hint #11 in the material on this chapter in Appendix A.

c. Which of the following statements comes closest to your recommendation for further action to be taken in the estimation of this equation?
   i. No further specification changes are advisable (go to Section 8.7.3).
   ii. I would like to drop PREP from the equation (go to run #8.20).
   iii. I would like to add RACE to the equation (go to run #8.18).
   iv. I would like to replace the AP combination variable with APMATH and APENG (go to run #8.4).

If you need feedback on your answer, see hint #16 in the material on this chapter in Appendix A.

Regression Run 8.18

```
ORDINARY LEAST SQUARES                    DEPENDENT VARIABLE IS SAT
SAMPLE RANGE:              1-65

                         COEFFICIENT      STANDARD ERROR        T-SCORE
      CONST               522.4920          107.1073            4.878210
      GPA                 154.0768          34.42039            4.476323
      AP                  125.9048          45.75812            2.751529
      PREP               -41.06153          38.80679           -1.058102
      RACE               -61.63421          37.41938           -1.647120

   R-squared               0.537224     Mean of depend var      1075.538
   Adjusted R-squared      0.506372     Std dev depend var      191.3605
   Std err of regress      134.4472     Residual sum            1.69678E-11
   Durbin Watson stat      1.887634     Sum squared resid       1084563.
   F Statistic             17.41313
```

Answer each of the following questions for the above regression run.

a. Evaluate this result with respect to its economic meaning, overall fit, and the signs and significance of the individual coefficients.

b. What econometric problems (out of omitted variables, irrelevant variables, or multicollinearity) does this regression have? Why? If you need feedback on your answer, see hint #11 in the material on this chapter in Appendix A.

c. Which of the following statements comes closest to your recommendation for further action to be taken in the estimation of this equation?

   i. No further specification changes are advisable (go to Section 8.7.3).

   ii. I would like to drop RACE from the equation (go to run #8.17).

   iii. I would like to add ESL to the equation (go to run #8.19).

   iv. I would like to replace the AP combination variable with APMATH and APENG (go to run #8.11).

If you need feedback on your answer, see hint #16 in the material on this chapter in Appendix A.

Regression Run 8.19

```
ORDINARY LEAST SQUARES                    DEPENDENT VARIABLE IS SAT
SAMPLE RANGE:            1-65

                        COEFFICIENT       STANDARD ERROR        T-SCORE
     CONST               524.8762           108.0514            4.857655
     GPA                 153.7341           34.67841            4.433136
     AP                  122.3201           47.01130            2.601930
     ESL                 26.00898           67.33954            0.386236
     PREP               -43.55594           39.61488           -1.099484
     RACE               -84.43699           70.04203           -1.205519

  R-squared                0.538391        Mean of depend var     1075.538
  Adjusted R-squared       0.499272        Std dev depend var     191.3605
  Std err of regress       135.4107        Residual sum           9.54969E-12
  Durbin Watson stat       1.894863        Sum squared resid      1081828.
  F Statistic              13.76280
```

Answer each of the following questions for the above regression run.

a. Evaluate this result with respect to its economic meaning, overall fit, and the signs and significance of the individual coefficients.

b. What econometric problems (out of omitted variables, irrelevant variables, or multicollinearity) does this regression have? Why? If you need feedback on your answer, see hint #11 in the material on this chapter in Appendix A.

c. Which of the following statements comes closest to your recommendation for further action to be taken in the estimation of this equation?
   i. No further specification changes are advisable (go to Section 8.7.3).
   ii. I would like to drop ESL from the equation (go to run #8.18).
   iii. I would like to replace the AP combination variable with APMATH and APENG (go to run #8.14).

If you need feedback on your answer, see hint #16 in the material on this chapter in Appendix A.

Regression Run 8.20

```
ORDINARY LEAST SQUARES                    DEPENDENT VARIABLE IS SAT
SAMPLE RANGE:               1-65

                         COEFFICIENT        STANDARD ERROR        T-SCORE
     CONST                457.2010             101.7863           4.491773
     GPA                  161.2106             34.19889           4.713912
     AP                   112.7129             44.46296           2.534985

   R-squared               0.511275        Mean of depend var      1075.538
   Adjusted R-squared      0.495510        Std dev depend var      191.3605
   Std err of regress      135.9185        Residual sum            1.87867E-11
   Durbin Watson stat      1.917047        Sum squared resid       1145378.
   F Statistic             32.43043
```

Answer each of the following questions for the above regression run.

a. Evaluate this result with respect to its economic meaning, overall fit, and the signs and significance of the individual coefficients.

b. What econometric problems (out of omitted variables, irrelevant variables, or multicollinearity) does this regression have? Why? If you need feedback on your answer, see hint #13 in the material on this chapter in Appendix A.

c. Which of the following statements comes closest to your recommendation for further action to be taken in the estimation of this equation?
   i. No further specification changes are advisable (go to Section 8.7.3).
   ii. I would like to add PREP to the equation (go to run #8.17).
   iii. I would like to replace the AP combination variable with APMATH and APENG (go to run #8.1).

If you need feedback on your answer, see hint #13 in the material on this chapter in Appendix A.

8.7.3 Evaluating the Results from Your Interactive Exercise

Congratulations! If you've reached this section, you must have found a specification that met your theoretical and econometric goals. Which one did you pick? Our experience is that most beginning econometricians end up with either regression run #8.3, #8.6, or #8.10, but only after looking at three or more regression results (or a hint or two) before settling on that choice.

In contrast, we've found that most experienced econometricians gravitate to regression run #8.6, usually after inspecting, at most, one other specification. What lessons can we learn from this difference?

1. *Learn that a variable isn't irrelevant simply because its t-score is low.* In our opinion, ESL belongs in the equation for strong theoretical reasons, and a slightly insignificant t-score in the expected direction isn't enough evidence to get us to rethink the underlying theory.

2. *Learn to spot redundant (multicollinear) variables.* ESL and RACE wouldn't normally be redundant, but in this high school, with its limited ethnic diversity, they are. Once one is included in the equation, the other shouldn't even be considered.

3. *Learn to spot false omitted variables.* At first glance, PREP is a tempting variable to include because prep courses almost surely improve the SAT scores of the students who choose to take them. The problem is that a student's decision to take a prep course isn't independent of his or her previous SAT scores (or expected scores). We trust the judgment of students who feel a need for a prep course, and we think that all the course will do is bring them up to the level of their colleagues who didn't feel they needed a course. As a result, we wouldn't expect a significant effect in either direction.

If you enjoyed and learned from this interactive regression learning exercise, you'll be interested to know that there are three more in future sections. We'll present more difficult interactive exercises in Section 11.4 and 11.6 and a computerized interactive exercise in Section 11.7. Good luck!

9

Serial Correlation

In the next two chapters, we'll investigate the final component of the specification of a regression equation—choosing the correct form of the stochastic error term. Our first topic, serial correlation, is the violation of the classical assumption that different observations of the error term are uncorrelated with each other. Serial correlation, also called autocorrelation, can exist in any research study in which the order of the observations has some meaning. It therefore occurs most frequently in time-series data-sets. In essence, serial correlation implies that the error term from one time period depends in some systematic way on error terms from other time periods. Since time-series data are used in many applications of econometrics, it's important to understand serial correlation and its consequences for OLS estimators.

The approach of this chapter to the problem of serial correlation will be similar to that used in the previous chapter to study multicollinearity. We'll attempt to answer the same four questions:

1. What is the nature of the problem?
2. What are the consequences of the problem?

3. How is the problem diagnosed?
4. What remedies for the problem are available?

9.1 Pure vs. Impure Serial Correlation

The correlation of observations of the error term with others over time is what gives rise to the name serial correlation. This section gives a description of the nature of serial correlation and distinguishes between two forms of the disease, "pure" and "impure" serial correlation.

9.1.1 Pure Serial Correlation

Pure serial correlation occurs when Classical Assumption IV, which assumes uncorrelated observations of the error term, is violated in a correctly specified equation. Recall that Assumption IV states that:

$$E(\epsilon_i \cdot \epsilon_j) = 0 \qquad (i \neq j)$$

If the expected value of the multiple of any two error terms is not equal to zero, then the error terms are said to be serially correlated. When econometricians use the term serial correlation without any modifier, they are referring to pure serial correlation.

The most commonly assumed kind of serial correlation is **first order serial correlation,** in which the current observation of the error term is a function of the previous observation of the error term:

$$\epsilon_t = \rho\epsilon_{t-1} + u_t \tag{9.1}$$

where: ϵ is the error term of the equation in question,
ρ is the parameter depicting the functional relationship between observations of the error term
u is a classical (nonserially correlated) error term

The functional form in Equation 9.1 is called a first-order Markov scheme, and the new symbol, ρ (rho, pronounced "row"), is called the **first-order autocorrelation coefficient.** For this kind of serial correlation, all that is needed is for the value of one observation of the error term to affect directly the value of the next observation of the error term.

The magnitude of ρ indicates the strength of the serial correlation in an equation. If ρ is zero, then there is no serial correlation (because ε would equal u, a classical error term). As ρ approaches one in absolute value, the value of the previous observation of the error term becomes more important in determining the current value of ϵ_t, and a high degree of serial correlation exists. For ρ to be greater than one in absolute value is unreasonable because it implies that the error term has a tendency to continually increase in absolute value over time ("explode"). As a result of the above, we can state that:

$$-1 < \rho < +1$$

The sign of ρ indicates the nature of the serial correlation in an equation. A positive value for ρ implies that the error term tends to have the same sign from one time period to the next. Such a tendency means that if ϵ_t happens by chance to take on a large value in one time period, subsequent observations would tend to retain a portion of this original large value and would have the same sign as the original. For example, in time-series models, a large external shock to an economy in one period may linger on for several time periods. If this occurs now and then, the error term will tend to be positive for a number of observations, then negative for several more, and then back again. This is called **positive serial correlation.** Figure 9.1 shows two different examples.

The error term observations plotted in Figure 9.1 are arranged in chronological order, with the first observation being the first period for which data are available, the second being the second, and so on. To see the difference between error terms with and without positive serial correlation, compare the patterns in Figure 9.1 with the depiction of no serial correlation ($\rho = 0$) in Figure 9.2.

A negative value of ρ implies that the error term has a tendency to switch signs from negative to positive and back again in consecutive observations. This is called **negative serial correlation** and implies that there is some sort of cycle (like a pendulum) behind the drawing of stochastic disturbances. Figure 9.3 shows two different examples of negative serial correlation. For instance, negative serial correlation might exist in the error term of a semiannual equation for the demand for some seasonal item (like Christmas lights) that had no seasonal dummy. In most time-series applications, however, negative pure serial correlation is much less likely than positive pure serial correlation. As a result, most econometricians analyzing pure serial correlation concern themselves primarily with positive serial correlation.

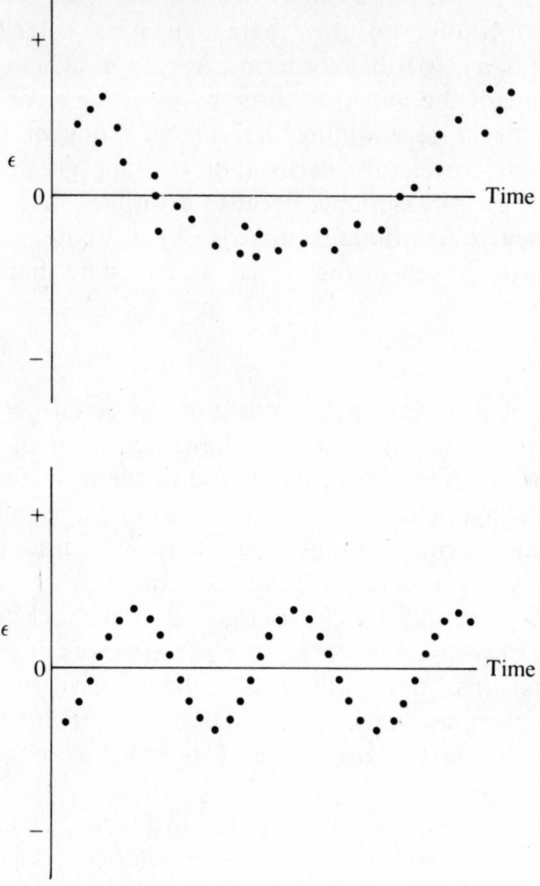

Figure 9.1 POSITIVE SERIAL CORRELATION
With positive first-order serial correlation, the current observation of the error term tends to have the same sign as the previous observation of the error term. An example of positive serial correlation would be external shocks to an economy that take more than one time period to completely work through the system.

Serial correlation can take on many forms other than first-order serial correlation. For example, in a quarterly model, the current quarter's error term observation may be functionally related to the observation of the error term from the same quarter in the previous year. This is called seasonally based serial correlation:

$$\epsilon_t = \rho\epsilon_{t-4} + u_t$$

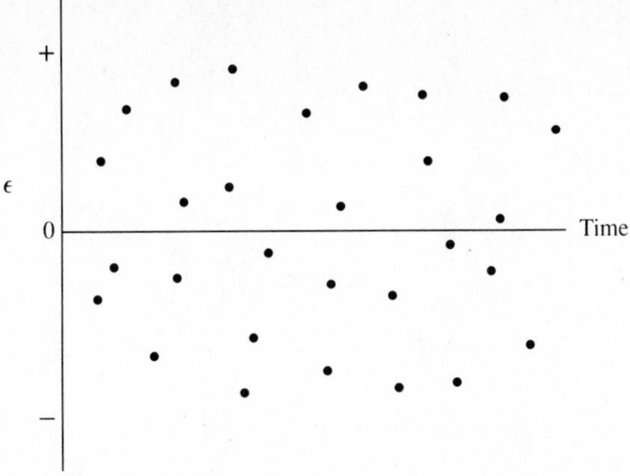

Figure 9.2 NO SERIAL CORRELATION
With no serial correlation, different observations of the error term are completely uncorrelated with each other. Such error terms would conform to Classical Assumption IV.

Similarly, it is possible that the error term in an equation might be a function of more than one previous observation of the error term:

$$\epsilon_t = \rho_1\epsilon_{t-1} + \rho_2\epsilon_{t-2} + u_t$$

Such a formulation is called second-order serial correlation. Higher-order expressions are similarly formed, but the justifications for assuming these higher-order forms are usually weaker than the justification for the second-order form, which itself is not always all that strong.

9.1.2 Impure Serial Correlation

By **impure serial correlation** we mean serial correlation that is caused by a specification error such as an omitted variable or an incorrect functional form. While pure serial correlation is caused by the underlying distribution of the error term of the true specification of an equation (which cannot be changed by the researcher), impure serial correlation is caused by a specification error that often can be corrected.

How is it possible for a specification error to cause serial correlation? Recall that the error term can be thought of as the effect of omitted variables, nonlinearities, measurement errors, and pure stochastic disturbances on the dependent variable. This means that if we

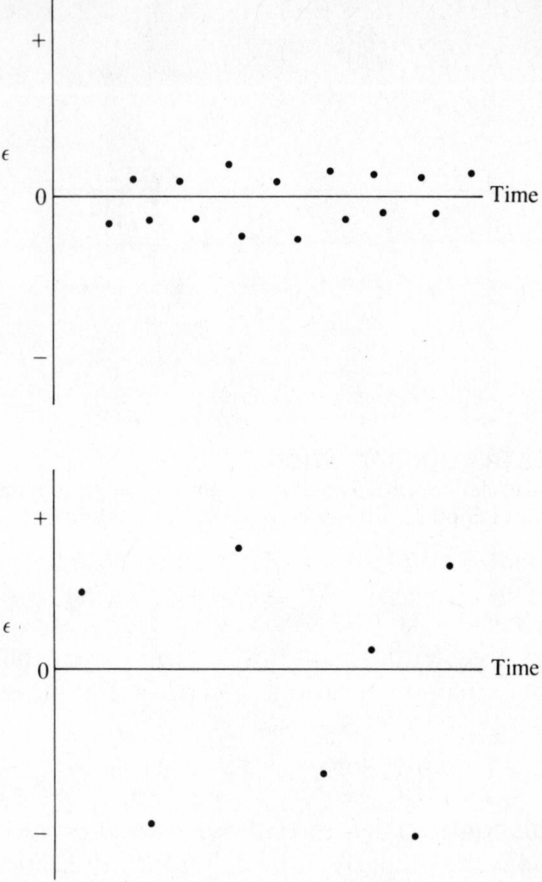

Figure 9.3 NEGATIVE SERIAL CORRELATION
With negative first-order serial correlation, the current observation of the error term tends to have the opposite sign from the previous observation of the error term. In most time series applications, negative serial correlation is much less likely than positive serial correlation.

omit a relevant variable or use the wrong functional form, then the portion of that omitted effect that cannot be represented by the included explanatory variables must be absorbed by the error term. The error term for an incorrectly specified equation thus includes a portion of the effect of any omitted variables and/or a portion of the effect of the difference between the proper functional form and the one chosen by the researcher. This new error term might be serially correlated even if the true one is not. If this is the case, the serial correlation has been caused by the researcher's choice of a specification and not by the pure error term associated with the correct specification.

As we'll see in Section 9.4, the proper remedy for serial correlation depends on whether the serial correlation is likely to be pure or impure. Not surprisingly, the best remedy for impure serial correlation usually is to attempt to find the omitted variable (or at least a good proxy) or the correct functional form for the equation. As a result, most econometricians try to make sure they have the best specification possible before they spend too much time worrying about pure serial correlation.

To see how a left-out variable can cause the error term to be serially correlated, suppose that the true equation is:

$$Y_t = \beta_0 + \beta_1 X_{1t} + \beta_2 X_{2t} + \epsilon_t \quad (9.2)$$

where ϵ_t is a classical error term. As shown in Section 6.1, if X_2 is accidentally omitted from the equation (or if data for X_2 are unavailable), then:

$$Y_t = \beta_0 + \beta_1 X_{1t} + \epsilon_t^* \qquad \text{where } \epsilon_t^* = \beta_2 X_{2t} + \epsilon_t \quad (9.3)$$

Thus the error term being used in the omitted variable case is not the classical error term ϵ. Instead, it's also a function of one of the independent variables, X_2. As a result, the new error term, ϵ^*, can be serially correlated even if the true error term ϵ is not. In particular, the new error term ϵ^* will tend to be serially correlated when:

1. X_2 itself is serially correlated (this is quite likely in a time-series).

2. The size of ϵ is small compared to the size[1] of $\beta_2\overline{X}_2$.

These tendencies hold even if there are a number of included and/or omitted variables.

1. If typical values of ϵ are significantly larger in absolute value than $\beta_2\overline{X}_2$, then even a serially correlated omitted variable (X_2) will not change ϵ very much. In addition, recall that the omitted variable, X_2, will cause bias in the estimate of β_1 depending on the correlation between the two Xs. If $\hat{\beta}_1$ is biased because of the omission of X_2, then a portion of the $\beta_2 X_2$ effect must have been absorbed by $\hat{\beta}_1$ and will not end up in the residuals. As a result, tests for serial correlation based on those residuals may give incorrect readings. Just as important, such residuals may leave misleading clues as to possible specification errors. This is only one of many reasons why an analysis of the residuals should not be the only procedure used to determine the nature of possible specification errors.

Before we leave the topic of Equation 9.3, we should mention two other items of interest. First, note that the error term ϵ^* appears to have a nonzero mean because of the impure serial correlation. The OLS estimate of the constant term, $\hat{\beta}_0$, will adjust to offet this problem. Second, since impure serial correlation implies a specification error such as an omitted variable, impure serial correlation is likely to be associated with biased coefficient estimates. Both the bias and the impure serial correlation will disappear if the specification error is corrected.

An example of how an omitted variable might cause serial correlation in the error term of an incorrectly specified equation involves the fish-demand equation of Section 8.5:

$$F_t = \beta_0 + \beta_1 RP_t + \beta_2 \ln Yd_t + \beta_3 D_t + \epsilon_t \tag{9.4}$$

where F_t is per capita pounds of fish consumed in year t, RP_t is the price of fish relative to beef in year t, Yd_t is real per capita disposable income in year t, D_t is a dummy variable equal to zero in years before the Pope's decision and one thereafter, and ϵ_t is a classical (nonserially correlated) error term. Assume that Equation 9.4 is the "correct" specification. What would happen to this equation if disposable income, Yd, were omitted?

$$F_t = \beta_0 + \beta_1 RP_t + \beta_3 D_t + \epsilon_t^* \tag{9.5}$$

The most obvious effect would be that the estimated coefficients of RP and D would be biased depending on the correlation of RP and D with Yd. A secondary effect would be that the error term would now include a large portion of the left-out effect of disposable income on the consumption of fish. That is, ϵ_t^* would equal $\epsilon_t + \beta_2 \ln Yd_t$. It's reasonable to expect that disposable income (and therefore its log) might follow a fairly serially correlated pattern:

$$\ln Yd_t = f(\ln Yd_{t-1}) + u_t \tag{9.6}$$

Why is this likely? Observe Figure 9.4, which plots the log of U.S. disposable income over time. Note that the continual rise of disposable income over time makes it (and its log) act in a serially correlated or *autoregressive* manner. But if disposable income is serially correlated (and if its impact is not small relative to ϵ) then ϵ^* is likely to also be serially correlated, which can be expressed as:

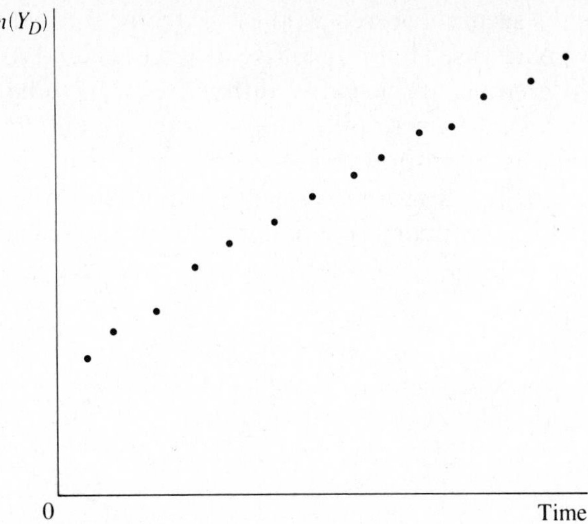

Figure 9.4 U.S. DISPOSABLE INCOME AS A FUNCTION OF TIME
U.S. disposable income (and most other national aggregates) tends to increase steadily over time. As a result, such variables are serially correlated (or autocorrelated), and the omission of such a variable from an equation could potentially introduce impure serial correlation into the error term of that equation.

$$\epsilon_t^* = \rho\epsilon_{t-1}^* + u_t$$

where ρ is the coefficient of serial correlation and u is a classical error term. This example has shown that it is indeed possible for a left-out variable to introduce "impure" serial correlation into an equation. For more on this example, see Exercise 10.

Another common kind of impure serial correlation is that caused by an incorrect functional form. In this kind of situation, the choice of the wrong functional form can cause the error terms to be serially correlated. Let's suppose that the true equation is double-log in nature:

$$\ln Y_t = \beta_0 + \beta_1 \ln X_{1t} + \epsilon_t \tag{9.7}$$

but that instead a linear regression is run:

$$Y_t = \alpha_0 + \alpha_1 X_{1t} + \epsilon_t^* \tag{9.8}$$

The new error term ϵ^* is now a function of the true error term ϵ and of the differences between the linear and the double-log functional forms. As can be seen in Figure 9.5, these differences often follow fairly autoregressive patterns. That is, positive differences tend to be followed by positive differences, and negative differences tend to be followed by negative differences. As a result, using a linear functional form when a nonlinear one is appropriate will usually result in positive impure serial correlation. For a more complete example of impure serial correlation caused by an incorrect functional form, see Exercise 14.

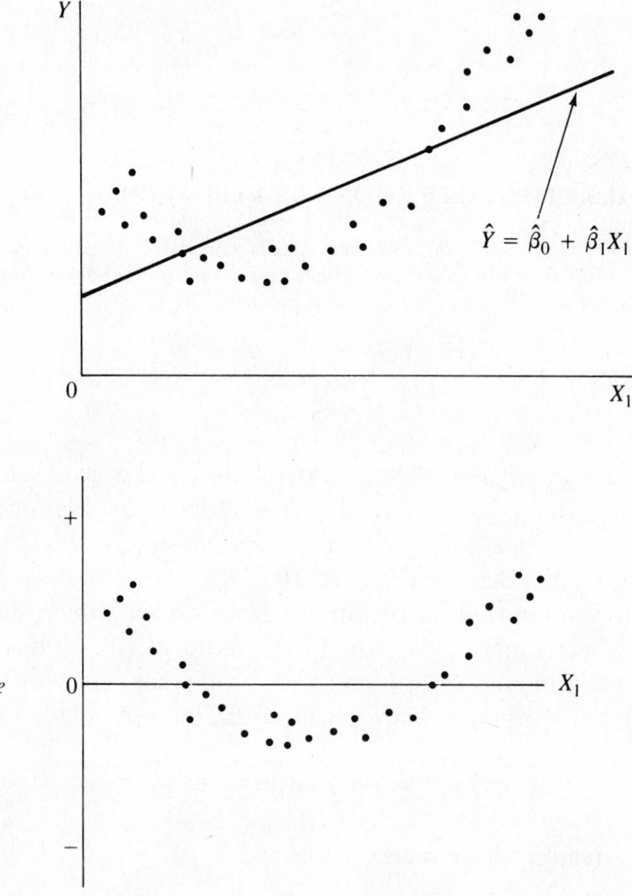

Figure 9.5 INCORRECT FUNCTIONAL FORM AS A SOURCE OF IMPURE SERIAL CORRELATION

The use of an incorrect functional form tends to group positive and negative residuals together, causing positive impure serial correlation.

9.2 The Consequences of Serial Correlation

The consequences of serial correlation are quite different in nature from the consequences of the problems discussed so far in this text. Omitted variables, irrelevant variables, and multicollinearity all have fairly recognizable external symptoms. Each problem changes the estimated coefficients and standard errors in a particular way, and an examination of these changes (and the underlying theory) often provides enough information for the problem to be detected. As we shall see, serial correlation is more likely to have internal symptoms; it affects the estimated equation in a way that is not easily observable from an examination of just the results themselves.

There are three major consequences of serial correlation:

1. Pure serial correlation does not cause bias in the coefficient estimates.
2. Serial correlation increases the variances of the $\hat{\beta}$ distributions.[2]
3. Serial correlation causes OLS to underestimate the variances (and standard errors) of the coefficients.

Let's now go on to explain these consequences in more detail and to then work through a hypothetical example of how serially correlated errors affect the estimation of an equation. In the process we will focus mainly on positive pure first-order serial correlation because it's the kind of autocorrelation most frequently assumed in economic analysis.

9.2.1 An Overview of the Consequences of Serial Correlation

The existence of serial correlation in the error terms of an equation violates Classical Assumption IV, and the estimation of the equation with OLS will have at least three consequences:

1. *Pure serial correlation does not cause bias in the coefficient estimates.* Recall that the most important property of the OLS estimation technique is that it is minimum variance for the class of linear unbiased estimators. If the errors are serially correlated, one of the

2. This holds as long as the serial correlation is positive, as is typically the case in economic examples. In addition, if the regression includes a lagged dependent variable as an independent variable, then the problems worsen significantly. For more on this topic (called distributed lags), see Chapter 12.

assumptions of the Gauss-Markov Theorem is violated, but this violation does not cause the coefficient estimates to be biased. Suppose that the error term of the following equation:

$$Y_t = \beta_0 + \beta_1 X_{1t} + \beta_2 X_{2t} + \epsilon_t \quad (9.9)$$

is known to have pure first-order serial correlation:

$$\epsilon_t = \rho\epsilon_{t-1} + u_t \quad (9.10)$$

where u_t is a classical (nonserially correlated) error term.

If Equation 9.9 is correctly specified and is estimated with OLS, then the estimates of the coefficients of the equation obtained from the OLS estimation will be unbiased. That is,

$$E(\hat{\beta}_1) = \beta_1 \quad \text{and} \quad E(\hat{\beta}_2) = \beta_2$$

Pure serial correlation introduces no bias into the estimation procedure. This conclusion doesn't depend on whether the serial correlation is positive or negative or first-order. If the serial correlation is impure, however, bias may be introduced by the use of an incorrect specification.

This lack of bias does not necessarily mean that the OLS estimates of the coefficients of a serially correlated equation will be close to the true coefficient values; the single estimate observed in practice can come from a wide range of possible values. In addition, the standard errors of these estimates will typically be increased by the serial correlation. This increase will raise the probability that a $\hat{\beta}$ will differ significantly from the true β value. What unbiased means in this case is that the distribution of the $\hat{\beta}$s is still centered around the true β.

2. *Serial correlation increases the variances of the $\hat{\beta}$ distributions.* While the violation of Classical Assumption IV causes no bias, it does affect the main conclusion of the Gauss-Markov Theorem, that of minimum variance. In particular, we cannot prove that the OLS $\hat{\beta}$s have minimum variance when Assumption IV is violated. As a result, if the error terms are serially correlated, then OLS no longer provides minimum variance estimates of the coefficients.

The serially correlated error terms cause the dependent variable to fluctuate in a way that the OLS estimation procedure attributes to the independent variables. Thus OLS is more likely to mis-estimate the true β in the face of serial correlation. On balance, the $\hat{\beta}$s are still unbiased because overestimates are just as likely as underestimates;

however, these errors increase the variance of the distribution of the estimates, increasing the amount that any given estimate is likely to differ from the true β. Indeed, it can be shown that if the error terms are distributed as in Equation 9.10, then the variance of the $\hat{\beta}$s is a function of ρ. The larger the ρ, the larger the variance of the $\hat{\beta}$s.

The effect of serial correlation on the distribution of the coefficient estimates is shown in Figure 9.6, which shows that the distribution of $\hat{\beta}$s from a serially correlated equation is centered around the true β but is much wider than the distribution from an equation without serial correlation.

3. *Serial correlation causes OLS to underestimate the variances (and standard errors) of the coefficients.* If serial correlation increases the variances (and also the standard deviations) of the $\hat{\beta}$s, then one might guess that the OLS SE($\hat{\beta}$)s would also increase, but this isn't usually the case. Instead, these SE($\hat{\beta}$)s tend to be too low. As a result,

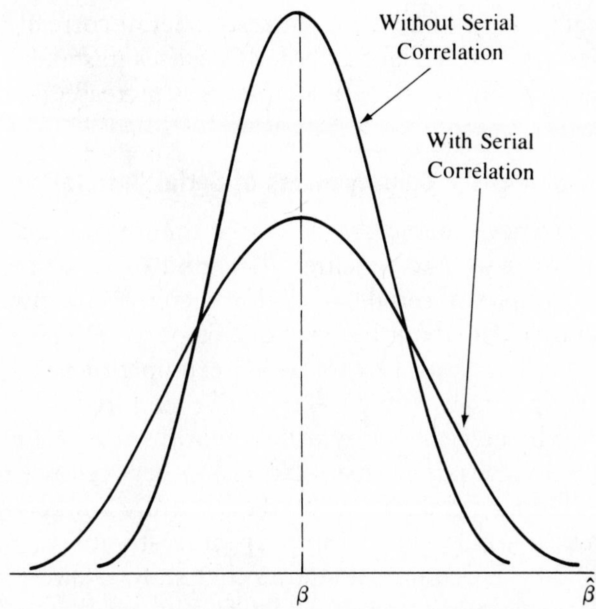

Figure 9.6 DISTRIBUTION OF $\hat{\beta}$s WITH AND WITHOUT SERIAL CORRELATION
The distribution of $\hat{\beta}$s from a serially correlated equation is centered around the true β, but it is often much wider than the distribution from an equation without serial correlation because serial correlation increases the variances of the $\hat{\beta}$ distributions. Unfortunately, OLS underestimates these variances, masking this effect.

serial correlation increases the standard deviations of the estimated coefficients, but it does so in a way that is masked by the OLS estimates.

OLS tends to underestimate the standard errors of the coefficients of serially correlated equations, because serial correlation usually results in a pattern of observations that allows a better fit than the actual nonserially correlated observations would justify. This better fit results in underestimates not only of the standard errors of the $\hat{\beta}$s but also of the standard error of the residuals, so neither the t-statistic nor the F-statistic can be relied on in the face of uncorrected serial correlation.

In particular, the tendency of OLS to underestimate the SE($\hat{\beta}$)s will cause it to overestimate the t-statistics of the estimated coefficients since:

$$t = \frac{(\hat{\beta} - \beta_{H_0})}{SE(\hat{\beta})} \tag{9.11}$$

If a too low SE($\hat{\beta}$) causes a high t-score for a particular coefficient, then it becomes more likely that we will reject a null hypothesis ($\beta = 0$) when it's true. In a sense, then, OLS misleads the researcher about the significance of a particular result. Serial correlation not only increases the standard deviations but also causes mistaken conclusions by making it difficult for OLS to capture this increase.

9.2.2 An Example of the Consequences of Serial Correlation

Error terms can never be observed, so one cannot examine an existing "real world" data-set and be sure what kind of serial correlation, if any, exists. Examples of serial correlation therefore are always clouded by lack of knowledge about the true degree of serial correlation in the error terms. Let's look at a hypothetical example of serial correlation and its effects on OLS estimates. After we've explored this hypothetical example, we'll be able to deal with real world situations much better.

Suppose you're studying the relationship between the real interest rate and the budget deficit. You read the literature on the topic, finding many theoretical articles mostly in favor (at least until recently) of such a link but finding no empirical studies that show a direct, significant, positive relationship between the deficit and real interest rates in the U.S.[3]

3. See, for example, Martin S. Feldstein and Otto Eckstein, "The Fundamental Determinants of the Interest Rate," *Review of Economics and Statistics*, Nov. 1970, pp. 363–375, or Gregory P. Hoelscher, "Federal Borrowing and Short-Term Interest Rates," *Southern Economic Journal*, Oct. 1983, pp. 319–333.

After some mainly Keynesian consideration, you decide to specify the following equation:

$$r_t = f(\overset{+}{D_t},\overset{-}{M_t}) = \beta_0 + \beta_D D_t + \beta_M M_t + \epsilon_t \tag{9.12}$$

where: r_t = the short-term real interest rate in year t
D_t = the budget deficit in year t (percent of GNP)
M_t = the nominal money growth rate in year t
ϵ_t = a classical error term

You then estimate Equation 9.12 on annual U.S. data from 1960 through 1990:

$$\begin{aligned} \hat{r}_t = 0.050 + &\ 0.008D_t - 0.002M_t \\ &\ (0.002) \qquad (0.001) \\ t = &\ 4.00 \qquad -2.00 \qquad \bar{R}^2 = .60 \end{aligned} \tag{9.13}$$

You're excited because it appears that you've shown that the deficit is a significant positive factor in the determination of short-term real interest rates. You worry, however, that serial correlation might invalidate your results. Your concern is in part due to the possibility that many of the involved relationships might take at least a year to work through the macroeconomy. This would mean that shocks to the system would work their way through slowly, causing positive first-order serial correlation. In addition, it seems likely that any omitted variables (of which there could be many in such a simplistic equation) in a time-series would have some "autocorrelated" pattern over time, causing impure serial correlation.

Since your concern is with the estimated coefficient and standard error of the deficit variable, let's take a look at the consequences of serial correlation for your hypothetical results (with respect to the D variable only):

A. With no serial correlation:

$\hat{\beta}_D$ = 0.008
$SE(\hat{\beta}_D)$ = 0.002
t-score = 4.00

With no serial correlation, valid inferences about the statistical significance of $\hat{\beta}$ can be drawn from Equation 9.13's t-scores. Unfortunately, if the error term is serially correlated, we will not obtain these results.

B. With serial correlation but a correct estimate of the standard error:

$\hat{\beta}_D = 0.008$ $SE(\hat{\beta}_D) = 0.006$ t-score $= 1.33$	With serial correlation, the standard deviation increases, and a correct estimate of it would decrease the t-score. This is the result that would be printed out if there were a computer program capable of estimating "correct" $SE(\hat{\beta})$s and t-scores. Since no such program exists, we can never observe these results.

C. With serial correlation and the OLS underestimate of the standard error:

$\hat{\beta}_D = 0.008$ $SE(\hat{\beta}_D) = 0.003$ t-score $= 2.66$	OLS will underestimate the standard error, giving an unrealistically high t-score. This is the result that will actually be printed out by the OLS computer program. It masks what should be a decreased t-score.

In a real case of serial correlation, we would never see result B, only result C. In this hypothetical example, however, we've been able to do what can never be done in an actual regression. That is, we have separated the increase in the standard deviation due to serial correlation from the simultaneous underestimate of that standard deviation by OLS. As a result, we can see that the OLS result (t = 2.66) is not a good indication of the actual significance of a particular coefficient in the face of serial correlation (t= 1.33). In order to decide what to do, it's clear that we need to be able to test for the existence of serial correlation.

9.3 The Durbin-Watson d Test

The test for serial correlation that is most widely used is the Durbin-Watson d test.

9.3.1 The Durbin-Watson d Statistic

The **Durbin-Watson d statistic**[4] is used to determine if there is first-order serial correlation in the error term of an equation by examining the *residuals* of a particular estimation of that equation. It's important

4. J. Durbin and G. S. Watson, "Testing for Serial Correlation in Least-Squares Regression," *Biometrika*, 1951, pp. 159–177.

to use the Durbin-Watson d statistic only when the assumptions that underlie its derivation are met:

1. The regression model includes an intercept term.
2. The serial correlation is first-order in nature:

$$\epsilon_t = \rho\epsilon_{t-1} + u_t$$

   where ρ is the coefficient of serial correlation and u is a classical (nonserially correlated) error term.
3. The regression model does not include a lagged dependent variable (discussed in Chapter 12) as an independent variable.[5]

The equation for the *Durbin-Watson d statistic* for T observations is:

$$d = \sum_{2}^{T}(e_t - e_{t-1})^2 / \sum_{1}^{T} e_t^2 \qquad (9.14)$$

where the e_ts are the OLS residuals.[6] Note that the numerator has one fewer observation than the denominator because an observation must be used to calculate e_{t-1}. The Durbin-Watson d statistic equals zero if there is extreme positive serial correlation, two if there is no serial correlation, and four if there is extreme negative serial correlation. To see this, put appropriate residual values into Equation 9.14 for these cases:

1. Extreme Positive Serial Correlation: $d = 0$

   In this case, $e_t = e_{t-1}$, so $(e_t - e_{t-1}) = 0$ and $d = 0$.

2. Extreme Negative Serial Correlation: $d \approx 4$

   In this case, $e_t = -e_{t-1}$, and $(e_t - e_{t-1}) = (2e_t)$. Substituting into Equation 9.14, we obtain $d = \Sigma(2e_t)^2/\Sigma(e_t)^2$ and $d \approx 4$.

5. In such a circumstance, the Durbin-Watson d is biased towards 2, but the Durbin h test or other tests can be used instead; see Section 12.2.

6. Another approach to the calculation of the Durbin-Watson statistic is:

$$d = 2(1 - \hat{\rho})$$

where $\hat{\rho}$ is the coefficient of a regression of the residuals as a function of their values lagged one time period. This $\hat{\rho}$ is also a rough estimate of rho, the coefficient of serial correlation.

3. No Serial Correlation: d ≈ 2

When there is no serial correlation, the mean of the distribution of d is equal to two.[7] That is, if there is no serial correlation, d ≈ 2.

9.3.2 Using the Durbin-Watson d Test

The Durbin-Watson d test is unusual in two respects. First, econometricians almost never test the one-sided null hypothesis that there is negative serial correlation in the residuals because negative serial correlation, as mentioned above, is quite difficult to explain theoretically in economic or business analysis. Its existence means that impure serial correlation probably has been caused by some error of specification.

Second, the Durbin-Watson test is sometimes inconclusive. While previously explained decision rules always have had only "acceptance" regions and rejection regions, the Durbin-Watson test has a third possibility, called the inconclusive region.[8] We'll discuss what to do when the test is inconclusive in Section 9.4.

With these exceptions, the use of the Durbin-Watson d test is quite similar to the use of the *t*- and *F*-tests. In order to test for positive serial correlation, the following steps are required:

1. Obtain the OLS residuals from the equation to be tested and calculate the d statistic by using Equation 9.14.
2. Determine the sample size and the number of explanatory variables and then consult Statistical Tables B-4, B-5, or B-6 in Appendix B to find the upper critical d value, d_U, and the lower critical d value, d_L, respectively. Instructions for the use of these tables are also in that appendix.
3. Given the null hypothesis of no positive serial correlation and a one-sided alternative hypothesis:

7. To see this, multiply out the numerator of Equation 9.14, obtaining

$$d = [\sum_2^T e_t^2 - 2\sum_2^T(e_t e_{t-1}) + \sum_2^T e_{t-1}^2]/\sum_1^T e_t^2 \approx [\sum_2^T e_t^2 + \sum_2^T e_{t-1}^2]/\sum_1^T e_t^2 \approx 2$$

If there is no serial correlation, then e_t and e_{t-1} are not related, and, on average, $\Sigma(e_t e_{t-1}) = 0$.

8. This inconclusive region is troubling, but the development of exact Durbin-Watson tests may eliminate this problem in the near future. See William H. Greene, *Econometric Analysis* (New York: Macmillan, 1990), pp. 449–452. Some computer programs, including recent versions of SHAZAM, allow the user the option of calculating an exact Durbin-Watson probability (of first-order serial correlation).

$$H_0: \rho \leq 0 \quad \text{(no positive serial correlation)}$$
$$H_A: \rho > 0 \quad \text{(positive serial correlation)}$$

the appropriate decision rule is:

$$\begin{aligned} \text{if } d < d_L &\quad \text{Reject } H_0 \\ \text{if } d > d_U &\quad \text{Do Not Reject } H_0 \\ \text{if } d_L \leq d \leq d_U &\quad \text{Inconclusive} \end{aligned}$$

In some circumstances, a two-sided d test will be appropriate. In such a case, steps 1 and 2 above are still used, but step 3 is now:

3A. Given the null hypothesis of no serial correlation and a two-sided alternative hypothesis:

$$H_0: \rho = 0 \quad \text{(no serial correlation)}$$
$$H_A: \rho \neq 0 \quad \text{(serial correlation)}$$

the appropriate decision rule is:

$$\begin{aligned} \text{if } d < d_L &\quad \text{Reject } H_0 \\ \text{if } d > 4 - d_L &\quad \text{Reject } H_0 \\ \text{if } 4 - d_U > d > d_U &\quad \text{Do Not Reject } H_0 \\ \text{otherwise} &\quad \text{Inconclusive} \end{aligned}$$

Figure 9.7 presents a graph of the rejection, "acceptance," and inconclusive regions for a two-sided hypothesis of no serial correlation. A graph of these regions for one-sided hypotheses is given in the example that follows.

9.3.3 Examples of the Use of the Durbin-Watson d Statistic

Let's work through some applications of the Durbin-Watson test. First, turn to Statistical Tables B-4, B-5, and B-6. Note that the upper and lower critical d values (d_U and d_L) depend on the number of explanatory variables (do not count the constant term), the sample size, and the level of significance of the test.

Now set up a one-sided 95 percent confidence test for a regression with three explanatory variables and 25 observations. As can be seen from the 5 percent table (B-4), the critical d values are $d_L = 1.12$ and $d_U = 1.66$. As a result, if the hypotheses are:

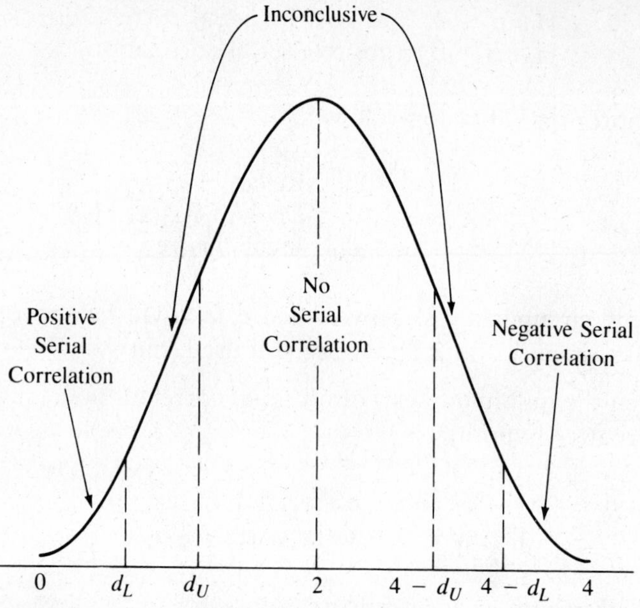

Figure 9.7 A TWO-SIDED DURBIN-WATSON d TEST
The Durbin-Watson d statistic is symmetrically distributed around its mean of 2. In a two-sided test, the farther that the observed d is from 2, the more likely is serial correlation. Note the inconclusive regions between the upper and lower critical values of d.

$$H_0: \rho \leq 0 \quad \text{(no positive serial correlation)}$$
$$H_A: \rho > 0 \quad \text{(positive serial correlation)}$$

the appropriate decision rule is:

$$\begin{aligned} \text{if } d &< 1.12 \quad \text{Reject } H_0 \\ \text{if } d &> 1.66 \quad \text{Do Not Reject } H_0 \\ \text{if } 1.12 \leq d &\leq 1.66 \quad \text{Inconclusive} \end{aligned}$$

A computed d statistic of 1.78, for example, would indicate that there is no evidence of positive serial correlation, a value of 1.28 would be inconclusive, and a value of 0.60 would imply positive serial correlation. Figure 9.8 provides a graph of the "acceptance," rejection, and inconclusive regions for this example.

For a more familiar example, we return to the chicken demand model of Equation 6.5. As can be confirmed with the data provided in

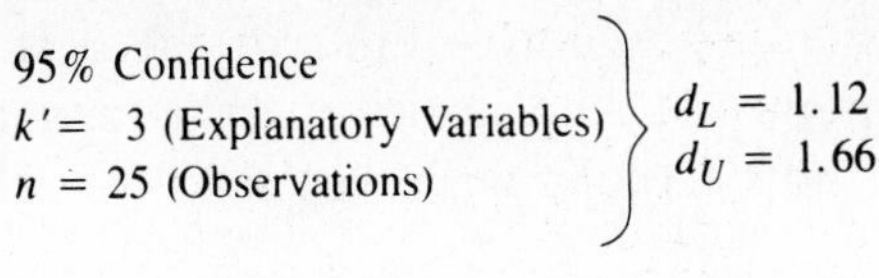

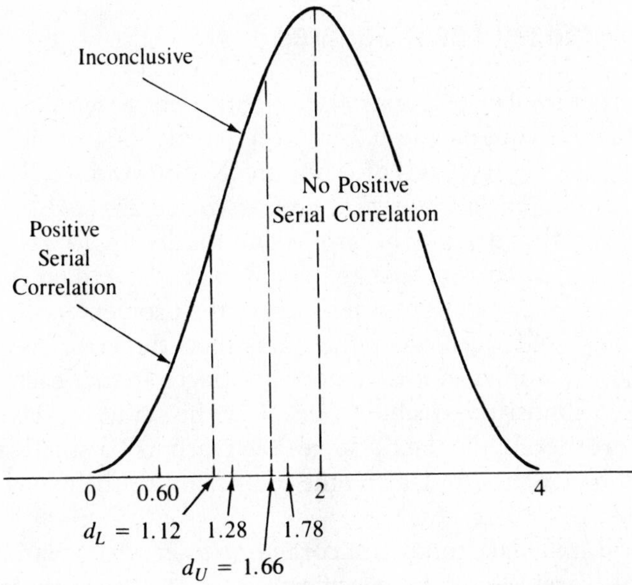

Figure 9.8 AN EXAMPLE OF A ONE-SIDED DURBIN-WATSON d TEST
In a one-sided Durbin-Watson test for positive serial correlation, only values of d significantly below 2 cause the null hypothesis of no positive serial correlation to be rejected. In this example, a d of 1.78 would indicate no positive serial correlation, a d of 0.60 would indicate positive serial correlation, and a d of 1.28 would be inconclusive.

Table 6.3, the Durbin-Watson statistic from Equation 6.5 is 0.94. Is that cause to be concerned about serial correlation? What would be the result of a one-sided 95 percent test of the null hypothesis of no positive serial correlation? Our first step would be to consult Statistical Table B-4. In that table, with k′ (the number of explanatory variables, K) equal to 3 and n (the number of observations) equal to 35, we would find the critical d values $d_L = 1.28$ and $d_U = 1.65$.

The decision rule would thus be:

$$\begin{aligned} &\text{if } d < 1.28 && \text{Reject } H_0 \\ &\text{if } d > 1.65 && \text{Do Not Reject } H_0 \\ &\text{if } 1.28 \leq d \leq 1.65 && \text{Inconclusive} \end{aligned}$$

Since 0.94 is less than the critical lower limit of the d statistic, we would reject the null hypothesis of no positive serial correlation, and we would have to decide how to cope with that serial correlation.

9.4 Generalized Least Squares

Suppose that the Durbin-Watson d statistic detects serial correlation in the residuals of your equation. Is there a remedy? Some students suggest reordering the observations of Y and the Xs to avoid serial correlation. That is, if this time's error term appears to be affected by last time's error term, why not reorder the data randomly to get rid of the problem? The answer is that the reordering of the data does not get rid of the serial correlation; it just makes the problem harder to detect. If $\epsilon_2 = f(\epsilon_1)$ and we reorder the data, then the error terms are still related to each other, but they now no longer follow each other, and it becomes almost impossible to discover the serial correlation. Interestingly, reordering the data changes the Durbin-Watson d statistic but does not change the estimates of the coefficients or their standard errors at all.[9]

Instead, the place to start in correcting a serial correlation problem is to look carefully at the specification of the equation for possible errors that might be causing impure serial correlation. Is the functional form correct? Are you sure that there are no omitted variables? In particular, are there specification errors that might have some pattern over time that could have introduced impure serial correlation into the residuals? Only after the specification of the equation has been reviewed carefully should the possibility of an adjustment for pure serial correlation be considered.

It's worth noting that if one or more of the variables increases or decreases steadily over time, as is often the case, or if the data-set is logically reordered (say, according to the magnitude of one of the variables), then the Durbin-Watson statistic can help detect impure serial correlation. A significant Durbin-Watson statistic can easily be caused by a left-out variable or an incorrect functional form. In such circumstances, the Durbin-Watson test does not distinguish between pure and impure serial correlation, but the detection of negative serial

9. This can be proven mathematically, but it is usually more instructive to estimate a regression yourself, change the order of the observations, and then reestimate the regression. See Exercise 3.

correlation is often a strong hint that the serial correlation is impure. If you conclude that you have pure serial correlation, then the appropriate response is to consider the application of generalized least squares.

9.4.1 What Is Generalized Least Squares?

Generalized Least Squares (GLS) is a method of ridding an equation of pure first-order serial correlation and in the process restoring the minimum variance property to its estimation. Also called the Aitken estimator, GLS starts with an equation that does not meet the Classical Assumptions (due in this case to the pure serial correlation in the error term) and transforms it into one (Equation 9.19) that does meet those assumptions.

At this point, you could skip directly to Equation 9.19, but it's easier to understand the GLS estimator by examining the transformation from which it comes. Start with an equation that has first-order serial correlation:

$$Y_t = \beta_0 + \beta_1 X_{1t} + \epsilon_t \tag{9.15}$$

which, if $\epsilon_t = \rho\epsilon_{t-1} + u_t$ (due to pure serial correlation) also equals:

$$Y_t = \beta_0 + \beta_1 X_{1t} + \rho\epsilon_{t-1} + u_t \tag{9.16}$$

where ϵ is the serially correlated error term, ρ is the coefficient of serial correlation, and u is a classical (nonserially correlated) error term.

If we could get the $\rho\epsilon_{t-1}$ term out of Equation 9.16, the serial correlation would be gone because the remaining portion of the error term (u_t) has no serial correlation in it. In order to rid $\rho\epsilon_{t-1}$ from Equation 9.16, multiply Equation 9.15 by ρ and then lag the new equation by one time period, obtaining

$$\rho Y_{t-1} = \rho\beta_0 + \rho\beta_1 X_{1t-1} + \rho\epsilon_{t-1} \tag{9.17}$$

Notice that we now have an equation with a $\rho\epsilon_{t-1}$ term in it. If we now subtract Equation 9.17 from Equation 9.16, the equivalent equation that remains no longer contains the serially correlated component of the error term:

$$Y_t - \rho Y_{t-1} = \beta_0(1 - \rho) + \beta_1(X_{1t} - \rho X_{1t-1}) + u_t \tag{9.18}$$

Equation 9.18 can be rewritten as:

$$Y_t^* = \beta_0^* + \beta_1 X_{1t}^* + u_t \tag{9.19}$$

where:

$$Y_t^* = Y_t - \rho Y_{t-1} \tag{9.20}$$
$$X_{1t}^* = X_{1t} - \rho X_{1t-1}$$
$$\beta_0^* = \beta_0 - \rho\beta_0$$

Equation 9.19 is called a GLS (or "quasi-differenced") version of Equation 9.16; notice that:

1. The error term is not serially correlated. As a result, OLS estimation of Equation 9.19 will be minimum variance.
2. The slope coefficient β_1 is the same as the slope coefficient of the original serially correlated equation, Equation 9.15.

As a result, if we substitute $\hat{\rho}$ for ρ, we can estimate Equation 9.19 to obtain estimates of Equation 9.15 or Equation 9.16. One typical question at this point is, "If the slope coefficients are the same for both equations, will the estimated coefficients also be the same?" The answer is that, as always, there is no reason for different unbiased estimates to be identical. In addition, the variances of the two estimates are different, further increasing the likelihood of different $\hat{\beta}$s.

9.4.2 How to Obtain Estimates of ρ

To get rid of pure serial correlation you have to redefine your variables as in Equation 9.20, substitute $\hat{\rho}$ for ρ, and run OLS on Equation 9.19. This will give you unbiased, minimum variance (if $\rho = \hat{\rho}$) estimates of the βs. There's only one problem with all this: Where does the $\hat{\rho}$ come from? Since it's quite unlikely that the true ρ will be known theoretically, how do you go about obtaining estimates of the coefficient of serial correlation?

The simplest way to obtain estimates of ρ is to use an approximation formula roughly based on the Durbin-Watson statistic:

$$\hat{\rho} \approx 1 - \frac{d}{2} \tag{9.21}$$

where d is the Durbin-Watson statistic in Equation 9.14 and the "approximately equal" sign indicates that the relationship is inexact. That this equation works for at least some points can be seen by plugging in the three best-known values of d:

With extreme positive serial correlation, d = 0, and $\hat{\rho} \approx 1$.

With extreme negative serial correlation, d = 4, and $\hat{\rho} \approx -1$.

With no serial correlation, d = 2, and $\hat{\rho} \approx 0$.

In each of these three cases, Equation 9.21 gives exactly the right value for $\hat{\rho}$.

To obtain more accurate estimates of ρ, particularly for small samples, there are a number of more complicated methods, including the *Cochrane-Orcutt iterative method,*[10] a two-step iterative procedure that provides an estimate of ρ and estimates the GLS regression.

Step 1: Estimate $\hat{\rho}$ by running a regression based on the residuals of the equation suspected of having serial correlation:

$$e_t = \rho e_{t-1} + u_t \tag{9.22}$$

where the e_ts are the OLS residuals from the equation suspected of having pure serial correlation and u is a classical (nonserially correlated) error term.

Step 2: Use this $\hat{\rho}$ to run (estimated) GLS by substituting it into Equation 9.20 and then using OLS to estimate Equation 9.19 with the adjusted data.

In practice, these two steps are repeated (iterated) until further iteration results in little change in $\hat{\rho}$. Once $\hat{\rho}$ has converged (usually in only a few iterations), the results of the last estimate of Step 2 are printed out (often along with a listing of all the intermediate $\hat{\rho}$s computed in the process). While other methods for estimating ρ are available (Hildreth-Lu, Theil-Nagar, Durbin, among others) the Cochrane-Orcutt method is commonly used and meets our needs quite well, so we'll limit our coverage to that method.

Let's examine the application of (estimated) GLS (using the Cochrane-Orcutt method) to the chicken demand example that was found to have positive serial correlation in the previous section. Recall what Equation 6.5 looked like:

10. D. Cochrane and G. H. Orcutt, "Application of Least Squares Regression to Relationships Containing Autocorrelated Error Terms," *Journal of the American Statistical Association,* 1949, pp. 32–61.

$$\hat{Y}_t = -60.5 - \underset{(0.07)}{0.45PC_t} + \underset{(0.05)}{0.12PB_t} + \underset{(1.2)}{12.2LYD_t} \qquad (6.5)$$
$$t = \quad -6.4 \qquad 2.5 \qquad 10.6$$
$$\bar{R}^2 = .984 \qquad n = 35 \qquad DW\ d = 0.94$$

where: Y = annual per capita chicken consumption (in pounds)
PC = the price of chicken (cents per pound)
PB = the price of beef (cents per pound)
LYD = the log of per capita disposable income (dollars)

Note that we have added the Durbin-Watson d statistic to the documentation with the notation "DW." All future time-series results will include the DW statistic, but cross-sectional documentation of the DW is not required unless the observations are ordered in some logical manner.

If Equation 6.5 is reestimated with GLS, we obtain:

$$\hat{Y}_t = -67.2 - \underset{(0.09)}{0.29PC_t} + \underset{(0.05)}{0.13PB_t} + \underset{(1.3)}{12.7LYD_t} \qquad (9.23)$$
$$t = \quad -3.3 \qquad 2.6 \qquad 9.6$$
$$\bar{R}^2 = .989 \qquad n = 35 \qquad \hat{\rho} = 0.69$$

Compare these two results. First, note that the $\hat{\rho}$ used in Equation 9.23 was 0.69. That means that Y was actually run as $Y_t^* = Y_t - 0.69Y_{t-1}$, PC as $PC_t^* = PC_t - 0.69PC_{t-1}$, etc. Second, note that the t-scores have decreased (mainly due to increased estimates of the standard errors). This makes sense since one of the consequences of serial correlation is that OLS underestimates the standard errors of the $\hat{\beta}$s. Indeed, one reason for adjusting for serial correlation is to avoid making mistakes of inference because of t-scores that are too high. Finally, note that even though serial correlation causes no bias in the estimates of the βs, the GLS estimates of the slope coefficients still differ somewhat from the OLS ones. To compare intercepts, we would divide $\hat{\beta}_0^*$ by $[1 - \hat{\rho}]$.

With respect to documentation, note that the $\hat{\rho}$ replaces the DW in the documentation of a GLS result since the DW statistic of Equation 9.23 is not strictly comparable with non-GLS DWs (it is biased towards 2). In fact, if any Durbin-Watson statistic should be presented, it should be the one from the OLS estimation (in order to document the need for GLS). Also note that the estimated model is typically presented without the asterisks attached to the transformed Y* and X* variables. This sometimes causes confusion if the reader cannot tell whether the variables in Equation 9.23 are actually Xs or X*s. Since the only difference between the OLS and GLS results is the constant term (remember, the slope coefficients are estimates of the same βs), and since the constant term is usually of little interest to researchers, the common practice is to present the results of GLS with the variables in their untransformed state. To forecast with GLS, adjustments like those discussed in Section 15.2 need to be made.

9.4.3 Why Generalized Least Squares Shouldn't Automatically Be Used

There are a number of reasons why GLS should not be applied every time that the Durbin-Watson test indicates the likelihood of serial correlation in the residuals of an equation. As a result of these reasons, many econometricians try to avoid the use of GLS when the Durbin-Watson test is inconclusive.

1. *The significant DW may be caused by impure serial correlation.* When autocorrelation is detected, the cause may be an omitted variable or a poor choice of functional form. In such a case, the best solution is to find the missing variable or the proper form. Even if these easy answers cannot be found, the application of GLS to the misspecified equation is not necessarily superior to OLS. Impure serial correlation justifies using GLS only when the cause is an omitted variable that is at least reasonably correlated with one of the included variables. In this case, if the left-out variable cannot be found, GLS will reduce the bias somewhat, because the procedure proxies for the autocorrelated portion of the omitted variable. In cases of uncorrelated omitted variables or improper functional form, it can be shown that OLS is superior to GLS for estimating an incorrectly specified equation. In all cases, of course, the best course of action to use the correct specification.

2. *The serial correlation may be fixed by timewise aggregation.* In general, we should expect pure autocorrelation to be more likely in weekly, monthly, or quarterly data than in annual or cross-sectional data. This is because the lag between a shock and its final impact on a system is more likely to be included in a particular time period the longer that time period is. For example, it's possible to reduce serial correlation in a data-set by aggregating quarterly data into annual data, called "timewise aggregation." While this new data-set will have fewer degrees of freedom, it will not necessarily contain substantially less

information. In particular, if the reason for running the regression is to measure the effect of a single policy change (like the Pope's decision to allow Catholics to eat meat on Fridays), then timewise aggregation will better capture that effect because short-run "noise" is less likely to interfere with the estimation of a longer-run phenomenon.

3. *The consequences of the serial correlation may be minor.* GLS works well if $\hat{\rho}$ is close to the actual ρ, but $\hat{\rho}$ is biased in small samples, potentially causing estimation problems. Since serial correlation causes no bias, it's possible that the harm done to the equation by the serial correlation may be less than the damage done by attempting to fix that problem with a biased $\hat{\rho}$. In particular, when coefficient estimates seem theoretically reasonable, and when the t-scores of the various coefficients are not being relied on for the retention of independent variables, the harm caused by serial correlation may be minor.

9.4.4 A Complete Example of the Application of Generalized Least Squares

To illustrate the GLS technique, let's make up an example with a serial correlation problem. We will choose a "true" equation, force the error terms to be serially correlated, and generate a data-set from the equation. We will first estimate the equation with OLS to see if the Durbin-Watson statistic will detect the serial correlation. We will then apply GLS to the equation to see if GLS can estimate the actual ρ, rid the equation of the serial correlation, and thus come close to the true values we used to generate the observations.

Think how you might act if you received occasional gifts of money from an uncle with the instruction that you should "go out and have a nice dinner." Let's build a model of the amount of money (in dollars) you would spend on dinner (Y) as a function of the size of the gift you received from your uncle (X). In such a case, the following equation might make sense:

$$Y_t = 4.00 + 0.90X_t + \epsilon_t \qquad (9.24)$$

Let's further assume that the error terms are serially correlated

$$\epsilon_t = 0.80\epsilon_{t-1} + u_t \qquad (9.25)$$

That is, your behavior follows a pattern of positive serial correlation with an autocorrelation coefficient of 0.80. If one time you spend more than normal, then you tend to spend more than normal the next time, and if you spend less than normal on a gift dinner, you tend to spend less subsequently. (If your behavior instead followed a pattern of negative serial correlation, then you would tend to correct your overspending one time by underspending the next, and the error terms would be negatively related.)

To create the data-set, we need to create a set of error terms and then use these error terms to generate observations of Y. For this example's error term observations, we randomly selected values u_t from a normal distribution with mean = 0 and variance = 3, signified by N(0,3). We then plugged the randomly picked *us* into Equation 9.25 above. For the first time period, for example, we randomly drew a u of −3.72, and then we calculated ϵ_1 (assuming $\epsilon_0 = 0$):

$$\epsilon_1 = 0.80\epsilon_0 + u_1 = 0.80(0) + (-3.72) = -3.72$$

For the second time period, we randomly drew u = −2.09, and ϵ_2 was:

$$\epsilon_2 = 0.80\epsilon_1 + u_2 = 0.80(-3.72) + (-2.09) = -5.07$$

For the generation of the rest of the observations of the error term (ϵ) see Table 9.1.

TABLE 9.1 THE GENERATION OF THE HYPOTHETICAL DATA SET USED IN THE SERIAL CORRELATION EXAMPLE

X_t	$u_t \sim N(0,3)$	$\epsilon_t = u_t + 0.8\epsilon_{t-1}$	$Y_t = 4.0 + 0.9X_t + \epsilon_t$
1	−3.72	−3.72	1.18
2	−2.09	−5.07	0.73
3	−0.28	−4.34	2.36
4	1.99	−1.48	6.12
5	1.06	−0.12	8.38
6	−1.07	−1.17	8.23
7	0.06	−0.88	9.42
8	1.37	0.67	11.87
9	1.22	1.76	13.86
10	0.35	1.76	14.76
11	−1.31	0.10	14.00
12	−1.56	−1.48	13.32
13	−3.25	−4.43	11.27
14	−1.55	−5.09	11.51
15	0.59	−3.48	14.02

We then selected a set of fixed Xs (1 to 15) and combined these with the now serially correlated ϵs to generate the Ys from Equation 9.24. For example, Y_1 was calculated by plugging $X_1 = 1.0$ and $\epsilon_1 = -3.72$ (as calculated above) into Equation 9.24:

$$Y_1 = 4.00 + 0.90X_1 + \epsilon_1 = 4.00 + 0.90(1.0) + (-3.72) = 1.18$$

For the generation of the rest of the observations of the dependent variable (Y), see Table 9.1.

Our next step was to estimate the serially correlated equation with OLS:

$$\begin{aligned} \hat{Y} &= 1.91 + 0.937X \\ &\qquad\quad (0.149) \\ t &= 6.31 \\ DW &= 0.425 \end{aligned} \tag{9.26}$$

While we shouldn't draw conclusions from a single trial of a Monte Carlo experiment, this seems like a reasonable result, since the estimates (1.91 and 0.937) are fairly close to the true values (4.0 and 0.90). Does the Durbin-Watson statistic indicate likely serial correlation? For a 99 percent one-sided test, Table B-6 indicates that d_L is 0.81. Since the observed DW is far less than 0.81, the test indicates the presence of positive serial correlation. (This comes as no real surprise, since we introduced the serial correlation ourselves.)

Given this, and given that there are no specification errors, the next step was to reestimate Equation 9.26 with GLS using the Cochrane-Orcutt method:

$$\begin{aligned} \hat{Y} &= 1.48 + 0.924X \\ &\qquad\quad (0.230) \\ t &= 4.016 \qquad\qquad \hat{\rho} = 0.762 \end{aligned} \tag{9.27}$$

How did GLS do? First of all, the $\hat{\rho}$ of 0.762 is very close to the actual value of 0.80. Second, the GLS estimate of the standard error is higher than the OLS estimate, which is logical when you recall that the OLS estimate of the standard error is biased. Because of this increased standard error, the t-score falls; in general, this decreased t-score will help lower the chance of errors of inference. Third, the GLS estimated coefficient of X in Equation 9.27 is slightly closer to the true value of 0.90 than the OLS estimated coefficient in Equation 9.26. Finally, note that while the estimate of the slope coefficient changed hardly at all in this example, such changes do sometimes occur when GLS is used.

9.5 Summary

1. Serial correlation, or autocorrelation, is the violation of the classical assumption that the observations of the error term are uncorrelated with each other. Usually, econometricians focus on first-order serial correlation, in which the current observation of the error term is assumed to be a function of the previous observation of the error term and a nonserially correlated error term (u):

$$\epsilon_t = \rho\epsilon_{t-1} + u_t \qquad -1 < \rho < 1$$

where ρ is "rho," the coefficient of serial correction.

2. Pure serial correlation is serial correlation that is a function of the error term of the correctly specified regression equation. Impure serial correlation is caused by specification errors such as an omitted variable or an incorrect functional form. While impure serial correlation can be positive ($0 < \rho < 1$) or negative ($-1 < \rho < 0$), pure serial correlation in economics or business situations is almost always positive.

3. The major consequence of serial correlation is an increase in the variances of the $\hat{\beta}$ distributions that is masked by an underestimation of those variances (and the standard errors) by OLS. Pure serial correlation does not cause bias in the estimates of the βs.

4. The most commonly used method of detecting first-order serial correlation is the Durbin-Watson d test, which uses the residuals of an estimated regression to test the possibility of serial correlation in the error terms. A d value of 0 indicates extreme positive serial correlation, a d value of 2 indicates no serial correlation, and a d value of 4 indicates extreme negative serial correlation.

5. The first step in ridding an equation of serial correlation is to check for possible specification errors. Only once the possibility of impure serial correlation has been reduced to a minimum should remedies for pure serial correlation be considered.

6. Generalized least squares (GLS) is a method of transforming an equation to rid it of pure first-order serial correlation. The use of GLS requires the estimation of ρ, which is most commonly accomplished through an iterative method devised by Cochrane and Orcutt. GLS should not be automatically applied every time the Durbin-Watson test indicates the possibility of serial correlation in an equation.

Exercises

(Answers to even-numbered exercises are in Appendix A.)

1. Write the meaning of each of the following terms without reference to the book (or your notes), and compare your definition with the version in the text for each:
   a. impure serial correlation
   b. first-order serial correlation
   c. first-order autocorrelation coefficient
   d. Durbin-Watson d statistic
   e. Generalized Least Squares
   f. positive serial correlation

2. Use Statistical Tables B-4, B-5, and B-6 to test for serial correlation given the following Durbin-Watson d statistics for serial correlation.
   a. $d = 0.81, k' = 3, n = 21$, 95 percent, one-sided positive test
   b. $d = 3.48, k' = 2, n = 15$, 99 percent, one-sided positive test
   c. $d = 1.56, k' = 5, n = 30$, 90 percent, two-sided test
   d. $d = 2.84, k' = 4, n = 35$, 95 percent, two-sided test
   e. $d = 1.75, k' = 1, n = 45$, 95 percent, one-sided positive test
   f. $d = 0.91, k' = 2, n = 28$, 98 percent, two-sided test
   g. $d = 1.03, k' = 6, n = 26$, 95 percent, one-sided positive test

3. Recall from Section 9.4 that switching the order of a data-set will not change its coefficient estimates. A changed order will change the Durbin-Watson statistic, however. To see both these points, run regressions ($HS = \beta_0 + \beta_1 P + \epsilon$) and compare the coefficient estimates and DW d statistics for this data-set:

Year	Housing Starts	Population
1	9090	2200
2	8942	2222
3	9755	2244
4	10327	2289
5	10513	2290

in the following three orders (in terms of year):

a. 1,2,3,4,5
b. 5,4,3,2,1
c. 2,4,3,5,1

4. After GLS has been run on an equation, the $\hat{\beta}$s are still good estimates of the original (nontransformed) equation except for the constant term:
   a. What must be done to the estimate of the constant term generated by GLS to compare it with the one estimated by OLS?
   b. Why is such an adjustment necessary?
   c. Return to Equation 9.23 and calculate the $\hat{\beta}_0$ that would be comparable to the one in Equation 6.5.
   d. Return to Equation 9.27 and calculate the $\hat{\beta}_0$ that would be comparable to the one in Equation 9.26.
   e. In both cases, the two estimates are different. Why? Would such a difference concern you?

5. Carefully distinguish between the following concepts:
   a. positive and negative serial correlation
   b. pure and impure serial correlation
   c. serially correlated observations of the error term and serially correlated residuals
   d. GLS and the Cochrane-Orcutt method

6. In Statistical Table B-4, column $k' = 5$, d_U is greater than two for the five smallest sample sizes in the table. What does it mean if $d_U > 2$?

7. Recall the example of the relationship between the short-term real interest rate and the budget deficit discussed at the end of Section 9.2. The hypothetical results in that section were extrapolated from

a cross-sectional study by Hutchinson and Pyle[11] that found at least some evidence of such a link in a sample that pools annual time series and cross-sectional data from six countries.

a. Suppose you were told that the Durbin-Watson d from their best regression was 0.81. Test this DW for indications of serial correlation ($n = 70$, $k' = 4$, 95 percent one-sided test for positive serial correlation).
b. Based on this result, would you conclude that serial correlation existed in their study? Why or why not? (Hint: the six countries were the U.K., France, Japan, Canada, Italy, and the U.S.; assume that the order of the data was U.K. 1973–82, followed by France 1973–82, etc.)
c. How would you use GLS to correct for serial correlation in this case?

8. Suppose the data in a time-series study were entered in reverse chronological order. Would this change in any way the testing or adjusting for serial correlation? How? In particular:

a. What happens to the Durbin-Watson statistic's ability to detect serial correlation if the order is reversed?
b. What happens to the GLS method's ability to adjust for serial correlation if the order is reversed?
c. What is the intuitive economic explanation of reverse serial correlation?

9. Suppose that a plotting of the residuals of a regression with respect to time indicates a significant outlier in the residuals. (Be careful here, this is not an outlier in the original data but is an outlier in the *residuals* of a regression.)

a. How could such an outlier occur? What does it mean?
b. Is the Durbin-Watson d statistic applicable in the presence of such an outlier? Why or why not?

10. Recall the discussion of impure serial correlation caused by leaving out the log of disposable income variable from the fish-demand equation of the previous chapter (see Equations 8.29 and 9.4–9.6).

a. Return to that data-set and estimate Equation 9.5; that is, leave out the lnYd variable and estimate:

11. M. M. Hutchinson and D. H. Pyle, "The Real Interest Rate/Budget Deficit Link: International Evidence, 1973–82," *Federal Reserve Bank of San Francisco Economic Review,* Fall 1984, pp. 26–35.

$$F = \beta_0 + \beta_1 RP_t + \beta_3 D_t + \epsilon_t^*$$

(If you do not have access to a computer, or if you do not have time to estimate the equation yourself, look up the result in the answer to this question in Appendix A and then attempt to do the rest of the question on your own.)

b. Analyze the results. In particular, test the coefficients for five percent statistical significance, test for serial correlation, and decide whether or not the result confirms our original expectation that the Pope's decision did indeed decrease per capita fish consumption.

c. How would you have gone about analyzing this problem if you had not known that the omission of the lnYd variable was the cause? In particular, how would you have determined whether the potential serial correlation was pure or impure?

11. Your friend is just finishing a study of attendance at Los Angeles Laker regular season home basketball games when she hears that you've read a chapter on serial correlation and asks your advice. Before running the equation on last season's data, she "reviewed the literature" by interviewing a number of basketball fans. She found out that fans like to watch winning teams and that the Lakers almost always play quite well. In addition, she learned that while some fans like to watch games throughout the season, others are most interested in games played late in the season. Her estimated equation (standard errors in parentheses) was:

$$\hat{A}_t = 14123 + \underset{(500)}{20L_t} + \underset{(1000)}{2600P_t} + \underset{(300)}{900W_t}$$

$$DW = 0.85 \qquad n = 40 \qquad \bar{R}^2 = .46$$

where: A_t = the attendance at game t

L_t = the winning percentage (games won divided by games played) of the Lakers before game t

P_t = the winning percentage before game t of the Lakers' opponent in that game

W_t = a dummy variable equal to one if game t was on Friday, Saturday, or Sunday, 0 otherwise

a. Test for serial correlation using the Durbin-Watson d test at the 5 percent level.

b. Make and test appropriate hypotheses about the slope coefficients at the 1 percent level.
c. Compare the size and significance of the estimated coefficient of L with that for P. Is this difference surprising? Is L an irrelevant variable? Explain your answer.
d. If serial correlation exists, would you expect it to be pure or impure serial correlation? Why?
e. Your friend omitted the first game of the year from the sample because the first game is always a sellout and because neither team had a winning percentage yet. Was this a good decision?

12. In Section 2.2.3, we considered an equation for U.S. per capita consumption of beef for the years 1960–1987 that was kept simple for the purpose of that chapter. A more complete specification of the equation estimated on the same data produces:

$$\hat{B}_t = -330.3 + 49.1 \ln Y_t - 0.34 PB_t + 0.33 PP_t - 15.4 D_t$$
$$\qquad (7.4) \qquad (0.13) \qquad (0.12) \qquad (4.1)$$
$$t = \quad 6.6 \qquad -2.6 \qquad 2.7 \qquad -3.7$$
$$\bar{R}^2 = .700 \quad n = 28 \quad DW = 0.94 \qquad (9.28)$$

where: B_t = the annual per capita pounds of beef consumed in the U.S. in year t
$\ln Y$ = log of real per capita disposable income in the U.S. in 1982 dollars
PB = average annualized real wholesale price of beef (in cents per pound)
PP = average annualized real wholesale price of pork (in cents per pound)
D = a dummy variable equal to 1 for years after 1981, 0 otherwise (an attempt to capture the increased consumer awareness of the health dangers of red meat)

a. Develop and test your own hypotheses with respect to the individual estimated slope coefficients.
b. Test for serial correlation in Equation 9.28 using the Durbin-Watson d test at the 5 percent level.
c. What econometric problem(s) (if any) does Equation 9.28 appear to have? What remedy would you suggest?
d. You take your own advice, and apply GLS to Equation 9.28, obtaining:

$$\hat{B}_t = -193.3 + 35.2\ln Y_t - 0.38PB_t + 0.10PP_t - 5.7D_t$$
$$\qquad (14.1) \qquad (0.10) \qquad (0.09) \qquad (3.9)$$
$$t = \quad 2.5 \qquad -3.7 \qquad 1.1 \qquad -1.5$$
$$\bar{R}^2 = .857 \quad n = 28 \quad \hat{\rho} = 0.82 \qquad (9.29)$$

Compare Equations 9.28 and 9.29. Which do you prefer? Why?

13. You're hired by "Farmer Vin," a famous producer of bacon and ham, to test the possibility that feeding pigs at night allows them to grow faster than feeding them during the day. You take 200 pigs (from newborn piglets to extremely old porkers) and randomly assign them to feeding only during the day or feeding only at night and, after six months, end up with the following (admittedly very hypothetical) equation:

$$\hat{W}_i = 12 + 3.5G_i + 7.0D_i - 0.25F_i$$
$$\qquad (1.0) \qquad (1.0) \qquad (0.10)$$
$$t = \quad 3.5 \qquad 7.0 \qquad -2.5$$
$$\bar{R}^2 = .70 \quad n = 200 \quad DW\ d = 0.50$$

where: W_i = the percentage weight gain of the *i*th pig
G_i = a dummy variable equal to 1 if the *i*th pig is a male, 0 otherwise
D_i = a dummy variable equal to 1 if the *i*th pig was fed only at night, 0 if only during the day
F_i = the amount of food (pounds) eaten per day by the *i*th pig

a. Test for serial correlation at the 5 percent level in this equation.
b. What econometric problems appear to exist in this equation? (Hint: Be sure to make and test appropriate hypotheses about the slope coefficients.)
c. The goal of your experiment is to determine whether feeding at night represents a significant improvement over feeding during the day. What can you conclude?
d. The observations are ordered from the youngest pig to the oldest pig. Does this information change any of your answers to the previous parts of this question? Is this ordering a mistake? Explain your answer.

14. As an example of impure serial correlation caused by an incorrect functional form, let's return to the equation for the percentage of

putts made (P_i) as a function of the length of the putt in feet (L_i) that we discussed originally in Exercise 8 in Chapter 1. The complete documentation of that equation is:

$$\hat{P}_i = 83.6 - \underset{\substack{(0.4) \\ t = -10.6}}{4.1L_i} \quad (9.30)$$

$$n = 19 \quad \bar{R}^2 = .861 \quad DW = 0.48$$

a. Test Equation 9.30 for serial correlation using the Durbin-Watson d test at the 1 percent level.
b. Why might the linear functional form be inappropriate for this study? Explain your answer.
c. If we now re-run Equation 9.30 using a double-log functional form, we obtain:

$$\widehat{\ln P_i} = 5.50 - \underset{\substack{(0.07) \\ t = -13.0}}{0.92 \ln L_i} \quad (9.31)$$

$$n = 19 \quad \bar{R}^2 = .903 \quad DW = 1.22$$

Test Equation 9.31 for serial correlation using the Durbin-Watson d test at the 1 percent level.
d. Compare Equations 9.30 and 9.31. Which equation do you prefer? Why?

10

Heteroskedasticity

Heteroskedasticity[1] is the violation of Classical Assumption V, which states that the observations of the error term are drawn from a distribution that has a constant variance. The assumption of constant variances for different observations of the error term (homoskedasticity) is not always realistic. For example, in a model explaining heights, compare a one-inch error in measuring the height of a basketball player with a one-inch error in measuring the height of a mouse. It's likely

1. Various authors spell this "heteros*c*edasticity," but a recent article by Huston McCulloch appears to settle this controversy in favor of "heteros*k*edasticity" because of the word's Greek origin. See J. Huston McCulloch, "On Heteros*edasticity," *Econometrica*, March 1985, p. 483. While heteroskedasticity is a difficult word to spell, at least it's an impressive comeback when parents ask, "What'd you learn for all that money?"

that error terms associated with the height of a basketball player would come from distributions with larger variances than those associated with the height of a mouse. As we'll see, the distinction between heteroskedasticity and homoskedasticity is important because OLS, when applied to heteroskedastic models, is no longer the minimum variance estimator (it still is unbiased, however).

Heteroskedasticity often occurs in data-sets in which there is a wide disparity between the largest and smallest observed values. The larger the disparity between the size of observations in a sample, the larger the likelihood that the error term observations associated with them will have different variances and therefore be heteroskedastic. That is, we'd expect that the error term distribution for very large observations might have a large variance, while the error term distribution for small observations might have a small variance.

One can easily get such a large range between the highest and lowest values of the variables in cross-sectional data-sets. Recall that in cross-sectional models, the observations are all from the same time period (a given month or year, for example) but are from different entities (individuals, states, or countries, for example). The difference between the size of California's labor force and Rhode Island's, for instance, is quite large (comparable in percentage terms to the difference between the heights of a basketball player and a mouse). Since cross-sectional models often include observations of widely different size in the same sample (cross-state studies of the U.S. usually include California and Rhode Island as individual observations, for example), heteroskedasticity is hard to avoid if economic topics are going to be studied cross-sectionally.

This focus on cross-sectional models is not to say that heteroskedasticity in time-series models is impossible, nor to deny the possibility that an omitted variable could cause impure heteroskedasticity in any kind of data. In general, though, heteroskedasticity is more likely to take place in cross-sectional models than in time-series models.

Within this context, we'll attempt to answer the same four questions for heteroskedasticity that we answered for multicollinearity and serial correlation in the previous two chapters:

1. What is the nature of the problem?
2. What are the consequences of the problem?
3. How is the problem diagnosed?
4. What remedies for the problem are available?

10.1 Pure vs. Impure Heteroskedasticity

Heteroskedasticity, like serial correlation, can be divided into pure and impure versions. Pure heteroskedasticity is that caused by the error term of the correctly specified equation, while impure heteroskedasticity is caused by a specification error such as an omitted variable.

10.1.1 Pure Heteroskedasticity

Pure heteroskedasticity refers to heteroskedasticity that is a function of the error term of a correctly specified regression equation. As with serial correlation, use of the word "heteroskedasticity" without any modifier (like pure or impure) implies *pure* heteroskedasticity.

Such **pure heteroskedasticity** occurs when Classical Assumption V, which assumes that the variance of the error term is constant, is violated in a correctly specified equation. Recall that Assumption V assumes that:

$$\mathrm{VAR}(\epsilon_i) = \sigma^2 \qquad (i = 1,2,\ldots,n) \tag{10.1}$$

If this assumption is met, all the observations of the error term can be thought of as being drawn from the same distribution: a distribution with a mean of zero and a variance of σ^2. This σ^2 does not change for different observations of the error term; this property is called homoskedasticity. A homoskedastic error term distribution is pictured in the top half of Figure 10.1; note that the variance of the distribution is constant (even though individual observations drawn from that sample will vary quite a bit).

With heteroskedasticity, this error term variance is not constant; instead, the variance of the distribution of the error term depends on exactly which observation is being discussed:

$$\mathrm{VAR}(\epsilon_i) = \sigma_i^2 \qquad (i = 1,2,\ldots,n) \tag{10.2}$$

Note that the only difference between Equation 10.1 and Equation 10.2 is the subscript "i" attached to σ^2, which implies that instead of being constant over all the observations, a heteroskedastic error term's variance can change depending on the observation (hence the subscript).

Another way to visualize heteroskedasticity is to picture a world

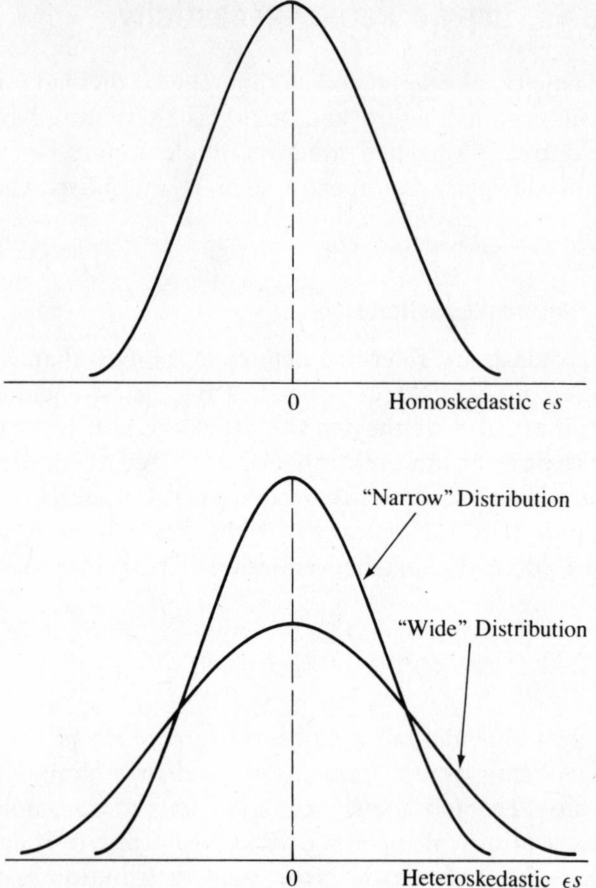

Figure 10.1 HOMOSKEDASTICITY VS. DISCRETE HETEROSKEDASTICITY
In homoskedasticity, the distribution of the error term has a constant variance, so the observations are continually drawn from the same distribution (shown in the top panel). In the simplest heteroskedastic case, discrete heteroskedasticity, there would be two different error term variances and therefore two different distributions (one wider than the other, as in the bottom panel) from which the observations of the error term could be drawn.

in which some of the observations of the error term are drawn from much wider distributions than are others. The simplest situation would be that the observations of the error term could be grouped into just two different distributions, "wide" and "narrow." We will call this simple version of the problem *discrete heteroskedasticity.* Here, both distributions would be centered around zero, but one would have a larger variance than the other, as indicated in the bottom half of Figure

10.1. Note the difference between the two halves of the figure. With homoskedasticity, all the error term observations come from the same distribution; with heteroskedasticity, they come from different distributions.

Heteroskedasticity takes on many more complex forms, however; the number of different models of heteroskedasticity is virtually limitless, and an analysis of even a small percentage of these alternatives would be a huge task. Instead, we'd like to address the general principles of heteroskedasticity by focusing on the most frequently specified model of pure heteroskedasticity, just as we focused on positive first-order serial correlation in the previous chapter. However, don't let this focus mislead you into concluding that econometricians are concerned only with one kind of heteroskedasticity.

In this model of heteroskedasticity, the variance of the error term is related to an exogenous variable Z_i. For a typical regression equation:

$$Y_i = \beta_0 + \beta_1 X_{1i} + \beta_2 X_{2i} + \epsilon_i \tag{10.3}$$

the variance of the otherwise classical error term ϵ might be equal to:

$$VAR(\epsilon_i) = \sigma^2 Z_i^2 \tag{10.4}$$

where Z may or may not equal one of the Xs in the equation. The variable Z is called a **proportionality factor** because the variance of the error term changes proportionally to the square of Z_i. The higher the value of Z_i, the higher the variance of the distribution of the *i*th observation of the error term. There could be *n* different distributions, one for each observation, from which the observations of the error term could be drawn depending on the number of different values that Z takes. To see what homoskedastic and heteroskedastic distributions of the error term look like with respect to Z, examine Figures 10.2 and 10.3. Note that the heteroskedastic distribution gets wider as Z increases but that the homoskedastic distribution maintains the same width no matter what value Z takes.

What is an example of a proportionality factor Z? How is it possible for an exogenous variable such as Z to change the whole distribution of an error term? Think about a function that relates the consumption of a household to its income. The expenditures of a low-income household are not likely to be as variable in absolute value as the expenditures of a high-income one because a ten-percent change in spending for a high-income family involves a lot more money than a ten-percent change for a low-income one. In addition, the proportion

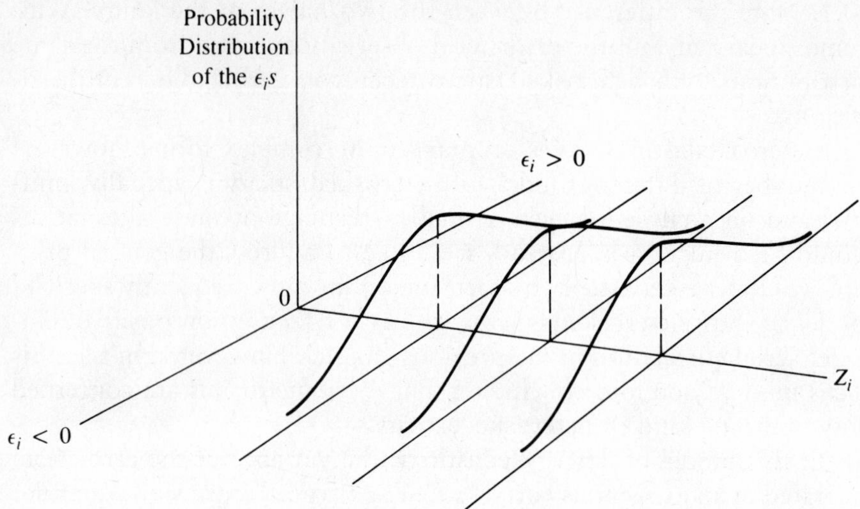

Figure 10.2 A HOMOSKEDASTIC ERROR TERM WITH RESPECT TO Z_i
If an error term is homoskedastic with respect to Z_i, the variance of the distribution of the error term is the same (constant) no matter what the value of Z_i is: $VAR(\epsilon_i) = \sigma^2$.

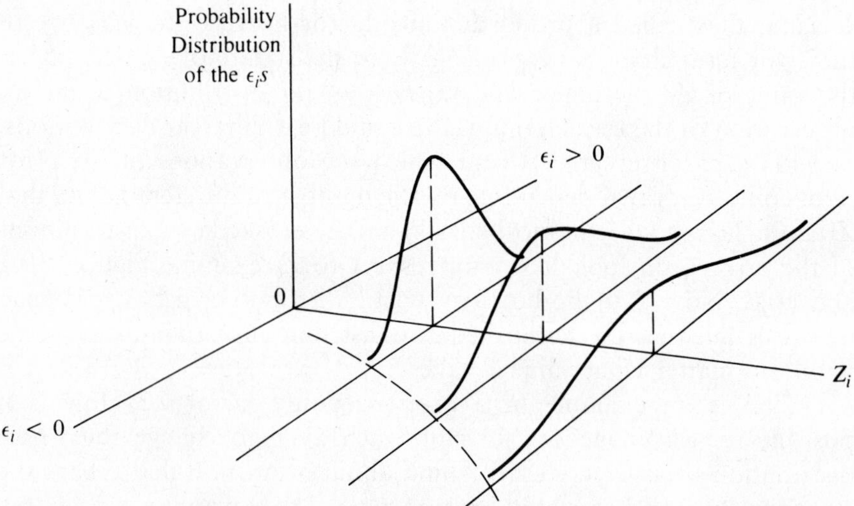

Figure 10.3 A HETEROSKEDASTIC ERROR TERM WITH RESPECT TO Z_i
If an error term is heteroskedastic with respect to Z_i, the variance of the distribution of the error term changes systematically as a function of Z_i. In this example, the variance is an increasing function of Z_i, as in $VAR(\epsilon_i) = \sigma^2 Z_i^2$.

of the low-income budget that must be spent on necessities is much higher than that of the high-income budget. In such a case, the Y_i would be consumption expenditures and the proportionality factor, Z, would be household income. As household income rose, so too would the variance of the error term of an equation built to explain expenditures. The error term distributions would look something like those in Figure 10.3, where the Z in Figure 10.3 is household income, one of the independent variables in the function.

This example helps emphasize that heteroskedasticity is likely to occur in cross-sectional models because of the large variation in the size of the dependent variable involved. An exogenous disturbance that might seem huge to a low-income household could seem miniscule to a high-income one, for instance.

Heteroskedasticity can occur in at least two situations other than a cross-sectional data-set with a large amount of variation in the size of the dependent variable:

1. Heteroskedasticity can occur in a time-series model with a significant amount of change in the dependent variable. If you were modeling sales of VCRs from 1970 to 1990, it's quite possible that you would have a heteroskedastic error term. As the phenomenal growth of the industry took place, the variance of the error term probably increased as well. Such a possibility is unlikely in time-series that have low rates of change, however.

2. Heteroskedasticity can occur in any model, time-series or cross-sectional, where the quality of data collection changes dramatically within the sample. As data collection techniques get better, the variance of the error term should fall because measurement errors are included in the error term. As measurement errors decrease in size, so should the variance of the error term. For more on this topic (called "errors in the variables"), see Section 14.6.

10.1.2 Impure Heteroskedasticity

Heteroskedasticity that is caused by an error in specification, such as an omitted variable, is referred to as **impure heteroskedasticity.** While improper functional form is less likely to cause impure heteroskedasticity than it is to cause impure serial correlation, the two concepts are similar in most other ways.

An omitted variable can cause a heteroskedastic error term because the portion of the omitted effect not represented by one of the included explanatory variables must be absorbed by the error term. If this effect has a heteroskedastic component, the error term of the misspecified equation might be heteroskedastic even if the error term of the true equation is not. This distinction is important because with impure heteroskedasticity the correct remedy is to attempt to find the left-out variable and include it in the regression. It's therefore important to be sure that your specification is correct before trying to detect or remedy pure heteroskedasticity.[2]

For example, consider a cross-sectional study of the 1990 imports of a number of variously sized nations. For simplicity, assume that the best model of a nation's imports in such a cross-sectional setting includes a positive function of its GNP and a positive function of the relative price ratio (including the impact of exchange rates) between it and the rest of the world. In such a case, the "true" model would look like:

$$M_i = f(\overset{+}{GNP},\overset{+}{PR}) = \beta_0 + \beta_1 GNP_i + \beta_2 PR_i + \epsilon_i \tag{10.5}$$

where: M_i = the imports (in dollars) of the *i*th nation
GNP_i = the Gross National Product (in dollars) of the *i*th nation
PR_i = the ratio of the domestic price of normally traded goods (converted to dollars by the exchange rate) to the world price or those goods (measured in dollars) for the *i*th nation
ϵ_i = a classical error term

Now suppose that the equation is run without GNP. Since GNP is left out, the equation would become:

$$M_i = \beta_0 + \beta_2 PR_i + \epsilon_i^* \tag{10.6}$$

where the error term of the misspecified equation, ϵ_i^*, is a function of the left-out variable (GNP) and a nonheteroskedastic error term ϵ:

$$\epsilon_i^* = \epsilon_i + \beta_1 GNP_i$$

2. If this paragraph sounds vaguely familiar, that's because our discussion of impure heteroskedasticity parallels our previous discussion of impure serial correlation. If this paragraph doesn't sound at all familiar (or if you skipped Section 9.1), then you might consider briefly reading section 9.1.2 for a fuller explanation of these ideas.

To the extent that the relative price ratio does not act as a proxy for GNP, the error term has to incorporate the effect of the omitted variable. If this new effect has a larger variance for larger values of GNP, which seems likely, the new error term, ϵ_i^*, is heteroskedastic. The impact of such an effect also depends on the size of the $\beta_1\overline{GNP}$ component compared with the absolute value of the typical ϵ component. The larger the omitted variable portion of ϵ_i^*, the more likely it is to have impure heteroskedasticity. In such a case, the error term observations, ϵ_i^*, when plotted with respect to GNP_i, appear as in Figure 10.4. As can be seen, the larger the GNP, the larger the variance of the error term.

10.2 The Consequences of Heteroskedasticity

If the error term of your equation is known to be heteroskedastic, what does that mean for the estimation of your coefficients? It turns out that the consequences of heteroskedasticity are almost identical in

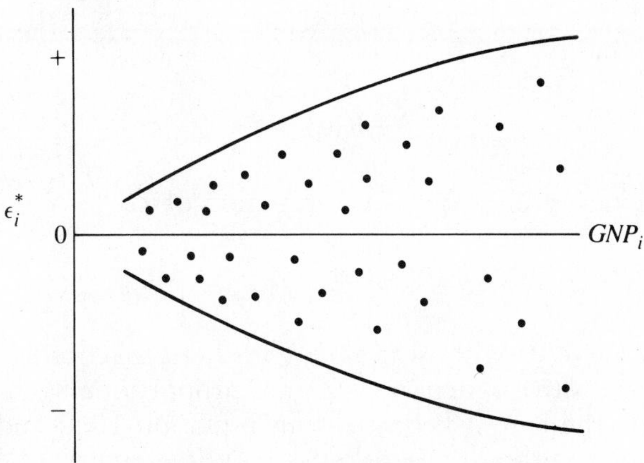

Figure 10.4 IMPURE HETEROSKEDASTICITY CAUSED BY THE OMISSION OF GNP

Impure heteroskedasticity is a nonconstant variance of the distribution of the error term that is caused by an incorrect specification. In this case, the omission of GNP from the equation has forced the error term to incorporate the impact of GNP, causing the distribution of the error term to be wider (higher variance) for large values of GNP than for small ones.

general framework to those of serial correlation,[3] though the two problems are quite different.

If the error term of an equation is heteroskedastic, there are three major consequences:

1. *Pure heteroskedasticity does not cause bias in the coefficient estimates.*
   Even if the error term of an equation is known to be purely heteroskedastic, that heteroskedasticity will not cause bias in the OLS estimates of the coefficients. As a result, we can say that an otherwise correctly specified equation that has pure heteroskedasticity still has the property that:

$$E(\hat{\beta}) = \beta \text{ for all } \beta\text{s}$$

   Lack of bias does not guarantee "accurate" coefficient estimates, especially since heteroskedasticity increases the variance of the estimates, but the distribution of the estimates is still centered around the true β. Equations with impure heteroskedasticity caused by an omitted variable, of course, will have possible specification bias.

2. *Heteroskedasticity increases the variances of the $\hat{\beta}$ distributions.*
   Pure heteroskedasticity causes no bias in the estimates of the OLS coefficients, but it does affect the minimum variance property. If the error term of an equation is heteroskedastic with respect to a proportionality factor Z:

$$VAR(\epsilon_i) = \sigma^2 Z_i^2 \tag{10.7}$$

   then the variance of the $\hat{\beta}$s is a function of Z:

$$VAR^{**}(\hat{\beta}_k) = f(Z^2) \cdot [VAR(\hat{\beta}_k)] \tag{10.8}$$

   where $VAR^{**}(\hat{\beta})$ is the variance with heteroskedasticity; $f(Z^2)$ indicates a positive function of Z, the proportionality factor that is "causing" the heteroskedasticity in Equation 10.7, and $VAR(\hat{\beta})$ is the variance without heteroskedasticity (see Figure 10.5). Not surprisingly, the minimum-variance portion of the Gauss-Markov Theorem cannot be proven if Classical Assumption V (homoskedasticity) is violated.

3. The consequences of heteroskedasticity are so similar to those of serial correlation that the following section assumes at least some familiarity with them. As long as you've read Section 9.2, the following discussion can stand on its own. If you didn't read Section 9.2, you should do so before continuing with this chapter.

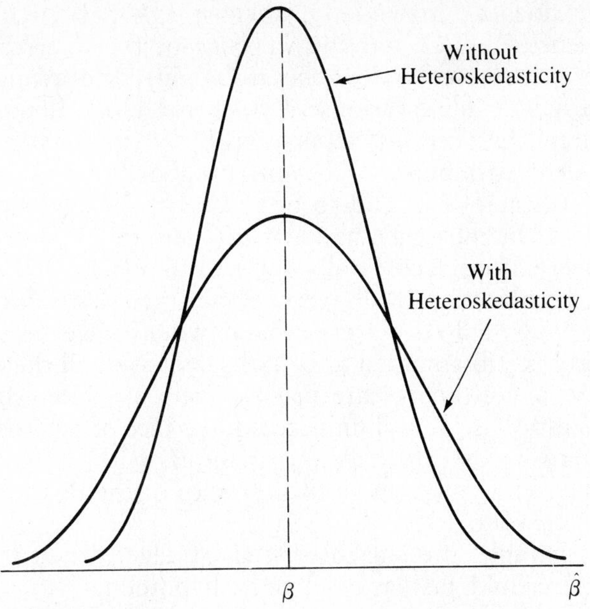

Figure 10.5 $\hat{\beta}$ DISTRIBUTION WITH AND WITHOUT HETEROSKEDASTICITY
Heteroskedasticity increases the variance of the $\hat{\beta}$s, widening the $\hat{\beta}$ distribution. It does not cause bias, however, so the $\hat{\beta}$ distribution is centered around the true β whether or not there is heteroskedasticity.

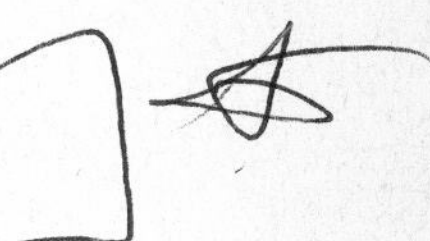

3. *Heteroskedasticity causes OLS to underestimate the variances (and standard errors) of the coefficients.*
Heteroskedasticity turns out to increase the variances of the $\hat{\beta}$s in a way that is masked by the OLS estimates of them, and OLS nearly always underestimates[4] those variances. As a result, neither the t-statistic nor the F-statistic can be relied on in the face of uncorrected heteroskedasticity. In fact, OLS usually ends up with higher t-scores than would be obtained if the error terms were homoskedastic, sometimes leading researchers to reject null hypotheses that shouldn't be rejected.

4. Actually, the OLS estimates of the variance and the standard error of the coefficient estimates $\hat{\beta}_k$ are biased, but the bias is negative as long as σ_i^2 and $(X_{ki} - \bar{X}_k)^2$ are positively correlated. The SE($\hat{\beta}$)s will be underestimated as long as an increase in X is related to an increase in the variance of the error terms. In economic examples, such a positive correlation would almost always be expected in cases when a sizable correlation is likely to exist. For some variables, no correlation at all might exist, but a negative correlation would occur quite infrequently. As a result, the statement that OLS underestimates the variances, while a simplification, is almost always true.

Why does heteroskedasticity cause this particular pattern of consequences? As Z and the variance of the distribution of the error term increase, so does the probability of drawing a large (in absolute value) observations of the error term. If the pattern of these large observations happens to be positive when one of the independent variables is substantially above average, the OLS $\hat{\beta}$ for that variable will tend to be greater than it would have been otherwise. On the other hand, if the pattern of these large error term observations accidentally happens to be negative when one of the Xs is substantially above average, then the OLS $\hat{\beta}$ for that variable will tend to be less than it would have been. Since the error term is still assumed to be independent of all the explanatory variables, overestimates are just as likely as underestimates, and the OLS estimator is still unbiased in the face of heteroskedasticity. The heteroskedasticity has caused the $\hat{\beta}$s to be farther from the true value, however, and so the variance of the distribution of the $\hat{\beta}$s has increased.

For example, the real interest rate/budget deficit study of Section 9.2.2 could just as well be a hypothetical example of the consequences of heteroskedasticity. The OLS-estimated t-scores are too high, leading to potential mistakes of inference whenever *t*-tests are used in heteroskedastic equations. Before we can get rid of heteroskedasticity, however, we must detect it.

10.3 Testing for Heteroskedasticity

Econometricians do not all use the same test for heteroskedasticity because heteroskedasticity takes a number of different forms, and its precise manifestation in a given equation is almost never known. The "Z_i proportionality factor" approach of this chapter, for example, is only one of many specifications of the form of heteroskedasticity. As a result, there is no universally agreed-upon method of testing for heteroskedasticity; econometrics textbooks list as many as eight different methods of such testing.

Because of this wide variety, we'll describe the use of four different tests for heteroskedasticity. Our primary focus will be on the *Park test* because it tests precisely the functional form that we use in this chapter to illustrate the problem of pure heteroskedasticity. At least three other tests are more generally used than the Park test. They are the *Goldfeld-Quandt test,* the *Breusch-Pagan test,* and the *White test,* which will also be described later in this section. No test for heteroskedasticity can "prove" that heteroskedasticity exists in an equation, though, so the best we can do is to get a general indication of its likelihood.

10.3.1 The Park Test

How do we test for pure heteroskedasticity of the form that we assumed in the previous section? That form, as we outlined in the previous section, is:

$$VAR(\epsilon_i) = \sigma^2 Z_i^2$$

where ϵ is the error term of the equation being estimated, σ^2 is the variance of the homoskedastic error term, and Z is the proportionality factor.

The **Park Test**[5] is a formal procedure that attempts to test the residuals for this heteroskedasticity in a manner similar to the way that the Durbin-Watson d statistic tests residuals for serial correlation. The Park test has three basic steps. First, the regression equation is estimated by OLS and the residuals are calculated. Second, the log of the squared residuals is used as the dependent variable of an equation whose sole explanatory variable is the log of the proportionality factor Z. Finally, the results of this second regression are tested to see if there is any evidence of heteroskedasticity.

There's no need to run a Park test for every equation estimated, however, so before using the Park test it's a good idea to ask the following preliminary questions:

1. Are there any obvious specification errors? If the estimated equation is suspected of having an omitted variable or is about to be re-run for some other specification reason, the Park test should be delayed until the specification is as good as possible.
2. Is the subject of the research often afflicted with heteroskedasticity? Not only are cross-sectional studies the most likely source of heteroskedasticity, but some cross-sectional studies (with large variations in the size of the dependent variable, for instance) are more susceptible to heteroskedasticity than others.
3. Finally, does a graph of the residuals show any evidence of heteroskedasticity? It sometimes saves time to plot the residuals with respect to a potential Z proportionality factor. In such cases, the graphs can often show that heteroskedasticity is or is not likely without a Park test. Figure 10.4 above shows an example of what to look for: an expanding (or contracting) *range* of the residuals.

5. R. E. Park, "Estimation with Heteroscedastic Error Terms," *Econometrica*, Oct. 1966, p. 888.

If there is some reason to suspect heteroskedasticity, it's appropriate to run a Park test. Since the Park test is not run automatically by computer regression packages, you should know how to run the test yourself:

1. *Obtain the residuals of the estimated regression equation.* The first step is to estimate the equation with OLS and then find the residuals from that estimation:

$$e_i = Y_i - \hat{\beta}_0 - \hat{\beta}_1 X_{1i} - \hat{\beta}_2 X_{2i} \tag{10.9}$$

These residuals, which are printed out by most computer regression packages, are the same used to calculate the Durbin-Watson d statistic to test for serial correlation.

2. *Use these residuals to form the dependent variable in a second regression.* In particular, the Park test suggests that you run the following double-log regression:

$$\ln(e_i^2) = \alpha_0 + \alpha_1 \ln Z_i + u_i \tag{10.10}$$

where: e_i = the residual from the *i*th observation from Equation 10.9

Z_i = your best choice as to the possible proportionality factor (Z)

u_i = a classical (homoskedastic) error term[6]

3. *Test the significance of the coefficient of Z in Equation 10.10 with a t-test.* The last step is to use the t-statistic to test the significance of lnZ in explaining $\ln(e^2)$ in Equation 10.10. If the coefficient of Z is significantly different from zero, this is evidence of heteroskedastic patterns in the residuals with respect to Z; otherwise, heteroskedasticity related to this particular Z is not supported by the evidence in these residuals. However, it's impossible to prove that a particular equation's error terms are homoskedastic.

The Park test is not always easy to use. The major problem with it and most other methods of testing for heteroskedasticity is the identification of the proportionality factor Z. Although Z is often an explanatory variable in the original regression equation, there is no guarantee of that. A particular Z should be chosen for your Park test only after investigating the type of potential heteroskedasticity in your

6. One criticism of the Park test is that this error term is not necessarily homoskedastic. See S. M. Goldfeld and R. E. Quandt, *Nonlinear Methods in Econometrics* (Amsterdam: North-Holland Publishing Company, 1972), pp. 93–94.

equation.[7] A good Z is a variable that seems likely to vary with the variance of the error term.

For example, in a cross-sectional model of countries or states, a good Z would be one that measured the size of the observation relative to the dependent variable in question. For a dependent variable such as gallons of gasoline consumed, the number of registered drivers or automobiles might be a better measure of size than the population. While it's difficult to identify the best Z for a particular equation, it's often easier to distinguish good Zs from bad Zs. In the gasoline consumption equation, for example, a bad Z might be the speed limit in the state because, while the speed limit might be important in determining how much gasoline is used, it is unlikely to "cause" any heteroskedasticity. This is because the speed limit in a state does not vary in size in the same way that gasoline consumption does. That is, the states likely to have large error term variances are not also likely to have high speed limits. For more on this fairly thorny issue, see Exercise 2.

10.3.2 An Example of the Use of the Park Test

Let's return to the Woody's Restaurants example of Section 3.4 and test for heteroskedasticity in the residuals of Equation 3.14. Recall that regression explained the number of customers, as measured by the check volume (Y) at a cross section of 33 different Woody's restaurants as a function of the number of nearby competitors (C), the nearby population (P), and the average household income of the local area (I):

$$\begin{aligned} \hat{Y}_i = 102{,}192 - &\ 9075C_i + 0.354P_i + 1.288I_i \qquad (3.14) \\ &(2053) \qquad (0.073) \qquad (0.543) \\ t = &-4.42 \qquad 4.88 \qquad 2.37 \\ n = 33 \quad &\bar{R}^2 = .579 \quad F = 15.65 \end{aligned}$$

This equation is cross-sectional, so heteroskedasticity is a theoretical possibility, but the dependent variable does not change much in size from restaurant to restaurant, so heteroskedasticity is not likely to be a major problem. As a result, the assumption of a constant variance of the error term (homoskedasticity) seems to be reasonable.

7. Some econometricians suggest using the Park test for insight into the form of the heteroskedasticity. If $VAR(\epsilon_i) = \sigma^2 Z_1{}^2$, then $\ln(e_i^2)$ should be equal to $\alpha_0 + 2\ln Z_i$ plus an error term. Thus the estimate of the coefficient of lnZ in the Park test implies whether the proportionality factor should be squared or raised to some other power. For more on this and its implications for deciding which form to use to adjust for heteroskedasticity, see R. S. Pindyck and D. L. Rubinfeld, *Econometric Models and Economic Forecasts* (New York: McGraw-Hill, 1981), pp. 150–152.

To judge whether this tentative conclusion is correct, let's use the Park test to see if the residuals from Equation 3.14 give any indication of heteroskedasticity.

1. *Calculate the residuals:* First, obtain the residuals from the equation you want to test. In the Woody's example, these residuals have already been calculated. They're at the end of Section 3.4.

2. *Use these residuals as the dependent variable in a second regression:* Run a regression with the log of the squared residual as the dependent variable as a function of the log of the suspected proportionality factor Z as first outlined in Equation 10.10:

$$\ln(e_i^2) = \alpha_0 + \alpha_1 \ln Z_i + u_i \tag{10.10}$$

It's possible that no Z exists, but if one does, it seems likely that it would somehow be related to the size of the market that the particular Woody's restaurant serves. Since larger error term variances might exist in more heavily populated areas, population (P) is a reasonable choice as a Z to try in our Park test. Any other variable related to the size of the market or of the particular restaurant would also be a reasonable possibility.

If the logged and squared residuals from Equation 3.14 are regressed as a function of the log of P, we obtain:

$$\widehat{\ln(e_i^2)} = 21.05 - 0.2865 \ln P_i \tag{10.11}$$

$$(0.6263)$$

$$t = -0.457$$

$$n = 33 \qquad \bar{R}^2 = .0067 \qquad F = 0.209$$

3. *Test the significance of $\hat{\alpha}_1$ in Equation 10.10:* As can be seen from the calculated t-score, there is virtually no measurable relationship between the squared residuals of Equation 3.14 and population. The calculated t-score of -0.457 is quite a bit smaller in absolute value than 2.750, the critical t-value (from Statistical Table B-1) for a two-tailed, one percent test. As a result, we would not be able to reject[8] the null hypothesis of homoskedasticity:

8. Recall that not being able to reject the null hypothesis of homoskedasticity doesn't prove that the error terms are homoskedastic. In addition, note that *this* Park test says nothing about *other* proportionality factors or other forms of heteroskedasticity. While heteroskedasticity of any kind is unlikely in this example because of the nature of the dependent variable, it's possible to find homoskedasticity with respect to one proportionality factor (or form) but heteroskedasticity with respect to some other proportionality factor (or form). Careful thinking is necessary before a potential Z can be chosen. Running a Park test on every conceivable variable would do little but increase the chance of Type I Error (rejecting the null hypothesis of homoskedasticity when it's true).

$$H_0: \alpha_1 = 0$$
$$H_A: \alpha_1 \neq 0$$

For more practice in the use of the Park test, see Exercise 4.

10.3.3 The Goldfeld-Quandt Test

Perhaps the most commonly used test for heteroskedasticity is the **Goldfeld-Quandt test,**[9] which reorders the data according to the value of a potential proportionality factor Z (which might or might not be an independent variable in the equation) and then tests to see if the variance of the residuals from the first third of the newly reordered data-set differs significantly from that of the residuals from the last third. This test is fairly easy to use and is particularly useful to test for heteroskedasticity related to a proportionality factor when the specific form of that relationship is unknown.

To use the Goldfeld-Quandt test:

1. Reorder the observations according to the size of Z, the particular proportionality factor suspected of being related to any possible heteroskedasticity.
2. Omit the middle third of the reordered observations, and use OLS to estimate separate regressions on the first and last thirds of the data-set. Use the specification of the original equation.
3. Calculate $GQ = RSS_3/RSS_1$, where RSS_1 is the residual sum of squares from the first third of the reordered data-set, and RSS_3 is the RSS from the last third.
4. Use the *F*-test to test the null hypothesis of homoskedasticity by comparing GQ to the critical F-value for K and ("n" − K − 1) degrees of freedom in the numerator and denominator, respectively, and where "*n*" is the sample size of each of the new smaller data-sets.

   If GQ is greater than the critical F-value, the summed squared residuals from the last third of the estimated equation are significantly greater than those from the first third, and you can reject the null hypothesis of homoskedasticity. If GQ is less than the

9. S. M. Goldfeld and R. F. Quandt, "Some Tests for Homoscedasticity," *Journal of the American Statistical Association,* Sept. 1965, pp. 539–547. Goldfeld and Quandt do not explicitly specify that exactly a third of the sample should be omitted, but since the power of the test increases as more observations are omitted, the use of one-third has become almost standard practice. The Goldfeld-Quandt test avoids the potential heteroskedasticity (and, it turns out, serial correlation) of the Park test referred to in footnote 6.

critical F-value, you cannot reject the null hypothesis. For an example of and practice in the use of the Goldfeld-Quandt test, see Exercises 7 and 8.

10.3.4 The Breusch-Pagan and White Tests

Unfortunately, to use either the Park test or the Goldfeld-Quandt test we must know Z_i, the variable suspected of being proportional to the possible heteroskedasticity. Quite often, however, we may want to test the possibility that more than one proportionality factor is involved simultaneously. Less frequently, we might not be able to decide which of a number of possible Z factors to test. In either of these situations, it's unadvisable to run a series of Park or Goldfeld-Quandt tests (one for each possible proportionality factor). Instead, it is appropriate to use either the Breusch-Pagan test or the White test.[10]

The **Breusch-Pagan test**[11] attempts to detect heteroskedasticity by measuring the overall significance of a secondary regression that specifies the original residuals, squared, to be a function of *more than one* Z proportionality factor. In essence, it is a linear Park test with more than one Z.

To run a Breusch-Pagan test:

1. Obtain the residuals of the estimated regression equation. As we did with the Park test, we can use Equation 10.9 to calculate the residuals of the equation to be tested for heteroskedasticity:

$$e_i = Y_i - \hat{\beta}_0 - \hat{\beta}_1 X_{1i} - \hat{\beta}_2 X_{2i} \qquad (10.9)$$

2. Use the squared residuals as the dependent variable in a secondary equation that includes as independent variables all those variables suspected of being related to the variance of the error term of the original equation:

$$(e_i)^2 = \alpha_0 + \alpha_1 Z_{1i} + \alpha_2 Z_{2i} + \cdots + \alpha_p Z_{pi} + u_i \qquad (10.12)$$

While these Zs are often Xs in the original equation, not every original independent variable needs to be included as a possible

10. Both the Breusch-Pagan test and the White test belong to a general group of tests based on the Lagrange Multiplier (LM). For more on LM tests, see Section 12.2.

11. T. S. Breusch and A. R. Pagan, "A Simple Test for Heteroscedasticity and Random Coefficient Variation," *Econometrica*, 1979, pp. 1287–1294. The version of the test we present is that suggested by G. S. Maddala, *Introduction to Econometrics* (New York: Macmillan, 1988), pp. 159–167.

proportionality factor. In addition, variables and/or functional forms not in the original equation can be used in the secondary equation.

3. Test the overall significance of Equation 10.12 by testing the null hypothesis:

$$H_0: \alpha_1 = \alpha_2 = \cdots = \alpha_p = 0$$
$$H_A: \text{otherwise}$$

The appropriate test statistic to use is:

$$L = \frac{ESS}{2[\Sigma(e_i^2/n)]^2} \tag{10.13}$$

where ESS is the explained sum of squares from Equation 10.12, e_i is the residual from the *i*th observation of the original equation, and n is the sample size. For large samples, L has a Chi-square distribution with p degrees of freedom, where p is the number of proportionality factors in the null hypothesis. (Most researchers treat L as a Chi-square variable for small samples as well.) If L is larger than the critical Chi-square value from Statistical Table B-8, then we reject the null hypothesis that the slope coefficients jointly equal zero and conclude that Equation 10.12 is indeed significant. Such a conclusion is evidence that the variance of the residuals of the original equation is not a constant, and it's likely that we have heteroskedasticity. If L is less than the critical Chi-square value, then we cannot reject the null hypothesis of homoskedasticity.

The Breusch-Pagan test's biggest advantage is that more than one proportionality factor can be tested simultaneously. On the other hand, the researcher still must specify the proportionality factors and the form of heteroskedasticity to be tested, and a fairly large sample is required. For an example of and practice in the use of the Breusch-Pagan test, see Exercises 7 and 8.

The **White test**[12] also approaches the detection of heteroskedasticity by running a regression with the squared residuals as the dependent variable. This time, though, the right-hand side of the secondary equation includes all the original independent variables, the squares of all the original independent variables, and the cross products of all the

12. Halbert White, "A Heteroskedasticity-Consistent Covariance Matrix Estimator and a Direct Test for Heteroskedasticity," *Econometrica*, 1980, pp. 817–838.

original independent variables with each other. The White test thus has the distinct advantage of not assuming any particular form of heteroskedasticity. As a result, it's rapidly gaining support as the best[13] test yet devised to apply to all types of heteroskedasticity.

To run a White test:

1. Obtain the residuals of the estimated regression equation. This first step is identical to the first steps in the Park and Breusch-Pagan tests.
2. Use these residuals (squared) as the dependent variable in a second equation that includes as explanatory variables each X from the original equation, the square of each X, and the product of each X times every other X. For example, if the original equation's independent variables are X_1, X_2, and X_3, the appropriate White test equation is:

$$\begin{aligned}(e_i)^2 = \alpha_0 + \alpha_1 X_{1i} + \alpha_2 X_{2i} + \alpha_3 X_{3i} + \\ \alpha_4 X_{1i}^2 + \alpha_5 X_{2i}^2 + \alpha_6 X_{3i}^2 + \\ \alpha_7 X_{1i} X_{2i} + \alpha_8 X_{1i} X_{3i} + \alpha_9 X_{2i} X_{3i} + u_i\end{aligned} \tag{10.14}$$

3. Test the overall significance of Equation 10.14 with the Chi-square test. The appropriate test statistic here is nR^2, or the sample size (n) times the coefficient of determination (the unadjusted R^2) of Equation 10.14. This test statistic has a Chi-square distribution with degrees of freedom equal to the number of slope coefficients in Equation 10.14. The decision rule is virtually identical to that used in the Breusch-Pagan test. If nR^2 is larger than the critical Chi-square value found in Statistical Table B-8, then we reject the null hypothesis and conclude that it's likely that we have heteroskedasticity. If nR^2 is less than the critical Chi-square value, then we cannot reject the null hypothesis of homoskedasticity.

One problem with the White test is that, in some situations, the secondary equation cannot be estimated because it has negative degrees of freedom. This can happen when the original equation has such a small sample size and/or so many variables that the secondary equation

13. For time-series data, the best test is Engle's Autoregressive Conditional Heteroskedasticity (ARCH) test. An ARCH model considers the variance of the current error term to be a function of (to be conditional on) the variances of previous time periods' error terms. Thus testing for heteroskedasticity consists of measuring the fit of an equation that specifies e_t^2 as a function of e_{t-1}^2, e_{t-2}^2, e_{t-3}^2, etc. See Robert F. Engle, "Autoregressive Conditional Heteroscedasticity with Estimates of Variance of United Kingdom Inflation," *Econometrica*, July 1982, pp. 987–1007.

has more independent variables (including the squared and cross-product terms) than observations. Sometimes the difficulty can be avoided if there are dummy independent variables in the original equation because we must drop the squares of all dummies from Equation 10.14 (since $0^2 = 0$ and $1^2 = 1$, they're perfectly collinear). For an example of and practice in the use of the White test, see Exercises 7 and 8.

10.4 Remedies for Heteroskedasticity

This section presents some remedies for heteroskedasticity, but it will also imply that there are times when the problem should be left unadjusted. Part of the art of econometrics is learning to tell one situation from the other.

The first step to take to rid an equation of heteroskedasticity is to try and figure out whether the heteroskedasticity is pure or impure. If the heteroskedasticity might be impure, determine and include the omitted variable that is causing the impure heteroskedasticity. If the heteroskedasticity is pure, there are two general remedies to consider:

1. *Use Weighted Least Squares.* If there is pure heteroskedasticity, weighted least squares (a form of generalized least squares) should be considered. Divide the equation through by the proportionality factor Z (or a function of Z) that appears to be related to the heteroskedasticity. After this division, reestimate the equation with the newly adjusted dependent and independent variables.
2. *Redefine the variables.* The effects of heteroskedastic residuals can often be negated by redefining the variables. This is a direct approach to correcting heteroskedasticity rather than the indirect approach of weighted least squares. The redefinition of the variables should be based on the underlying theory and refocusing the equation on the basic behavior it is supposed to explain.

Thus, the first thing to do if the Park test (or another test of heteroskedasticity) indicates the possibility of heteroskedasticity is to examine the equation carefully for specification errors. Although you should never include an explanatory variable simply because the Park test indicates the possibility of heteroskedasticity, you ought to rigorously think through the specification of the equation. If this rethinking allows you to discover a variable that should have been in the regression from the beginning, then that variable should be added to the equation. However, if there are no obvious specification errors, the heteroskedasticity is probably pure in nature, and one of the other remedies of this section should be considered.

10.4.1 Weighted Least Squares

Take an equation with pure heteroskedasticity caused by a proportionality factor Z:

$$Y_i = \beta_0 + \beta_1 X_{1i} + \beta_2 X_{2i} + \epsilon_i \qquad (10.15)$$

where the variance of the error term, instead of being constant, is

$$\text{VAR}(\epsilon_i) = \sigma_i^2 = \sigma^2 Z_i^2 \qquad (10.16)$$

where σ^2 is the constant variance of a classical (homoskedastic) error term u_i and Z_i is the proportionality factor. Given that pure heteroskedasticity exists, then Equation 10.15 can be shown[14] to be equal to:

$$Y_i = \beta_0 + \beta_1 X_{1i} + \beta_2 X_{2i} + Z_i u_i \qquad (10.17)$$

The error term in Equation 10.17, $Z_i u_i$, is heteroskedastic because $\sigma^2 Z_i^2$, its variance, is not constant. How could we adjust Equation 10.17 to make the error term homoskedastic? That is, what should be done to $Z_i u_i$ to make it turn into u_i? The easiest method is to divide the entire equation through by the proportionality factor Z_i, resulting in an error term, u_i, that has a constant variance σ^2. The new equation satisfies the Classical Assumptions, and a regression run on this new equation would no longer be expected to have heteroskedastic error terms. This general remedy to heteroskedasticity is called weighted least squares, which is actually a version of GLS.

Weighted least squares involves dividing Equation 10.17 through by whatever will make the error term once again homoskedastic and then rerunning the regression on the transformed variables. Given the commonly assumed form of heteroskedasticity in Equation 10.16, this means that the technique consists of three steps:

1. Divide Equation 10.17 through by the proportionality factor Z, obtaining:

$$Y_i/Z_i = \beta_0/Z_i + \beta_1 X_{1i}/Z_i + \beta_2 X_{2i}/Z_i + u_i \qquad (10.18)$$

14. The key is to show that the error term $Z_i u_i$ has a variance equal to $\sigma^2 Z_i^2$. For more, see Exercise 6.

The error term of Equation 10.18 is now u_i, which is homoskedastic.

2. Recalculate the data for the variables to conform to Equation 10.18.

3. Estimate Equation 10.18 with OLS.

This third step in weighted least squares, the estimation of the transformed equation, is fairly tricky, because the exact details of how to complete this regression depend on whether the proportionality factor Z is also an explanatory variable in Equation 10.15. If Z is not an explanatory variable in Equation 10.15, then the regression to be run in step 3 might seem to be:

$$Y_i/Z_i = \beta_0/Z_i + \beta_1 X_{1i}/Z_i + \beta_2 X_{2i}/Z_i + u_i \qquad (10.19)$$

Note, however, that this equation has no constant term. Most OLS computer packages can run such a regression only if the equation is forced through the origin by specifically suppressing the intercept with an instruction to the computer.

As pointed out in Section 7.4, however, the omission of the constant term forces the constant effect of omitted variables, nonlinearities, and measurement error into the other coefficient estimates. To avoid having these constant term elements forced into the slope coefficient estimates, one alternative approach to Equation 10.19 is to add a constant term[15] before the transformed equation is estimated. Consequently, when Z is not identical to one of the Xs in the original equation, then we suggest that the following specification can be run as step 3 in weighted least squares:

$$Y_i/Z_i = \alpha_0 + \beta_0/Z_i + \beta_1 X_{1i}/Z_i + \beta_2 X_{2i}/Z_i + u_i \qquad (10.20)$$

If Z *is* an explanatory variable in Equation 10.15, then no constant term need be added because one already exists. Look again at Equation 10.18. If $Z = X_1$ (or, similarly, if $Z = X_2$), then one of the slope coefficients becomes the constant term in the transformed equation because $X_1/Z = 1$:

$$Y_i/Z_i = \beta_0/Z_i + \beta_1 + \beta_2 X_{2i}/Z_i + u_i \qquad (10.21)$$

15. The suggestion of adding a constant term is also made by Potluri Rao and Roger LeRoy Miller, *Applied Econometrics* (Belmont, California: Wadsworth, 1971), p. 121; and others.

If this form of weighted least squares is used, however, coefficients obtained from an estimation of Equation 10.21 must be interpreted very carefully. Notice that β_1 is now the intercept term of Equation 10.21 even though it is a slope coefficient in Equation 10.15 and that β_0 is a slope coefficient in Equation 10.21, even though it is the intercept in Equation 10.15. As a result, a researcher interested in an estimate of the coefficient of X_1 in Equation 10.15 would have to examine the intercept of Equation 10.21, and a researcher interested in an estimate of the intercept term of Equation 10.15 would have to examine the coefficient of $1/Z_i$ in Equation 10.21. The computer will print out $\hat{\beta}_0$ as a "slope coefficient" and $\hat{\beta}_1$ as a "constant term" when in reality they are estimates of the opposite coefficients in the original Equation 10.15.

There are three other major problems with using weighted least squares:

1. The job of identifying the proportionality factor Z is, as has been pointed out, quite difficult.
2. The functional form that relates the Z factor to the variance of the error term of the original equation may not be the commonly assumed squared function of Equation 10.16. When some other functional relationship is involved, a different transformation is required. For more on these advanced transformations, see Exercise 12.
3. Sometimes weighted least squares is applied to an equation with impure heteroskedasticity. In such cases, it can be shown that the estimates reduce somewhat the bias from an omitted variable, but the estimates are inferior to those obtained from the correctly specified equation.

Given these uncertainties, it makes sense to consider returning to the original theory and attempting to formulate the project in a way that's not so susceptible to heteroskedasticity.

10.4.2 The Direct Approach: Redefining the Variables

Another approach to ridding an equation of heteroskedasticity is to go back to the basic underlying theory of the equation and redefine the variables in a way that avoids heteroskedasticity. A redefinition of the variables often is useful in allowing the estimated equation to focus more on the behavioral aspects of the relationship. Such a rethinking is a difficult and discouraging process because it appears to dismiss all the work already done. However, once the theoretical work has been reviewed, the alternative approaches that are discovered are often exciting in that they offer possible ways to avoid problems that had previously seemed insurmountable.

Unfortunately, it's difficult to specify procedures for something as general as "completely rethinking the research project," so we'd like to present a nonnumerical example of what we mean. When we get to the numerical example of the next section, the direct approach of redefining the variables will then be compared with the more formal method of weighted least squares.

Consider a cross-sectional model of the total expenditures by the governments of different cities. Logical explanatory variables to consider in such an analysis are the aggregate income, the population, and the average wage in each city. The larger the total income of a city's residents and businesses, for example, the larger the city government's expenditures (see Figure 10.6). In this case, it's not very enlightening to know that the larger cities have larger incomes and larger expenditures (in absolute magnitude) than the smaller ones.

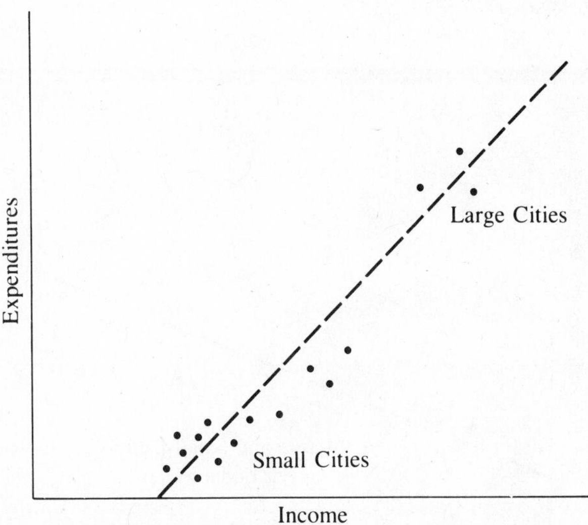

Figure 10.6 AN AGGREGATE CITY EXPENDITURES FUNCTION
If city expenditures are explained in an aggregate model, the larger cities play a major role in the determination of the coefficient values. Note how the slope would be somewhat lower without the heavy influence of the larger cities. In addition, heteroskedasticity is a potential problem in an aggregate model, because the wide range of sizes of the dependent variable makes different error term variances more likely.

Fitting a regression line to such data (see the line in Figure 10.6) also gives undue weight to the larger cities because they would otherwise give rise to large squared residuals. That is, since OLS minimizes the summed squared residuals, and since the residuals from the large cities are likely to be large due simply to the size of the city, the regression estimation will be especially sensitive to the residuals from the larger cities. This is often called "spurious correlation" due to size.

In addition, the residuals may indicate heteroskedasticity. The remedy for this kind of heteroskedasticity is not to automatically use weighted least squares, however, nor is it to throw out the observations from large cities. It makes sense to consider reformulating the model in a way that will discount the scale factor (the size of the cities) and emphasize the underlying behavior. In this case, per capita expenditures would be a logical dependent variable and per capita income a logical explanatory variable. Such a transformation is shown in Figure 10.7. This form of the equation places New York and Los Angeles on the same scale as, say, Pasadena or New Brunswick and thus gives them the same weight in estimation. If an explanatory variable happened not to be a function of the size of the city, however, it would not need to be adjusted to per capita terms. If the equation included the average

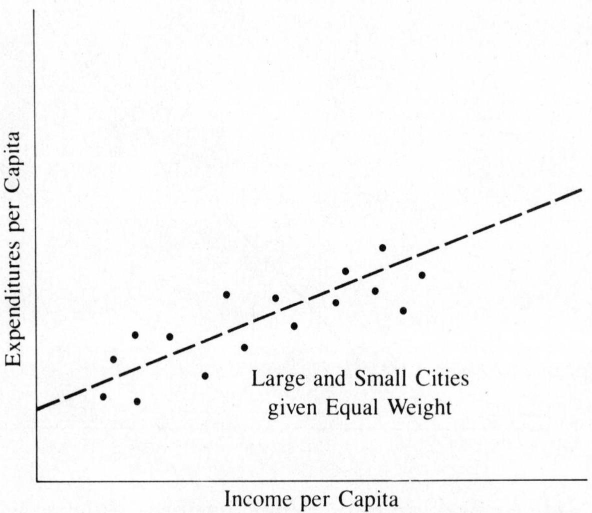

Figure 10.7 A PER CAPITA CITY EXPENDITURES FUNCTION
If city expenditures are explained in a per capita model, then large and small cities have equal weights. In addition, heteroskedasticity is less likely, because the dependent variable does not vary over a wide range of sizes.

wage of city workers, for example, that wage would not be divided through by population in the transformed equation.

Note that this transformation is similar in some ways to weighted least squares, since both independent and dependent variables have been divided by population. The difference is that there is no term equal to the reciprocal of population (as there is in weighted least squares) and that not all explanatory variables are divided by population. For the original equation,

$$EXP_i = \beta_0 + \beta_1 POP_i + \beta_2 INC_i + \beta_3 WAGE_i + \epsilon_i \qquad (10.22)$$

the weighted least squares version would be:

$$EXP_i/POP_i = \beta_1 + \beta_0/POP_i + \beta_2 INC_i/POP_i + \beta_3 WAGE_i/POP_i + u_i \qquad (10.23)$$

while the directly transformed equation would be

$$EXP_i/POP_i = \alpha_0 + \alpha_1 INC_i/POP_i + \alpha_2 WAGE_i + u_i \qquad (10.24)$$

where EXP_i refers to the expenditures, INC_i refers to the income, $WAGE_i$ refers to the average wage, and POP_i refers to population of the *i*th city. As can be seen, the weighted least squares Equation 10.23 divides through the entire equation by population, while the theoretically transformed one divides only expenditures and income by population. While the directly transformed Equation 10.24 does indeed solve any potential heteroskedasticity in the model, such a solution should be considered incidental to the benefits of rethinking the equation in a way that focuses on the basic behavior being examined.

Note that it's possible that the *reformulated* Equation 10.24 could have heteroskedasticity; the error variances might be larger for the observations having the larger per capita values for income and expenditures than they are for smaller per capita values. Thus it is legitimate to suspect and test for heteroskedasticity even in this transformed model. Such heteroskedasticity in the transformed equation is unlikely, however, because there will be little of the variation in size normally associated with heteroskedasticity.

A thoughtful transformation of the variables that corrects for heteroskedasticity while at the same time avoiding the spurious correlation due to size may sometimes be the best approach to solving these problems. Note however, that not every variable in the equation is treated the same (unlike weighted least squares). Each variable in a cross-sectional model can be examined for possible transformations that will yield a meaningful and properly interpreted regression equation.

10.5 A More Complete Example

Let's work through a more complete example that involves a cross-sectional data-set, potential heteroskedasticity, and the use of the Park test and weighted least squares. Back in the mid-1970s, the U.S. Department of Energy attempted to allocate gasoline to regions, states, and even individual retailers on the basis of past usage, changing demographics, and other factors. Underlying these allocations must have been some sort of model of the usage of petroleum by state (or region) as a function of a number of factors. It seems likely that such a cross-sectional model, if ever estimated, would have had to cope with the problem of heteroskedasticity.

In a model where the dependent variable is petroleum consumption by state, possible explanatory variables include functions of the size of the state (such as the number of miles of roadway, the number of motor vehicle registrations, or the population) and variables that are *not* functions of the size of the state (such as the gasoline tax *rate* or the speed limit). Since there is little to be gained by including more than one variable that measures the size of the state (because such an addition would be theoretically redundant and likely to cause needless multicollinearity), and since the speed limit was the same for all states (it would be a useful variable in a time-series model, however) a reasonable model to consider might be:

$$PCON_i = f(\overset{+}{REG}, \overset{-}{TAX}) = \beta_0 + \beta_1 REG_i + \beta_2 TAX_i + \epsilon_i \quad (10.25)$$

where: $PCON_i$ = petroleum consumption in the *i*th state (trillions of BTUs)
REG_i = motor vehicle registrations in the *i*th state (thousands)
TAX_i = gasoline tax rate in the *i*th state (cents per gallon)
ϵ_i = a classical error term

The more cars registered in a state, we would think, the more petroleum consumed, while a high tax rate on gasoline would decrease aggregate gasoline purchases in that state. If we now collect the data for this example (see Table 10.1) we can estimate Equation 10.25, obtaining

$$\widehat{PCON_i} = 551.7 + \underset{(0.0117)}{0.1861}REG_i - \underset{(16.86)}{53.59}TAX_i \qquad (10.26)$$

$$t = \qquad 15.88 \qquad\qquad -3.18$$

$$\bar{R}^2 = .861 \qquad n = 50$$

This equation seems to have no problems—the coefficients are significant in the hypothesized directions, and the overall equation is statistically significant. No Durbin-Watson d statistic is shown because there is no "natural" order of the observations to test for serial correlation (if you're curious, the DW for the order in Table 10.1 is 2.19). Given the discussion in the previous sections, let's investigate the possibility of heteroskedasticity caused by variation in the size of the states.

To test this possibility, we obtain the residuals from Equation 10.26, which are listed in Table 10.1, and run a Park test on them. Before we can run a Park test, we must decide what possible proportionality factor Z to investigate.

Almost any variable related to market size would be appropriate, but motor vehicle registrations (REG) is certainly a reasonable choice. Note that to run a Park test with the gasoline tax rate (TAX) as the proportionality factor Z would be a mistake, since there is little evidence that the rate varies significantly with the size of the state. Total tax receipts, on the other hand, would be a possible alternative to REG. To see what the residuals look like if plotted against REG, see Figure 10.8; note that the residuals do indeed look potentially heteroskedastic. The next step would be to run a Park test:

$$\ln(e_i^2) = \alpha_0 + \alpha_1 \ln REG_i + u_i \qquad (10.27)$$

where e_i is the residual for the *i*th state from Equation 10.26, and u_i is a classical (homoskedastic) error term.

If we run this Park test regression, we obtain:

$$\widehat{\ln(e_i^2)} = 1.650 + \underset{(0.308)}{0.952}\ln REG_i \qquad (10.28)$$

$$t = \quad 3.09$$

$$\bar{R}^2 = .148 \qquad n = 50 \qquad F = 9.533$$

TABLE 10.1 DATA FOR THE PETROLEUM CONSUMPTION EXAMPLE (1982)

PCON	UHM	TAX	REG	POP	e	STATE
270	2.2	9	743	1136	62.335	Maine
122	2.4	14	774	948	176.52	New Hampshire
58	0.7	11	351	520	30.481	Vermont
821	20.6	9.9	3750	5750	101.87	Massachusetts
98	3.6	13	586	953	133.92	Rhode Island
450	10.1	11	2258	3126	67.527	Connecticut
1819	36.4	8	8235	17567	163.24	New York
1229	22.2	8	4917	7427	190.83	New Jersey
1200	27.9	11	6725	11879	−13.924	Pennsylvania
1205	29.2	11.7	7636	10772	−140.98	Ohio
650	17.6	11.1	3884	5482	−29.764	Indiana
1198	30.3	7.5	7242	11466	−299.72	Illinois
760	25.1	13	6250	9116	−258.33	Michigan
460	13.8	13	3162	4745	16.446	Wisconsin
503	13.0	13	3278	4133	37.855	Minnesota
371	8.1	13	2346	2906	79.330	Iowa
571	13.9	7	3412	4942	−240.63	Missouri
136	1.6	8	653	672	−108.50	North Dakota
109	1.6	13	615	694	139.52	South Dakota
203	4.3	13.9	1215	1589	170.08	Nebraska
349	8.4	8	2061	2408	−157.58	Kansas
118	1.4	11	415	600	78.568	Delaware
487	9.8	13.5	2893	4270	120.31	Maryland
628	12.4	11	3705	5485	−23.806	Virginia
192	2.9	10.5	1142	1961	−9.5451	West Virginia
642	17.1	12	4583	6019	−119.64	North Carolina
320	7.1	13	1975	3227	97.385	South Carolina
677	15.6	7.5	3916	5648	−201.65	Georgia
1459	28.5	8	8335	10466	−215.37	Florida
434	6.9	10	2615	3692	−68.513	Kentucky
482	11.9	9	3381	4656	−216.68	Tennessee
457	13.7	11	3039	3941	−70.842	Alabama
325	6.3	9	1593	2569	−40.877	Mississippi
300	7.4	9.5	1481	2307	−18.235	Arkansas
1417	10.1	8	2800	4383	772.87	Louisiana
451	11.4	6.58	2780	3226	−265.51	Oklahoma
3572	59.9	5	11388	15329	1168.6	Texas
131	2.3	9	758	805	−79.457	Montana
105	2.2	7.5	873	977	−207.25	Idaho
163	1.5	8	508	509	−54.515	Wyoming
323	9.2	9	2502	3071	−212.07	Colorado
192	4.4	11	1193	1367	7.7577	New Mexico
291	8.9	10	2216	2892	−137.25	Arizona
169	5.0	11	1038	1571	13.608	Utah
133	2.4	12	710	876	92.250	Nevada
562	14.8	12	3237	4276	50.895	Washington
364	8.4	8	2075	2668	−145.18	Oregon
2840	62.5	9	17130	24697	−417.81	California
155	1.2	8	319	444	−27.336	Alaska
214	1.3	8.5	586	997	8.7623	Hawaii

Source: *1985 Statistical Abstract* (U.S. Department of Commerce), except the residual.

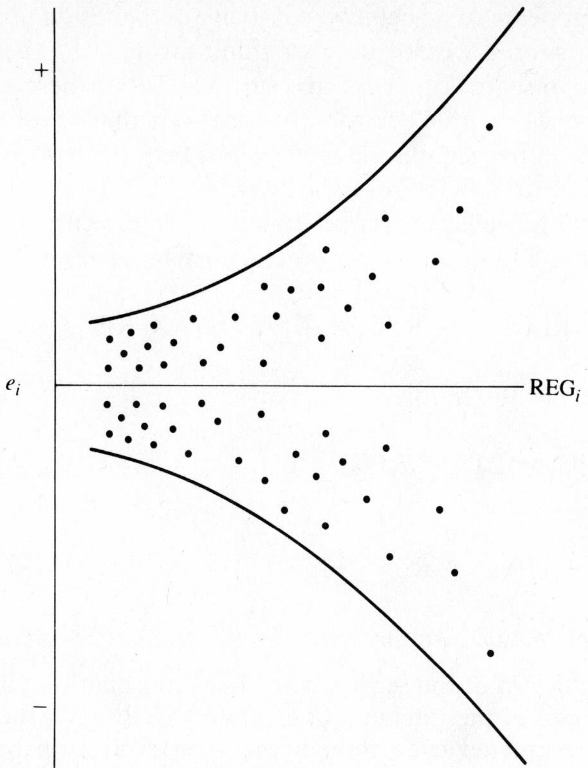

Figure 10.8 HETEROSKEDASTIC RESIDUALS FROM EQUATION 10.26
If the residuals from Equation 10.26 and Table 10.1 are plotted with respect to Motor Vehicle Registrations by state (REG), they appear to follow a wider distribution for large values of REG than for small values of REG. Such a pattern is preliminary evidence of heteroskedasticity that should be tested more formally with a test like the Park test.

Since the critical t-value for a 99 percent two-tailed *t*-test is about 2.7 in Statistical Table B-1,[16] we can reject the null hypothesis of homoskedasticity because the appropriate decision rule is:

$$\text{Reject } H_0\text{: } \alpha_1 = 0 \quad \text{if } |t_{PARK}| > 2.7$$
$$\text{Do Not Reject } H_0 \quad \text{if } |t_{PARK}| \leq 2.7$$

16. Note that Statistical Table B-1 doesn't list critical values for exactly 50 observations. In such cases, most researchers either use the number of observations closest to the value they're looking for or interpolate to find the critical values.

Since there appears to be heteroskedasticity in the residuals of Equation 10.26, what should we do? First we think through the specification of the equation in search of an omitted variable. While there are a number of possible ones for this equation, it turns out that all of the ones we tried involve either significant multicollinearity (as shown in Chapter 8) or do not cure the heteroskedasticity.

As a result, we'll reestimate Equation 10.25 with weighted least squares, using REG[17] as the proportionality factor Z:

$$PCON_i/REG_i = \beta_0/REG_i + \beta_1 + \beta_2 TAX_i/REG_i + u_i \qquad (10.29)$$

which results in the following estimates:

$$\widehat{PCON_i/REG_i} = 218.54/REG_i + 0.168 - 17.389TAX_i/REG_i \qquad (10.30)$$
$$(0.014) \quad (4.682)$$
$$t = \quad 12.27 \quad -3.71$$
$$\bar{R}^2 = .333 \qquad n = 50$$

Compare this result carefully with Equation 10.26. Note that:

1. The coefficient of the reciprocal of REG in Equation 10.30 is really an estimate of the intercept of Equation 10.26, and therefore no *t*-test is conducted even though the OLS regression program will indicate that it is a slope coefficient.
2. What appears to be the intercept of Equation 10.30 is an estimate of the coefficient of REG in Equation 10.26. Note that this particular estimate is quite close in magnitude and significance to the original results in Equation 10.26.
3. The t-score of the coefficient of the proportionality factor, REG, is lower in the weighted least squares estimate than it is in the potentially heteroskedastic Equation 10.26. The overall fit is also worse, but this has no particular importance because the dependent variables are different in the two equations.

However, as mentioned in Section 10.4.2, an alternative is to rethink the purpose of the regression and reformulate the variables of

17. Note that we've divided the equation through by REG_i. This assumes that the error term $\epsilon_i = Z_i u_i$. The coefficient of lnREG in the Park test is approximately one, which is evidence that the appropriate functional form may be $\epsilon_i = u_i\sqrt{Z}$, but such a transformation should not be adopted simply on the basis of the Park test coefficient alone. If the underlying theory gives evidence in support of such a change, which is not the case in this example, then we would divide the equation through by the square root of Z_i. For more on this, reread footnote 7.

the equation to try to avoid heteroskedasticity resulting from spurious correlation due to size. If we were to rethink Equation 10.25, we might decide to attempt to explain per capita petroleum consumption, coming up with:

$$PCON_i/POP_i = \beta_0 + \beta_1 REG_i/POP_i + \beta_2 TAX_i + \epsilon_i \qquad (10.31)$$

where POP_i is the population of the *i*th state in thousands of people.

We've reformulated the equation in a way similar to weighted least squares, but we now have an equation that can stand on its own from a theoretical point of view. If we estimate Equation 10.31, we obtain:

$$\begin{aligned} \widehat{PCON_i/POP_i} = 0.168 &+ 0.1082REG_i/POP_i - 0.0103TAX_i \qquad (10.32)\\ &\quad (0.0716) \qquad\qquad\qquad (0.0035)\\ t = &\quad 1.51 \qquad\qquad\qquad\quad -2.95\\ \bar{R}^2 &= .165 \qquad n = 50 \end{aligned}$$

If we compare Equation 10.32 with Equations 10.30 and 10.26, we see that this third approach is not necessarily better but quite different. The statistical properties of Equation 10.32, though not directly comparable to the other equations, do not appear as strong as they ought to be, but this is not necessarily an important factor.

Which is better, the unadjusted potentially heteroskedastic equation, the one derived from weighted least squares, or the reformulated one? It depends on the purposes of your research. If your goal is to determine the impact of tax rates on gasoline consumption, all three models give virtually the same results in terms of the sign and significance of the coefficient, but the latter two models avoid the heteroskedasticity. If your goal is to allocate petroleum in aggregate amounts to states, then the original equation may be just fine. In most cases of severe heteroskedasticity, some remedial action is necessary, but whether weighted least squares or a reformulation is called for depends on the particular equation in question. We generally find that if reformulation makes intuitive sense, it's usually the best remedy to apply, since it more easily avoids the arbitrary process of choosing a Z.

10.6 Summary

1. Heteroskedasticity is the violation of the classical assumption that the observations of the error term are drawn from a distribution with a constant variance. While homoskedastic error term obser-

vations are drawn from a distribution that has a constant variance for all observations, heteroskedastic error term observations are drawn from distributions whose variances differ with different observations. Heteroskedasticity occurs most frequently in cross-sectional data-sets.

2. The variance of a heteroskedastic error term is not equal to σ^2, a constant. Instead, it equals σ_i^2, where the subscript *i* indicates that the variance can change from observation to observation. Many different kinds of heteroskedasticity are possible, but a common model is one in which the variance changes systematically as a function of some other variable, a proportionality factor Z:

$$\mathrm{VAR}(\epsilon_i) = \sigma^2 Z_i^2$$

The proportionality factor Z is usually a variable related in some way to the size or accuracy of the dependent variable.

3. Pure heteroskedasticity is a function of the error term of the correctly specified regression equation. Impure heteroskedasticity is caused by a specification error such as an omitted variable.

4. The major consequence of heteroskedasticity is an increase in the variance of the $\hat{\beta}$s that is masked by an underestimation of the standard errors by OLS. As a result, OLS tends to overestimate t-scores in the face of heteroskedasticity, sometimes leading to errors of inference. Pure heteroskedasticity does not cause bias in the estimates of the βs themselves.

5. Many tests use the residuals of an equation to test for the possibility of heteroskedasticity in the error terms. The Park test uses a function of these residuals as the dependent variable of a second regression whose explanatory variable is a function of the suspected proportionality factor Z:

$$\ln(e_i^2) = \alpha_0 + \alpha_1 \ln Z_i + u_i$$

If $\hat{\alpha}_1$ is significantly different from zero, then we reject the null hypothesis of homoskedasticity.

6. The first step in correcting heteroskedasticity is to check for an omitted variable that might be causing impure heteroskedasticity.

If the specification is as good as possible, then solutions such as weighted least squares or a reformulation of the variables of the equation should be considered.

7. Weighted least squares is a method of ridding an equation of heteroskedasticity by dividing it through by a function of the proportionality factor Z and then reestimating the equation with OLS. In some circumstances, it's more appropriate to rethink the underlying theory of the equation completely and to reformulate the variables, for example by converting them to a per capita basis, before rerunning the equation.

Exercises

(Answers to even-numbered exercises are in Appendix A.)

1. Write the meaning of each of the following terms without reference to the book (or your notes), and compare your definition with the version in the text for each:
   a. impure heteroskedasticity
   b. proportionality factor Z
   c. the Park test
   d. the Goldfeld-Quandt test
   e. weighted least squares

2. In the common model of heteroskedasticity ($VAR(\epsilon_i) = \sigma^2 Z_i^2$), one of the major difficulties is making good choices for potential proportionality factors (Zs). In each of the following equations, separate the listed explanatory variables into those that are likely or unlikely to be proportionality factors.
   a. The number of economics majors in a cross section of various sized colleges and universities as a function of the number of undergraduates attending a school, the number of required courses in that school's economics major, the average GPA in the major, and the number of economics professors there.
   b. GNP in a cross section of various sized countries as a function of the aggregate gross investment in a nation, the percentage growth of its money supply, the maximum marginal tax rate on capital gains there, and its population.
   c. The demand for carrots in a time-series model of the U.S. as a function of the real price of carrots, U.S. disposable income,

U.S. per capita disposable income, population, the percentage error in carrot sales measurement, and the real price of celery.

3. Of all the econometric problems we've encountered, heteroskedasticity is the one that seems the most difficult to understand. Close your book and attempt to write an explanation of heteroskedasticity in your own words. Be sure to include a diagram in your description.

4. Use the Park test to test the null hypothesis of homoskedasticity in each of the following situations (one percent level of significance):
   a. The calculated t-score of your suspected proportionality factor Z is 3.561 from a Park test regression with 25 degrees of freedom.
   b. The following residuals and values for the potential Z:

Observation	Residual	Proportionality Factor Z
1	3.147	120
2	9.394	240
3	−2.344	900
4	−1.034	50
5	5.678	600
6	2.113	20
7	−4.356	200

   c. How would your answer to the above change if the seventh observation of Z was minus 200? How would you even take the log of Z?
   d. Test the first column of error terms in Table 9.1. Use X as the potential proportionality factor Z.

5. Ando and Modigliani collected the following data on the income and consumption of non-self-employed homeowners:[18]

18. Albert Ando and Franco Modigliani, "The 'Permanent Income' and 'Life Cycle' Hypotheses of Saving Behavior: Comparisons and Tests," in I. Friend and R. Jones, eds., *Consumption and Saving*, Vol. II., 1960, p. 154.

Income Bracket ($)	Average Income ($)	Average Consumption ($)
0–999	556	2760
1000–1999	1622	1930
2000–2999	2664	2740
3000–3999	3587	3515
4000–4999	4535	4350
5000–5999	5538	5320
6000–7499	6585	6250
7500–9999	8582	7460
10000–above	14033	11500

a. Run a regression attempting to explain average consumption as a function of average income.
b. Use the Park test to test the residuals from the equation you ran in part a for heteroskedasticity, using income as the potential proportionality factor Z (five percent).
c. If there is only one explanatory variable, what does the equation for weighted least squares look like? Does running weighted least squares have any effect on the estimation? Why or why not?
d. If the Park test run in part b above shows evidence of heteroskedasticity, then what, if anything, should be done about it?

6. Show that Equation 10.17 is true by showing that the variance of an error term that equals a classical error term multiplied times a proportionality constant Z is that shown in Equation 10.16. That is, show that if $\epsilon_i = u_i Z_i$, then $\text{VAR}(\epsilon_i) = \sigma^2 Z_i^2$ if $\text{VAR}(u_i) = \sigma^2$ (a constant). (Hint: Start with the definition of a variance and then calculate the variance for an error term $Z_i u_i$.)

7. The best way to feel comfortable with the various tests for heteroskedasticity that we've introduced is to apply these tests to a set of residuals to which we've already applied the Park test. Using the residuals from the Woody's example in Table 3.1, conduct the following tests. (Hint: Use the one percent level of significance, and be sure to compare your result with the Park test result in Section 10.3.2.)

a. the Goldfeld-Quandt test (with P as your proportionality factor)
b. the Breusch-Pagan test (Zs = C, P, I)
c. the White test

8. As further practice with the different tests for heteroskedasticity, use the residuals from the state gasoline model (Equation 10.26) in Table 10.1 to conduct the following tests. (Hint: Use the one percent level of significance, and be sure to compare your result with the Park test result in Section 10.5.)
   a. the Goldfeld-Quandt test (with POP as your proportionality factor)
   b. the Breusch-Pagan test (Zs = POP, REG, TAX)
   c. the White test

9. Is it really possible to decide whether to adjust an equation for heteroskedasticity if we can never actually observe the error terms (just the residuals), have no way of knowing whether we have the right proportionality factor, and have no way of knowing whether we have the right functional relationship between the proportionality factor and the error term? Are there any circumstances in which you could feel confident about having used weighted least squares? Are there any circumstances in which you could feel confident about having reformulated the equation? Why?

10. Consider the following double-log equation (standard errors in parentheses):[19]

$$\hat{Y}_i = \underset{(0.058)}{0.442X_{1i}} + \underset{(0.042)}{0.092X_{2i}} + \underset{(0.014)}{0.045X_{3i}} + \underset{(0.034)}{0.259X_{4i}}$$

$$R^2 = .620 \qquad n = 430$$

where: Y_i = the log of the gross value of agricultural output (in drachmas) of the *i*th Greek farm in a given year

X_{1i} = the log of farmer-workdays in a year on the *i*th farm

X_{2i} = the log of the amount of land on the *i*th farm (in stremmata, equal to a quarter of an acre)

19. Adapted from Pan A. Yotopoulos and Jeffrey B. Nugent, *Economics of Development* (New York: Harper & Row, 1976), p. 82. No estimate of the intercept was reported.

X_{3i} = the log of the value of the plant and equipment (plus operating expenses for plant and equipment) on the *i*th farm that year (in drachmas)

X_{4i} = the log of the value of livestock (including trees) plus operating expenses on livestock (in drachmas) in the *i*th farm that year

a. Create hypotheses about the signs of the various coefficients and then calculate the t-scores to test those hypotheses at the five percent level of significance.

b. Suppose you were now told that the Park test, using X_1 as a potential proportionality factor Z, indicated the likelihood of heteroskedasticity in the residuals of this equation. Is it likely that there actually is heteroskedasticity in such a double-log equation? Why or why not?

c. Is there is a logical reformulation of the equation that might rid the model of heteroskedasticity?

d. If you decided to apply weighted least squares to the equation, what equation would you estimate?

11. Consider the following estimated regression equation for average annual hours worked (per capita) for young black men (16–21 years) in 94 standard metropolitan statistical areas (standard errors in parentheses):

$$\hat{B}_i = 300.0 + 0.50W_i - 7.5U_i - 18.3\ln P_i$$
$$(0.05) \qquad (7.5) \qquad (6.1)$$
$$DW = 2.00 \qquad n = 94 \qquad \bar{R}^2 = .64$$

where: B_i = average annual hours worked (per capita) by young black men (age 16 to 21) in the *i*th city

W_i = average annual hours worked by young white men in the *i*th city

U_i = black unemployment rate in the *i*th city

$\ln P_i$ = natural log of the black population of the *i*th city

a. Develop and test (five percent level) your own hypotheses with respect to the individual estimated slope coefficients.

b. How would you respond to the claim that having the estimated coefficient of W be far less than one is proof that racial discrimination exists in the labor markets in this country?

c. Since this is a cross-sectional model, is it reasonable to worry about heteroskedasticity? What variable would you choose as a possible Z? Explain your choice.
d. Suppose you ran a Park test with respect to your chosen Z and found a t-score of 1.91. Does this support or refute your answer to part c above? (Hint: Be sure to complete the Park test.)
e. Suppose you were asked to make one change in the specification of this equation. What would you suggest? Explain your choice.

12. Given the most commonly used functional form for the relationship between the proportionality factor Z and the error term ($\epsilon_i = Z_i u_i$ where u_i is a homoskedastic error term), we can derive the appropriate weighted least squares equation of:

$$Y/Z_i = \beta_0/Z_i + \beta_1 + \beta_2 X_{2i}/Z_i + u_i$$

when $Z = X_1$, an explanatory variable already in the equation. This is accomplished by dividing the equation by the precise value (Z_i) necessary to make the error term homoskedastic in nature. Find the appropriate weighted least squares equations to be used in the following situations:

a. $\epsilon_i = u_i\sqrt{Z_i}$ where $Z = X_1$, an explanatory variable already in the equation
b. $\epsilon_i = u_i Z_i$ where $Z = X_3$, a variable not in the equation
c. $\epsilon_i = u_i\hat{Y}_i$, where $\hat{Y}_i$ is the estimated value of the dependent variable obtained from the regression equation

13. Bucklin, Caves, and Lo[20] estimated the following double-log model to explain the yearly circulation of metropolitan newspapers (standard errors in parentheses):

$$\begin{aligned}\hat{C}_i = -\,8.2 \;-\; &\underset{(0.58)}{0.56P_i} + \underset{(0.14)}{0.90I_i} + \underset{(0.21)}{0.76Q_i} + \underset{(0.14)}{0.27A_i}\\ + \;&\underset{(0.05)}{0.08S_i} - \underset{(0.27)}{0.77T_i}\\ &n = 50\end{aligned}$$

20. R. E. Bucklin, R. E. Caves, and A. W. Lo, "Games of Survival in the U.S. Newspaper Industry," *Applied Economics*, May 1989, pp. 631–650. Note that the Park test results are hypothetical and that the equation was originally estimated with Two-Stage Least Squares (to be discussed in Chapter 14). These facts don't change the equation's usefulness as an exercise in this chapter.

where: C_i = yearly circulation of the *i*th newspaper
P_i = the weighted average single copy price of the *i*th newspaper
I_i = the total disposable income of the metropolitan area of the *i*th newspaper
Q_i = number of personnel in editorial positions for the *i*th newspaper
A_i = volume of retail advertising in the *i*th newspaper
S_i = amount of competition from suburban dailies in the *i*th newspaper's region
T_i = number of television stations in the *i*th newspaper's region
(All variables are in logarithmic form.)

a. Hypothesize signs and run *t*-tests on each of the individual slope coefficients.
b. Does heteroskedasticity seem theoretically likely? Test for heteroskedasticity at the 99 percent level assuming that a Park test with I_i as Z produces a t-score of 3.13.
c. Given your responses to parts a and b above, what econometric problems (out of omitted variables, irrelevant variables, incorrect functional form, multicollinearity, serial correlation, and heteroskedasticity) appear to exist in this equation?
d. If you could suggest just one change in the specification of this equation, what would that change be? Carefully explain your answer.

14. Think back to the "Farmer Vin" pig-growing equation of Exercise 13 in Chapter 9:

$$\hat{W}_i = 12 + \underset{(1.0)}{3.5G_i} + \underset{(1.0)}{7.0D_i} - \underset{(0.10)}{0.25F_i} \qquad (10.33)$$
$$t = \quad 3.5 \qquad 7.0 \qquad -2.5$$
$$\bar{R}^2 = .70 \qquad n = 200 \qquad DW = 0.50$$

where: W_i = the percentage weight gain of the *i*th pig during the six-month experiment
G_i = a dummy variable equal to 1 if the *i*th pig is a male, 0 otherwise
D_i = a dummy variable equal to 1 if the *i*th pig was fed only at night, 0 if the *i*th pig was fed only during the day

F_i = the amount of food (pounds) eaten per day by the *i*th pig

It turns out this study was estimated originally with the dependent variable equal to "pounds gained by the *i*th pig," but a Park test showed severe heteroskedasticity. (Recall that the sample pigs ranged all the way from pink piglets to ponderous porkers.) The dependent variable was converted to percentage terms in an effort to "rethink the equation" and eliminate the heteroskedasticity. Equation 10.33 was the result.

a. How theoretically likely is it that there is heteroskedasticity in the revised equation, Equation 10.33? Explain your answer.

b. Suppose that you run a Park test on the residuals of Equation 10.33 and find a t-score of −6.31 (using the weight of the *i*th pig as the potential proportionality factor). Do you appear to have heteroskedasticity?

c. The t-score from your Park test in part b above is negative, and yet we've never discussed "negative" or "positive" heteroskedasticity. What importance, if any, would you attach to the *sign* of the estimated slope coefficient in this Park test?

d. You appear to have heteroskedasticity in both the original and the revised versions of your equation. What should you do? Explain.

11

A Regression User's Handbook

The real world problems that a regression user encounters are not so neatly labeled and compartmentalized as the previous ten chapters might imply. Instead, researchers must consider all the possible difficulties in an equation in order to decide which specification and estimation techniques to use. As a result, it's useful to have summaries of the definitions, problems, solutions, and statistical tests that are central to basic single-equation linear regression models. The first two sections of this chapter contain such summaries; while you should certainly read these now, we also hope that you'll benefit from using them as a reference whenever you're undertaking econometric research.

Frequently, even the best planned and executed regression analysis will not produce the expected set of estimated coefficients and their accompanying statistics the first time it is applied—something is deficient. When this happens, econometrics becomes an art. The core of the problem is that we never know what the true model is. There's a fine line between curve fitting and searching for the truth by formulating and estimating alternative specifications or using alternative estimating techniques. Some regression results may very well be strictly a product of chance, because the researchers might have experimented with a number of models and estimators until they obtained the results that came closest to what they wanted; such an approach is unscientific. On the other hand, it would be foolhardy to go to the opposite extreme and ignore obvious estimation or specification errors in a first regression run; that would also be unscientific. The last five sections of this chapter help beginning researchers strike a balance between these two positions by providing advice on econometric ethics and by giving some hands-on regression experience. The interactive regression exercises in Sections 11.4, 11.6, and 11.7 are "half-way houses" between reading someone else's regression results (and having no input) and doing one's own regression analysis (and getting no feedback). We strongly encourage the reader to take the exercises seriously and work through the examples rather than just read them.

11.1 A Regression User's Checklist

Table 11.1 contains a list of the items that a researcher checks when reviewing the output from a computer regression package. Not every item in the checklist will be produced by your computer package, and not every item in your computer output will be in the checklist, but the checklist can be a very useful reference. In most cases, a quick glance at the checklist will remind you of the text sections that deal with the item, but if this is not the case, the fairly minimal explanations in the checklist should not be relied on to cover everything needed for complete analysis and judgment. Instead, you should look up the item in the index. In addition, note that the actions in the right-hand column are merely suggestions. The circumstances of each individual research project are much more reliable guides than any dogmatic list of actions.

There are two ways to use the checklist. First, you can refer to it as a "glossary of packaged computer output terms" when you encounter something in your regression result that you don't understand.

Second, you can work your way through the checklist in order, finding the items in your computer output and marking them. This latter use of the checklist will help ensure that you don't forget to note some important item (such as the Durbin-Watson statistic for a time-series or the simple correlation matrix for a model with likely collinearity). As with the Regression User's Guide (Table 11.2), the use of the Regression User's Checklist will be most helpful for beginning researchers, but we also find ourselves referring back to it once in a while even after years of experience.

Be careful. All simplified tables, like the two in this chapter, must trade completeness for ease of use. As a result, strict adherence to a set of rules is not recommended even if the rules come from one of our tables. Someone who understands the purpose of the research, the exact definitions of the variables, and the problems in the data is much more likely to make a correct judgment than is someone equipped with a set of rules created to apply to a wide variety of possible applications.

11.2 A Regression User's Guide

Table 11.2 contains a brief summary of the major econometric maladies discussed so far in this text. For each econometric problem, we list:

1. Its nature
2. Its consequences for OLS estimation
3. How to detect it
4. How to attempt to get rid of it

How might you use the guide? If an estimated equation has a particular problem, such as a significant unexpected sign for a coefficient estimate or a correct but insignificant coefficient estimate, a quick glance at the guide can give some idea of what econometric problems might be causing that symptom. Both multicollinearity and irrelevant variables can cause regression coefficients to have insignificant t-scores, for example, and someone who remembered only one of these potential causes might take the wrong corrective action. After some practice, the use of this guide will decrease until it eventually will seem fairly limiting and simplistic. Until then, however, our experience is that those about to undertake their first econometric research can benefit by referring to this guide.

TABLE 11.1 REGRESSION USER'S CHECKLIST

Symbol	Checkpoint	Reference	Decision
X,Y	Data observations	Check for data errors, especially outliers, in computer printout of the data. Spot check transformations of variables.	Correct any errors. If the quality of the data is poor, may want to avoid regression analysis or use just OLS.
d. f.	Degrees of freedom	$n - K - 1 > 0$ n = number of observations K = number of explanatory variables	If $n - K - 1 \leq 0$, equation cannot be estimated, and if the degrees of freedom are low, precision is low. In such a case, try to include more observations.
$\hat{\beta}$	Estimated coefficient	Compare signs and magnitudes to expected values.	If they are unexpected, respecify model if appropriate or assess other statistics for possible corrective procedures.
t	t-statistic $t_k = \frac{\hat{\beta}_k - \beta_{H_0}}{SE(\hat{\beta}_k)}$ or $t_k = \frac{\hat{\beta}_k}{SE(\hat{\beta}_k)}$ for computer-supplied t-scores or whenever $\beta_{H_0} = 0$	Two-sided test: H_0: $\beta_k = \beta_{H_0}$ H_A: $\beta_k \neq \beta_{H_0}$ One-sided test: H_0: $\beta_k \leq \beta_{H_0}$ H_A: $\beta_k > \beta_{H_0}$ β_{H_0}, the hypothesized β, is supplied by the researcher, and is often zero.	Reject H_0 if $\lvert t_k \rvert > t_c$ The estimate must be of the expected sign to reject H_0. t_c is the critical value for α level of significance and $n - K - 1$ degrees of freedom.
R^2	Coefficient of Determination	Measures the degree of overall fit of the model to the data.	A guide to the overall fit.
$\bar{R}^2$	R^2 adjusted for degrees of freedom	Same as R^2. Also attempts to show the contribution of an additional explanatory variable.	One indication that an explanatory variable is irrelevant is if the $\bar{R}^2$ falls when it is included.

TABLE 11.1 REGRESSION USER'S CHECKLIST *(continued)*

Symbol	Checkpoint	Reference	Decision
F	F-statistic	To test $H_0: \beta_1 = \beta_2 = \cdots = \beta_K = 0$ H_A: H_0 not true Calculate special F-statistic to test joint hypotheses.	Reject H_0 if $F > F_c$, the critical value for α level of significance and K numerator and $n - K - 1$ denominator d.F.
DW	Durbin-Watson d statistic	Tests: H_0: $p \leq 0$ H_A: $p > 0$ For positive serial correlation	Reject H_0 if DW $< d_L$. Inconclusive if $d_L \leq DW \leq d_u$. (d_L and d_u are critical DW values.)
e_i	Residual	Check for transcription errors. Check for heteroskedasticity by examining the pattern of the residuals.	Correct the data. May take appropriate corrective action, but test first.
SEE	Standard error of the equation	An estimate of σ. Compare with $\bar{Y}$ for a measure of overall fit.	A guide to the overall fit.
TSS	Total sum of squares	$TSS = \sum_i (Y_i - \bar{Y})^2$	Used to compute F, R^2, and $\bar{R}^2$.
ESS	Explained sum of squares	$ESS = \sum_i (\hat{Y}_i - \bar{Y})^2$	Same as above.
RSS	Residual sum of squares	$RSS = \sum_i (Y_i - \hat{Y}_i)^2$	Same as above. Also used in hypothesis testing.
$SE(\hat{\beta}_k)$	Standard error of $\hat{\beta}_k$	Used in t-statistic.	A "rule of thumb" is if $\lvert\hat{\beta}_k\rvert > 2SE(\hat{\beta}_k)$ reject H_0: $\hat{\beta}_k = 0$.
$\hat{p}$	Estimated first-order autocorrelation coefficient	Usually provided by an autoregressive routine.	If negative, implies a specification error.
r_{12}	Simple correlation coefficient between X_1 and X_2	Used to detect collinearity.	Suspect severe multicollinearity if *t*-test shows r_{12} is significant.
VIF	Variance inflation factor	Used to detect multicollinearity.	Suspect severe multicollinearity if VIF > 5 or > 10.

TABLE 11.2 A REGRESSION USER'S GUIDE

What Can Go Wrong?	What Are the Consequences?	How Can It Be Detected?	How Can It Be Corrected?
Omitted Variable			
The omission of a relevant independent variable.	Bias in the coefficient estimates (the $\hat{\beta}$s) of the included Xs.	Theory, significant unexpected signs or surprisingly poor fits.	Include the left-out variable or a proxy.
Irrelevant Variable			
The inclusion of a variable that does not belong in the equation.	Decreased precision in the form of lower $\bar{R}^2$, higher standard errors, and lower t-scores.	1. Theory 2. *t*-test on $\hat{\beta}$ 3. $\bar{R}^2$ 4. Impact on other coefficients if X is dropped	Delete the variable if its inclusion is not required by the underlying theory.
Incorrect Functional Form			
The functional form is inappropriate.	Biased and inconsistent estimates, poor fit, and difficult interpretation.	Examine the theory carefully; think about the relationship between X and Y.	Transform the variable or the equation to a different functional form.
Multicollinearity			
Some of the independent variables are (imperfectly) correlated.	No biased $\hat{\beta}$s, but estimates of the separate effects of the Xs are not reliable, i.e., high SEs (and low t-scores).	No universally accepted rule or test is available. Use the *t*-test on r_{12} or the VIF test.	Drop redundant variables, but to drop others might introduce bias. A combination variable may be useful, but often doing nothing is best.
Serial Correlation			
The error terms for different observations are correlated, as in: $\epsilon_t = \rho\epsilon_{t-1} + u_t$	No biased $\hat{\beta}$s, but the variances of the $\hat{\beta}$s increase (and t-scores fall) in a way not captured by OLS.	Use Durbin-Watson d test; if significantly less than 2, positive serial correlation exists.	If impure, add the omitted variable or change the functional form. Otherwise, consider generalized least squares.
Heteroskedasticity			
The variance of the error term is not constant for all observations, as in: $VAR(\epsilon_i) = \sigma^2 Z_i^2$	Same as for serial correlation.	Plot the spread or contraction of the residuals (against one of the Xs, for example) or use the Park or Goldfeld-Quandt tests.	If impure, add the omitted variable. Otherwise, redefine the variables or apply a weighted least squares correction.

11.3 The Ethical Econometrician

In Section 11.4, we'll present an opportunity to practice the art of econometrics. The interactive regression learning exercise of that section will give the reader some "hands-on" experience with regression analysis without requiring the use of a computer and will provide the kind of feedback that usually is possible only in one-on-one sessions with an experienced econometrician. Unfortunately, the format of the section also has the potential to mislead because the interactive exercise includes a large number of alternative specifications of the same regression project. One conclusion that a casual reader might draw from this large number of specifications is that we encourage the estimation of numerous regression results as a way of insuring the discovery of the best possible estimates.

Nothing could be further from the truth!

As every reader of this book should know by now, our opinion is that the best models are those on which much care has been spent to develop the theoretical underpinnings and only a short time is spent pursuing alternative estimations of that equation. Many econometricians, ourselves included, would hope to be able to estimate only *one* specification of an equation for each data-set. Econometricians are fallible and our data are sometimes imperfect, however, so it is unusual for a first attempt at estimation to be totally problem-free. As a result, two or even more regressions are often necessary to rid an estimation of fairly simple difficulties that perhaps could have been avoided in a world of perfect foresight.

Unfortunately, a beginning researcher usually has little motivation to stop running regressions until he or she likes the way the result looks. If running another regression provides a result with a better fit, why shouldn't one more specification be tested?

The reason is a compelling one. Every time an extra regression is run and a specification choice is made on the basis of fit or statistical significance, the chances of making a mistake of inference increase dramatically. This can happen in at least two ways:

1. If you consistently drop a variable when its coefficient is insignificant but keep it when it is significant, it can be shown, as discussed in Section 6.4, that you bias your estimates of the coefficients of the equation and of the t-scores.

2. If you choose to use a lag structure, or a functional form, or an estimation procedure other than OLS on the basis of fit[1] rather than on the basis of previously theorized hypotheses, you run the risk that your equation will work poorly when it's applied to data outside your sample. If you restructure your equation to work well on one data-set, you might decrease the chance of it working well on another.

What might be thought of as "ethical econometrics" is also in reality "good econometrics." That is, the real reason to avoid running too many different specifications is that the fewer regressions you run, the more reliable and more consistently trustworthy are your results. The instance in which professional ethics come into play is when a number of changes are made (different variables, lag structures, functional forms, estimation procedures, data-sets, dropped outliers, and so on), but the regression results are presented to colleagues, clients, editors, or journals as if the final and best equation had been the first and only one estimated. Our recommendation is that all estimated equations be reported even if footnotes or an appendix have to be added to the documentation.

We think that there are two reasonable goals for econometricians when estimating models:

1. Run as few different specifications as possible while still attempting to avoid the major econometric problems.[2]
2. Report honestly the number and type of different specifications estimated so that readers of the research can evaluate how much weight to give to your results.

Therefore, the art of econometrics boils down to attempting to find the best possible equation in the fewest possible number of regression

1. Choices on the basis of fit are only valid if two different data-sets are used. One data-set is used to develop hypotheses by testing the fits of various specifications (this is often called "scanning"). The second data-set then is used to test these hypotheses by estimating the equation involved. The use of such dual data-sets is easiest when there is a plethora of data. This is sometimes the case in cross-sectional research projects but rarely for time-series research. If the same data-set is used both to scan competing hypotheses and also to test them, our typical statistical tests have little meaning in the second use because the researcher knows ahead of time what the result will be.

2. The only exception to our recommendation to run as few specifications as possible is an advanced technique called "sensitivity analysis." Sensitivity analysis consists of purposely altering specifications to ensure that particular results are "robust." (Not statistical flukes). Researchers who use sensitivity analysis report all specifications estimated and tend to discount a result that appears significant in some specifications and insignificant in others.

runs. Only careful thinking and reading before the first regression can bring this about. An ethical econometrician is honest and complete in reporting the different specifications and/or data-sets used.

11.4 The Passbook Deposits Interactive Regression Learning Exercise

This section contains an interactive exercise similar in format to the SAT interactive exercise in Section 8.7. If you worked through that exercise and are familiar with the procedures involved, skip directly to Section 11.4.1.

Econometrics is difficult to learn by reading examples, no matter how expertly done and explained they are. Most econometricians, the authors included, had trouble understanding how to use econometrics, particularly in the area of specification choice, until they had a chance to run their own regression projects. There is an element of econometric understanding that is better learned by doing than by reading about someone else's doing.

Mastering the art of econometrics by running your own regression projects without feedback is also difficult, unfortunately, because it takes quite a while to consistently avoid some fairly simple mistakes. Probably the best way to learn is to work on your own regression project, analyzing your own problems and making your own decisions, but with a more experienced econometrician nearby to give you one-on-one feedback on exactly which of your decisions were inspired and which were flawed (and why).

This section is an attempt to give the reader the opportunity to make independent specification decisions and to then get feedback on the advantages or disadvantages of those decisions. Using the interactive learning exercise of this section requires neither a computer nor a tutor, although either would certainly be useful. Instead, we've designed an example that can be used on its own to help bridge the gap between the typical econometrics examples (which require no decision making) and the typical econometrics projects (which give little feedback). Two more interactive learning exercises are in the appendices to this chapter.

STOP!

In order to get the most out of the example, *you should not read any portion of this section until you are ready to work through at least one specification, if not the entire exercise.* Reading the pages "in order" as with any other example will waste your time because, once you have seen even a few of the results, the benefits to you of making specification decisions will diminish. *Until you are ready to follow the instructions of the example, you should stop now and come back to this section later.* The same warning applies to the interactive learning exercises in Sections 11.6 and 11.7.

11.4.1 Building a Model of U.S. Savings and Loan Passbook Deposits

For this interactive learning exercise, let's assume that you've been hired by the President to solve the savings and loan crisis. As a first step, you decide to build a model of how the industry worked before the crisis, so your dependent variable is total deposits in passbook accounts (savings accounts) in Savings and Loan Associations (S & Ls) in the U.S. The data are quarterly for the 1970s, so there are 40 observations. For each quarter, the dependent variable, QDPASS, measures the quarterly (hence Q) aggregate current (nominal) dollars on deposit (D) in passbook (PASS) accounts in S & Ls in the U.S. Deposits in checking accounts, money market accounts, certificates of deposit, NOW accounts, commercial bank accounts, or brokerage house accounts are not included. Instead, QDPASS just measures the stock of deposits in traditional savings accounts in S & Ls. (The term "passbook" comes from the small books that often document such accounts.)

What is the basic financial theory on which a model of passbook deposits can be built? What kinds of variables should be considered for inclusion as explanatory variables? It's important to note that while they are called savings accounts by many people, passbook accounts should not be considered a measure of aggregate savings. Instead, the stock of passbook deposits is only one component of wealth. As such, the main theoretical work to be done before choosing variables should investigate how people decide what portion of their assets to keep in passbook accounts. The different levels of sophistication of such models in the literature are impressive, and any reader interested in reviewing a portion of the applicable articles should do so now before continuing on with the section.[3]

3. Background articles include Peter Fortune, "The Effectiveness of Recent Policies to Maintain Thrift-Deposit Flows," *Journal of Money, Credit and Banking,* Aug. 1975, pp. 297–316, and A. Thomas King, "Thrift Institution Deposits: The Influence of MMCs and MMMFs," *Journal of Money, Credit and Banking,* Aug. 1984, pp. 328–333. See also a number of excellent articles in Edward M. Gramlich and Dwight M. Jaffee, eds., *Savings Deposits, Mortgages, and Housing* (Lexington, Mass: Lexington Books, 1972).

If you were going to build a single-equation linear model of passbook deposits, what factors should you consider? During any quarter, new deposits can be generated by individuals wishing to save a portion of their income in that quarter, so any reasonable specification ought to include some measure of income or wealth.

Two such variables are quarterly disposable income in the U.S. (QYDUS) and permanent income (QYPERM), defined below. Passbook savings accounts bear a lower rate of interest and are more liquid than almost any other financial asset except demand deposits (checking accounts), so there is some reason to believe that people might treat passbook holdings as interim transaction accounts. If this is true, then the income/wealth variables should be quite important.

A second set of important factors refers to competition from other assets and includes the interest rate on passbook accounts (QRDPASS), the interest rate on three-month treasury bills (QRTB3Y), and a dummy variable (MMCDUM) equal to zero before the legalization in 1978 of money market certificates account paying higher "money market" rates. In addition, a fourth variable, equal to the difference (SPREAD) between the two interest rate variables, can also be specified.

A third set of factors measures the environment in which the deposits are operating. If high inflation is expected, for example, holdings in all low-interest accounts might fall, so an "adaptive expectations" measure of expected inflation (EXPINF) is a possible explanatory variable. This measures last quarter's level of inflation; it's based on the theory that individuals form their expectation of future inflation rates by extrapolating from past inflation rates. In addition, the convenience of depositing funds may be relevant, so the total number of branches of S & Ls open nationwide (BRANCH) is also a possible variable.

To sum, the variables available for your model are:

$QDPASS_t$ — the aggregate stock of deposits held in passbook accounts in S & Ls in the U.S. in quarter t (millions of nominal dollars)

$QYDUS_t$ — U.S. disposable income in quarter t (millions of nominal dollars)

$QYPERM_t$ — U.S. "permanent" income in quarter t (millions of nominal dollars) (This variable was formed by taking a four-quarter declining weighted moving average of disposable income in previous quarters.)

$QRDPASS_t$ — the average rate of return (in percentage points) on passbook accounts in S & Ls in quarter t

$QRTB3Y_t$	the interest rate on three-month treasury bills in quarter t
$SPREAD_t$	$QRDPASS_t - QRTB3Y_t$
$MMCDUM_t$	a dummy variable equal to zero before the third-quarter 1978 legalization of money market certificates and equal to one thereafter
$EXPINF_t$	the expected percentage rate of inflation in quarter t (equal to the previous quarter's inflation rate)
$BRANCH_t$	the number of S & L branches operating in the U.S. in quarter t

The data for these variables are contained in Table 11.3, but a few comments are in order. First, the data are quarterly from 1970 through 1979. More recent data are available, but the deregulation of the financial services industry makes that data not easily comparable to the data from the 1970s. Second, the interest rate on passbook accounts was controlled by "Regulation Q" and did not change much during the 1970s. In particular, QRDPASS was 5 percent until the fourth quarter of 1973, when it changed to 5.25 percent, where it remained until the third quarter of 1979 when it rose to 5.50 percent. Third, two of the variables were formed using data from previous quarters. QYPERM, a measure of "permanent" income, is a function of the four previous quarters' disposable income; this gives more weight to the last quarter than to quarters a year ago. EXPINF uses a simplified adaptive expectations model of the formation of inflationary expectations by using last quarter's inflation rate (specifically, the percentage rate of change of the fixed-weight GNP deflator) as the rate that will be expected. Finally, data on S & L branches (technically, total facilities) were not available for every quarter. While a smooth series could have been formed by interpolating between known observations to calculate the missing observations, we simply used the most recent known value as a proxy for the missing observations. As a result of the made-up nature of some of the observations of this variable, it will be potentially unreliable.

Now:

1. Hypothesize expected signs for the coefficients of all these variables in an equation for passbook deposits in the 1970s. Examine each variable carefully; what is the economic content of your hypothesis?
2. Choose carefully the best set of explanatory variables. Assume every model should have QYDUS and either the two interest rate variables or the spread variable. Don't simply include all the vari-

TABLE 11.3 DATA FOR THE PASSBOOK DEPOSITS INTERACTIVE LEARNING EXERCISE

OBSERV	QDPASS	QYDUS	QYPERM	BRANCH	QRTB3Y	EXPINF
1970:1	84312	671.5	646.39	8498	7.501241	4.8
1970:2	83141	692.4	659.41	8498	6.964994	5.8
1970:3	82754	705.8	675.30	8498	6.569049	4.9
1970:4	84120	711.5	690.33	8372	5.507346	3.5
1971:1	85525	732.7	701.97	8722	3.955616	5.9
1971:2	89286	749.3	716.93	8722	4.310242	5.7
1971:3	90618	757.6	732.41	8722	5.186687	5.1
1971:4	92310	767.4	745.52	8862	4.339589	4.1
1972:1	95112	782.2	757.37	8862	3.513554	3.2
1972:2	97361	794.5	769.55	8862	3.836743	5.4
1972:3	99317	815.6	781.70	8862	4.346840	2.6
1972:4	101634	849.0	797.77	9584	4.979777	3.6
1973:1	103884	878.9	821.40	9977	5.800682	4.8
1973:2	104905	903.5	848.83	9977	6.813935	6.3
1973:3	102782	925.3	876.43	10547	8.689076	7.2
1973:4	103231	950.3	901.85	10547	7.710729	7.7
1974:1	104492	963.9	926.30	11324	7.856835	8.6
1974:2	104675	988.6	946.06	11324	8.561761	10.8
1974:3	102871	1012.7	967.20	12123	8.581159	11.2
1974:4	104386	1028.1	989.47	12123	7.578421	12.4
1975:1	109321	1035.2	1009.16	13056	6.044646	12.6
1975:2	114970	1105.2	1023.91	13056	5.551466	8.2
1975:3	117018	1109.4	1059.53	13839	6.529253	6.6
1975:4	119006	1134.5	1085.17	13839	5.847297	7.8
1976:1	124515	1163.7	1111.18	14479	5.085805	6.6
1976:2	127351	1180.8	1138.23	14479	5.309831	4.2
1976:3	129752	1203.3	1159.27	15108	5.309831	4.4
1976:4	132282	1229.6	1181.75	15108	4.820496	5.5
1977:1	136636	1255.2	1205.36	15835	4.743668	6.7
1977:2	139378	1291.9	1229.70	15835	4.956230	7.4
1977:3	142246	1335.5	1259.57	16439	5.625816	6.6
1977:4	143466	1373.5	1295.77	16439	6.301799	5.1
1978:1	146005	1405.7	1333.95	16955	6.603971	7.2
1978:2	144368	1451.3	1370.62	16955	6.680457	6.8
1978:3	141777	1496.2	1410.48	17492	7.556666	9.6
1978:4	134460	1542.7	1452.36	17492	8.998331	8.3
1979:1	130017	1587.5	1496.77	18100	9.717486	8.9
1979:2	129480	1624.0	1542.18	18100	9.737695	9.9
1979:3	125824	1674.3	1584.01	18676	10.00877	9.5
1979:4	116100	1714.9	1628.69	18676	12.33570	10.0

ables, intending to drop the insignificant ones. Instead, think through the problem carefully and find the best possible equation you can. For example, using SPREAD instead of both of the rate variables imposes a constraint on the interest rate coefficients; what is that constraint? Does the constraint make sense in our case? Why or why not? Or what about the permanent income hypothesis? Does it mean we should include QYPERM? Why or why not?

Once you have specified your equation, you're ready to begin the interactive example in Section 11.4.2. Keep following the instructions in the example until you've completely specified your equation and have been instructed to go to the discussion of serial correlation and heteroskedasticity in Section 11.4.3. You may take some time to think over the questions contained in Section 11.4.2 or take a break, but when you return to the interactive example, make sure to return to the exact point from which you left rather than starting over again. To the extent you can do it, you should avoid looking at the hints until after completing the entire project. The hints are there to help you if you get stuck, not to check every decision you make.

One final bit of advice: Take the time to answer all the questions. Rushing through this interactive example will lessen its effectiveness.

11.4.2 The Passbook Deposits Interactive Regression Exercise

Start by turning to the regression result that corresponds to the specification you would like to estimate (note that the simple correlation coefficient matrix for this data-set is in Table 11.4 just before the results begin). To find your regression result, carefully follow these instructions:

All the equations include QDPASS as the dependent variable, QYDUS as one of the explanatory variables, and either SPREAD or both interest rate variables (QRDPASS and QRTB3Y) as explanatory variables. If you chose SPREAD as one of your variables, continue to question number 1; if you instead chose both interest rate variables, jump to question number 4 below.

1. Do you want to include QYPERM? If yes, continue to question number 2; if no, skip to question number 3.
2. MMCDUM is in all QYPERM equations. Do you want also to include EXPINF? If yes, go to regression 11.9; if no, go to regression 11.10.
3. Find below the combination of explanatory variables (from EXPINF, BRANCH, and MMCDUM) that you wish to include and go to the indicated regression:

None of them, go to regression 11.8
MMCDUM only, go to regression 11.4
EXPINF only, go to regression 11.7
BRANCH only, go to regression 11.6
MMCDUM and BRANCH, go to regression 11.2
MMCDUM and EXPINF, go to regression 11.3
BRANCH and EXPINF, go to regression 11.5
All three, go to regression 11.1

4. Do you want to include QYPERM? If yes, continue to question number 5; if no, skip to question number 6.

5. MMCDUM is in all QYPERM equations. Do you want also to include EXPINF? If yes, go to regression 11.17; if no, go to regression 11.18.

6. Find below the combination of explanatory variables (from EXPINF, BRANCH, and MMCDUM) that you wish to include and go to the indicated regression:

None of them, go to regression 11.16
MMCDUM only, go to regression 11.12
EXPINF only, go to regression 11.15
BRANCH only, go to regression 11.14
MMCDUM and BRANCH, go to regression 11.20
MMCDUM and EXPINF, go to regression 11.11
BRANCH and EXPINF, go to regression 11.13
All three, go to regression 11.19

TABLE 11.4 MEANS, VARIANCES, AND SIMPLE CORRELATION COEFFICIENTS FOR THE PASSBOOK DEPOSITS INTERACTIVE LEARNING EXERCISE

Means, Variances, and Correlations

Sample Range: 1-40

Variable	Mean	Standard Dev	Variance
QDPASS	113017.2	19655.37	3.8633E + 08
QYDUS	1089.880	303.5133	92120.32
QYPERM	1038.246	284.1240	80726.44
BRANCH	12824.10	3521.420	1.24004E + 07
QRDPASS	5.175000	0.139194	0.019375
QRTB3Y	6.509237	1.944990	3.782986
SPREAD	−1.334237	1.861988	3.466998
EXPINF	6.887500	2.476710	6.134094
MMCDUM	0.150000	0.357071	0.127500

	Correlation Coeff		Correlation Coeff
QYPERM,QYDUS	0.999	QYDUS,QDPASS	0.867
BRANCH,QDPASS	0.911	QYPERM,QDPASS	0.867
BRANCH,QYPERM	0.987	BRANCH,QYDUS	0.985
QRDPASS,QDPASS	0.694	QRDPASS,QYDUS	0.836
QRDPASS,QYPERM	0.841	QRDPASS,BRANCH	0.844
QRTB3Y,QYDUS	0.565	QRTB3Y,QDPASS	0.191
QRTB3Y,BRANCH	0.488	QRTB3Y,QYPERM	0.560
SPREAD,QYDUS	−0.528	QRTB3Y,QRDPASS	0.619
SPREAD,BRANCH	−0.447	SPREAD,QDPASS	−0.148
SPREAD,QRTB3Y	−0.998	SPREAD,QYPERM	−0.522
EXPINF,QDPASS	0.270	SPREAD,QRDPASS	−0.572
EXPINF,QYPERM	0.489	EXPINF,QYDUS	0.486
EXPINF,QRDPASS	0.660	EXPINF,BRANCH	0.472
EXPINF,SPREAD	−0.634	EXPINF,QRTB3Y	0.654
MMCDUM,QDPASS	0.354	MMCDUM,QYDUS	0.715
MMCDUM,QYPERM	0.710	MMCDUM,BRANCH	0.628
MMCDUM,QRDPASS	0.477	MMCDUM,QRTB3Y	0.694
MMCDUM,SPREAD	−0.689	MMCDUM,EXPINF	0.420

Regression Run 11.1

```
ORDINARY LEAST SQUARES                    DEPENDENT VARIABLE IS QDPASS
SAMPLE RANGE:     1970:1-1979:4

                      COEFFICIENT        STANDARD ERROR         T-SCORE
      CONST            34318.26            4577.597            7.497004
      QYDUS            37.71483            25.19174            1.497111
      SPREAD           2085.763            759.5224            2.746150
      MMCDUM          -16836.48            4539.872           -3.708579
      BRANCH           3.621914            1.966156            1.842129
      EXPINF          -514.7155            457.7060           -1.124555

   R-squared               0.941624      Mean of depend var       113017.2
   Adjusted R-squared      0.933039      Std dev depend var       19905.77
   Std err of regress      5150.947      Residual sum          -5.96629E-10
   Durbin Watson stat      0.573681      Sum squared resid      9.02096E+08
   F Statistic             109.6873
```

Answer each of the following questions for the above regression run.

a. Evaluate this result with respect to its economic meaning, overall fit, and the signs and significance of the individual coefficients.

b. What econometric problems does this regression have? Why? If you need feedback on your answer, see hint #17 in the material on this chapter in Appendix A.

c. Which of the following statements comes closest to your recommendation for further action to be taken in the estimation of this equation?

- **i.** No further specification changes are advisable (go to Section 11.4.3).
- **ii.** No further variable changes are advisable, but I am concerned about heteroskedasticity or serial correlation (go to Section 11.4.3).
- **iii.** I would like to drop BRANCH from the equation (go to run #11.3).
- **iv.** I would like to drop EXPINF from the equation (go to run #11.2).
- **v.** I would like to change away from the spread interest rate formulation (go to run #11.19).

If you need feedback on your answer, see hint #18 in the material on this chapter in Appendix A.

Regression Run 11.2

```
ORDINARY LEAST SQUARES                  DEPENDENT VARIABLE IS QDPASS
SAMPLE RANGE:      1970:1-1979:4

                    COEFFICIENT        STANDARD ERROR          T-SCORE
     CONST           32355.77            4247.861             7.616956
     QYDUS           42.54051            24.91723             1.707273
     SPREAD          2563.729            631.8402             4.057559
     MMCDUM         -16532.77            4548.932            -3.634429
     BRANCH          3.134560            1.925036             1.628313

   R-squared                0.939453       Mean of depend var      113017.2
   Adjusted R-squared       0.932533       Std dev depend var      19905.77
   Std err of regress       5170.382       Residual sum            1.09139E-10
   Durbin Watson stat       0.524571       Sum squared resid       9.35650E+08
   F Statistic              135.7664
```

Answer each of the following questions for the above regression run.

a. Evaluate this result with respect to its economic meaning, overall fit, and the signs and significance of the individual coefficients.

b. What econometric problems does this regression have? Why? If you need feedback on your answer, see hint #17 in the material on this chapter in Appendix A.

c. Which of the following statements comes closest to your recommendation for further action to be taken in the estimation of this equation?

  i. No further specification changes are advisable (go to Section 11.4.3).

  ii. No further variable changes are advisable, but I am concerned about heteroskedasticity or serial correlation (go to Section 11.4.3).

  iii. I would like to drop BRANCH from the equation (go to run #11.4).

  iv. I would like to add EXPINF to the equation (go to run #11.1).

  v. I would like to change away from the spread interest rate formulation (go to run #11.20).

If you need feedback on your answer, see hint #18 in the material on this chapter in Appendix A.

Regression Run 11.3

```
ORDINARY LEAST SQUARES                     DEPENDENT VARIABLE IS QDPASS
SAMPLE RANGE:     1970:1-1979:4

                        COEFFICIENT      STANDARD ERROR       T-SCORE
     CONST               30836.71          4309.422           7.155648
     QYDUS               83.52150          4.172854           20.01544
     SPREAD              2565.976          737.3738           3.479885
     MMCDUM             -21061.36          4049.704          -5.200717
     EXPINF             -328.8701          461.4616          -0.712670

  R-squared               0.935798     Mean of depend var      113017.2
  Adjusted R-squared      0.928460     Std dev depend var      19905.77
  Std err of regress      5324.155     Residual sum            1.40062E-09
  Durbin Watson stat      0.729307     Sum squared resid       9.92132E+08
  F Statistic             127.5391
```

Answer each of the following questions for the above regression run.

a. Evaluate this result with respect to its economic meaning, overall fit, and the signs and significance of the individual coefficients.

b. What econometric problems does this regression have? Why? If you need feedback on your answer, see hint #17 in the material on this chapter in Appendix A.

c. Which of the following statements comes closest to your recommendation for further action to be taken in the estimation of this equation?
   i. No further specification changes are advisable (go to Section 11.4.3).
   ii. No further variable changes are advisable, but I am concerned about heteroskedasticity or serial correlation (go to Section 11.4.3).
   iii. I would like to drop EXPINF from the equation (go to run #11.4).
   iv. I would like to add BRANCH to the equation (go to run #11.1).
   v. I would like to change away from the spread interest rate formulation (go to run #11.11).

If you need feedback on your answer, see hint #15 in the material on this chapter in Appendix A.

Regression Run 11.4

```
ORDINARY LEAST SQUARES                      DEPENDENT VARIABLE IS QDPASS
SAMPLE RANGE:      1970:1-1979:4

                          COEFFICIENT       STANDARD ERROR        T-SCORE
      CONST                29833.37           4045.035            7.375306
      QYDUS                82.62301           3.950564            20.91423
      SPREAD               2843.567           621.8082            4.573061
      MMCDUM              -20475.63           3938.230           -5.199196

   R-squared                0.934866     Mean of depend var      113017.2
   Adjusted R-squared       0.929438     Std dev depend var      19905.77
   Std err of regress       5287.641     Residual sum            4.58385E-10
   Durbin Watson stat       0.676536     Sum squared resid       1.00653E+09
   F Statistic              172.2372
```

Answer each of the following questions for the above regression run.

a. Evaluate this result with respect to its economic meaning, overall fit, and the signs and significance of the individual coefficients.

b. What econometric problems does this regression have? Why? If you need feedback on your answer, see hint #13 in the material on this chapter in Appendix A.

c. Which of the following statements comes closest to your recommendation for further action to be taken in the estimation of this equation?

   i. No further specification changes are advisable (go to Section 11.4.3).

   ii. No further variable changes are advisable, but I am concerned about heteroskedasticity or serial correlation (go to Section 11.4.3).

   iii. I would like to add BRANCH to the equation (go to run #11.2).

   iv. I would like to add EXPINF to the equation (go to run #11.3).

   v. I would like to change away from the spread interest rate formulation (go to run #11.12).

If you need feedback on your answer, see hint #15 in the material on this chapter in Appendix A.

Regression Run 11.5

```
ORDINARY LEAST SQUARES                    DEPENDENT VARIABLE IS QDPASS
SAMPLE RANGE:     1970:1-1979:4

                         COEFFICIENT         STANDARD ERROR         T-SCORE
      CONST               44000.71              4392.003            10.01837
      QYDUS              -16.54446              23.95441           -0.690664
      SPREAD              2840.254              854.7564            3.322881
      BRANCH              7.305554              1.981996            3.685958
      EXPINF             -413.7377              533.6858           -0.775246

   R-squared               0.918010         Mean of depend var      113017.2
   Adjusted R-squared      0.908640         Std dev depend var      19905.77
   Std err of regress      6016.666         Residual sum           -3.96540E-
   Durbin Watson stat      0.319973         Sum squared resid       1.26701E+09
   F Statistic             97.97121
```

Answer each of the following questions for the above regression run.

a. Evaluate this result with respect to its economic meaning, overall fit, and the signs and significance of the individual coefficients.

b. What econometric problems does this regression have? Why? If you need feedback on your answer, see hint #12 in the material on this chapter in Appendix A.

c. Which of the following statements comes closest to your recommendation for further action to be taken in the estimation of this equation?

- **i.** No further specification changes are advisable (go to Section 11.4.3).
- **ii.** No further variable changes are advisable, but I am concerned about heteroskedasticity or serial correlation (go to Section 11.4.3).
- **iii.** I would like to drop EXPINF from the equation (go to run #11.6).
- **iv.** I would like to add MMCDUM to the equation (go to run #11.1).
- **v.** I would like to change away from the spread interest rate formulation (go to run #11.13).

If you need feedback on your answer, see hint #8 in the material on this chapter in Appendix A.

Regression Run 11.6

```
ORDINARY LEAST SQUARES                         DEPENDENT VARIABLE IS QDPASS
SAMPLE RANGE:      1970:1-1979:4

                    COEFFICIENT        STANDARD ERROR         T-SCORE
     CONST           42276.73            3766.360             11.22482
     QYDUS          -11.86216            23.05168            -0.514590
     SPREAD          3214.838            701.1670             4.584981
     BRANCH          6.858817            1.885838             3.637013

  R-squared             0.916602       Mean of depend var      113017.2
  Adjusted R-squared    0.909653       Std dev depend var      19905.77
  Std err of regress    5983.231       Residual sum           -6.25732E-10
  Durbin Watson stat    0.337198       Sum squared resid       1.28877E+09
  F Statistic           131.8897
```

Answer each of the following questions for the above regression run.

a. Evaluate this result with respect to its economic meaning, overall fit, and the signs and significance of the individual coefficients.

b. What econometric problems does this regression have? Why? If you need feedback on your answer, see hint #12 in the material on this chapter in Appendix A.

c. Which of the following statements comes closest to your recommendation for further action to be taken in the estimation of this equation?

i. No further specification changes are advisable (go to Section 11.4.3).

ii. No further variable changes are advisable, but I am concerned about heteroskedasticity or serial correlation (go to Section 11.4.3).

iii. I would like to drop BRANCH from the equation (go to run #11.8).

iv. I would like to add MMCDUM to the equation (go to run #11.2).

v. I would like to change away from the spread interest rate formulation (go to run #11.14).

If you need feedback on your answer, see hint #8 in the material on this chapter in Appendix A.

Regression Run 11.7

```
ORDINARY LEAST SQUARES                    DEPENDENT VARIABLE IS QDPASS
SAMPLE RANGE:     1970:1-1979:4

                          COEFFICIENT       STANDARD ERROR        T-SCORE
      CONST               41152.02            5022.709           8.193192
      QYDUS               70.63577            4.408044           16.02429
      SPREAD              4653.520            812.0529           5.730563
      EXPINF              158.1950            593.2161           0.266673

   R-squared               0.886184        Mean of depend var      113017.2
   Adjusted R-squared      0.876699        Std dev depend var      19905.77
   Std err of regress      6989.748        Residual sum            1.09867E-09
   Durbin Watson stat      0.293023        Sum squared resid       1.75884E+09
   F Statistic             93.43341
```

Answer each of the following questions for the above regression run.

a. Evaluate this result with respect to its economic meaning, overall fit, and the signs and significance of the individual coefficients.

b. What econometric problems does this regression have? Why? If you need feedback on your answer, see hint #12 in the material on this chapter in Appendix A.

c. Which of the following statements comes closest to your recommendation for further action to be taken in the estimation of this equation?
   i. No further specification changes are advisable (go to Section 11.4.3).
   ii. No further variable changes are advisable, but I am concerned about heteroskedasticity or serial correlation (go to Section 11.4.3).
   iii. I would like to drop EXPINF from the equation (go to run #11.8).
   iv. I would like to add MMCDUM to the equation (go to run #11.3).
   v. I would like to change away from the spread interest rate formulation (go to run #11.15).

If you need feedback on your answer, see hint #8 in the material on this chapter in Appendix A.

Regression Run 11.8

```
ORDINARY LEAST SQUARES                        DEPENDENT VARIABLE IS QDPASS
SAMPLE RANGE:    1970:1-1979:4

                         COEFFICIENT       STANDARD ERROR        T-SCORE
     CONST                41799.30           4341.727            9.627344
     QYDUS                70.90675           4.235142            16.74247
     SPREAD               4543.382           690.3493            6.581281

  R-squared                0.885959        Mean of depend var      113017.2
  Adjusted R-squared       0.879794        Std dev depend var      19905.77
  Std err of regress       6901.451        Residual sum            3.00133E-10
  Durbin Watson stat       0.280912        Sum squared resid       1.76231E+09
  F Statistic              143.7227
```

Answer each of the following questions for the above regression run.

a. Evaluate this result with respect to its economic meaning, overall fit, and the signs and significance of the individual coefficients.

b. What econometric problems does this regression have? Why? If you need feedback on your answer, see hint #14 in the material on this chapter in Appendix A.

c. Which of the following statements comes closest to your recommendation for further action to be taken in the estimation of this equation?
   i. No further specification changes are advisable (go to Section 11.4.3).
   ii. No further variable changes are advisable, but I am concerned about heteroskedasticity or serial correlation (go to Section 11.4.3).
   iii. I would like to add EXPINF to the equation (go to run #11.7).
   iv. I would like to add MMCDUM to the equation (go to run #11.4).
   v. I would like to change away from the spread interest rate formulation (go to run #11.16).

If you need feedback on your answer, see hint #8 in the material on this chapter in Appendix A.

Regression Run 11.9

```
ORDINARY LEAST SQUARES                    DEPENDENT VARIABLE IS QDPASS
SAMPLE RANGE:     1970:1-1979:4

                         COEFFICIENT      STANDARD ERROR         T-SCORE
      CONST               32416.73          4288.033            7.559814
      QYDUS               224.5774          81.32013            2.761647
      QYPERM             -150.3951          86.59623           -1.736739
      SPREAD              2859.399          736.6466            3.881644
      MMCDUM             -21205.30          3938.738           -5.383780
      EXPINF             -147.9428          460.6521           -0.321159

   R-squared                0.941029     Mean of depend var      113017.2
   Adjusted R-squared       0.932357     Std dev depend var      19905.77
   Std err of regress       5177.123     Residual sum            1.30967E-09
   Durbin Watson stat       0.995422     Sum squared resid       9.11288E+08
   F statistic              108.5123
```

Answer each of the following questions for the above regression run.

a. Evaluate this result with respect to its economic meaning, overall fit, and the signs and significance of the individual coefficients.

b. What econometric problems does this regression have? Why? If you need feedback on your answer, see hint #1 in the material on this chapter in Appendix A.

c. Which of the following statements comes closest to your recommendation for further action to be taken in the estimation of this equation?

  i. No further specification changes are advisable (go to Section 11.4.3).

  ii. No further variable changes are advisable, but I am concerned about heteroskedasticity or serial correlation (go to Section 11.4.3).

  iii. I would like to drop QYPERM from the equation (go to run #11.3).

  iv. I would like to drop EXPINF from the equation (go to run #11.10).

  v. I would like to change away from the spread interest rate formulation (go to run #11.17).

If you need feedback on your answer, see hint #3 in the material on this chapter in Appendix A.

Regression Run 11.10

```
ORDINARY LEAST SQUARES                      DEPENDENT VARIABLE IS QDPASS
SAMPLE RANGE:      1970:1-1979:4

                        COEFFICIENT        STANDARD ERROR          T-SCORE
     CONST               32054.54            4083.725             7.849338
     QYDUS               230.0929            78.46106             2.932574
     QYPERM             -156.6846            83.26498            -1.881759
     SPREAD              2990.158            605.9888             4.934346
     MMCDUM             -20961.30            3814.933            -5.494540

  R-squared               0.940850      Mean of depend var       113017.2
  Adjusted R-squared      0.934090      Std dev depend var       19905.77
  Std err of regress      5110.362      Residual sum           -6.03904E-10
  Durbin Watson stat      0.983859      Sum squared resid       9.14053E+08
  F Statistic             139.1810
```

Answer each of the following questions for the above regression run.

a. Evaluate this result with respect to its economic meaning, overall fit, and the signs and significance of the individual coefficients.

b. What econometric problems does this regression have? Why? If you need feedback on your answer, see hint #1 in the material on this chapter in Appendix A.

c. Which of the following statements comes closest to your recommendation for further action to be taken in the estimation of this equation?
   i. No further specification changes are advisable (go to Section 11.4.3).
   ii. No further variable changes are advisable, but I am concerned about heteroskedasticity or serial correlation (go to Section 11.4.3).
   iii. I would like to drop QYPERM from the equation (go to run #11.4).
   iv. I would like to add EXPINF to the equation (go to run #11.9).
   v. I would like to change away from the spread interest rate formulation (go to run #11.18).

If you need feedback on your answer, see hint #3 in the material on this chapter in Appendix A.

Regression Run 11.11

```
ORDINARY LEAST SQUARES                    DEPENDENT VARIABLE IS QDPASS
SAMPLE RANGE:     1970:1-1979:4

                          COEFFICIENT        STANDARD ERROR        T-SCORE
     CONST                 184591.2            68093.77            2.710838
     QYDUS                 97.20543            7.223235            13.45733
     QRDPASS              -31233.99            14958.53           -2.088039
     QRTB3Y               -1987.620            742.8786           -2.675565
     MMCDUM               -26680.35            4565.580           -5.843803
     EXPINF                153.8465            485.8750            0.316638

  R-squared                 0.944196       Mean of depend var      113017.2
  Adjusted R-squared        0.935990       Std dev depend var      19905.77
  Std err of regress        5036.200       Residual sum          -7.76708E-10
  Durbin Watson stat        1.038720       Sum squared resid      8.62353E+08
  F Statistic               115.0559
```

Answer each of the following questions for the above regression run.

a. Evaluate this result with respect to its economic meaning, overall fit, and the signs and significance of the individual coefficients.

b. What econometric problems does this regression have? Why? If you need feedback on your answer, see hint #2 in the material on this chapter in Appendix A.

c. Which of the following statements comes closest to your recommendation for further action to be taken in the estimation of this equation?
   i. No further specification changes are advisable (go to Section 11.4.3).
   ii. No further variable changes are advisable, but I am concerned about heteroskedasticity or serial correlation (go to Section 11.4.3).
   iii. I would like to drop EXPINF from the equation (go to run #11.12).
   iv. I would like to add BRANCH to the equation (go to run #11.19).
   v. I would like to change to the spread interest rate formulation (go to run #11.3).

If you need feedback on your answer, see hint #5 in the material on this chapter in Appendix A.

Regression Run 11.12

```
ORDINARY LEAST SQUARES                        DEPENDENT VARIABLE IS QDPASS
SAMPLE RANGE:     1970:1-1979:4

                       COEFFICIENT        STANDARD ERROR        T-SCORE
      CONST             175517.2            60970.86            2.878707
      QYDUS             96.70339            6.955899            13.90236
      QRDPASS          -29260.79            13422.73           -2.179944
      QRTB3Y           -1918.369            700.7701           -2.737516
      MMCDUM           -26556.05            4489.825           -5.914718

   R-squared             0.944031         Mean of depend var       113017.2
   Adjusted R-squared    0.937635         Std dev depend var       19905.77
   Std err of regress    4971.046         Residual sum             3.20142E-10
   Durbin Watson stat    1.034733         Sum squared resid        8.64896E+08
   F Statistic           147.5888
```

Answer each of the following questions for the above regression run.

a. Evaluate this result with respect to its economic meaning, overall fit, and the signs and significance of the individual coefficients.

b. What econometric problems does this regression have? Why? If you need feedback on your answer, see hint #2 in the material on this chapter in Appendix A.

c. Which of the following statements comes closest to your recommendation for further action to be taken in the estimation of this equation?
   i. No further specification changes are advisable (go to Section 11.4.3).
   ii. No further variable changes are advisable, but I am concerned about heteroskedasticity or serial correlation (go to Section 11.4.3).
   iii. I would like to add BRANCH to the equation (go to run #11.20).
   iv. I would like to add EXPINF to the equation (go to run #11.11).
   v. I would like to change to the spread interest rate formulation (go to run #11.4).

If you need feedback on your answer, see hint #6 in the material on this chapter in Appendix A.

Regression Run 11.13

```
ORDINARY LEAST SQUARES                      DEPENDENT VARIABLE IS QDPASS
SAMPLE RANGE:     1970:1-1979:4

                  COEFFICIENT         STANDARD ERROR        T-SCORE
      CONST        75571.63             75526.37            1.000599
      QYDUS       -18.30103             24.60198           -0.743884
      QRDPASS     -3918.494             16163.94           -0.242422
      QRTB3Y      -2756.596             887.7795           -3.105046
      BRANCH       7.625268             2.146174            3.552960
      EXPINF      -315.6677             588.6806           -0.536229

   R-squared              0.918431     Mean of depend var      113017.2
   Adjusted R-squared     0.906435     Std dev depend var      19905.77
   Std err of regress     6088.825     Residual sum          -1.48430E-09
   Durbin Watson stat     0.317306     Sum squared resid      1.26051E+09
   F Statistic            76.56534
```

Answer each of the following questions for the above regression run.

a. Evaluate this result with respect to its economic meaning, overall fit, and the signs and significance of the individual coefficients.

b. What econometric problems does this regression have? Why? If you need feedback on your answer, see hint #7 in the material on this chapter in Appendix A.

c. Which of the following statements comes closest to your recommendation for further action to be taken in the estimation of this equation?
   i. No further specification changes are advisable (go to Section 11.4.3).
   ii. No further variable changes are advisable, but I am concerned about heteroskedasticity or serial correlation (go to Section 11.4.3).
   iii. I would like to drop EXPINF from the equation (go to run #11.14).
   iv. I would like to add MMCDUM to the equation (go to run #11.19).
   v. I would like to change to the spread interest rate formulation (go to run #11.5).

If you need feedback on your answer, see hint #10 in the material on this chapter in Appendix A.

Regression Run 11.14

```
ORDINARY LEAST SQUARES                    DEPENDENT VARIABLE IS QDPASS
SAMPLE RANGE:     1970:1-1979:4

                      COEFFICIENT        STANDARD ERROR        T-SCORE
     CONST             90549.10            69453.96            1.303728
     QYDUS            -16.18897            24.03620           -0.673524
     QRDPASS          -7121.353            14866.23           -0.479028
     QRTB3Y           -2954.533            799.1416           -3.697133
     BRANCH            7.501259            2.111849            3.551986

  R-squared              0.917741     Mean of depend var       113017.2
  Adjusted R-squared     0.908340     Std dev depend var       19905.77
  Std err of regress     6026.534     Residual sum          -1.68438E-09
  Durbin Watson stat     0.333103     Sum squared resid      1.27117E+09
  F Statistic            97.62198
```

Answer each of the following questions for the above regression run.

a. Evaluate this result with respect to its economic meaning, overall fit, and the signs and significance of the individual coefficients.

b. What econometric problems does this regression have? Why? If you need feedback on your answer, see hint #7 in the material on this chapter in Appendix A.

c. Which of the following statements comes closest to your recommendation for further action to be taken in the estimation of this equation?
   i. No further specification changes are advisable (go to Section 11.4.3).
   ii. No further variable changes are advisable, but I am concerned about heteroskedasticity or serial correlation (go to Section 11.4.3).
   iii. I would like to drop BRANCH from the equation (go to run #11.16).
   iv. I would like to add MMCDUM to the equation (go to run #11.20).
   v. I would like to change to the spread interest rate formulation (go to run #11.6).

If you need feedback on your answer, see hint #10 in the material on this chapter in Appendix A.

Regression Run 11.15

```
ORDINARY LEAST SQUARES                        DEPENDENT VARIABLE IS QDPASS
SAMPLE RANGE:     1970:1-1979:4

                           COEFFICIENT        STANDARD ERROR         T-SCORE
      CONST                -22324.94            81162.10            -0.275066
      QYDUS                 66.48008            6.911253             9.619107
      QRDPASS               18136.45            17225.33             1.052894
      QRTB3Y               -4662.207            816.5163            -5.709877
      EXPINF               -90.29220            675.4795            -0.133671

   R-squared                0.888146       Mean of depend var       113017.2
   Adjusted R-squared       0.875363       Std dev depend var       19905.77
   Std err of regress       7027.519       Residual sum            -8.25821E-10
   Durbin Watson stat       0.312751       Sum squared resid        1.72851E+09
   F Statistic              69.47733
```

Answer each of the following questions for the above regression run.

a. Evaluate this result with respect to its economic meaning, overall fit, and the signs and significance of the individual coefficients.

b. What econometric problems does this regression have? Why? If you need feedback on your answer, see hint #7 in the material on this chapter in Appendix A.

c. Which of the following statements comes closest to your recommendation for further action to be taken in the estimation of this equation?
   i. No further specification changes are advisable (go to Section 11.4.3).
   ii. No further variable changes are advisable, but I am concerned about heteroskedasticity or serial correlation (go to Section 11.4.3).
   iii. I would like to add BRANCH to the equation (go to run #11.13).
   iv. I would like to add MMCDUM to the equation (go to run #11.11).
   v. I would like to change to the spread interest rate formulation (go to run #11.7).

If you need feedback on your answer, see hint #9 in the material on this chapter in Appendix A.

Regression Run 11.16

```
ORDINARY LEAST SQUARES                    DEPENDENT VARIABLE IS QDPASS
SAMPLE RANGE:     1970:1-1979:4

                         COEFFICIENT      STANDARD ERROR       T-SCORE
      CONST              -17529.78          71804.10          -0.244133
      QYDUS               66.69228          6.634067           10.05300
      QRDPASS             17105.76          15191.61           1.126000
      QRTB3Y             -4710.521          722.1070          -6.523300

   R-squared                0.888089     Mean of depend var      113017.2
   Adjusted R-squared       0.878763     Std dev depend var      19905.77
   Std err of regress       6930.995     Residual sum         -7.38510E-10
   Durbin Watson stat       0.314723     Sum squared resid     1.72939E+09
   F Statistic              95.22845
```

Answer each of the following questions for the above regression run.

a. Evaluate this result with respect to its economic meaning, overall fit, and the signs and significance of the individual coefficients.

b. What econometric problems does this regression have? Why? If you need feedback on your answer, see hint #7 in the material on this chapter in Appendix A.

c. Which of the following statements comes closest to your recommendation for further action to be taken in the estimation of this equation?

- i. No further specification changes are advisable (go to Section 11.4.3).
- ii. No further variable changes are advisable, but I am concerned about heteroskedasticity or serial correlation (go to Section 11.4.3).
- iii. I would like to add BRANCH to the equation (go to run #11.14).
- iv. I would like to add MMCDUM to the equation (go to run #11.12).
- v. I would like to change to the spread interest rate formulation (go to run #11.8).

If you need feedback on your answer, see hint #9 in the material on this chapter in Appendix A.

Regression Run 11.17

```
ORDINARY LEAST SQUARES                    DEPENDENT VARIABLE IS QDPASS
SAMPLE RANGE:     1970:1-1979:4

                     COEFFICIENT       STANDARD ERROR        T-SCORE
      CONST           158889.6           72022.07           2.206123
      QYDUS           185.0812           82.05795           2.255494
      QYPERM         -96.22873           89.51066          -1.075053
      QRDPASS        -25174.00           15953.44          -1.577967
      QRTB3Y         -2275.845           788.1823          -2.887460
      MMCDUM         -25796.23           4628.803          -5.572981
      EXPINF          185.7460           485.6733           0.382450

   R-squared              0.946084     Mean of depend var       113017.2
   Adjusted R-squared     0.936281     Std dev depend var       19905.77
   Std err of regress     5024.705     Residual sum          -8.62201E-10
   Durbin Watson stat     1.118237     Sum squared resid      8.33173E+08
   F Statistic            96.51171
```

Answer each of the following questions for the above regression run.

a. Evaluate this result with respect to its economic meaning, overall fit, and the signs and significance of the individual coefficients.

b. What econometric problems does this regression have? Why? If you need feedback on your answer, see hint #1 in the material on this chapter in Appendix A.

c. Which of the following statements comes closest to your recommendation for further action to be taken in the estimation of this equation?
   i. No further specification changes are advisable (go to Section 11.4.3).
   ii. No further variable changes are advisable, but I am concerned about heteroskedasticity or serial correlation (go to Section 11.4.3).
   iii. I would like to drop QYPERM from the equation (go to run #11.11).
   iv. I would like to drop EXPINF from the equation (go to run #11.18).
   v. I would like to change to the spread interest rate formulation (go to run #11.9).

If you need feedback on your answer, see hint #3 in the material on this chapter in Appendix A.

Regression Run 11.18

```
ORDINARY LEAST SQUARES                        DEPENDENT VARIABLE IS QDPASS
SAMPLE RANGE:     1970:1-1979:4

                        COEFFICIENT          STANDARD ERROR          T-SCORE
      CONST              148533.7              65894.91              2.254099
      QYDUS              182.5674              80.76083              2.260593
      QYPERM             -94.13722             88.21462             -1.067139
      QRDPASS            -22932.27             14650.16             -1.565325
      QRTB3Y             -2186.282             743.0808             -2.942186
      MMCDUM             -25665.93             4557.923             -5.631059

   R-squared               0.945845       Mean of depend var        113017.2
   Adjusted R-squared      0.937881       Std dev depend var        19905.77
   Std err of regress      4961.220       Residual sum           -8.07631E-10
   Durbin Watson stat      1.116034       Sum squared resid       8.36866E+08
   F Statistic             118.7670
```

Answer each of the following questions for the above regression run.

a. Evaluate this result with respect to its economic meaning, overall fit, and the signs and significance of the individual coefficients.

b. What econometric problems does this regression have? Why? If you need feedback on your answer, see hint #1 in the material on this chapter in Appendix A.

c. Which of the following statements comes closest to your recommendation for further action to be taken in the estimation of this equation?
   i. No further specification changes are advisable (go to Section 11.4.3).
   ii. No further variable changes are advisable, but I am concerned about heteroskedasticity or serial correlation (go to Section 11.4.3).
   iii. I would like to drop QYPERM from the equation (go to run #11.12).
   iv. I would like to add EXPINF to the equation (go to run #11.17).
   v. I would like to change to the spread interest rate formulation (go to run #11.10).

If you need feedback on your answer, see hint #3 in the material on this chapter in Appendix A.

Regression Run 11.19

```
ORDINARY LEAST SQUARES                        DEPENDENT VARIABLE IS QDPASS
SAMPLE RANGE:     1970:1-1979:4

                        COEFFICIENT        STANDARD ERROR         T-SCORE
        CONST            205021.5            64868.81             3.160556
        QYDUS            46.06039            23.45340             1.963911
        QRDPASS         -35397.74            14231.12            -2.487347
        QRTB3Y          -1374.301            750.7925            -1.830467
        MMCDUM          -22428.71            4694.232            -4.777930
        BRANCH           4.159648            1.825168             2.279049
        EXPINF          -8.002371            463.8907            -0.017250

    R-squared                0.951785      Mean of depend var       113017.2
    Adjusted R-squared       0.943018      Std dev depend var       19905.77
    Std err of regress       4751.651      Residual sum            -1.91721E-09
    Durbin Watson stat       0.858614      Sum squared resid        7.45080E+08
    F Statistic              108.5728
```

Answer each of the following questions for the above regression run.

a. Evaluate this result with respect to its economic meaning, overall fit, and the signs and significance of the individual coefficients.

b. What econometric problems does this regression have? Why? If you need feedback on your answer, see hint #2 in the material on this chapter in Appendix A.

c. Which of the following statements comes closest to your recommendation for further action to be taken in the estimation of this equation?
   i. No further specification changes are advisable (go to Section 11.4.3).
   ii. No further variable changes are advisable, but I am concerned about heteroskedasticity or serial correlation (go to Section 11.4.3).
   iii. I would like to drop EXPINF from the equation (go to run #11.20).
   iv. I would like to drop BRANCH from the equation (go to run #11.11).
   v. I would like to change to the spread interest rate formulation (go to run #11.1).

If you need feedback on your answer, see hint #5 in the material on this chapter in Appendix A.

Regression Run 11.20

```
ORDINARY LEAST SQUARES                          DEPENDENT VARIABLE IS QDPASS
SAMPLE RANGE:     1970:1-1979:4

                       COEFFICIENT         STANDARD ERROR          T-SCORE
    CONST               205458.8             58827.37              3.492571
    QYDUS               46.14516             22.59320              2.042436
    QRDPASS            -35493.15             12918.26             -2.747517
    QRTB3Y             -1378.529             699.1426             -1.971743
    MMCDUM             -22439.95             4579.932             -4.899625
    BRANCH              4.154828             1.776940              2.338193

  R-squared              0.951784          Mean of depend var       113017.2
  Adjusted R-squared     0.944694          Std dev depend var       19905.77
  Std err of regress     4681.273          Residual sum         -5.56611E-10
  Durbin Watson stat     0.859207          Sum squared resid     7.45087E+08
  F Statistic            134.2342
```

Answer each of the following questions for the above regression run.

a. Evaluate this result with respect to its economic meaning, overall fit, and the signs and significance of the individual coefficients.

b. What econometric problems does this regression have? Why? If you need feedback on your answer, see hint #2 in the material on this chapter in Appendix A.

c. Which of the following statements comes closest to your recommendation for further action to be taken in the estimation of this equation?
   i. No further specification changes are advisable (go to Section 11.4.3).
   ii. No further variable changes are advisable, but I am concerned about heteroskedasticity or serial correlation (go to Section 11.4.3).
   iii. I would like to drop BRANCH from the equation (go to run #11.12).
   iv. I would like to add EXPINF to the equation (go to run #11.19).
   v. I would like to change to the spread interest rate formulation (go to run #11.2).

If you need feedback on your answer, see hint #4 in the material on this chapter in Appendix A.

11.4.3 Heteroskedasticity and Serial Correlation in the Interactive Exercise

Congratulations! You've arrived at this section, which implies that you have completed the specification of your model of passbook deposits. Now that you've chosen your specification, the next step is to consider the possibility that pure heteroskedasticity and/or serial correlation might exist in your residuals. We also need to discuss a few other topics that the strict limitations of the interactive exercise did not allow; these topics will be covered in the exercises at the end of the chapter.

Heteroskedasticity. Heteroskedasticity is extremely unlikely in this example for a number of reasons. First, it is a time-series model, meaning that huge cross-sectional differences in size do not exist. Second, while there was certainly growth in passbook deposits during the 1970s, that growth was not exceptional. Indeed, competition from money market certificates and other assets actually decreased passbook deposits in the late 1970s. Finally, there is no indication of any dramatic change in the quality of measurement of the data.

As a result, many econometricians would not run a Park test on this example, feeling that even a significant result would indicate impure, rather than pure, heteroskedasticity. However, it would be good to get into the habit of examining the residuals of the final specification plotted against a suspected proportionality factor and then running a Park test if the plot indicates heteroskedasticity. Look over your independent variables; which of them would be candidates to be a potential proportionality factor Z? One good choice is to use either income variable because income is the best measure of size available to us. The interest rate variables are much less likely to be proportionality factors, mainly because of their fluctuations. Fortunately, plots of the residuals of the final specifications most frequently chosen by previous users of this example give no indications of pure heteroskedasticity with respect to disposable income, and so we should not seriously consider adjusting for heteroskedasticity. As it turns out, a Park test using the log of QYDUS to explain the log of the squared residuals from regression run #11.4 shows no evidence of heteroskedasticity at all ($t_z = 0.80$).

Serial Correlation. As you have surely noted, almost all of the regression runs produce Durbin-Watson statistics that indicate positive serial correlation (or inconclusive DW statistics quite close to the lower limit). Does this mean that there is serial correlation in the error terms? How do we tell whether we have pure or impure serial correlation? Should we adjust the equation with generalized least squares?

One way to determine whether the serial correlation indicated by the Durbin-Watson statistic is pure or impure is to examine the degree to which the model matches the underlying theory. The last part of the decade was marked by turmoil in the S & L industry (deregulation, high interest rates, some closures of S & Ls, and increasing competition for passbook accounts from other kinds of savings vehicles), but only the MMC dummy attempts to account for this turmoil. To most observers, such a set of circumstances is an indication that the residual pattern picked up by the Durbin-Watson statistic might be due to impure serial correlation. That is, the model has not properly captured all the factors that were influencing passbook deposits in the last few years of the decade. Such a result is a strong indication that the introduction of a more sophisticated explanatory variable (somehow measuring increasing competition for passbook accounts from other kinds of savings assets) should be attempted before GLS is run.

Since such a variable is not available in the present data-set, we went ahead and ran GLS on the most frequently chosen "best" equation, regression run #11.4. The results of that estimation are in Table 11.5. Note that the estimated coefficient of the spread variable is very similar in both the GLS and the OLS versions of run #11.4. In addition, note that even when the supposed serial correlation is "corrected," the Durbin-Watson d is only 0.97, even though it is biased toward 2 after GLS is run. This is evidence of an omitted variable or higher-order serial correlation. Finally, note that all the slope coefficients are still significantly different from zero in the hypothesized directions after the GLS adjustment except for the coefficient of MMCDUM, which is just barely insignificant in the expected direction. As a result, we would prefer sticking with the OLS estimate at present and attempting to find the possible omitted variable. Even if that variable is not found, it appears that serial correlation is doing little harm to the OLS results. Document such situations, including both the OLS and GLS results.

TABLE 11.5 GENERALIZED LEAST SQUARES ESTIMATION OF REGRESSION RUN 11.4

```
FIRST ORDER AUTOCORRELATION              DEPENDENT VARIABLE IS QDPASS
SAMPLE RANGE:     1970:1-1979:4

RHO=  0.847330
```

	COEFFICIENT	STANDARD ERROR	T-SCORE
CONST	54991.04	8023.691	6.853584
QYDUS	55.79906	7.370787	7.570300
SPREAD	2641.605	607.8764	4.345628
MMCDUM	-4683.828	3423.625	-1.368090

R-squared	0.977043	Mean of depend var	113017.2
Adjusted R-squared	0.975130	Std dev depend var	19905.77
Std err of regress	3139.164	Residual sum	-522.1689
Durbin Watson stat	0.969755	Sum squared resid	3.54757E+08
F Statistic	510.7248		

Reporting Your Final Equation. When reporting your "final" equation, remember also to report (say, in an appendix or a footnote) the equations estimated previously and subsequently (the order of estimation doesn't matter) to the one you chose. This practice allows readers to make up their own minds as to whether your choice was one with which they would agree. Such complete reporting will also let readers be able to judge for themselves the extent to which the reported t-scores are likely to follow the t-distribution and be comparable to the critical t-values found in the t-table.

11.5 Summary

1. Table 11.1 contains a listing of terms that should be checked when reviewing the output from a computer regression package.

2. Table 11.2 contains a summary of the nature, consequences, detection, and correction procedures for the various econometric problems covered so far in this text. A review of this table is a good way to prepare for the first few attempts at applied regression analysis.

3. The art of econometrics involves finding the best possible equation in the fewest possible number of regression runs. The only way to do this is to spend quite a bit of time thinking through the underlying principles of every research project before the first regression is run.

4. An ethical econometrician is always honest and complete in reporting all the different regressions estimated and/or data-sets used before the final results were chosen.

5. The interactive exercises of Sections 11.4, 11.6, and 11.7 are meant to provide a bridge between the structured examples of the previous chapters and the unstructured world of running regressions completely on your own.

Exercises

(Answers to even-numbered exercises are in Appendix A.)

1. Is there a problem with dominant variables in the interactive learning example? Don't the high $\bar{R}^2$s and high t-scores associated with QYDUS indicate the possibility of it being a dominant variable?

Examine regression run #11.4, for example. Why are we not worried about a dominant variable?

2. Some of the regression results you may not have encountered in working through the interactive exercise contain important lessons to be learned. Return to the following regression runs and answer the questions attached to those results for each one. (You will gain the most from this exercise if you answer the questions in writing without reference to the hints in Appendix A.)
   a. Regression run #11.1
   b. Regression run #11.9
   c. Regression run #11.12
   d. Regression run #11.16

3. What about real income in the interactive exercise? Since both QDPASS and QYDUS are measured in nominal terms, some of the correlation between them actually measures changes in prices.
   a. Think through the theory behind the relationship between holding assets in particular accounts and nominal vs. real income. Should a measure of real income be included?
   b. Analyze the following regression result, which replaces QYDUS with a new variable, $QYDUSR_t$, in regression run #11.3, where:

$QYDUSR_t$ = real disposable income in the U.S. in quarter t calculated by dividing QYDUS through by the consumer price index in quarter t

$$\widehat{QDPASS}_t = -175680 + \underset{\substack{(2504) \\ t = 17.01}}{42599QYDUSR_t} + \underset{\substack{(1067) \\ 3.54}}{3774SPREAD_t} - \underset{\substack{(4192) \\ t = -1.51}}{6329MMCDUM_t} + \underset{\substack{(610.5) \\ 0.71}}{435.1EXPINF_t}$$

$$\bar{R}^2 = .909 \quad n = 40 \quad DW = 0.50$$

   c. Compare this equation to regression run #11.3. Which specification do you prefer, and why? Be explicit.
   d. Would it make a difference if the real disposable income variable had been multiplied through by 100 as is typically done after dividing by a price index? How would the estimated equation have been changed?

11.6 Appendix: The Demand for Pork Interactive Exercise

This interactive exercise is another attempt to bridge the gap between textbook and computer. It's a complete exercise that helps you make independent specification decisions and get feedback on those decisions without requiring you to use a computer to do the estimation or have access to an experienced econometrician. When you complete this interactive exercise, you should be able to do econometric analysis on your own. As with the other interactive examples, we strongly encourage you to:

1. Look over a portion of the reading list before deciding on your specification.
2. Try to estimate as few regression runs as possible. There's nothing wrong with looking at only one regression result.
3. Avoid looking at the "hints" until after you reached what you believe is your best specification.

As before, we urge you to put this interactive exercise aside until you have the time to take it seriously. We believe that the benefits from completing it will be directly proportional to the effort you put into it. In other words, stop reading this section if you don't have the time it will take to get out of it what we have intended.

11.6.1 Building a "Demand-Side" Model for Pork

The dependent variable for this interactive exercise is the quantity of pork consumed (pounds per person per year) in the U.S. The data[4] are quarterly for 1975 through 1984, so there are 40 observations. For each quarter, the dependent variable, *CONPK*, measures the annual per capita consumption (CON) of pork (PK).

4. This data-set was obtained from Prof. William G. Tomek of Cornell University, who has created an eight-part series of hands-on problem sets (of up to six exercises per problem set) on the demand for pork. Most of the tasks to which he sets his students go beyond what we have covered so far in this text. They include specifications with price as the dependent variable (which is more difficult than the quantity dependent varible model we use), distributed lag equations, forecasting, and a simultaneous model of the pork industry. We are extremely grateful to Professor Tomek for his help in preparing this interactive exercise, but we hasten to point out that since the approach we have taken is different from his, any errors in this example should not be blamed on him. (Don't blame us either, they're probably stochastic errors!)

This text has discussed demand-side equations for consumption goods frequently, so the underlying theory will be fairly familiar. In the past, all of our equations have included at least the price of the product, in this case, *PRIPK,* the price (PRI) of pork (PK). We usually added some measure of consumer buying power, and in this example we will use *YDUSP,* which is disposable income (YD) in the U.S. (US) per capita (P). The price of a substitute, PRIBF, the price of beef (BF), is also available as an explanatory variable.

Since many of the questions about inclusion of variables have been considered already, we now go beyond that choice to include somewhat more complex items than were studied in the first interactive exercises. You'll choose whether the functional form of the demand/income relationship should be linear (using YDUSP) or semi-log (using the log of that variable, LYDUSP). In addition, you'll have to figure out whether to adjust the intercept of the quarterly model for seasonal variation with the inclusion of quarterly seasonal dummies (D1, D2, and D3). You also will have to decide on the extent to which simultaneity can be dealt with by including a production variable, *PROPK,* the production (PRO) of pork, in the equation. Finally, serial correlation or heteroskedasticity may be problems.

As even a little reading will show you, attention paid in the literature[5] to the demand for pork is impressive. One question is whether the typical consumer decides to buy a product based on real prices and incomes or on nominal prices and incomes. The more sophisticated the consumer behavior is assumed to be, the more logical it is to use real prices and incomes since the dependent variable in this example (pounds of pork) is in real terms. The variables in this exercise are nominal; however, they could be converted to real prices and incomes by dividing through by the consumer price index (CPI) or GNP deflator of the quarter in question and multiplying by 100:

$$\text{Real } X_t = \text{Nominal } X_t(100/\text{CPI}_t)$$

5. The place to start is to review the demand for chicken example of Section 6.1 and to consult a typical intermediate microeconomics text regarding the demand for a product. More advanced readings, among many suggested by Professor Tomek, include E. R. Arzac and M. Wilkinson, "A Quarterly Econometric Model of United States Livestock and Feed Grain Markets and Some of Its Policy Implications," *American Journal of Agricultural Economics,* May 1979, pp. 297–308; Giancarlo Moschini and Karl D. Meilke, "Parameter Stability and the U.S. Demand for Beef," *Western Journal of Agricultural Economics,* Dec. 1984, pp. 271–282; and Z. A. Hassen and S. R. Johnson, "Structural Stability and Recursive Residuals: Quarterly Demand for Meat," *Agricultural Economics Research,* Oct. 1979, pp. 20–29.

The variables available for your model are:

$CONPK_t$	per capita pounds of pork consumed in the U.S. in quarter t
$PRIPK_t$	the price of a pound of pork (in dollars per 100 pounds) in quarter t
$PRIBF_t$	the price of a pound of beef (in dollars per 100 pounds) in quarter t
$YDUSP_t$	per capita disposable income in the U.S. in quarter t (current dollars)
$LYDUSP_t$	the log of per capita disposable income
$PROPK_t$	pounds of pork produced (billions) in the U.S. in quarter t
$D1_t$	dummy equal to 1 in the first quarter of the year and 0 otherwise
$D2_t$	dummy equal to 1 in the second quarter of the year and 0 otherwise
$D3_t$	dummy equal to 1 in the third quarter of the year and 0 otherwise

The data for the variables are in Table 11.6.

Now:

1. Hypothesize expected signs for these variables in an equation for the consumption of pork. Consider each variable carefully; what is the economic content of each hypothesis?

2. Choose the best combination of explanatory variables for this model. Assume that every model has at least the price of pork variable and one of the two income variables. Don't take the attitude that you'll see what a particular specification looks like before making up your mind. As you surely remember from Section 6.4 on specification searches, such an attitude leads to possible bias and ruins the applicability of the *t*- and *F*-tests. In addition, this example has been set up in the hope that you will be capable of getting the "perfect" equation after looking at only one regression estimate. Take it as a challenge; try to make your first equation your final equation.

Once you've chosen your best equation, you'll be ready to begin. As with previous interactive exercises, keep following the instructions until you've completely specified the equation and have been instructed to go to the discussion of serial correlation and heteroskedasticity in Section 11.6.3. To the extent that you can,

TABLE 11.6 DATA FOR THE DEMAND FOR PORK INTERACTIVE EXERCISE

OBSERV	CONPK	PRIPK	PRIBF	PROPK	YDUSP	D1	D2	D3
1975:1	13.98	114.1	137.2	3.142	4.182	1	0	0
1975:2	12.90	122.7	155.3	2.992	5.125	0	1	0
1975:3	11.29	148.8	166.0	2.555	5.129	0	0	1
1975:4	12.49	152.9	160.9	2.896	5.232	0	0	0
1976:1	13.01	141.2	151.3	2.958	5.335	1	0	0
1976:2	12.08	138.2	150.8	2.847	5.422	0	1	0
1976:3	12.95	137.1	145.3	3.014	5.511	0	0	1
1976:4	15.68	119.6	145.4	3.669	5.617	0	0	0
1977:1	14.28	120.5	144.6	3.294	5.721	1	0	0
1977:2	13.48	121.7	146.4	3.185	5.873	0	1	0
1977:3	13.25	131.0	149.0	3.073	6.055	0	0	1
1977:4	14.81	128.2	153.4	3.500	6.209	0	0	0
1978:1	13.94	137.0	162.7	3.243	6.340	1	0	0
1978:2	13.60	142.4	185.7	3.265	6.529	0	1	0
1978:3	13.61	144.7	189.4	3.160	6.711	0	0	1
1978:4	14.70	150.1	189.7	3.541	6.900	0	0	0
1979:1	14.50	156.1	215.4	3.395	7.082	1	0	0
1979:2	15.55	148.2	235.5	3.754	7.226	0	1	0
1979:3	16.01	138.0	226.6	3.775	7.427	0	0	1
1979:4	17.75	134.3	227.7	4.346	7.584	0	0	0
1980:1	17.30	133.9	235.2	4.125	7.814	1	0	0
1980:2	17.69	124.4	231.4	4.299	7.871	0	1	0
1980:3	16.04	144.2	241.6	3.756	8.095	0	0	1
1980:4	17.25	154.3	242.3	4.252	8.345	0	0	0
1981:1	16.81	148.7	237.5	4.073	8.606	1	0	0
1981:2	15.87	144.7	234.7	3.881	8.732	0	1	0
1981:3	15.38	157.5	243.1	3.605	9.023	0	0	1
1981:4	16.90	158.7	239.5	4.157	9.134	0	0	0
1982:1	15.33	160.1	237.3	3.693	9.209	1	0	0
1982:2	14.76	169.3	247.2	3.550	9.295	0	1	0
1982:3	13.87	185.0	248.3	3.240	9.439	0	0	1
1982:4	15.06	187.1	237.2	3.638	9.593	0	0	0
1983:1	14.51	183.0	237.9	3.483	9.675	1	0	0
1983:2	15.33	171.1	245.1	3.771	9.832	0	1	0
1983:3	15.42	165.4	238.4	3.657	10.082	0	0	1
1983:4	16.91	159.8	231.1	4.206	10.318	0	0	0
1984:1	15.34	161.5	242.6	3.738	10.608	1	0	0
1984:2	15.14	159.4	242.1	3.670	10.806	0	1	0
1984:3	14.77	164.0	236.2	3.355	11.000	0	0	1
1984:4	16.46	163.3	237.3	3.957	11.133	0	0	0

avoid looking at the hints until after completing the project. (If you need to look, you'll find that the hints are somewhat more sparse than in previous examples in the hope that you'll eventually be able to do without such hints.) Try to take the time to actually write out the answers to the questions after each result. Good luck!

11.6.2 The Demand for Pork Interactive Learning Exercise

To start the example, write out the specification you chose in Section 11.6.1 above. To find that equation's estimates, carefully follow these instructions. (Note that the means and simple correlation coefficients for this data-set are in Table 11.7 just before the results begin.)

All the equations include CONPK as the dependent variable and PRIPK as one of the explanatory variables. In addition, all the equations available to you at the beginning of the example include either YDUSP or LYDUSP. If you chose YDUSP as one of your variables, go to question number one, but if you instead chose LYDUSP, jump to question number two below.

1. Find below the combination of explanatory variables (from PROPK, PRIBF, and the seasonal dummies) that you wish to include in your regression and then go to the indicated estimated regression equation:
None of them, go to regression run #11.21
The seasonal dummies only, go to regression run #11.22
PROPK only, go to regression run #11.23
PRIBF only, go to regression run #11.25
The seasonal dummies and PROPK, to go regression run #11.24
The seasonal dummies and PRIBF, go to regression run #11.26
PROPK and PRIBF, go to regression run #11.27
All three, go to regression run #11.28

2. Find below the combination of explanatory variables (from PROPK, PRIBF, and the seasonal dummies) that you wish to include in your regression and then go to the indicated estimated regression equation:
None of them, go to regression run #11.29
PROPK only, go to regression run #11.31
The seasonal dummies only, go to regression run #11.30
PRIBF only, go to regression run #11.33
The seasonal dummies and PROPK, go to regression run #11.32
The seasonal dummies and PRIBF, go to regression run #11.34
PROPK and PRIBF, go to regression run #11.35
All three, go to regression run #11.36

TABLE 11.7 MEANS, VARIANCES, AND CORRELATIONS FOR THE DEMAND FOR PORK INTERACTIVE EXERCISE

Variable	Mean	Standard Dev	Variance
CONPK	14.90000	1.555892	2.420800
PRIPK	148.0550	18.06813	326.4575
PRIBF	205.6075	40.29446	1623.643
PROPK	3.542750	0.435895	0.190004
YDUSP	7.745500	1.928367	3.718598
LYDUSP	2.014656	0.258263	0.066700
D1	0.250000	0.433013	0.187500
D2	0.250000	0.433013	0.187500
D3	0.250000	0.433013	0.187500

	Correlation Coeff		Correlation Coeff
PRIBF,PRIPK	0.708	PRIPK,CONPK	0.110
PROPK,CONPK	0.988	PRIBF,CONPK	0.682
PROPK,PRIBF	0.707	PROPK,PRIPK	0.165
YDUSP,CONPK	0.573	YDUSP,PRIPK	0.758
YDUSP,PRIBF	0.884	YDUSP,PROPK	0.604
LYDUSP,PRIPK	0.752	LYDUSP,CONPK	0.611
LYDUSP,PROPK	0.643	LYDUSP,PRIBF	0.909
D1,PRIPK	−0.078	LYDUSP,YDUSP	0.993
D1,PROPK	−0.037	D1,CONPK	0.000
D1,LYDUSP	−0.096	D1,PRIBF	−0.077
D2,CONPK	−0.096	D1,YDUSP	−0.086
D2,PRIBF	0.025	D2,PRIPK	−0.122
D2,YDUSP	−0.022	D2,PROPK	−0.028
D2, D1	−0.333	D2,LYDUSP	−0.016
D3,CONPK	−0.237	D3,PRIPK	0.112
D3,PRIBF	0.039	D3,PROPK	−0.296
D3,YDUSP	0.030	D3,LYDUSP	0.033
D3,D1	−0.333	D3,D2	−0.333

Regression Run 11.21

```
ORDINARY LEAST SQUARES                    DEPENDENT VARIABLE IS CONPK
SAMPLE RANGE:     1975:1-1984:4

                  COEFFICIENT          STANDARD ERROR         T-SCORE
     CONST         17.44424              1.481554            11.77429
     PRIPK        -0.065922              0.014110            -4.671866
     YDUSP         0.931629              0.132211             7.046528

   R-squared              0.578212      Mean of depend var      14.90000
   Adjusted R-squared     0.555413      Std dev depend var      1.575713
   Std err of regress     1.050643      Residual sum            1.52101E-13
   Durbin Watson stat     0.891839      Sum squared resid       40.84251
   F Statistic            25.36097
```

Answer each of the following questions for the above regression run.

a. Evaluate this result with respect to its economic meaning, overall fit, and the signs and significance of the individual coefficients.

b. What econometric problems does this regression have? Why? If you need feedback on your answer, see hint #19 in the material on this chapter in Appendix A.

c. Which of the following statements comes closest to your recommendation for further action to be taken in the estimation of this equation?

  i. No specification changes are advisable (go to Section 11.6.3).

  ii. No variable changes are advisable, but I am concerned about heteroskedasticity or serial correlation (go to Section 11.6.3).

  iii. I would like to add PRIBF to the equation (go to run #11.25).

  iv. I would like to add seasonal dummies to the equation (go to run #11.22).

  v. I would like to change to LYDUSP from YDUSP (go to run #11.29).

If you need feedback on your answer, see hint #21 in the material on this chapter in Appendix A.

Regression Run 11.22

```
ORDINARY LEAST SQUARES                    DEPENDENT VARIABLE IS CONPK
SAMPLE RANGE:      1975:1-1984:4

                    COEFFICIENT           STANDARD ERROR          T-SCORE
     CONST           18.52019               1.373476              13.48418
     PRIPK          -0.067431               0.012656             -5.327895
     YDUSP           0.930688               0.117379              7.928913
     D1             -0.741767               0.416293             -1.781835
     D2             -1.295246               0.418102             -3.097919
     D3             -1.343842               0.414675             -3.240705

  R-squared                   0.699346      Mean of depend var      14.90000
  Adjusted R-squared          0.655132      Std dev depend var      1.575713
  Std err of regress          0.925344      Residual sum            1.42553E-13
  Durbin Watson stat          0.282212      Sum squared resid       29.11292
  F Statistic                 15.81736
```

Answer each of the following questions for the above regression run.

a. Evaluate this result with respect to its economic meaning, overall fit, and the signs and significance of the individual coefficients.

b. What econometric problems does this regression have? Why? If you need feedback on your answer, see hint #19 in the material on this chapter in Appendix A.

c. Which of the following statements comes closest to your recommendation for further action to be taken in the estimation of this equation?
   i. No specification changes are advisable (go to Section 11.6.3).
   ii. No variable changes are advisable, but I am concerned about heteroskedasticity or serial correlation (go to Section 11.6.3).
   iii. I would like to add PRIBF to the equation (go to run #11.26).
   iv. I would like to add PROPK to the equation (go to run #11.24).
   v. I would like to change to LYDUSP from YDUSP (go to run #11.30).

If you need feedback on your answer, see hint #21 in the material on this chapter in Appendix A.

Regression Run 11.23

```
ORDINARY LEAST SQUARES                    DEPENDENT VARIABLE IS CONPK
SAMPLE RANGE:     1975:1-1984:4

                         COEFFICIENT       STANDARD ERROR       T-SCORE
      CONST               3.546665           0.584502           6.067834
      PRIPK              -8.56798E-03        3.61098E-03       -2.372757
      YDUSP               0.053112           0.041880           1.268189
      PROPK               3.446612           0.122413           28.15540

   R-squared               0.981677       Mean of depend var      14.90000
   Adjusted R-squared      0.980150       Std dev depend var      1.575713
   Std err of regress      0.221998       Residual sum            1.19960E-13
   Durbin Watson stat      2.344786       Sum squared resid       1.774204
   F Statistic             642.9325
```

Answer each of the following questions for the above regression run.

a. Evaluate this result with respect to its economic meaning, overall fit, and the signs and significance of the individual coefficients.

b. What econometric problems does this regression have? Why? If you need feedback on your answer, see hint #22 in the material on this chapter in Appendix A.

c. Which of the following statements comes closest to your recommendation for further action to be taken in the estimation of this equation?
   i. No specification changes are advisable (go to Section 11.6.3).
   ii. No variable changes are advisable, but I am concerned about heteroskedasticity or serial correlation (go to Section 11.6.3).
   iii. I would like to add PRIBF to the equation (go to run #11.27).
   iv. I would like to drop PROPK from the equation (go to run #11.21).
   v. I would like to change to LYDUSP from YDUSP (go to run #11.31).

If you need feedback on your answer, see hint #24 in the material on this chapter in Appendix A.

Regression Run 11.24

```
ORDINARY LEAST SQUARES                   DEPENDENT VARIABLE IS CONPK
SAMPLE RANGE:     1975:1-1984:4

                      COEFFICIENT         STANDARD ERROR        T-SCORE
      CONST            3.261090            0.543027            6.005385
      PRIPK           -8.88974E-03         2.94923E-03        -3.014258
      YDUSP            0.041744            0.035263            1.183812
      PROPK            3.549741            0.112242            31.62553
      D1               0.146837            0.080576            1.822349
      D2              -0.159385            0.083920           -1.899245
      D3               0.236159            0.090303            2.615171

   R-squared               0.990397         Mean of depend var     14.90000
   Adjusted R-squared      0.988651         Std dev depend var     1.575713
   Std err of regress      0.167863         Residual sum           8.67639E-14
   Durbin Watson stat      1.491153         Sum squared resid      0.929878
   F Statistic             567.2373
```

Answer each of the following questions for the above regression run.

a. Evaluate this result with respect to its economic meaning, overall fit, and the signs and significance of the individual coefficients.

b. What econometric problems does this regression have? Why? If you need feedback on your answer, see hint #22 in the material on this chapter in Appendix A.

c. Which of the following statements comes closest to your recommendation for further action to be taken in the estimation of this equation?
   i. No specification changes are advisable (go to Section 11.6.3).
   ii. No variable changes are advisable, but I am concerned about heteroskedasticity or serial correlation (go to Section 11.6.3).
   iii. I would like to add PRIBF to the equation (go to run #11.28).
   iv. I would like to drop PROPK rom the equation (go to run #11.22).
   v. I would like to change to LYDUSP from YDUSP (go to run #11.32).

If you need feedback on your answer, see hint #24 in the material on this chapter in Appendix A.

Regression Run 11.25

```
ORDINARY LEAST SQUARES                    DEPENDENT VARIABLE IS CONPK
SAMPLE RANGE:     1975:1-1984:4

                      COEFFICIENT       STANDARD ERROR        T-SCORE
     CONST             15.81965           1.143847            13.83021
     PRIPK            -0.073081           0.010607            -6.889909
     PRIBF             0.036680           6.64475E-03         5.520233
     YDUSP             0.304512           0.150450            2.024009

  R-squared              0.771571        Mean of depend var     14.90000
  Adjusted R-squared     0.752535        Std dev depend var     1.575713
  Std err of regress     0.783851        Residual sum           4.65183E-14
  Durbin Watson stat     1.895964        Sum squared resid      22.11922
  F Statistic            40.53278
```

Answer each of the following questions for the above regression run.

a. Evaluate this result with respect to its economic meaning, overall fit, and the signs and significance of the individual coefficients.

b. What econometric problems does this regression have? Why? If you need feedback on your answer, see hint #25 in the material on this chapter in Appendix A.

c. Which of the following statements comes closest to your recommendation for further action to be taken in the estimation of this equation?
   i. No specification changes are advisable (go to Section 11.6.3).
   ii. No variable changes are advisable, but I am concerned about heteroskedasticity or serial correlation (go to Section 11.6.3).
   iii. I would like to add PROPK to the equation (go to run #11.27).
   iv. I would like to add seasonal dummies to the equation (go to run #11.26).
   v. I would like to change to LYDUSP from YDUSP (go to run #11.33).

If you need feedback on your answer, see hint #27 in the material on this chapter in Appendix A.

Regression Run 11.26

```
ORDINARY LEAST SQUARES                      DEPENDENT VARIABLE IS CONPK
SAMPLE RANGE:     1975:1-1984:4

                      COEFFICIENT        STANDARD ERROR        T-SCORE
      CONST            16.99718            0.627607           27.08250
      PRIPK           -0.076755            5.71353E-03       -13.43405
      PRIBF            0.041556            3.54856E-03        11.71067
      YDUSP            0.225024            0.079902            2.816248
      D1              -0.917087            0.186697           -4.912169
      D2              -1.633960            0.189128           -8.639402
      D3              -1.529973            0.186052           -8.223335

   R-squared              0.941685        Mean of depend var     14.90000
   Adjusted R-squared     0.931083        Std dev depend var     1.575713
   Std err of regress     0.413656        Residual sum           3.70814E-14
   Durbin Watson stat     1.085841        Sum squared resid      5.646691
   F Statistic            88.81648
```

Answer each of the following questions for the above regression run.

a. Evaluate this result with respect to its economic meaning, overall fit, and the signs and significance of the individual coefficients.

b. What econometric problems does this regression have? Why? If you need feedback on your answer, see hint #23 in the material on this chapter in Appendix A.

c. Which of the following statements comes closest to your recommendation for further action to be taken in the estimation of this equation?
   i. No specification changes are advisable (go to Section 11.6.3).
   ii. No variable changes are advisable, but I am concerned about heteroskedasticity or serial correlation (go to Section 11.6.3).
   iii. I would like to add PROPK to the equation (go to run #11.28).
   iv. I would like to drop the dummies from the equation (go to run #11.25).
   v. I would like to change to LYDUSP from YDUSP (go to run #11.34).

If you need feedback on your answer, see hint #26 in the material on this chapter in Appendix A.

Regression Run 11.27

```
ORDINARY LEAST SQUARES                    DEPENDENT VARIABLE IS CONPK
SAMPLE RANGE:     1975:1-1984:4

                         COEFFICIENT       STANDARD ERROR        T-SCORE
      CONST               4.016224           0.652050            6.159372
      PRIPK              -0.011906           4.17235E-03        -2.853550
      PRIBF               3.70474E-03        2.43832E-03         1.519383
      YDUSP               0.029828           0.043901            0.679452
      PROPK               3.289467           0.158608            20.73957

   R-squared               0.982811     Mean of depend var      14.90000
   Adjusted R-squared      0.980846     Std dev depend var      1.575713
   Std err of regress      0.218070     Residual sum            3.89966E-14
   Durbin Watson stat      2.329738     Sum squared resid       1.664422
   F Statistic             500.3034
```

Answer each of the following questions for the above regression run.

a. Evaluate this result with respect to its economic meaning, overall fit, and the signs and significance of the individual coefficients.

b. What econometric problems does this regression have? Why? If you need feedback on your answer, see hint #22 in the material on this chapter in Appendix A.

c. Which of the following statements comes closest to your recommendation for further action to be taken in the estimation of this equation?
   i. No specification changes are advisable (go to Section 11.6.3).
   ii. No variable changes are advisable, but I am concerned about heteroskedasticity or serial correlation (go to Section 11.6.3).
   iii. I would like to add seasonal dummies to the equation (go to run #11.28).
   iv. I would like to drop PROPK from the equation (go to run #11.25).
   v. I would like to drop YDUSP from the equation (go to run #11.40).

If you need feedback on your answer, see hint #24 in the material on this chapter in Appendix A.

Regression Run 11.28

```
ORDINARY LEAST SQUARES                    DEPENDENT VARIABLE IS CONPK
SAMPLE RANGE:     1975:1-1984:4

                 COEFFICIENT          STANDARD ERROR          T-SCORE
    CONST         4.981253              0.891042              5.590364
    PRIPK        -0.017923              4.73211E-03          -3.787536
    PRIBF         6.66840E-03           2.83469E-03           2.352423
    YDUSP         0.042957              0.033068              1.299023
    PROPK         3.092723              0.220951             13.99731
    D1            4.29957E-03           0.096848              0.044394
    D2           -0.359976              0.116029             -3.102451
    D3            2.87148E-03           0.130400              0.022020

  R-squared                0.991812         Mean of depend var      14.90000
  Adjusted R-squared       0.990021         Std dev depend var      1.575713
  Std err of regress       0.157398         Residual sum            1.18322E-13
  Durbin Watson stat       1.560405         Sum squared resid       0.792779
  F Statistic              553.7938
```

Answer each of the following questions for the above regression run.

a. Evaluate this result with respect to its economic meaning, overall fit, and the signs and significance of the individual coefficients.

b. What econometric problems does this regression have? Why? If you need feedback on your answer, see hint #22 in the material on this chapter in Appendix A.

c. Which of the following statements comes closest to your recommendation for further action to be taken in the estimation of this equation?
   i. No specification changes are advisable (go to Section 11.6.3).
   ii. No variable changes are advisable, but I am concerned about heteroskedasticity or serial correlation (go to Section 11.6.3).
   iii. I would like to drop PROPK from the equation (go to run #11.26).
   iv. I would like to drop the dummies from the equation (go to run #11.27).
   v. I would like to change to LYDUSP from YDUSP (go to run #11.36).

If you need feedback on your answer, see hint #28 in the material on this chapter in Appendix A.

Regression Run 11.29

```
ORDINARY LEAST SQUARES                    DEPENDENT VARIABLE IS CONPK
SAMPLE RANGE:     1975:1-1984:4

                     COEFFICIENT        STANDARD ERROR          T-SCORE
     CONST            10.39432            1.296442             8.017572
     PRIPK           -0.069246            0.012628            -5.483233
     LYDUSP           7.325314            0.883513             8.291116

  R-squared                0.654355      Mean of depend var       14.90000
  Adjusted R-squared       0.635671      Std dev depend var       1.575713
  Std err of regress       0.951094      Residual sum             1.09301E-13
  Durbin Watson stat       1.141043      Sum squared resid        33.46947
  F Statistic              35.02317
```

Answer each of the following questions for the above regression run.

a. Evaluate this result with respect to its economic meaning, overall fit, and the signs and significance of the individual coefficients.

b. What econometric problems does this regression have? Why? If you need feedback on your answer, see hint #19 in the material on this chapter in Appendix A.

c. Which of the following statements comes closest to your recommendation for further action to be taken in the estimation of this equation?
   i. No specification changes are advisable (go to Section 11.6.3).
   ii. No variable changes are advisable, but I am concerned about heteroskedasticity or serial correlation (go to Section 11.6.3).
   iii. I would like to add PRIBF to the equation (go to run #11.33).
   iv. I would like to add seasonal dummies to the equation (go to run #11.30).
   v. I would like to change to YDUSP from LYDUSP (go to run #11.21).

If you need feedback on your answer, see hint #21 in the material on this chapter in Appendix A.

Regression Run 11.30

```
ORDINARY LEAST SQUARES                    DEPENDENT VARIABLE IS CONPK
SAMPLE RANGE:     1975:1-1984:4

                      COEFFICIENT         STANDARD ERROR          T-SCORE
      CONST            11.45430             1.137026             10.07392
      PRIPK            -0.071056            0.010748             -6.610806
      LYDUSP            7.347388            0.744910              9.863450
      D1               -0.691991            0.357754             -1.934261
      D2               -1.316661            0.359186             -3.665681
      D3               -1.337528            0.356145             -3.755564

   R-squared              0.778169      Mean of depend var      14.90000
   Adjusted R-squared     0.745547      Std dev depend var      1.575713
   Std err of regress     0.794842      Residual sum            1.24345E-14
   Durbin Watson stat     0.440595      Sum squared resid       21.48032
   F Statistic            23.85399
```

Answer each of the following questions for the above regression run.

a. Evaluate this result with respect to its economic meaning, overall fit, and the signs and significance of the individual coefficients.

b. What econometric problems does this regression have? Why? If you need feedback on your answer, see hint #19 in the material on this chapter in Appendix A.

c. Which of the following statements comes closest to your recommendation for further action to be taken in the estimation of this equation?

   i. No specification changes are advisable (go to Section 11.6.3).
   ii. No variable changes are advisable, but I am concerned about heteroskedasticity or serial correlation (go to Section 11.6.3).
   iii. I would like to add PRIBF to the equation (go to run #11.34).
   iv. I would like to add PROPK to the equation (go to run #11.32).
   v. I would like to change to YDUSP from LYDUSP (go to run #11.22).

If you need feedback on your answer, see hint #21 in the material on this chapter in Appendix A.

Regression Run 11.31

```
ORDINARY LEAST SQUARES                    DEPENDENT VARIABLE IS CONPK
SAMPLE RANGE:     1975:1-1984:4

                    COEFFICIENT        STANDARD ERROR         T-SCORE
      CONST          3.174375            0.417704             7.599574
      PRIPK         -8.34067E-03         3.82251E-03         -2.181986
      LYDUSP         0.380609            0.344662             1.104294
      PROPK          3.441876            0.136468             25.22103

   R-squared             0.981486       Mean of depend var       14.90000
   Adjusted R-squared    0.979943       Std dev depend var       1.575713
   Std err of regress    0.223155       Residual sum             1.83964E-13
   Durbin Watson stat    2.325275       Sum squared resid        1.792740
   F Statistic           636.1611
```

Answer each of the following questions for the above regression run.

a. Evaluate this result with respect to its economic meaning, overall fit, and the signs and significance of the individual coefficients.

b. What econometric problems does this regression have? Why? If you need feedback on your answer, see hint #22 in the material on this chapter in Appendix A.

c. Which of the following statements comes closest to your recommendation for further action to be taken in the estimation of this equation?

- **i.** No specification changes are advisable (go to Section 11.6.3).
- **ii.** No variable changes are advisable, but I am concerned about heteroskedasticity or serial correlation (go to Section 11.6.3).
- **iii.** I would like to add PRIBF to the equation (go to run #11.35).
- **iv.** I would like to drop LYDUSP from the equation (go to run #11.39).
- **v.** I would like to change to YDUSP from LYDUSP (go to run #11.23).

If you need feedback on your answer, see hint #24 in the material on this chapter in Appendix A.

Regression Run 11.32

```
ORDINARY LEAST SQUARES                    DEPENDENT VARIABLE IS CONPK
SAMPLE RANGE:      1975:1-1984:4

                COEFFICIENT        STANDARD ERROR         T-SCORE
      CONST       2.936173           0.400830             7.325220
      PRIPK      -8.27414E-03        3.28513E-03         -2.518660
      LYDUSP      0.253455           0.309607             0.818634
      PROPK       3.561963           0.133387            26.70395
      D1          0.150814           0.082635             1.825064
      D2         -0.154851           0.088159            -1.756489
      D3          0.240370           0.096289             2.496344

   R-squared                0.990188      Mean of depend var      14.90000
   Adjusted R-squared       0.988404      Std dev depend var      1.575713
   Std err of regress       0.169676      Residual sum            2.26485E-14
   Durbin Watson stat       1.480375      Sum squared resid       0.950073
   F Statistic              555.0630
```

Answer each of the following questions for the above regression run.

a. Evaluate this result with respect to its economic meaning, overall fit, and the signs and significance of the individual coefficients.

b. What econometric problems does this regression have? Why? If you need feedback on your answer, see hint #22 in the material on this chapter in Appendix A.

c. Which of the following statements comes closest to your recommendation for further action to be taken in the estimation of this equation?

- **i.** No specification changes are advisable (go to Section 11.6.3).
- **ii.** No variable changes are advisable, but I am concerned about heteroskedasticity or serial correlation (go to Section 11.6.3).
- **iii.** I would like to add PRIBF to the equation (go to run #11.36).
- **iv.** I would like to drop LYDUSP from the equation (go to run #11.37).
- **v.** I would like to drop PROPK from the equation (go to run #11.30).

If you need feedback on your answer, see hint #24 in the material on this chapter in Appendix A.

Regression Run 11.33

```
ORDINARY LEAST SQUARES                     DEPENDENT VARIABLE IS CONPK
SAMPLE RANGE:     1975:1-1984:4

                    COEFFICIENT        STANDARD ERROR        T-SCORE
      CONST          13.21439            1.222610           10.80835
      PRIPK         -0.073307            0.010273           -7.135437
      PRIBF          0.032975            7.31164E-03         4.510048
      LYDUSP         2.858571            1.222098            2.339069

   R-squared              0.779142      Mean of depend var      14.90000
   Adjusted R-squared     0.760738      Std dev depend var      1.575713
   Std err of regress     0.770750      Residual sum            3.74145E-14
   Durbin Watson stat     1.949536      Sum squared resid       21.38604
   F Statistic            42.33376
```

Answer each of the following questions for the above regression run.

a. Evaluate this result with respect to its economic meaning, overall fit, and the signs and significance of the individual coefficients.

b. What econometric problems does this regression have? Why? If you need feedback on your answer, see hint #25 in the material on this chapter in Appendix A.

c. Which of the following statements comes closest to your recommendation for further action to be taken in the estimation of this equation?
   i. No specification changes are advisable (go to Section 11.6.3).
   ii. No variable changes are advisable, but I am concerned about heteroskedasticity or serial correlation (go to Section 11.6.3).
   iii. I would like to add seasonal dummies to the equation (go to run #11.34).
   iv. I would like to add PROPK to the equation (go to run #11.35).
   v. I would like to change to YDUSP from LYDUSP (go to run #11.25).

If you need feedback on your answer, see hint #27 in the material on this chapter in Appendix A.

Regression Run 11.34

```
ORDINARY LEAST SQUARES                    DEPENDENT VARIABLE IS CONPK
SAMPLE RANGE:     1975:1-1984:4

                    COEFFICIENT        STANDARD ERROR        T-SCORE
      CONST          15.04305            0.671939           22.38752
      PRIPK         -0.076857            5.42260E-03       -14.17362
      PRIBF          0.038757            3.83568E-03        10.10445
      LYDUSP         2.121158            0.638094            3.324208
      D1            -0.891380            0.180553           -4.936920
      D2            -1.616533            0.182617           -8.851999
      D3            -1.516465            0.179540           -8.446363

   R-squared                0.945814      Mean of depend var     14.90000
   Adjusted R-squared       0.935962      Std dev depend var     1.575713
   Std err of regress       0.398742      Residual sum           7.21645E-14
   Durbin Watson stat       1.191319      Sum squared resid      5.246862
   F Statistic              96.00372
```

Answer each of the following questions for the above regression run.

a. Evaluate this result with respect to its economic meaning, overall fit, and the signs and significance of the individual coefficients.

b. What econometric problems does this regression have? Why? If you need feedback on your answer, see hint #19 in the material on this chapter in Appendix A.

c. Which of the following statements comes closest to your recommendation for further action to be taken in the estimation of this equation?
   i. No specification changes are advisable (go to Section 11.6.3).
   ii. No variable changes are advisable, but I am concerned about heteroskedasticity or serial correlation (go to Section 11.6.3).
   iii. I would like to add PROPK to the equation (go to run #11.36).
   iv. I would like to drop the dummies from the equation (go to run #11.33).
   v. I would like to change to YDUSP from LYDUSP (go to run #11.26).

If you need feedback on your answer, see hint #26 in the material on this chapter in Appendix A.

Regression Run 11.35

```
ORDINARY LEAST SQUARES                        DEPENDENT VARIABLE IS CONPK
SAMPLE RANGE:     1975:1-1984:4

                      COEFFICIENT        STANDARD ERROR        T-SCORE
      CONST            3.804671            0.579937            6.560486
      PRIPK           -0.011335            4.22807E-03        -2.681007
      PRIBF            3.88151E-03         2.52570E-03         1.536806
      LYDUSP           0.141817            0.372298            0.380923
      PROPK            3.299646            0.162819            20.26569

   R-squared                0.982656      Mean of depend var      14.90000
   Adjusted R-squared       0.980674      Std dev depend var      1.575713
   Std err of regress       0.219050      Residual sum            3.60822E-14
   Durbin Watson stat       2.321783      Sum squared resid       1.679414
   F Statistic              495.7593
```

Answer each of the following questions for the above regression run.

a. Evaluate this result with respect to its economic meaning, overall fit, and the signs and significance of the individual coefficients.

b. What econometric problems does this regression have? Why? If you need feedback on your answer, see hint #22 in the material on this chapter in Appendix A.

c. Which of the following statements comes closest to your recommendation for further action to be taken in the estimation of this equation?
   i. No specification changes are advisable (go to Section 11.6.3).
   ii. No variable changes are advisable, but I am concerned about heteroskedasticity or serial correlation (go to Section 11.6.3).
   iii. I would like to add seasonal dummies to the equation (go to run #11.36).
   iv. I would like to drop LYDUSP from the equation (go to run #11.40).
   v. I would like to drop PROPK from the equation (go to run #11.33).

If you need feedback on your answer, see hint #24 in the material on this chapter in Appendix A.

Regression Run 11.36

```
ORDINARY LEAST SQUARES                      DEPENDENT VARIABLE IS CONPK
SAMPLE RANGE:     1975:1-1984:4

                    COEFFICIENT       STANDARD ERROR        T-SCORE
      CONST          4.659579           0.830264            5.612163
      PRIPK         -0.017415           4.99238E-03        -3.488345
      PRIBF          6.68425E-03        2.87026E-03         2.328791
      LYDUSP         0.271936           0.290843            0.934991
      PROPK          3.100112           0.234565           13.21640
      D1             7.14724E-03        0.099132            0.072097
      D2            -0.357210           0.120017           -2.976327
      D3             4.91692E-03        0.135639            0.036249

   R-squared              0.991610      Mean of depend var      14.90000
   Adjusted R-squared     0.989775      Std dev depend var      1.575713
   Std err of regress     0.159333      Residual sum            6.08957E-14
   Durbin Watson stat     1.545712      Sum squared resid       0.812391
   F Statistic            540.3143
```

Answer each of the following questions for the above regression run.

a. Evaluate this result with respect to its economic meaning, overall fit, and the signs and significance of the individual coefficients.

b. What econometric problems does this regression have? Why? If you need feedback on your answer, see hint #22 in the material on this chapter in Appendix A.

c. Which of the following statements comes closest to your recommendation for further action to be taken in the estimation of this equation?
   i. No specification changes are advisable (go to Section 11.6.3).
   ii. No variable changes are advisable, but I am concerned about heteroskedasticity or serial correlation (go to Section 11.6.3).
   iii. I would like to drop LYDUSP from the equation (go to run #11.38).
   iv. I would like to drop the dummies from the equation (go to run #11.35).
   v. I would like to drop PROPK from the equation (go to run #11.34).

If you need feedback on your answer, see hint #24 in the material on this chapter in Appendix A.

Regression Run 11.37

ORDINARY LEAST SQUARES — DEPENDENT VARIABLE IS CONPK

SAMPLE RANGE: 1975:1-1984:4

	COEFFICIENT	STANDARD ERROR	T-SCORE
CONST	2.739559	0.319347	8.578610
PRIPK	-5.89585E-03	1.52615E-03	-3.863198
PROPK	3.655655	0.068175	53.62105
D1	0.171500	0.078294	2.190458
D2	-0.122343	0.078327	-1.561955
D3	0.279954	0.082862	3.378542

R-squared	0.989989	Mean of depend var	14.90000
Adjusted R-squared	0.988517	Std dev depend var	1.575713
Std err of regress	0.168851	Residual sum	3.27516E-14
Durbin Watson stat	1.500170	Sum squared resid	0.969367
F Statistic	672.4652		

Answer each of the following questions for the above regression run.

a. Evaluate this result with respect to its economic meaning, overall fit, and the signs and significance of the individual coefficients.

b. What econometric problems does this regression have? Why? If you need feedback on your answer, see hint #22 in the material on this chapter in Appendix A.

c. Which of the following statements comes closest to your recommendation for further action to be taken in the estimation of this equation?

i. No specification changes are advisable (go to Section 11.6.3).
ii. No variable changes are advisable, but I am concerned about heteroskedasticity or serial correlation (go to Section 11.6.3).
iii. I would like to add LYDUSP to the equation (go to run #11.32).
iv. I would like to drop the dummies from the equation (go to run #11.39).
v. I would like to replace PROPK with YDUSP (go to run #11.22).

If you need feedback on your answer, see hint #29 in the material on this chapter in Appendix A.

Regression Run 11.38

```
ORDINARY LEAST SQUARES                    DEPENDENT VARIABLE IS CONPK
SAMPLE RANGE:     1975:1-1984:4

                    COEFFICIENT        STANDARD ERROR         T-SCORE
      CONST          4.429906           0.791581             5.596272
      PRIPK         -0.014765           4.10206E-03         -3.599465
      PRIBF          6.61102E-03        2.86372E-03          2.308541
      PROPK          3.205620           0.205246            15.61843
      D1             0.030898           0.095640             0.323075
      D2            -0.320141           0.113063            -2.831516
      D3             0.049935           0.126564             0.394544

   R-squared             0.991381       Mean of depend var     14.90000
   Adjusted R-squared    0.989814       Std dev depend var     1.575713
   Std err of regress    0.159029       Residual sum           6.63358E-14
   Durbin Watson stat    1.524391       Sum squared resid      0.834585
   F Statistic           632.6324
```

Answer each of the following questions for the above regression run.

a. Evaluate this result with respect to its economic meaning, overall fit, and the signs and significance of the individual coefficients.

b. What econometric problems does this regression have? Why? If you need feedback on your answer, see hint #22 in the material on this chapter in Appendix A.

c. Which of the following statements comes closest to your recommendation for further action to be taken in the estimation of this equation?
   i. No specification changes are advisable (go to Section 11.6.3).
   ii. No variable changes are advisable, but I am concerned about heteroskedasticity or serial correlation (go to Section 11.6.3).
   iii. I would like to add LYDUSP to the equation (go to run #11.36).
   iv. I would like to drop the dummies from the equation (go to run #11.40).
   v. I would like to replace PROPK with YDUSP (go to run #11.26).

If you need feedback on your answer, see hint #29 in the material on this chapter in Appendix A.

Regression Run 11.39

```
ORDINARY LEAST SQUARES                    DEPENDENT VARIABLE IS CONPK
SAMPLE RANGE:      1975:1-1984:4

                   COEFFICIENT        STANDARD ERROR         T-SCORE
      CONST         2.980064            0.379955             7.843186
      PRIPK        -4.73003E-03         1.98603E-03         -2.381641
      PROPK         3.562273            0.082322             43.27210

   R-squared              0.980858      Mean of depend var      14.90000
   Adjusted R-squared     0.979824      Std dev depend var      1.575713
   Std err of regress     0.223816      Residual sum            7.26363E-14
   Durbin Watson stat     2.412963      Sum squared resid       1.853467
   F Statistic            948.0087
```

Answer each of the following questions for the above regression run.

a. Evaluate this result with respect to its economic meaning, overall fit, and the signs and significance of the individual coefficients.

b. What econometric problems does this regression have? Why? If you need feedback on your answer, see hint #22 in the material on this chapter in Appendix A.

c. Which of the following statements comes closest to your recommendation for further action to be taken in the estimation of this equation?
   i. No specification changes are advisable (go to Section 11.6.3).
   ii. No variable changes are advisable, but I am concerned about heteroskedasticity or serial correlation (go to Section 11.6.3).
   iii. I would like to add LYDUSP to the equation (go to run #11.31).
   iv. I would like to add seasonal dummies to the equation (go to run #11.37).
   v. I would like to replace PROPK with YDUSP (go to run #11.21).

If you need feedback on your answer, see hint #30 in the material on this chapter in Appendix A.

Regression Run 11.40

```
ORDINARY LEAST SQUARES                    DEPENDENT VARIABLE IS CONPK
SAMPLE RANGE:     1975:1-1984:4

                   COEFFICIENT         STANDARD ERROR        T-SCORE
      CONST         3.810085             0.572838           6.651243
      PRIPK        -0.010534             3.62398E-03       -2.906824
      PRIBF         4.28305E-03          2.26779E-03        1.888643
      PROPK         3.321979             0.150083          22.13423

   R-squared              0.982584      Mean of depend var      14.90000
   Adjusted R-squared     0.981133      Std dev depend var      1.575713
   Std err of regress     0.216434      Residual sum            5.44009E-14
   Durbin Watson stat     2.327164      Sum squared resid       1.686376
   F Statistic            677.0419
```

Answer each of the following questions for the above regression run.

a. Evaluate this result with respect to its economic meaning, overall fit, and the signs and significance of the individual coefficients.

b. What econometric problems does this regression have? Why? If you need feedback on your answer, see hint #22 in the material on this chapter in Appendix A.

c. Which of the following statements comes closest to your recommendation for further action to be taken in the estimation of this equation?
   i. No specification changes are advisable (go to Section 11.6.3).
   ii. No variable changes are advisable, but I am concerned about heteroskedasticity or serial correlation (go to Section 11.6.3).
   iii. I would like to add YDUSP to the equation (go to run #11.27).
   iv. I would like to add seasonal dummies to the equation (go to run #11.38).
   v. I would like to replace PROPK with LYDUSP (go to run #11.33).

If you need feedback on your answer, see hint #30 in the material on this chapter in Appendix A.

11.6.3 Heteroskedasticity and Serial Correlation in the Interactive Exercise

Now that you've chosen your "best" specification, we can discuss the possibility that pure heteroskedasticity or serial correlation might exist in your residuals.

Heteroskedasticity. This dependent variable is already per capita, is time-series, and does not change substantially during the sample period. As a result, the possibility of pure heteroskedasticity is so low that most econometricians would not even bother testing for it. (If a Park test were to be run, however, a logical proportionality factor Z might be per capita disposable income.)

Serial Correlation. Serial correlation is quite another matter, however, since the data are time-series and a number of the Durbin-Watson statistics are below the critical d_L value for positive serial correlation. In addition, it seems possible that short-run fads in the consumption of pork or, alternatively, supply shocks, might cause swings in consumption from year to year that would not be captured completely by the price variables or by coefficients estimated over the entire sample.

In particular, if your final equation was either regression run #11.26 or run #11.34, then there is a possibility of serial correlation. Their Durbin-Watson d's of 1.09 and 1.19, respectively, are right at the edge of the critical d_L for positive serial correlation for 40 observations and 6 explanatory variables.

The application of GLS to regression runs #11.26 and #11.34 are contained in regression runs #11.41 and #11.42, respectively; we include these mainly for instructional purposes, since the inconclusive Durbin-Watson statistics do not justify GLS on their own. As can be seen by comparing the GLS result with the OLS result, the correction with a $\hat{p}$ in the range of 0.43 does indeed decrease the significance of the price and income explanatory variables as would have been expected. The increase in significance of the seasonal dummies with GLS estimation raises the possibility that the seasonal pattern is not as simple as that implied by the three intercept dummies.

Note that most of the lowest Durbin-Watson statistics come with the seasonal dummy group included, supporting the hypothesis that this set of dummies doesn't properly capture the actual seasonality of the demand for pork; the introduction of the seasonal dummies might have deleted some but not all of the seasonal variation, leaving a serially correlated pattern in the residuals. While the equation is probably better off with the seasonal dummies than without, a better knowledge of the meat industry before estimation might have allowed a more sophisticated seasonal pattern to be chosen. For example, a better

Regression Run #11.41

```
FIRST ORDER AUTOCORRELATION                  DEPENDENT VARIABLE IS CONPK
SAMPLE RANGE:     1975:1-1984:4

RHO=  0.456086

                        COEFFICIENT        STANDARD ERROR          T-SCORE
      CONST              17.41717            0.801390              21.73368
      PRIPK             -0.079422            7.27183E-03          -10.92200
      PRIBF              0.040679            5.19787E-03            7.826139
      YDUSP              0.244969            0.108005               2.268130
      D1                -0.939656            0.137482              -6.834739
      D2                -1.648857            0.158989             -10.37086
      D3                -1.524040            0.134927             -11.29529

   R-squared                0.953952      Mean of depend var       14.90000
   Adjusted R-squared       0.945579      Std dev depend var        1.575713
   Std err of regress       0.367584      Residual sum              0.023041
   Durbin Watson stat       1.975891      Sum squared resid         4.458908
   F Statistic            113.9409
```

overall fit might have been obtained by using only one seasonal dummy (D_4, equal to one in the fourth quarter and zero otherwise). However, to make such a switch (or to drop one of the seasonal dummies while keeping the others) on the sole basis of the attached estimations would be a mistake because the hypothesis would be tested on the same dataset from which it was developed.

Finally, note that some of the Durbin-Watson statistics are greater than two when PROPK is included in the regression. This result almost surely occurs because the resulting equation includes both demand-side and supply-side variables, and the residuals are no longer the residuals of just the demand-side equation. This result is one of many reasons that most econometricians go to great lengths to avoid including such a production variable in a demand-side equation (except in some specific forecasting situations). In a sense, production acts like a dominant variable; its relationship to the dependent variable is strong but is definitional with little economic content.

Regression Run #11.42

```
FIRST ORDER AUTOCORRELATION                  DEPENDENT VARIABLE IS CONPK
SAMPLE RANGE:     1975:1-1984:4

RHO=  0.404512

                        COEFFICIENT        STANDARD ERROR          T-SCORE
      CONST              15.43464            0.907400              17.00973
      PRIPK             -0.078773            6.89552E-03          -11.42385
      PRIBF              0.038995            5.26234E-03            7.410364
      LYDUSP             2.050201            0.817806               2.506952
      D1                -0.924071            0.140062              -6.597590
      D2                -1.642813            0.159658             -10.28957
      D3                -1.522512            0.137046             -11.10945

   R-squared                0.954593      Mean of depend var       14.90000
   Adjusted R-squared       0.946338      Std dev depend var        1.575713
   Std err of regress       0.365014      Residual sum             -0.085264
   Durbin Watson stat       1.969933      Sum squared resid         4.396761
   F Statistic            115.6292
```

11.7 Appendix: The Housing Price Interactive Exercise

This final interactive regression learning exercise is somewhat different from the first three. Our goal is still to bridge the gap between textbook and computer, but we feel that if you've completed the previous interactive exercises you should be ready to do the computer work on your own. As a result, this interactive exercise will provide you with a short literature review and the data, but you'll be asked to calculate your own estimates. Feedback on your specification choices will once again be found in the hints in Appendix A.

Since the only difference between this interactive exercise and the other three is that this one requires you to estimate your chosen specification(s) with the computer, our guidelines for interactive exercises still apply:

1. Take the time to look over a portion of the reading list before choosing a specification.
2. Try to estimate as few regression runs as possible.
3. Avoid looking at the hints until after you've reached what you think is your best specification.

We believe that the benefits you get from an interactive exercise are directly proportional to the effort you put into it. If you have to delay this exercise until you have the time and energy to do your best, that's probably a good idea.

11.7.1 Building a Hedonic Model of Housing Prices

In the next section, we're going to ask you to specify the independent variables and functional form for an equation whose dependent variable is the price of a house in Southern California. Before making these choices, it's vital to review the housing price literature and to think through the theory behind such models. Such a review is especially important in this case because the model we'll be building will be *hedonic* in nature.

What is a hedonic model? Recall that in Section 1.4 we estimated an equation for the price of a house as a function of the size of that house. Such a model is called **hedonic** because it uses measures of the quality of a product as independent variables instead of measures of the market for that product (like quantity demanded, income, etc.). Hedonic models are most useful when the product being analyzed is heterogeneous in nature because we need to analyze what causes products to be different and therefore to have different prices. With a homogenous product, hedonic models are virtually useless.

Perhaps the most-cited early hedonic housing price study is that of Grether and Mieszkowski.[6] Grether and Mieszkowski collected a seven-year data-set and built a number of linear models of housing price using different combinations of variables. They included square feet of space, the number of bathrooms, and the number of rooms, although the latter turned out to be insignificant. They also included lot size and the age of the house as variables, specifying a quadratic function for the age variable. Most innovatively, they used several slope dummies in order to capture the interaction effects of various combinations of variables (like a hardwood floors dummy times the size of the house).

Peter Linneman[7] estimated a housing price model on data from Los Angeles, Chicago, and the entire U.S. His goal was to create a model that worked for the two individual cities and then to apply it to the nation to test the hypothesis of a national housing market. Linneman did not include any lot characteristics, nor did he use any interactive variables. His only measures of the size of the living space were the number of bathrooms and the number of nonbathrooms. Except for an age variable, the rest of the independent variables were dummies describing quality characteristics of the house and the neighborhood. While many of the dummy variables were quite fickle, the age, number of bathrooms, and the number of nonbathrooms were relatively stable and significant. Central air conditioning had a negative, insignificant coefficient for the Los Angeles regression.

Ihlanfeldt and Martinez-Vasquez[8] investigated sample bias in various methods of obtaining house price data and concluded that the house's sales price is the least biased of all measures. Unfortunately, they went on to estimate an equation by starting with a large number of variables and then dropping all those that had t-scores below one, almost surely introducing bias into their equation.

6. G. M. Grether and Peter Mieszkowski, "Determinants of Real Estate Values," *Journal of Urban Economics*, April 1974, pp. 127–146. Another classic article of the same era is J. Kain and J. Quigley, "Measuring the Value of Housing Quality," *Journal of American Statistical Association,* June 1970.

7. Peter Linneman, "Some Empirical Results on the Nature of Hedonic Price Functions for the Urban Housing Market," *Journal of Urban Economics*, July 1980, pp. 47–68.

8. Keith Ihlanfeldt and Jorge Martinez-Vasquez, "Alternative Value Estimates of Owner-Occupied Housing: Evidence on Sample Selection Bias and Systematic Errors," *Journal of Urban Economics*, Nov. 1986, pp. 356–369. Also see Eric Cassel and Robert Mendelsohn, "The Choice of Functional Forms for Hedonic Price Equations: Comment," *Journal of Urban Economics*, Sept. 1985, pp. 135–142.

Finally, Allen Goodman[9] added some innovative variables to a Box-Cox estimate on a national data-set. He included measures of specific problems like rats, cracks in the plaster, holes in the floors, plumbing breakdowns, and the level of property taxes. While the property tax variable showed the capitalization of low property taxes, as would be expected, the rats variable was insignificant, and the cracks variable's coefficient asserted that cracks significantly increase the value of a house.

11.7.2 The Housing Price Interactive Exercise

Now that we've reviewed at least a portion of the literature, it's time to build your own model. Recall that in Section 1.4, we built a simple model of the price of a house as a function of the size of that house, Equation 1.23:

$$\hat{P}_i = 40.1 + 0.138S_i \tag{1.23}$$

where: P_i = the price (in thousands of dollars) of the *i*th house
S_i = the size (in square feet) of the *i*th house

Equation 1.23 was estimated on a sample of 43 houses that were purchased in the same Southern California town (Monrovia) within a few weeks of each other. It turns out that we have a number of additional independent variables for the data-set we used to estimate equation 1.23. Also available are:

N_i = the quality of the neighborhood of the *i*th house (1 = best, 4 = worst) as rated by two local real estate agents
A_i = the age of the *i*th house in years
BE_i = the number of bedrooms in the *i*th house
BA_i = the number of bathrooms in the *i*th house
C_i = a dummy variable equal to 1 if the *i*th house has central air conditioning, 0 otherwise
SP_i = a dummy variable equal to 1 if the *i*th house has a pool, 0 otherwise
Y_i = the size of the yard around the *i*th house (in square feet)

9. Allen C. Goodman, "An Econometric Model of Housing Price, Permanent Income, Tenure Choice, and Housing Demand," *Journal of Urban Economics*, May 1988, pp. 327–353.

TABLE 11.8 DATA FOR THE HOUSING PRICE INTERACTIVE EXERCISE

P	S	N	A	BE	BA	C	SP	Y
107	736	4	39	2	1	0	0	3364
133	720	3	63	2	1	0	0	1780
141	768	2	66	2	1	0	0	6532
165	929	3	41	3	1	0	0	2747
170	1080	2	44	3	1	0	0	5520
173	942	2	65	2	1	0	0	6808
182	1000	2	40	3	1	0	0	6100
200	1472	1	66	3	2	0	0	5328
220	1200	1.5	69	3	1	0	0	5850
226	1302	2	49	3	2	0	0	5298
260	2109	2	37	3	2	1	0	3691
275	1528	1	41	2	2	0	0	5860
280	1421	1	41	3	2	0	1	6679
289	1753	1	1	3	2	1	0	2304
295	1528	1	32	3	2	0	0	6292
300	1643	1	29	3	2	0	1	7127
310	1675	1	63	3	2	0	0	9025
315	1714	1	38	3	2	1	0	6466
350	2150	2	75	4	2	0	0	14825
365	2206	1	28	4	2.5	1	0	8147
503	3269	1	5	4	2.5	1	0	10045
135	936	4	75	2	1	0	0	5054
147	728	3	40	2	1	0	0	1922
165	1014	3	26	2	1	0	0	6416
175	1661	3	27	3	2	1	0	4939
190	1248	2	42	3	1	0	0	7952
191	1834	3.5	40	3	2	0	1	6710
195	989	2	41	3	1	0	0	5911
205	1232	1	43	2	2	0	0	4618
210	1017	1	38	2	1	0	0	5083
215	1216	2	77	2	1	0	0	6834
228	1447	2	44	2	2	0	0	4143
242	1974	1.5	65	4	2	0	1	5499
250	1600	1.5	63	3	2	1	0	4050
250	1168	1.5	63	3	1	0	1	5182
255	1478	1	50	3	2	0	0	4122
255	1756	2	36	3	2	0	1	6420
265	1542	2	38	3	2	0	0	6833
265	1633	1	32	4	2	0	1	7117
275	1500	1	42	2	2	1	0	7406
285	1734	1	62	3	2	0	1	8583
365	1900	1	42	3	2	1	0	19580
397	2468	1	10	4	2.5	1	0	6086

Read through the list of variables again, developing your own analyses of the theory behind each variable. What are the expected signs of the coefficients? Which variables seem potentially redundant? Which variables *must* you include?

In addition, there are a number of functional form modifications that can be made. For example, you might consider a quadratic polynomial for age, as Grether and Mieszkowski did, or you might consider creating slope dummies such as S·PS or C·S. Finally, you might consider interactive variables that involve the neighborhood proxy variable such as N·S or N·BA. What hypotheses would each of these functional forms imply?

Develop your specification carefully. Think through each variable and/or functional form decision, and take the time to write out your expectations for the sign and size of each coefficient. Don't take the attitude that you should include *every* possible variable and functional form modification and then drop the insignificant ones. Instead, try to design the best possible hedonic model of housing prices you can the first time around.

Once you've chosen a specification, estimate your equation, using the data in Table 11.8, and analyze the result.

1. Test your hypotheses for each coefficient with the *t*-test. Pay special attention to any functional form modifications.
2. Test the overall significance of the equation with the *F*-test.
3. Decide what econometric problems exist in the equation, testing, if appropriate, for multicollinearity, serial correlation, or heteroskedasticity.
4. Decide whether to accept your first specification as the best one or to make a modification in your equation and estimate again. Make sure you avoid the temptation to estimate an additional specification "just to see what it looks like."

Once you've decided to make no further changes, you're finished—congratulations! Now turn to the hints for this Section in Appendix A for feedback on your choices.

12

Distributed Lag Models

Ever since we introduced the idea of lags in Section 3.1, we've used lagged independent variables whenever we've expected X to affect Y after a period of time. For example, if the underlying theory has suggested that X_1 affects Y with a one-time-period lag (but X_2 has an instantaneous impact on Y) we've used equations like:

$$Y_t = \beta_0 + \beta_1 X_{1t-1} + \beta_2 X_{2t} + \epsilon_t \qquad (12.1)$$

Such lags are called simple lags, and the estimation of β_1 with OLS is no more difficult than the estimation of the coefficients of nonlagged equations, except for possible impure serial correlation if the lag is misspecified. Remember, however, that the coefficients of such equations should be interpreted carefully. For example, β_2 in Equation 12.1 measures the effect of a one-unit change in this time's X_2 on this time's Y holding *last time's* X_1 constant.

A case more complicated than this simple lag model occurs when the impact of an independent variable is expected to be spread out over a number of time periods. For example, suppose we're interested in studying the impact of a change in the money supply on GNP. Theoretical and empirical studies have provided evidence that because of rigidities in the marketplace, it takes time for the economy to react completely to a change in the money supply. Some of the effect on GNP will take place the first quarter, some more in the second quarter, and so on. In such a case, the appropriate econometric model would be a distributed lag model:

$$Y_t = a_0 + \beta_0 X_t + \beta_1 X_{t-1} + \beta_2 X_{t-2} + \cdots + \beta_p X_{t-p} + \epsilon_t \quad (12.2)$$

A **distributed lag model** explains the current value of Y as a function of current and past values of X, thus "distributing" the impact of X over a number of time periods. Take a careful look at Equation 12.2. The coefficients β_0, β_1, and β_2 through β_p measure the effects of the various lagged values of X on the current value of Y. In most economic applications, including our money supply example, we'd expect the impact of X on Y to decrease as the length of the lag (indicated by the subscript of the β) increases. That is, while β_0 might be larger or smaller than β_1, we certainly would expect either β_0 or β_1 to be larger in absolute value than β_6 or β_7.

Unfortunately, the estimation of Equation 12.2 with OLS causes a number of problems:

1. The various lagged values of X are likely to be severely multicollinear, making coefficient estimates imprecise.
2. In large part because of this multicollinearity, there is no guarantee that the estimated βs will follow the smoothly declining pattern that economic theory would suggest. Instead, it's quite typical for the estimated coefficients of Equation 12.2 to follow a fairly irregular pattern, for example:

$$\hat{\beta}_0 = 0.26 \quad \hat{\beta}_1 = 0.07 \quad \hat{\beta}_2 = 0.17 \quad \hat{\beta}_3 = -0.03 \quad \hat{\beta}_4 = 0.08$$

3. The degrees of freedom tend to decrease, sometimes substantially, for two reasons. First, we have to estimate a coefficient for each lagged X, thus increasing K and lowering the degrees of freedom $(n-K-1)$. Second, unless data for lagged Xs outside the sample are available, we have to decrease the sample size by one for each lagged X we calculate, thus lowering the number of observations, n, and therefore the degrees of freedom.

As a result of these problems with OLS estimation of functions like Equation 12.2, called ad hoc distributed lag equations, it's standard practice to use a simplifying assumption to avoid these problems. The most commonly used simplification is the Koyck model.

12.1 Koyck Distributed Lag Models

12.1.1 What Are Koyck Lags?

A **Koyck distributed lag model** is one that assumes that the coefficients of the lagged variables decrease in a geometric fashion the longer the lag:[1]

$$\beta_i = \beta_0\lambda^i \tag{12.3}$$

where i is the length of the lag, 1,2,...,p and $0 < \lambda < 1$. For example, $\beta_3 = \beta_0\lambda^3$.

If we substitute Equation 12.3 into Equation 12.2 for each coefficient and factor out β_0, we obtain:

$$Y_t = a_0 + \beta_0(X_t + \lambda X_{t-1} + \lambda^2X_{t-2} + \lambda^3X_{t-3} + \cdots) + \epsilon_t \tag{12.4}$$

Since we have assumed that λ is between zero and one, λ to the $(n + 1)$th power is smaller than λ to the nth power. As a result, each successive lagged term has a smaller coefficient than the previous term. For example, if $\lambda = 0.2$, then Equation 12.4 becomes:

$$Y_t = a_0 + \beta_0(X_t + 0.2X_{t-1} + 0.04X_{t-2} + 0.008X_{t-3} + \cdots) + \epsilon_t \tag{12.5}$$

because $(0.2)^1 = 0.2$, $(0.2)^2 = 0.04$, etc. As can be seen in Figure 12.1, each successive lagged value does indeed have relatively less weight in determining the current value of Y.

How can we estimate Equation 12.4? As it stands, the equation is nonlinear in the coefficients and can't be estimated with OLS. However, with some manipulation, Equation 12.4 can be transformed into an equivalent equation that is linear in the coefficients α_0, β_0, and λ. To see this, multiply both sides of Equation 12.4 by λ and lag it once (that is, substitute $t-1$ for t in every instance that it appears):

$$\lambda Y_{t-1} = \lambda a_0 + \beta_0(\lambda X_{t-1} + \lambda^2X_{t-2} + \lambda^3X_{t-3} + \cdots) + \lambda\epsilon_{t-1} \tag{12.6}$$

1. L. M. Koyck, *Distributed Lags and Investment Analysis* (Amsterdam: North-Holland Publishing Company, 1954).

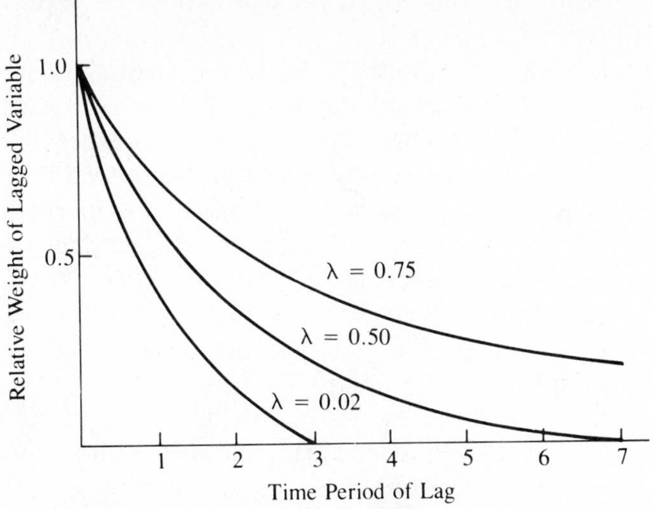

Figure 12.1 GEOMETRIC WEIGHTING SCHEMES FOR VARIOUS KOYCK LAG MODELS
No matter what λ is used, a Koyck distributed lag model has the impact of the independent variable declining as the length of the lag increases.

Now subtract the resulting equation (Equation 12.6) from Equation 12.4 and rewrite, producing:

$$Y_t = \alpha_0 + \beta_0 X_t + \lambda Y_{t-1} + u_t \tag{12.7}$$

where $u_t = \epsilon_t - \lambda\epsilon_{t-1}$ and where $\alpha_0 = a_0 - \lambda a_0$ (still a constant). Compare Equation 12.7 with Equation 12.4; the Koyck transformation has changed a distributed lag equation into an equation with a lagged dependent variable, often called an *autoregressive* equation.

This equation (Equation 12.7) then becomes the estimating equation. OLS can be applied to Equation 12.7 as long as the error term, u_t, meets the Classical Assumptions and as long as the sample is large enough. How large is "large enough?" Our recommendation, based more on experience than proof, is to aim for a sample size of at least 50 observations.[2] The smaller the sample, the more likely you are to encounter bias. Samples below 25 in size should be avoided entirely,

2. If u_t is well-behaved, OLS estimation of Equation 12.7 can be shown to have desirable properties for extremely large samples, but not enough is known about the small sample properties of this model to issue firm sample size guidelines. See H. Doran and W. Griffiths, "Inconsistency of the OLS Estimator of the Partial-Adjustment-Adaptive Expectations Model," *Journal of Econometrics*, 1978, pp. 133–146.

in part because of bias and in part because hypothesis testing becomes untrustworthy.

Koyck distributed lag models have other potentially serious problems. If the original error term ϵ_t satisfies all the Classical Assumptions, then the error term in the Koyck model, u_t, is almost sure to be serially correlated. If the error term u_t is serially correlated, OLS estimation of Equation 12.7 is biased no matter how large the sample. These problems will be discussed in Section 12.2. Also, the presence of a lagged dependent variable as an independent variable forces every other independent variable in the equation to be related to the dependent variable by a declining geometric (Koyck) distributed lag function.

As an example of this latter problem, suppose that we want to build a model of Y that includes a distributed lag of X_1 and also includes X_2 as an unlagged independent variable. Suppose further that we use the following equation:

$$Y_t = \alpha_0 + \beta_0 X_{1t} + \lambda Y_{t-1} + \beta_1 X_{2t} + u_t \qquad (12.8)$$

The use of the λY_{t-1} term forces both X_1 *and* X_2 to have Koyck distributed lag patterns with respect to Y. Thus if we don't expect X_2 to have lagged effects on Y, a Koyck function is not an appropriate model in this case.

12.1.2 An Example of Koyck Distributed Lags

As an example of a Koyck distributed lag model, let's look at an aggregate consumption function from a macroeconomic equilibrium GNP model. Many economists argue that in such a model consumption is not an instantaneous function of income. Instead, they maintain that current purchases of goods and services (C_t) are influenced by past levels of disposable income (Yd_{t-1}, Yd_{t-2}, etc.) as well as current levels of disposable income (Yd_t):

$$C_t = f(\overset{+}{Yd_t}, \overset{+}{Yd_{t-1}}, \overset{+}{Yd_{t-2}}, \text{etc.}) \qquad (12.9)$$

Such an equation fits well with simple models of consumption, but it only makes sense if the weights given past levels of income decrease as the length of the lag increases. That is, we would expect the coefficient of Yd_{t-2} to be less than the coefficient of Yd_{t-1}, and so on.

As a result, most econometricians would model Equation 12.9 with a Koyck distributed lag equation:

$$C_t = \alpha_0 + \beta_0 Yd_t + \lambda C_{t-1} + u_t \quad (12.10)$$

This equation not only fits a simple model of aggregate consumption, but it also is quite close to that suggested by Milton Friedman for his permanent income hypothesis.[3] In that hypothesis, Friedman suggested that consumption was based not on current income but instead on the consumer's perception of lifetime income. Consequently, changes in transitory income wouldn't affect consumption. Since it's reasonable to hypothesize that perceptions of permanent income are based on past levels of income, the simple consumption model and the more sophisticated permanent income model have similar equations.

To estimate Equation 12.10, we use data from Section 14.3, where we will build a small simultaneous equation macro model of the U.S. economy from 1964 through 1988. The OLS estimates of Equation 12.10 for this data-set are (standard errors in parentheses):

$$\begin{aligned} \hat{C}_t = & -55.29 + 0.64Yd_t + 0.34C_{t-1} \quad (12.11) \\ & \qquad\qquad (0.17) \qquad\quad (0.18) \\ & \qquad t = \ \ 3.8 \qquad\qquad 1.9 \\ \bar{R}^2 = .996 & \qquad n = 25 \text{ (annual 1964–1988)} \end{aligned}$$

One way to analyze the coefficients of this equation is to convert the estimates back into the format of Equation 12.9 by using the Koyck distributed lag definitions of the individual coefficients, Equation 12.3:

$$\beta_i = \beta_0 \lambda^i \quad (12.3)$$

If we substitute $\beta_0 = 0.64$ and $\lambda = 0.34$ into Equation 12.3 for $i = 1$, we obtain $\beta_1 = \beta_0\lambda^1 = (0.64)(0.34)^1 = 0.22$. If we continue this process, it turns out that Equation 12.11 is equivalent to:[4]

$$\hat{C}_t = -36.49 + 0.64Yd_t + 0.22Yd_{t-1} + 0.07Yd_{t-2} + 0.03Yd_{t-3} \quad (12.12)$$

In actuality, this equation keeps going out to the end of the data-set, but since the coefficient of Y_{t-4} is less than 0.01, it's close enough to zero to be disregarded.

3. Milton Friedman, *A Theory of the Consumption Function* (Princeton, N.J.: Princeton University Press/National Bureau of Economic Research, 1957). It's interesting to note, however, that Friedman's original function did not include a constant term because of the nature of his derivation of permanent income.

4. Note that the constant term equals $\hat{\alpha}_0/(1 - \hat{\lambda})$.

To compare this estimate with an OLS estimate of the same equation without the Koyck lag format, we'd need to estimate an ad hoc distributed lag equation with the same number of lagged variables:

$$C_t = \alpha_0 + \beta_0 Yd_t + \beta_1 Yd_{t-1} + \beta_2 Yd_{t-2} + \beta_3 Yd_{t-3} + \epsilon_t \quad (12.13)$$

As expected, the coefficients of the ad hoc estimate no longer follow the same smoothly declining pattern that both our theoretical expectations and the Koyck model follow:

$$\hat{C}_t = -115.1 + 1.13Yd_t - 0.02Yd_{t-1} - 0.06Yd_{t-2} - 0.11Yd_{t-3} \quad (12.14)$$

How do the coefficients of Equation 12.14 look? Note that the coefficient of Yd_t is greater than one, implying that in the short run, we'd increase spending by more than the causal increase in disposable income. In addition, all the coefficients of the lagged variables are negative but increase in absolute value as the lag increases. Neither economic theory nor common sense leads us to expect this pattern. Such a poor result is due to the severe multicollinearity between the lagged Xs. Most econometricians therefore estimate distributed lag models with a lagged dependent variable simplication scheme like the Koyck function in Equation 12.10.

An interesting interpretation of the results in Equation 12.11 concerns the long-run multiplier implied by the model. The long-run multiplier measures the total impact of a change in income on consumption after all the lagged effects have been felt. An estimate of the long-run multiplier can be obtained from Equation 12.11 by calculating $\hat{\beta}_0[1/(1 - \hat{\lambda})]$, which in this case equals $0.64[1/(1 - 0.34)]$ or 0.97. A sample of this size is likely to encounter small sample bias, however, so we shouldn't overanalyze the results. For more on this data-set and the other equations in the model, see Section 14.3. For more on testing and adjusting distributed lag equations like Equation 12.11 for serial correlation, let's move on to the next section.

12.2 Serial Correlation and Koyck Distributed Lags

Perhaps the most serious drawback of Koyck lags is the high probability of serial correlation. Equations with a lagged dependent variable are more likely to encounter serial correlation than are other equations.

To understand why, take a look at a typical Koyck error term from Equation 12.7:

$$u_t = \epsilon_t - \lambda\epsilon_{t-1}$$

Do you see the problem? This time's error term (u_t) is a moving average of ϵ_t, so ϵ_{t-1} affects both u_t and u_{t-1}. Since u_t and u_{t-1} are affected by the same variable (ϵ_{t-1}), they're almost sure to be correlated, violating Classical Assumption IV.[5]

Compounding this, the consequences, detection, and remedies for serial correlation that we discussed in Chapter 9 are all either incorrect or need to be modified in the presence of a lagged dependent variable. Since all Koyck distributed lag models include a lagged dependent variable, a discussion of serial correlation in Koyck models is critical to an understanding of distributed lags.

12.2.1 Serial Correlation Causes Bias in Koyck Lag Models

In Section 9.2, we stated that pure serial correlation does not cause bias in the estimates of the coefficients. Unfortunately, the use of a Koyck distributed lag model changes all that. More specifically, if an equation that contains a lagged dependent variable as an independent variable has a serially correlated error term, then OLS estimates of the coefficients of that equation will be biased, even in large samples.

To see where this bias comes from, let's look at an equation with a lagged dependent variable and a serially correlated error term of the type ($u_t = \epsilon_t - \lambda\epsilon_{t-1}$, called a moving average) found in Koyck distributed lag equations:

$$\overset{\downarrow}{Y_t} = \alpha_0 + \beta_0 X_t + \beta_1 \overset{\uparrow}{Y_{t-1}} + \epsilon_t - \lambda \overset{\uparrow}{\epsilon_{t-1}} \tag{12.15}$$

Let's also look at Equation 12.15 lagged one time period:

$$\overset{\uparrow}{Y_{t-1}} = \alpha_0 + \beta_0 X_{t-1} + \beta_1 Y_{t-2} + \overset{\uparrow}{\epsilon_{t-1}} - \lambda\epsilon_{t-2} \tag{12.16}$$

5. More formally, if we start with $u_t = \epsilon_t - \lambda\epsilon_{t-1}$, lag this equation one time period, solve for ϵ_{t-1}, and then substitute back into the original definition of u_t, we obtain: $u_t = \epsilon_t - \lambda u_{t-1} - \lambda^2\epsilon_{t-2}$. In other words, this time's error term u_t is a function of last time's error term, u_{t-1}.

What happens when last time period's error term (ϵ_{t-1}) is positive? In Equation 12.16, the positive ϵ_{t-1} causes Y_{t-1} to be larger than it would have been otherwise (these changes are marked by upward pointing arrows). In Equation 12.15, the positive ϵ_{t-1} causes Y_t to be smaller (also marked) because it is multiplied by $-\lambda$. In addition, note that Y_{t-1} appears on the right-hand side of Equation 12.15, and we already know that it has increased.

Take a look at the right-hand side of Equation 12.15. Every time ϵ_{t-1} is positive, Y_{t-1} will be larger than otherwise. Thus the two variables are correlated and, therefore, so are u_t and Y_{t-1}. Such a situation violates Classical Assumption III that the error term is not correlated with any of the explanatory variables. What happens if ϵ_{t-1} is negative? In this case, Y_{t-1} will be lower than it would have been otherwise, so they're still correlated.

The consequences of this correlation include biased estimates, in particular of the coefficient β_1. In essence, the uncorrected serial correlation acts like an omitted variable (ϵ_{t-1}). Since an omitted variable causes bias whenever it is correlated with one of the included independent variables, and since ϵ_{t-1} is correlated with Y_{t-1}, the combination of a lagged dependent variable and serial correlation causes bias in the coefficient estimates.[6]

Serial correlation in a Koyck lag model also causes estimates of the standard errors of the $\hat{\beta}$s and the residuals to be biased. The former bias means that hypothesis testing is invalid, even for large samples. The latter bias means that tests based on the residuals, like the Durbin-Watson d test, are potentially invalid.

12.2.2 Testing Koyck Lag Models for Serial Correlation

Until now, we've relied on the Durbin-Watson d test of Section 9.3 to test for serial correlation, but, as mentioned above, the Durbin-Watson d is potentially invalid for an equation that contains a lagged dependent variable as an independent variable. This is because the biased residuals described in the previous paragraph cause the DW d statistic to be biased toward 2. This bias toward 2 means that the Durbin-Watson test sometimes fails to detect the presence of serial correlation in a Koyck (or similar) lag model.[7]

6. The reason that pure serial correlation doesn't cause bias in the coefficient estimates of equations that don't include a lagged dependent variable is that the "omitted variable" ϵ_{t-1} isn't correlated with any of the included independent variables.

7. The opposite is not a problem. A Durbin-Watson d test that indicates serial correlation in the presence of a lagged dependent variable, despite the bias toward 2, is an even stronger affirmation of serial correlation.

The most-used alternative is **Durbin's h test,**[8] which is a large-sample method of adjusting the Durbin-Watson d statistic to test for first-order serial correlation in the presence of a lagged dependent variable. The equation for Durbin's h statistic is:

$$h = (1 - 0.5 \cdot d)\sqrt{\frac{n}{(1 - n \cdot [S_{\lambda}^2])}} \qquad (12.17)$$

where d is the Durbin-Watson statistic, n is the sample size, and S_{λ}^2 is the square of the estimated standard error of $\hat{\lambda}$, the estimated coefficient of Y_{t-1}.

Durbin's h is normally distributed, so a 95 percent two-tailed test implies a critical z-value of 1.96. Therefore, the decision rule is:

If the absolute value of h is greater than 1.96, reject the null hypothesis of no first-order serial correlation.

If the absolute value of h is less than 1.96, do not reject the null hypothesis of no first-order serial correlation.

As an example, let's test our Koyck distributed lag aggregate consumption function, Equation 12.11, for serial correlation:

$$\begin{aligned} \hat{C}_t = &- 55.29 + 0.64Yd_t + 0.34C_{t-1} \qquad (12.11)\\ &\qquad\quad (0.17) \qquad (0.18)\\ &\quad t = \;\; 3.8 \qquad\quad 1.9\\ \bar{R}^2 = .996 &\quad n = 25 \quad DW\ d = 0.69 \end{aligned}$$

Substituting into Equation 12.17, we obtain:

$$h = (1 - 0.5 \cdot 0.69)\sqrt{\frac{25}{(1 - 25 \cdot [0.18]^2)}} = 7.51 \qquad (12.18)$$

Since 7.51 > 1.96, we can reject the null hypothesis of no first-order serial correlation and conclude that, as expected, there is indeed significant serial correlation in our consumption function.

Durbin's h test has at least two problems, however. First, the test statistic is undefined in certain circumstances (when $n \cdot [S_{\lambda}^2] \geq 1$) because the value under the square root sign in Equation 12.17 is negative. Second, Durbin's h test cannot be used if the model in question has

8. J. Durbin, "Testing for Serial Correlation in Least Squares Regression When Some of the Regressors Are Lagged Dependent Variables," *Econometrica*, 1970, pp. 410–421.

more than one lagged dependent variable or if the serial correlation being tested for isn't first-order in nature. While some researchers adjust Durbin's h in these situations, a more satisfactory solution would be to find a test that avoids these two problems altogether while working just as well in the simple case.

One such test is the Lagrange Multiplier test. The **Lagrange Multiplier test (LM)** is a method that can be used to test for serial correlation in the presence of a lagged dependent variable by analyzing how well the lagged residuals explain the residuals of the original equation (in an equation that includes all the explanatory variables of the original model). If the lagged residuals are significant in explaining this time's residuals (as shown by the Chi-square test), then we can reject the null hypothesis of no serial correlation. The Lagrange Multiplier test also is useful as a specification test and as a test for heteroskedasticity and other econometric problems.[9]

Using the Lagrange Multiplier to test for serial correlation for a typical Koyck model involves three steps:

1. Obtain the residuals from the estimated equation:

$$e_t = Y_t - \hat{Y}_t = Y_t - \hat{\alpha}_0 - \hat{\beta}_0 X_{1t} - \hat{\lambda} Y_{t-1} \qquad (12.19)$$

2. Use these residuals as the dependent variable in an auxiliary equation that includes as independent variables all those on the right-hand side of the original equation as well as the lagged residuals:

$$e_t = a_0 + a_1 X_t + a_2 Y_{t-1} + a_3 e_{t-1} + u_t \qquad (12.20)$$

3. Estimate Equation 12.20 using OLS and then test the null hypothesis that $a_3 = 0$ with the following test statistic:

$$LM = n \cdot R^2 \qquad (12.21)$$

where n is the sample size and R^2 is the unadjusted coefficient of determination, both of the auxiliary equation, Equation 12.20. For large samples, LM has a Chi-square distribution with degrees of

9. For example, some readers may remember that the Breusch-Pagan and White tests of Section 10.3.4 are Lagrange Multiplier tests. For a survey of the various uses to which Lagrange Multiplier tests can be put and a discussion of the LM test's relationship to the Wald and Likelihood Ratio tests, see Robert F. Engle, "Wald, Likelihood Ratio, and Lagrange Multiplier Tests in Econometrics," in Z. Griliches and M. D. Intriligator, *Handbook of Econometrics,* Volume II (Amsterdam: Elsevier Science Publishers, 1984).

freedom equal to the number of restrictions in the null hypothesis (in this case, one). If LM is greater than the critical Chi-square value from Statistical Table B-8, then we reject the null hypothesis that
$a_3 = 0$ and conclude that there is indeed serial correlation in the original equation.

To run an LM test for second-order or higher-order serial correlation, add lagged residuals (e_{t-2} for second-order, e_{t-2} and e_{t-3} for third-order) to the auxiliary equation, Equation 12.20. This latter change makes the null hypothesis $a_3 = a_4 = a_5 = 0$. Such a null hypothesis raises the degrees of freedom in the Chi-square test to three because we have imposed three restrictions on the equation (three coefficients are jointly set equal to zero). To run an LM test with more than one lagged dependent variable, add the lagged variables (Y_{t-2}, Y_{t-3}, etc.) to the original equation. For practice with the LM test, see Exercise 10; for practice with testing for higher-order serial correlation, see Exercise 11.

12.2.3 Correcting for Serial Correlation in Koyck Models

There are three strategies for attempting to rid a Koyck lag model (or a similar model) of serial correlation: improving the specification, instrumental variables, and modified GLS.

The first strategy is to consider the possibility that the serial correlation could be impure, caused by either omitting a relevant variable or by failing to capture the actual distributed lag pattern accurately. Unfortunately, finding an omitted variable or an improved lag structure is easier said than done, as we've seen in previous chapters. Because of the dangers of sequential specification searches, this option should be considered only if an alternative specification exists that has a theoretically sound justification.

The second strategy, called instrumental variables, consists of substituting a proxy (an "instrument") for Y_{t-1} in the original equation, thus eliminating the correlation between Y_{t-1} and u_t. While using an instrument is a reasonable option that is straightforward in principle, it's not always easy to find a proxy that retains the distributed lag nature of the original equation. For a more complete discussion of instrumental variables, see Section 14.3.

The final solution to serial correlation in Koyck models (or in models with lagged dependent variables and similar error term structures) is to use an iterative maximum likelihood technique to estimate the components of the serial correlation and then to transform the original equation so that the serial correlation has been eliminated. This technique, which is similar to the GLS procedure outlined in Section 9.4, is not without its complications. In particular, the sample needs to be large, the standard errors of the estimated coefficients potentially need to be adjusted, and the estimation techniques are flawed under some circumstances.[10] In essence, serial correlation causes bias in Koyck distributed lag models, but ridding the equation of that serial correlation is not an easy task.

12.3 Almon Polynomial Distributed Lag Models

While Koyck lags avoid some of the problems inherent in ad hoc distributed lag models, there are situations where the Koyck formulation isn't appropriate. In particular, recall that the shape of the Koyck lag pattern over time is a steadily declining one (Panel A in Figure 12.2). While this often is exactly what we want, there are times when we'd expect some other kind of shape.

For example, suppose Coca-Cola announces that it is moving its corporate headquarters to your city, and the mayor asks you to build a distributed lag model of the impact of such announcements on the creation of local jobs. At first, the announcement would have little impact on jobs because it takes time to move a large company. After a while, the new corporation would start hiring and local firms would add workers to better serve the expanded local economy. Eventually, though, job creation would tail off, as the new firm finished its hiring and the "multiplier" effects started to die down.

So, we'd expect the number of new jobs created by the move to start out low, rise to a peak, and then decline steadily, as depicted in Panel B of Figure 12.2. While such a shape is difficult to model with a Koyck lag structure, it's exactly the kind of shape for which the Almon lag model was created.

12.3.1 What Are Almon Lags?

An **Almon Polynomial Distributed Lag Model** is an estimation procedure for distributed lags that allows the coefficients of the lagged

10. For more on these complications, see A.C. Harvey, *The Econometric Analysis of Time Series* (New York: Wiley, 1981), and R. Betancourt and H. Kelejian, "Lagged Endogenous Variables and the Cochrane-Orcutt Procedure," *Econometrica*, 1981, pp. 1073–1078.

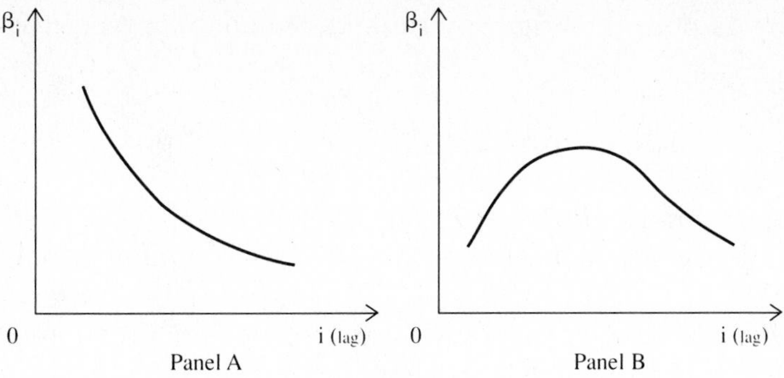

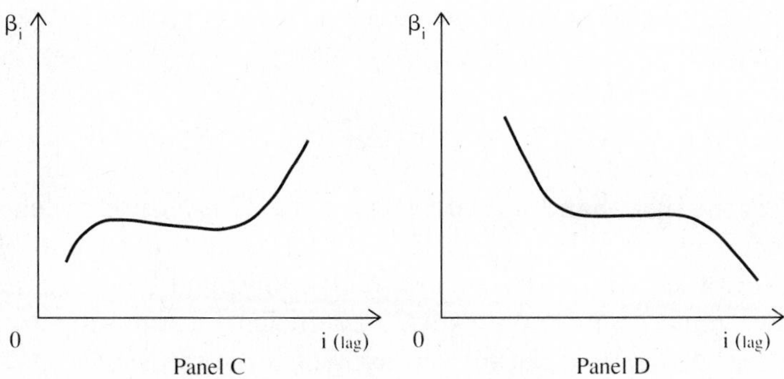

Figure 12.2 KOYCK VS. ALMON DISTRIBUTED LAGS
Panel A is best modeled with a Koyck lag, but Panels B, C, & D are most easily modeled with an Almon Polynomial Distributed Lag.

independent variables to follow a variety of continuous patterns as the length of the lag increases.[11] While coefficients from a simple Koyck model are limited to a steadily decreasing pattern (Panel A in Figure 12.2), the Almon model can allow coefficients to first increase and then decrease (Panel B) or to change direction more than once (Panels C and D).

To accomplish this, an Almon lag model starts with the same

11. Shirley Almon, "The Distributed Lag Between Capital Appropriations and Expenditures," *Econometrica,* 1965, pp. 178–196.

equation as does the original ad hoc distributed lag model, Equation 12.2:

$$Y_t = a_0 + \beta_0 X_t + \beta_1 X_{t-1} + \beta_2 X_{t-2} + \cdots + \beta_p X_{t-p} + \epsilon_t \quad (12.2)$$

An Almon model differs from an ad hoc one because the βs are estimated in a way that forces them to follow a continuous polynomial pattern.

Almon distributed lag models are "polynomial" in nature because the functional form that best describes Panels B (or C or D) is a polynomial. For example, recall from Section 7.1 that we used a quadratic polynomial functional form when we expected an inverted U shape similar to Panel B. The difference between the polynomials we've used before and the Almon polynomial is that instead of squaring or cubing the independent variables, it's the *coefficients* of the Almon model that contain squared or cubed terms. For example, a typical Almon lag coefficient, β_3, might look like:

$$\beta_3 = \alpha_0 + 3\alpha_1 + (3)^2\alpha_2 + (3)^3\alpha_3 \quad (12.22)$$

Note that the right-hand side of Equation 12.22 is a third-order (cubed) polynomial of the subscript of β! The coefficient α_1 is multiplied by the subscript (3) to the first power, α_2 is multiplied by 3 squared, etc.

How would you estimate such a coefficient? Estimating an Almon polynomial distributed lag model involves three steps:

1. Choose the number of lagged values of the independent variable to include (p in Equation 12.2), and choose the degree of the polynomial (for example, a quadratic polynomial would be a second degree polynomial). As discussed in Section 7.1, the researcher must specify the functional form before the equation can be estimated. Hence the exact shape of an Almon lag model depends not only on the data but also on the particular polynomial function chosen.

   We'll discuss how to choose the length of the lag and the degree of the polynomial at the end of the section, but for now let's assume that you've already chosen an equation with four lagged variables and a second degree polynomial:

$$Y_t = b + \beta_0 X_t + \beta_1 X_{t-1} + \beta_2 X_{t-2} + \beta_3 X_{t-3} + \beta_4 X_{t-4} + \epsilon_t \quad (12.23)$$

$$\text{where} \quad \beta_i = \alpha_0 + i \cdot \alpha_1 + i^2 \cdot \alpha_2 \quad (12.24)$$

2. Create and estimate with OLS a secondary equation that is a polynomial reformulation of Equation 12.23. This equation is obtained by substituting Equation 12.24 into Equation 12.23 for all the βs and then rearranging by factoring out the αs:

$$\begin{aligned} Y_t = a + \alpha_0\,(X_t + X_{t-1} + X_{t-2} + X_{t-3} + X_{t-4}) + \\ \alpha_1(X_{t-1} + 2X_{t-2} + 3X_{t-3} + 4X_{t-4}) + \\ \alpha_2(X_{t-1} + 4X_{t-2} + 9X_{t-3} + 16X_{t-4}) + \epsilon_t \end{aligned} \quad (12.25)$$

Equation 12.25 looks forbidding at first, but note that it is little more than the current value of Y as a function of past values of X only with the lagged values of X rearranged in a polynomial pattern similar to Equation 12.22. All of the expressions in parentheses have known, calculable values for all observations, so they can be used to estimate a and the αs.

3. Take the $\hat{\alpha}$s from Equation 12.25 above and use them to estimate the $\hat{\beta}$s in equation 12.23 by applying a system of auxiliary equations:

$$\begin{aligned} \hat{b} &= \hat{a} \\ \hat{\beta}_0 &= \hat{\alpha}_0 \\ \hat{\beta}_1 &= \hat{\alpha}_0 + \hat{\alpha}_1 + \hat{\alpha}_2 \\ \hat{\beta}_2 &= \hat{\alpha}_0 + 2\hat{\alpha}_1 + 4\hat{\alpha}_2 \\ \hat{\beta}_3 &= \hat{\alpha}_0 + 3\hat{\alpha}_1 + 9\hat{\alpha}_2 \\ \hat{\beta}_4 &= \hat{\alpha}_0 + 4\hat{\alpha}_1 + 16\hat{\alpha}_2 \end{aligned}$$

More generally, this system can be seen to equal the following polynomial auxiliary equation:

$$\hat{\beta}_i = \hat{\alpha}_0 + i\hat{\alpha}_1 + i^2\hat{\alpha}_2 + \cdots + i^r\hat{\alpha}_r \quad (12.26)$$

where i is the lag in question and r is the degree of the polynomial chosen in step one. Note that Equation 12.26 is a more general version of both the system of auxiliary equations and of Equation 12.24.

We thus get estimates of the coefficients of our distributed lag equation (Equation 12.23) by estimating a secondary equation (Equation 12.25) and then converting back to the original format using the auxiliary equations listed above. For practice with specifying this system for different numbers of lagged variables or for a different degree polynomial, see Exercise 6.

As mentioned, the first step in running an Almon model is to choose the degree of the polynomial and the length of the lag.[12] The need to make these choices is one of the major flaws of the Almon procedure. Some authors suggest choosing the degree of the polynomial and the length of the lag that provide the best fit, but, as readers of this text could have predicted, we discourage such a practice. Instead, we suggest a two-step procedure:

1. Choose the number of lags (p) to best describe the underlying theory of the particular variables involved. For each model, attempt to determine the number of time periods, on average, over which the impact of a change in the explanatory variable would be likely to continue, and then set the number of lagged variables in Equation 12.23 equal to this number. If theory provides little insight, then estimate the most sound theoretical alternatives and choose the length of the lag that minimizes Amemiya's PC (see Section 6.7).[13] For practice with this, see Exercise 8.

2. Choose the degree of the polynomial (r) based on the expected pattern of the lag structure; one good guess is that the degree of the polynomial is one more than the number of turning points in the diagram of the expected relationship. An inverted U shape, Panel B in Figure 12.2, indicates a second degree polynomial, while a shape with two turning points, such as Panels C and D, suggests a third degree polynomial in Equation 12.26. If you can't decide between two possibilities on theoretical grounds, estimate the higher degree polynomial and see if the last $\hat{\alpha}$ ($\hat{\alpha}_r$) is statistically significant; if not, then use the equation with the lower[14] polynomial degree $(r-1)$. For practice with this, see Exercise 8.

12.3.2 An Example of Almon Lags

Probably the most famous example of a polynomial distributed lag model is Almon's original *Econometrica* study. Almon estimated the relationship between capital appropriations and capital expenditures using quarterly data from 1953 to 1961 for U.S. manufacturing in-

12. Most Almon lag computer programs will also give the researcher the option of constraining one or both of the endpoints (β_{-1} and β_{r+1}) to zero. Our advice is to avoid constraining the endpoints unless there are sound theoretical reasons for doing so. Such theoretical reasons typically will be found in a thorough review of the literature on the subject being modeled.

13. Equivalently, you can choose the lag that optimizes any other specification criterion that tends to include fewer independent variables in the optimal specification than does maximizing $\bar{R}^2$; see P. A. Frost, "Some Properties of the Almon Lag Technique When One Searches for Degree of Polynomial and Lag," *Journal of the American Statistical Association,* 1975, pp. 606–612.

14. See T. W. Anderson, "The Choice of the Degree of a Polynomial Regression as a Multiple Decision Problem," *Annals of Mathematical Statistics,* 1966, pp. 255–265.

dustries. Almon felt that a Koyck distributed lag model was inappropriate for her topic, and she developed the polynomial distributed lag approach as a better model of the underlying relationship between capital expenditures and capital allocations.

To understand her reasoning, think about the time pattern of construction spending ("capital expenditures") you'd expect once the board of directors of a manufacturing company has authorized the building of a new factory ("capital appropriations"). In the first quarter, little money would be spent as blueprints are finalized and construction contracts are signed, but expenditures per quarter would increase in the second and third quarters as the foundation is laid and the building begins to take shape. By the end of a year or so, spending would start to tail off as major construction is completed and interior work begins, and a slowly decreasing spending pattern would continue through the finishing touches. In essence, we'd expect capital expenditures to rise and then fall something like Panel B in Figure 12.2, a shape that suggests a polynomial distributed lag model, not a Koyck model.

Almon specified the following polynomial distributed lag model for all U.S. manufacturing industries:

$$E_t = \beta_0 A_t + \beta_1 A_{t-1} + \beta_2 A_{t-2} + \beta_3 A_{t-3} + \beta_4 A_{t-4} + \beta_5 A_{t-5} + \beta_6 A_{t-6} + \beta_7 A_{t-7} + \alpha_1 S_{1t} + \alpha_2 S_{2t} + \alpha_3 S_{3t} + \alpha_4 S_{4t} + u_t \quad (12.27)$$

where: E_t = capital expenditures in quarter t (millions of dollars)
A_t = capital appropriations in quarter t (millions of dollars)
S_{kt} = a seasonal dummy equal to 1 in each kth quarter and 0 otherwise

As can be seen, Almon specified a lag of seven quarters. In addition, she specified a fourth degree polynomial with both endpoints restricted to zero. As a result, estimating Equation 12.27 involves instructing the computer to constrain β_{-1} and β_8 to be equal to zero.

The results were (standard errors in parentheses):

$$\begin{aligned}
\hat{E}_t = {} & \underset{(0.023)}{0.048A_t} + \underset{(0.016)}{0.099A_{t-1}} + \underset{(0.013)}{0.141A_{t-2}} + \underset{(0.023)}{0.165A_{t-3}} \\
& + \underset{(0.023)}{0.167A_{t-4}} + \underset{(0.013)}{0.146A_{t-5}} + \underset{(0.016)}{0.105A_{t-6}} + \underset{(0.024)}{0.053A_{t-7}} \\
& - 283S_{1t} + 13S_{2t} - 50S_{3t} + 320S_{4t} \qquad (12.28)
\end{aligned}$$

n = 36 (quarterly 1953–1961) $\bar{R}^2 = .92$ DW = 0.89

Take a look at the estimated coefficients; do they conform to our original expectations? The answer is yes, as they rise steadily from 0.048 to a peak of 0.167 (at a one year lag) before falling back down to 0.053. As can be seen in Figure 12.3, the shape of the plot of the estimated coefficients is virtually identical to the inverted U shape we hypothesized in Panel B of Figure 12.2.

A second item of analytical interest is the sum of the estimated coefficients of the As. In theory, what should this sum equal? If all capital appropriations eventually are expended, the sum should equal exactly 1.00, but if some firms spend more than is appropriated, the sum should be greater than 1.00. As you can verify, the sum of the slope coefficients in Equation 12.27 is 0.92, so it seems likely that some monies were approved but not spent (or not spent within two years).

Finally, note that Equation 12.28 has no constant term. Some readers might conclude that the equation has the equivalent of a constant, since there are four (not three) quarterly dummies in the equa-

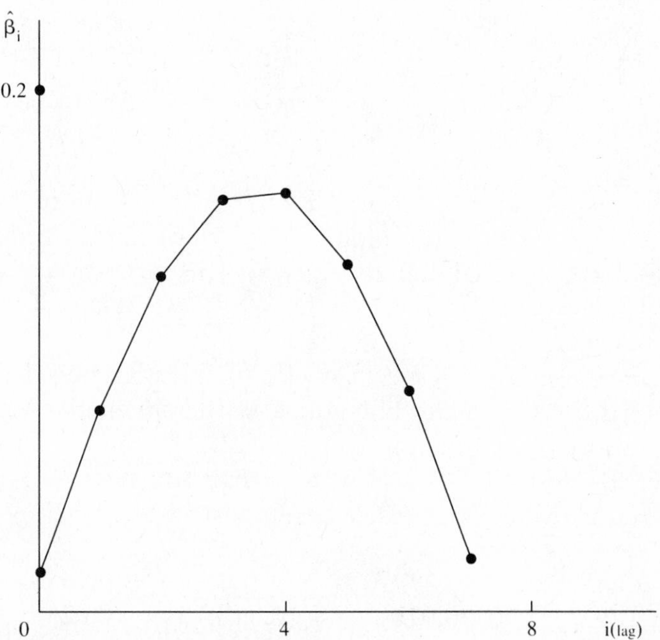

Figure 12.3 ALMON'S ESTIMATED MODEL
The coefficients on Almon's estimated model do indeed follow the inverted U shape hypothesized in Panel B of Figure 12.2.

tion. However, Almon estimated coefficients for just three of the dummies and then chose the fourth coefficient such that the coefficients of all four dummies would sum to zero, forcing the constant term to equal zero. Almon did this because of the underlying theory; she explicitly assumed "a zero intercept for the year as a whole"[15] because expenditures are likely to equal zero if appropriations equal zero. Unfortunately, the lack of a constant means that the expected value of the error term is no longer necessarily zero, potentially violating Classical Assumption II. For more on this topic, see Exercise 5.

12.4 Using the Distributed Lag Approach

There are many opportunities to use distributed lag models in economic and business applications. Two interesting examples of these uses are true lagged dependent variables and Granger causality.

12.4.1 True Lagged Dependent Variables

So far in this chapter, we've discussed only one model that has a lagged dependent variable, the Koyck model. The reader could thus get the impression that the only circumstance under which a lagged dependent variable could appear on the right-hand side of an equation is when the underlying theory calls for a geometric distributed lag of the independent variable(s). Not so.

Instead, there are a variety of situations in which a lagged dependent variable can appear on the right-hand side of the equation for theoretically sound reasons of its own. The two most well-known of these models are the adaptive expectations model and the partial stock adjustment model.

The Adaptive Expectations Model. In the **adaptive expectations model,** expectations about the future take past experience into account. For instance, an adaptive expectations model of inflation adjusts inflationary expectations up or down depending on the present level of inflation. If you expect a 6 percent rate of inflation and experience a 10 percent rate, you might raise the rate you expect to occur next year to 8 percent or 9 percent.

15. Shirley Almon, "The Distributed Lag Between Capital Appropriations and Expenditures," *Econometrica,* 1965, p. 183. It's possible that the lack of a constant is a second reason that the slope coefficients don't sum to 1.00.

How can we build a model of such adaptive expectations? We'd want our expected levels of inflation to change by some portion of the difference between what we expected and what we observed. Such a model could be written as:

$$E_t - E_{t-1} = \theta(I_t - E_{t-1}) \tag{12.29}$$

where: E_t = the expected rate of inflation in time period t
I_t = the actual rate of inflation in time period t
θ = the rate of adaptation $(0 < \theta \leq 1)$

It can be shown that Equation 12.29 is equivalent to the following econometric model:

$$E_t = \beta_0 + \beta_1\theta I_t + (1 - \theta)E_{t-1} + u_t \tag{12.30}$$

where $u_t = \epsilon_t - (1 - \theta)\epsilon_{t-1}$ and where ϵ_t is a classical stochastic error term.[16] Take a look at Equation 12.30 and note that it has the same basic functional form as a Koyck lag model, even though it has a dramatically different underlying model. Also note that the structure of the error term in Equation 12.30 is virtually identical to that of a Koyck lag model. This means that all our warnings in Section 12.2 about problems with serial correlation in the Koyck model also apply to the adaptive expectations model.

The Partial Stock Adjustment Model. In the **partial stock adjustment model,** the quantity of an item we decide to own depends on the amount of the item we held in the previous time period and on our long-run desired holding of the item. Thus a partial stock adjustment model of inventories of nuclear warheads would imply that we would move from the previous period's inventory to our desired inventory by making up some percentage, say 50 percent, of the difference between the two each year. Such a model would thus look like:

$$W_t - W_{t-1} = \Gamma(D_t - W_{t-1}) \tag{12.31}$$

16. For this derivation, the derivation of Equation 12.32, and an excellent discussion of other elements of the econometrics of expectations, see G. S. Maddala, *Introduction to Econometrics* (New York: Macmillan, 1988), pp. 340–374.

where: W_t = the stock of nuclear warheads in time period t
D_t = the desired stock of warheads in time period t
Γ = the rate of inventory adjustment $(0 < \Gamma < 1)$

Note that Equation 12.31 differs from Equation 12.29 in that the left-hand side measures actual, rather than expected, values. Given this difference, it can be shown that Equation 12.31 is equivalent to the following econometric model:

$$W_t = \beta_0 + \beta_1\Gamma D_t + (1 - \Gamma)W_{t-1} + \Gamma\epsilon_t \qquad (12.32)$$

where ϵ_t is a random error term. Note that once again we've derived an equation virtually identical to a Koyck model that's based on a lagged dependent variable acting as a true independent variable rather than as representing a geometric distributed lag of the independent variable(s).

There is one big difference between Equation 12.32 and the Koyck lag and adaptive expectations models we've discussed to date. The error term in Equation 12.32, $\Gamma\epsilon_t$, does *not* contain a lagged error term component (ϵ_{t-1}) as do the others. As a result, the estimation of Equation 12.32 is not as likely to encounter serial correlation as is Equation 12.30 or a true Koyck lag model. While such a situation might make you want to run out and declare all your distributed lag equations to be partial stock adjustment models, such a declaration would only mask the problem, not make it go away.

12.4.2 Granger Causality

One application of ad hoc distributed lag models is to test the direction of causality in economic relationships. Such a test is useful when we know that two variables are related but we don't know which variable causes the other to move. For example, most economists believe that increases in the money supply stimulate GNP, but others feel that increases in GNP eventually lead the monetary authorities to increase the money supply. Who's right?

One approach to such a question of indeterminate causality is to theorize that the two variables are determined simultaneously. We'll address the estimation of simultaneous equation models in Chapter 14. A second approach to the problem is to test for what is called "Granger" causality.

How can we claim to be able to test for causality? After all, didn't we say in Chapter 1 that even though most economic relationships are causal in nature, regression analysis cannot prove such causality? The answer is that we don't actually test for theoretical causality; instead, we test for Granger causality.

Granger causality, or precedence, is a circumstance where one time-series variable consistently and predictably changes before another variable does.[17] If one variable precedes ("Granger causes") another, we still can't be sure that the first variable "causes" the other to change, but we can be fairly sure that the opposite is not the case.

To see this, suppose event A always happens before event B. It's unlikely that B is the cause of A, isn't it? After all, how often does an event that will happen in the future cause an event that has already happened? In such a situation, we can reject the hypothesis that event B causes event A with a fairly high level of confidence.

On the other hand, we still have not shown that event A "causes" B. Proving that people carry umbrellas around before it rains doesn't prove that carrying the umbrellas actually causes the rain. All we've shown is that one event preceded or, more specifically, "Granger-caused" the other. Granger causality is important because it allows us to analyze which variable precedes or "leads" the other, and, as we shall see, such leading variables are extremely useful for forecasting purposes. Despite the value of Granger causality, however, we shouldn't let ourselves be lured into thinking that it allows us to prove economic causality in any rigorous way.

There are a number of different tests for Granger causality, and all the various methods involve distributed lag models in one way or another.[18] Our preference is to use an expanded version of a test originally developed by Granger. Granger suggested that to see if X "Granger-caused" Y, we should run:

$$Y_t = f(Y_{t-1}, Y_{t-2}, \ldots, Y_{t-p}, X_{t-1}, X_{t-2}, \ldots, X_{t-p}) \quad (12.33)$$

and test the null hypothesis that the coefficients of the lagged Xs jointly equal zero. If we can reject this null hypothesis using the *F*-test, then

17. See C. W. J. Granger, "Investigating Causal Relations by Econometric Models and Cross-Spectral Methods," *Econometrica,* 1969, pp. 24–36.

18. Perhaps the most famous of these tests is the Sims test: Christopher A. Sims, "Money, Causality, and Income," *American Economic Review,* 1972, pp. 540–552. Unfortunately, the Sims test has some problems, especially its inability to deal with serially correlated future values of X, and so it is no longer necessarily the test of choice for Granger causality. See John Geweke, R. Meese, and W. Dent, "Comparing Alternative Tests of Causality in Temporal Systems," *Journal of Econometrics,* 1982, pp. 161–194, and Rodney Jacobs, Edward Leamer, and Michael Ward, "Difficulties with Testing for Causation," *Economic Inquiry,* 1979, pp. 401–413.

we have evidence that X Granger-causes Y. Note that if $p = 1$, Equation 12.33 is similar to a Koyck distributed lag model.

For a number of reasons, we recommend running two Granger tests, one in each direction. That is, run Equation 12.33 and also run:

$$X_t = f(Y_{t-1}, Y_{t-2}, \ldots, Y_{t-p}, X_{t-1}, X_{t-2}, \ldots, X_{t-p}) \qquad (12.34)$$

testing for Granger causality in both directions. If the *F*-test is significant for Equation 12.33 but not for Equation 12.34, then we can conclude that X Granger-causes Y. For practice with this dual version of the Granger test, see Exercise 12.

12.5 Summary

1. A distributed lag model explains the current value of Y as a function of current and past values of X, thus "distributing" the impact of X over a number of lagged time periods. OLS estimation of distributed lag equations without any constraints (ad hoc distributed lags) encounters problems with multicollinearity, degrees of freedom, and a noncontinuous pattern of coefficients over time.

2. A Koyck distributed lag model avoids these problems by assuming that the coefficients of the lagged independent variables decrease in a geometric fashion the longer the lag ($\beta_i = \beta_0\lambda^i$), where i is the length of the lag and $0 < \lambda < 1$. Given this, the Koyck distributed lag equation can be simplified to:

$$Y_t = \alpha_0 + \beta_0 X_t + \lambda Y_{t-1} + u_t$$

where $u_t = \epsilon_t - \lambda\epsilon_{t-1}$ and where Y_{t-1} is a lagged dependent variable. Thus the use of a lagged dependent variable as an independent variable usually implies a Koyck distributed lag model.

3. In small samples, OLS estimates of a Koyck distributed lag model (or a similar model) are biased and have unreliable hypothesis testing properties. Even in large samples, OLS will produce biased estimates of the coefficients of a Koyck (or similar) model if the error term is serially correlated.

4. In a Koyck lag (or similar) model, the Durbin-Watson d test sometimes can fail to detect the presence of serial correlation because d is biased toward 2. The most-used alternative is Durbin's h test,

even though the Lagrange Multiplier test has similar properties and is more generally applicable.

5. An Almon polynomial distributed lag model is an estimation procedure for distributed lags that allows the coefficients of the lagged independent variables to follow a variety of continuous patterns as the lag increases. While coefficients from a Koyck model are limited to a steadily decreasing pattern, the Almon model can allow coefficients to first increase and then decrease, or to change direction more than once.

6. There are a variety of models in which a lagged dependent variable can appear on the right-hand side of the equation for sound theoretical reasons. The two most well-known of these models are the adaptive expectations model and the partial stock adjustment model.

7. Granger causality, or precedence, is a circumstance where one time-series variable consistently and predictably changes before another variable does. If one variable precedes (Granger-causes) another, we still can't be sure that the first variable "causes" the other to change, but we can be fairly sure that the opposite is not the case.

Exercises

(Answers to even-numbered exercises are in Appendix A.)

1. Write the meaning of each of the following terms without reference to the book (or your notes), and then compare your definition with the version in the text for each:
   a. Koyck distributed lag model
   b. ad hoc distributed lag model
   c. Durbin's h test
   d. Lagrange Multiplier test (LM)
   e. Almon polynomial distributed lag model
   f. adaptive expectations model
   g. partial stock adjustment model
   h. Granger causality

2. Calculate and graph the pattern of the impact of a lagged X on Y as the lag increases for each of the following estimated Koyck distributed lag equations.
   a. $Y_t = 13.0 + 12.0X_t + 0.04Y_{t-1}$
   b. $Y_t = 13.0 + 12.0X_t + 0.08Y_{t-1}$

c. $Y_t = 13.0 + 12.0X_t + 2.0Y_{t-1}$
d. $Y_t = 13.0 + 12.0X_t - 0.4Y_{t-1}$
e. Look over your graphs for parts c and d. What λ restriction do they combine to show the wisdom of?

3. Consider the following equation aimed at estimating the demand for real cash balances in Mexico (standard errors in parentheses):

$$\widehat{\ln M_t} = 2.00 - \underset{(0.10)}{0.10\ln R_t} + \underset{(0.35)}{0.70\ln Y_t} + \underset{(0.10)}{0.60\ln M_{t-1}}$$
$$\bar{R}^2 = .90 \quad \text{DW } d = 1.80 \quad \text{annual: 1959–1984}$$

where: M_t = money stock in year t (millions of pesos)
R_t = long-term interest rate in year t (percent)
Y_t = real GNP in year t (millions of pesos)

a. What economic relationship between Y and M is implied by the equation? (Hint: Note that the equation is double-log in form.)
b. How are Y and R similar in terms of their relationship to M?
c. Test for serial correlation in this equation.
d. Now suppose you learn that the estimated standard error of the estimated coefficient of $\ln M_{t-1}$ is actually 0.30. How does this change your answer to part c above?

4. Consider the following equation for the determination of wages in the United Kindgom (standard error in parentheses):[19]

$$\widehat{W_t} = 8.562 + \underset{(0.080)}{0.364P_t} + \underset{(0.072)}{0.004P_{t-1}} - \underset{(0.658)}{2.56U_t}$$
$$\bar{R}^2 = .87 \quad n = 19$$

where: W_t = wages and salaries per employee in year t
P_t = the price level in year t
U_t = the unemployment rate in year t

a. Develop and test your own hypotheses with respect to the individual slope coefficients at the 90 percent level.

19. *Prices and Earnings in 1951–1969; An Econometric Assessment*, United Kingdom Department of Employment, 1971, p. 35.

b. Discuss the theoretical validity of P_{t-1} and how your opinion of that validity has been changed by its statistical significance. Should P_{t-1} be dropped from the equation? Why or why not?
c. If P_{t-1} is dropped from the equation, the general functional form of the equation changes radically. Why?

5. Recall that Almon estimated her model with three quarterly seasonal dummies and then set the coefficient of the fourth season equal to the sum of the other three (to force the equation's intercept to equal zero). This is an example of the possibility that equations that appear to include a constant might not have one (and vice versa). Assume that D_i is defined as a seasonal dummy equal to one in the *i*th quarter and zero otherwise. Which of the following equations include(s) the equivalent of a constant term?

a. $Y_t = \alpha_0 + \alpha_1 D_1 + \alpha_2 D_2 + \alpha_3 D_3 + \alpha_4 X_t + \epsilon_t$
b. $Y_t = \alpha_0 \Sigma(D_1 + D_2 + D_3 + D_4) + \alpha_1 X_t + \epsilon_t$
c. $Y_t = \alpha_0 \Sigma(D_1 + D_2 + D_3) + \alpha_1 X_t + \epsilon_t$
d. $Y_t = \alpha_0 D_1 + \alpha_1 D_2 + \alpha_2 D_3 + \alpha_3 D_4 + \alpha_4 X_t + \epsilon_t$
e. $Y_t = \alpha_0 D_2 + \alpha_1 D_3 + \alpha_2 D_4 + \alpha_3 X_t + \epsilon_t$

6. Write out the specific secondary equation (similar to Equation 12.25) and the corresponding system of auxiliary equations for Almon polynomial distributed lag models with the following characteristics:
a. A second degree polynomial with length of lag = 3
b. A third degree polynomial with length of lag = 4
c. A third degree polynomial with length of lag = 5

7. Chalmers and Shelton[20] studied a number of major 1960s U.S. riots and estimated the following model of the intensity of a riot using a distributed lag approach:

$$\hat{I}_t = 0.024 + \hat{\beta}\Sigma C + 0.61I_{t-1} + 0.33I_{t-2} + 0.12I_{t-3} - 0.02I_{t-4} - 0.10I_{t-5} - 0.10I_{t-6} - 0.03I_{t-7} + 0.12I_{t-8}$$

where: I_t = the intensity of a riot (based on a scale of 1 to 30, 30 = high) in hour t
I_{t-1} = the intensity of that riot (based on the same scale) 3 hours prior to hour t

20. James A. Chalmers and Robert B. Shelton, "An Economic Analysis of Riot Participation," *Economic Inquiry,* 1975, pp. 322–336.

I_{t-i} = the intensity of that riot 3·i hours prior to hour t
C = a group of dummy variables measuring police activity and whether hour t was during day or night hours

a. Would you expect the coefficients of the lagged dependent variables to increase or decrease as the lag got longer? Explain your answer.
b. Study the estimated coefficients of the lagged I_ts. Do they follow the pattern you expected?
c. Which distributed lag estimation technique (ad hoc, Koyck, or Almon) do you think Chalmers and Shelton used? Explain your answer.
d. Suppose you go to the library and read that Chalmers and Shelton expected the intensity of a riot 24 hours before time t to have a positive effect on the intensity at time t. How does this research change your answer to part c above?
e. If your answer to part d above is "Almon," what degree polynomial do you think Chalmers and Shelton used? Why?

8. Based on the underlying theory of each of the following two research projects, choose the length of the lag and the degree of the polynomial that you would suggest to estimate a quarterly Almon polynomial distributed lag model. Explain your reasoning.
   a. The impact of the number of new housing building permits on GNP. (A permit to build new housing must be obtained before construction can begin. Hint: Assume that it takes between zero and two quarters to start building and between two and four quarters to complete the building process itself.)
   b. The impact of bathing suit advertising on sales of bathing suits in Alaska. (Hint: Assume that people buy bathing suits in direct proportion to the average temperature in a quarter and that they remember advertising for between zero minutes and five quarters.)

9. You've been hired to determine the impact of advertising on gross sales revenue for "Four Musketeers" candy bars. Four Musketeers has the same price and more or less the same ingredients as competing candy bars, so it seems likely that only advertising affects sales. You decide to build a distributed lag model of sales as a function of advertising, but you're not sure whether an ad hoc or a Koyck distributed lag model is more appropriate.

Using data on Four Musketeers candy bars from Table 12.1, estimate both of the following distributed lag equations from 1964–1988 and compare the lag structures implied by the estimated coefficients:

a. an ad hoc distributed lag model (4 lags)
b. a Koyck distributed lag model

10. Test for serial correlation in the estimated Koyck lag equation you got as your answer to Exercise #9b by using:

a. Durbin's h test
b. The Lagrange Multiplier test

TABLE 12.1 DATA FOR THE FOUR MUSKETEERS EXERCISE

Year	Sales	Advertising
1960	*	30
1961	*	35
1962	*	36
1963	320	39
1964	360	40
1965	390	45
1966	400	50
1967	410	50
1968	400	50
1969	450	53
1970	470	55
1971	500	60
1972	500	60
1973	490	60
1974	580	65
1975	600	70
1976	700	70
1977	790	60
1978	730	60
1979	720	60
1980	800	70
1981	820	80
1982	830	80
1983	890	80
1984	900	80
1985	850	75
1986	840	75
1987	850	75
1988	850	75

11. Suppose you're building a Koyck distributed lag model and are concerned with the possibility that serial correlation, instead of being first-order, is second-order: $u_t = f(u_{t-1}, u_{t-2})$.
   a. What is the theoretical meaning of such second-order serial correlation?
   b. Carefully write out the formula for the Lagrange Multiplier auxiliary equation (similar to Equation 12.20) that you would have to estimate to test such a possibility. How many degrees of freedom would there be in such a Lagrange Multiplier test?
   c. Test for second-order serial correlation in the estimated Koyck lag equation you got as your answer to Exercise #9b above.

12. Most economists consider investment and output to be jointly (simultaneously) determined. One test of this simultaneity would be to see whether one of the variables could be shown to Granger-cause the other. Take the data-set from the small macroeconomic model in Table 14.1 and test the possibility that investment (I) Granger-causes GNP (Y) (or vice-versa) with a two-sided Granger test with four lagged Xs.

13. Some farmers were interested in predicting inches of growth of corn as a function of rainfall on a monthly basis, so they collected data from the growing season and estimated an equation of the following form:

$$G_t = \beta_0 + \beta_1 R_t + \beta_2 G_{t-1} + \epsilon_t$$

where: G_t = inches of growth of corn in month t
R_t = inches of rain in month t
ϵ_t = a normally distributed classical error term

While the farmers expected a negative sign for β_2 (they felt that since corn can only grow so much, if it grows a lot in one month, it won't grow much in the next month), they got a positive estimate instead. What suggestions would you have for this problem?

13

Dummy Dependent Variable Techniques

Until now, our discussion of dummy variables has been restricted to dummy independent variables. However, there are many important research topics for which the *dependent* variable is appropriately treated as a dummy, equal only to zero or one.

In particular, researchers analyzing consumer choice often must cope with dummy dependent variables (also called qualitative dependent variables). For example, how do high school students decide whether to go to college? What distinguishes Pepsi drinkers from Coke drinkers? How can we convince people to commute to work using public transportation instead of driving? For an econometric study of these topics, or of any topic that involves a *discrete* choice of some sort, the dependent variable is typically a dummy variable.

In the first two sections of this chapter, we'll present two frequently used ways to estimate equations that have dummy dependent variables: the linear probability model and the binomial logit model. In the last section, we'll briefly discuss two other useful dummy dependent variable techniques: the binomial probit model and the multinomial logit model.

13.1 The Linear Probability Model

13.1.1 What Is a Linear Probability Model?

The most obvious way to estimate a model with a dummy dependent variable is to run OLS on a typical linear econometric equation. A **linear probability model** is just that, a linear-in-the-coefficients equation used to explain a dummy dependent variable:

$$D_i = \beta_0 + \beta_1 X_{1i} + \beta_2 X_{2i} + \epsilon_i \tag{13.1}$$

where D_i is a dummy variable and the Xs, βs, and ϵ are typical independent variables, regression coefficients, and an error term, respectively.

For example, suppose you're interested in understanding why some state legislatures voted to ratify the Equal Rights Constitutional Amendment (ERA) and others did not. In such a model, the appropriate dependent variable would be a dummy, for example D_i equal to one if the *i*th state ratified the ERA and equal to zero otherwise. If we hypothesize that states with a high percentage of females and a low percentage of Republicans would be likely to have approved the amendment, then a linear probability model of ERA voting by state would be:

$$D_i = \beta_0 + \beta_1 F_i + \beta_2 R_i + \epsilon_i$$

where: D_i = 1 if the *i*th state ratified the ERA, 0 otherwise
F_i = females as a percentage of the *i*th state's population
R_i = Republicans as a percentage of the *i*th state's registered voters

The phrase *linear probability model* comes from the fact that the right-hand side of the equation is linear, while the expected value of the left side is a probability. Let's discuss more thoroughly the concept that this equation measures a probability. It can be shown that the expected value of D_i equals the probability that D_i will equal one.[1] If

1. The expected value of a variable equals the sum of the products of each of the possible values the variable can take times the probability of that value occurring. If P_i is defined as the probability that D_i will equal one, then the probability that D_i will equal zero is $(1 - P_i)$, since D_i can take on only two values. Thus, the expected value of $D_i = P_i \cdot 1 + (1 - P_i) \cdot 0 = P_i$, the probability that D_i equals one.

we define P_i as the probability that D_i equals one, then this is the same as saying that the expected value of D_i equals P_i. Since Equation 13.1 specifies this choice as a function of X_{1i} and X_{2i}, this can be formally stated as:

$$E[D_i \mid X_{1i}, X_{2i}] = P_i \tag{13.2}$$

We can never observe the probability P_i, however, because it reflects the state of mind of a decision maker *before* a discrete choice is made. After a choice is made, we can observe only the outcome of that choice, and so the dependent variable D_i can take on the values of only zero or one. Thus, even though the expected value (P_i) can be anywhere from zero to one, we can only observe the two extremes (0 and 1) in our dependent variable (D_i).

13.1.2 Problems with the Linear Probability Model

Unfortunately, the use of OLS to estimate the coefficients of an equation with a dummy dependent variable encounters four major problems:

1. *The error term is not normally distributed.* Because the dependent variable takes on only two values, the error term is bimodal for small samples and approaches the normal distribution only for large samples. Classical Assumption VII thus is violated in all but the largest of samples.

2. *The error term is inherently heteroskedastic.* The variance of ϵ_i equals $P_i \cdot (1 - P_i)$, where P_i is the probability that D_i equals 1. Since P_i can vary from observation to observation, so too can the variance of ϵ_i. Thus the variance of ϵ_i is not constant, and Classical Assumption V is violated.

3. *$\bar{R}^2$ is not an accurate measure of overall fit.* For models with a dummy dependent variable, $\bar{R}^2$ tells us very little about how well the model explains the choices of the decision makers. To see why, take a look at Figure 13.1. D_i can equal only 1 or 0, but $\hat{D}_i$ must move in a continuous fashion from one extreme to the other. This means that $\hat{D}_i$ is likely to be quite different from D_i for some range of X_i. Thus $\bar{R}^2$ is likely to be much lower than 1 even if the model actually does an exceptional job of explaining the choices involved. As a result, $\bar{R}^2$ (or R^2) should not be relied on as a measure of the overall fit of a model with a dummy dependent variable.

4. *$\hat{D}_i$ is not bounded by 0 and 1.* Since the expected value of D_i is a probability, we'd expect $\hat{D}_i$ to be limited to a range of 0 to 1. After all, the prediction that a probability equals 2.6 (or -2.6, for that

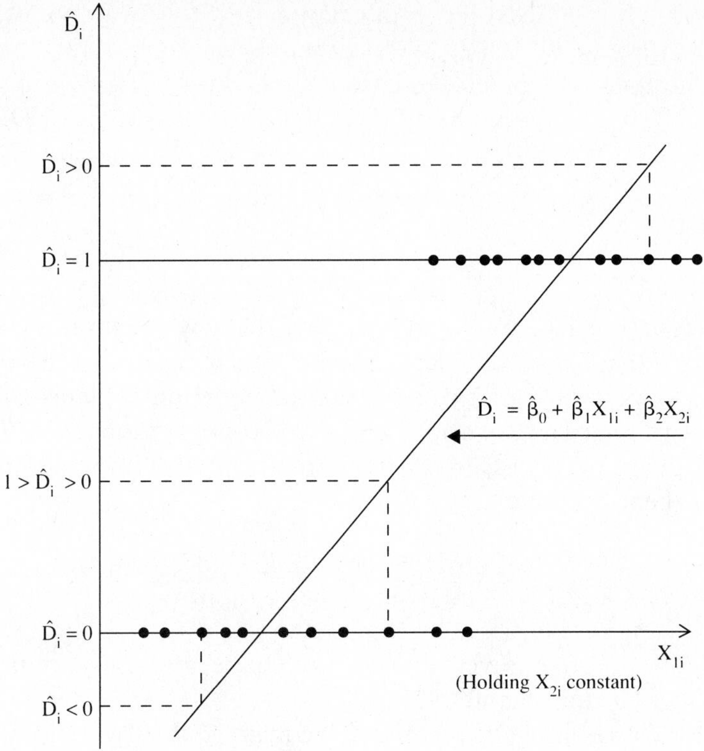

Figure 13.1 A LINEAR PROBABILITY MODEL
In a linear probability model, all the observed D_is equal either zero or one but $\hat{D}_i$ moves linearly from one extreme to the other. As a result, $\bar{R}^2$ is often quite low even if the model does an excellent job of explaining the decision maker's choice. In addition, exceptionally large or small values of X_{1i} (holding X_{2i} constant) can produce values of $\hat{D}_i$ outside the meaningful range of zero to one.

matter) is almost meaningless. However, take another look at Equation 13.1. Depending on the values of the Xs and the $\hat{\beta}$s, the right-hand side might well be outside the meaningful range. For instance, if all the Xs and $\hat{\beta}$s in Equation 13.1 equal 2.0, then $\hat{D}_i$ equals 10.0, substantially greater than 1.0.

Luckily, there are potential solutions to the first three problems cited above. First, as mentioned, a solution to the non-normality problem is to use an extremely large sample (or to avoid the use of hypothesis testing).

Second, a solution to the heteroskedasticity problem is to use weighted least squares. Recall that we know that the variance of ϵ_i

equals $P_i \cdot (1 - P_i)$. As shown in Chapter 10, if we were to divide the equation through by $\sqrt{P_i \cdot (1 - P_i)}$, then the variance of the error term would no longer be heteroskedastic. While we don't know the actual value of P_i, we do know that P_i equals the expected value of D_i. Thus, if we estimate Equation 13.1 and obtain $\hat{D}_i$, we can use $\hat{D}_i$ as an estimate of P_i. To run weighted least squares, we'd then calculate:

$$Z_i = \sqrt{\hat{D}_i \cdot (1 - \hat{D}_i)} \tag{13.3}$$

divide Equation 13.1 by Z_i, and estimate the new equation with OLS.[2]

Third, an alternative to $\bar{R}^2$ is $\mathbf{R_p^2}$, the percentage of the observations in the sample that a particular estimated equation explains correctly. To use this approach, consider a $\hat{D}_i \geq .5$ to predict that $D_i = 1$ and a $\hat{D}_i < .5$ to predict that $D_i = 0$, compare this prediction with the actual D_i, and then compute:

$$R_p^2 = \frac{\text{\# of observations "predicted" correctly}}{\text{total \# of observations (n)}} \tag{13.4}$$

Since R_p^2 is not used universally, we'll calculate and discuss both $\bar{R}^2$ and R_p^2 throughout this chapter.

For most researchers, therefore, the major difficulty with the linear probability model is the unboundedness of the predicted D_is. Take another look at Figure 13.1 for a graphical interpretation of the situation. Because of the linear relationship between the X_is and $\hat{D}_i$, $\hat{D}_i$ can fall well outside the relevant range of 0 to 1. Using the linear probability model, despite this unboundedness problem, may not cause insurmountable difficulties. In particular, the signs and general significance levels of the estimated coefficients of the linear probability model are often similar to those of the alternatives we will discuss later in this chapter.

One simplistic way to get around the unboundedness problem is to assign $\hat{D}_i = 1.0$ to all values of $\hat{D}_i$ above one and $\hat{D}_i = 0.0$ to all negative values. This approach copes with the problem by ignoring it, since an observation for which the linear probability model predicts a

2. Note that when $\hat{D}_i$ is quite close to 0 or 1, $\hat{D}_i \cdot (1 - \hat{D}_i)$ is extremely small and X_i/Z_i is huge. Also note that when $\hat{D}_i$ is outside the 0–1 range, $\hat{D}_i \cdot (1 - \hat{D}_i)$ is negative and Z_i is undefined. See R. G. McGilvray, "Estimating the Linear Probability Function," *Econometrica*, 1970, pp. 775–776. Some researchers arbitrarily drop all such observations to avoid the resulting estimation problems. A better alternative is to impose a floor of 0.02 on $\hat{D}_i \cdot (1 - \hat{D}_i)$. Either way, WLS is not efficient.

probability of 2.0 has been judged to be more likely to be equal to 1.0 than an observation for which the model predicts a 1.0, and yet they are lumped together. What is needed is a systematic method of forcing the $\hat{D}_i$s to range from 0 to 1 in a smooth and meaningful fashion. We'll present such a method, the binomial logit, in Section 13.2.

13.1.3 An Example of a Linear Probability Model

Before moving on to investigate the logit, however, let's take at a look at an example of a linear probability model: a disaggregate study of the labor force participation of women.

A person is defined as being in the labor force if she either has a job or is actively looking for a job. Thus a disaggregate (cross-sectional by person) study of women's labor force participation is appropriately modeled with a dummy dependent variable:

D_i = 1 if the *i*th woman has or is looking for a job, 0 otherwise (not in the labor force)

A review of the literature[3] reveals that there are many potentially relevant independent variables. Two of the most important are the marital status and the number of years of schooling of the woman. The expected signs for the coefficients of these variables are fairly straightforward, since a woman who is unmarried and well educated is much more likely to be in the labor force than her opposite:

$$D_i = f(\overset{-}{M_i}, \overset{+}{S_i})$$

where: M_i = 1 if the *i*th woman is married and 0 otherwise
S_i = the number of years of schooling of the *i*th woman

The data are presented in Table 13.1. The sample size is limited to 30 in order to make it easier for readers to estimate this example on their own. Unfortunately, such a small sample will make hypothesis testing fairly unreliable. Table 13.1 also includes the age of the *i*th woman for use in Exercises 8 and 9. Another typically used variable,

3. See Jerry Hausman, "Labor Supply," in *How Taxes Affect Economic Behavior* (Washington, D.C.: Brookings, 1981), pp. 27–84 and Malcolm S. Cohen, Samuel A. Rea, Jr., and Robert I. Lerman, *A Micro Model of Labor Supply* (Washington, D.C.: U. S. Bureau of Labor Statistics Staff Paper, 1970).

TABLE 13.1 DATA ON THE LABOR FORCE PARTICIPATION OF WOMEN

Observation #	D_i	M_i	A_i	S_i	$\hat{D}_i$	$\hat{D}_i(1 - \hat{D}_i)$	Z_i
1	1.0	0.0	31.0	16.0	1.20	0.020	0.141
2	1.0	1.0	34.0	14.0	0.63	0.231	0.481
3	1.0	1.0	41.0	16.0	0.82	0.146	0.382
4	0.0	0.0	67.0	9.0	0.55	0.247	0.497
5	1.0	0.0	25.0	12.0	0.83	0.139	0.374
6	0.0	1.0	58.0	12.0	0.45	0.247	0.497
7	1.0	0.0	45.0	14.0	1.01	0.020	0.141
8	1.0	0.0	55.0	10.0	0.64	0.228	0.478
9	0.0	0.0	43.0	12.0	0.83	0.139	0.374
10	1.0	0.0	55.0	8.0	0.45	0.248	0.498
11	1.0	0.0	25.0	11.0	0.73	0.192	0.439
12	1.0	0.0	41.0	14.0	1.01	0.020	0.141
13	0.0	1.0	62.0	12.0	0.45	0.247	0.497
14	1.0	1.0	51.0	13.0	0.54	0.248	0.498
15	0.0	1.0	39.0	9.0	0.17	0.141	0.376
16	1.0	0.0	35.0	10.0	0.64	0.228	0.478
17	1.0	1.0	40.0	14.0	0.63	0.231	0.481
18	0.0	1.0	43.0	10.0	0.26	0.194	0.440
19	0.0	1.0	37.0	12.0	0.45	0.247	0.497
20	1.0	0.0	27.0	13.0	0.92	0.069	0.263
21	1.0	0.0	28.0	14.0	1.01	0.020	0.141
22	1.0	1.0	48.0	12.0	0.45	0.247	0.497
23	0.0	1.0	66.0	7.0	−0.01	0.020	0.141
24	0.0	1.0	44.0	11.0	0.35	0.229	0.479
25	0.0	1.0	21.0	12.0	0.45	0.247	0.497
26	1.0	1.0	40.0	10.0	0.26	0.194	0.440
27	1.0	0.0	41.0	15.0	1.11	0.020	0.141
28	0.0	1.0	23.0	10.0	0.26	0.194	0.440
29	0.0	1.0	31.0	11.0	0.35	0.229	0.479
30	1.0	1.0	44.0	12.0	0.45	0.247	0.497

Note: $\hat{D}_i(1 - \hat{D}_i)$ has been set equal to 0.02 for all values of $\hat{D}_i$ less than 0.02 or greater than 0.98.

O_i = other income available to the *i*th woman, is not available for this sample, introducing possible omitted variable bias.

If we choose a linear functional form for both independent variables, we've got a linear probability model:

$$D_i = \beta_0 + \beta_1 M_i + \beta_2 S_i + \epsilon_i \tag{13.5}$$

where ϵ_i is an inherently heteroskedastic error term with variance = $P_i \cdot (1 - P_i)$. If we now estimate Equation 13.5 with the data on the labor force participation of women from Table 13.1, we obtain (standard errors in parentheses):

$$\hat{D}_i = -0.28 - 0.38M_i + 0.09S_i \qquad (13.6)$$
$$\qquad\qquad (0.15) \qquad (0.03)$$
$$n = 30 \qquad \bar{R}^2 = .32 \qquad R^2_p = .80$$

How do these results look? At first glance, they look terrific. Despite the small sample and the possible bias due to omitting O_i, both independent variables have estimated coefficients that are significant in the expected direction. In addition, the $\bar{R}^2$ of .32 is fairly high for a linear probability model (since D_i equals only 0 or 1, it's almost impossible to get a $\bar{R}^2$ much higher than .70). Further evidence of good fit is the fairly high R^2_p of .80, meaning that 80 percent of the choices were predicted "correctly" by Equation 13.6.

We need to be careful when we interpret the estimated coefficients in Equation 13.6, however. The slope coefficient in a linear probability model represents the change in the probability that D_i equals one caused by a one-unit change in the independent variable (holding the other independent variables constant). Viewed in this context, do the estimated coefficients still make economic sense? The answer is yes: the probability of a woman participating in the labor force falls by 38 percent if she is married (holding constant her schooling). In addition, each year of schooling increases the probability of labor force participation by 9 percent (holding constant marital status).

However, Equation 13.6 is far from perfect. Recall that the error term is inherently heteroskedastic, that hypothesis testing is unreliable in such a small sample, that $\bar{R}^2$ is not an accurate measure of fit, and that one or more relevant variables have been omitted. While we can do nothing about some of these problems, there is a solution to the heteroskedasticity problem: weighted least squares (WLS).

To use WLS, we take the $\hat{D}_i$ from Equation 13.6 and calculate $Z_i = \sqrt{\hat{D}_i \cdot (1 - \hat{D}_i)}$, as in Equation 13.3 [taking care to impose a floor of 0.02 on $\hat{D}_i \cdot (1 - \hat{D}_i)$ as suggested in footnote 2]. We then divide Equation 13.5 through by Z_i, obtaining:

$$D_i/Z_i = \alpha_0 + \beta_0(1/Z_i) + \beta_1 M_i/Z_i + \beta_2 S_i/Z_i + u_i \qquad (13.7)$$

where u_i is a nonheteroskedastic error term $= \epsilon_i/Z_i$. Note that since Z_i is not an independent variable in Equation 13.6, we have chosen to add α_0, a constant term, to Equation 13.7 to avoid violating Classical

Assumption II (as discussed in Chapter 9). If we now estimate Equation 13.7 with OLS, we obtain:

$$\widehat{D_i/Z_i} = 0.18 - 0.21(1/Z_i) - \underset{(0.15)}{0.39M_i/Z_i} + \underset{(0.02)}{0.08S_i/Z_i} \qquad (13.8)$$
$$n = 30 \qquad \bar{R}^2 = .86 \qquad R_p^2 = .83$$

Let's compare Equations 13.8 and 13.6. Surprisingly, the estimated standard errors of the estimated coefficients are almost identical in the two equations, indicating that at least for this sample the impact of the heteroskedasticity is minimal. The high $\bar{R}^2$ comes about in part because dividing the entire equation by the same number (Z_i) causes some spurious correlation, especially when some of the Z_i values are quite small. As evidence, note that R_p^2 is only slightly higher in Equation 13.8 than in Equation 13.6 even though $\bar{R}^2$ jumped from .32 to .86.

To make it easier for the reader to reproduce the WLS procedure, the values for $\hat{D}_i \cdot (1 - \hat{D}_i)$ and Z_i have been included in Table 13.1. Also included are the $\hat{D}_i$s from Equation 13.6; note that $\hat{D}_i$ is indeed often outside the meaningful range of 0 and 1, causing most of the problems cited earlier. To attack this problem of the unboundedness of $\hat{D}_i$, however, we need a new estimation technique, so let's take a look at one.

13.2 The Binomial Logit Model

Because of the difficulties cited earlier, most researchers avoid using the linear probability model. Instead, they attack dummy dependent variable research projects by using binomial (or binary) logit analysis.

13.2.1 What Is the Binomial Logit?

The **binomial logit** is an estimation technique for equations with dummy dependent variables that avoids the unboundedness problem of the linear probability model by using a variant of the cumulative logistic function:

$$\ln\left(\frac{D_i}{[1 - D_i]}\right) = \beta_0 + \beta_1 X_{1i} + \beta_2 X_{2i} + \epsilon_i \qquad (13.9)$$

where D_i is a dummy variable. The expected value of D_i continues to be P_i, the probability that the *i*th person will make the choice described by $D_i = 1$. Consequently, the dependent variable of Equation 13.9 can be thought of as the log of the odds[4] that the choice in question will be made.

It's important to be careful when analyzing the economic meaning of regression coefficients from a logit model. For instance, β_1 in Equation 13.9 measures the impact of a one-unit change in X_1 on the log of the odds of a given choice, holding X_2 constant. As a result, the absolute sizes of estimated logit coefficients tend to be quite different from the absolute sizes of estimated linear probability model coefficients for the same variables. Interestingly, as mentioned above, the signs and significance levels of the estimated coefficients from the two models often are similar.

How does the logit avoid the unboundedness problem of the linear probability model? It turns out that *both* sides of Equation 13.9 are unbounded. To see this, note that if $D_i = 1$, then the left-side of Equation 13.9 becomes:

$$\ln\left(\frac{D_i}{[1 - D_i]}\right) = \ln\left(\frac{1}{0}\right) = \infty \tag{13.10}$$

Similarly, if $D_i = 0$:

$$\ln\left(\frac{D_i}{[1 - D_i]}\right) = \ln\left(\frac{0}{1}\right) = -\infty \tag{13.11}$$

because the log of zero approaches negative infinity.

Are the $\hat{D}_i$s produced by a logit now limited by zero and one? The answer is yes, but to see why we need to solve Equation 13.9 for D_i. It can be shown[5] that Equation 13.9 is equivalent to:

$$D_i = \frac{1}{1 + e^{-[\beta_0 + \beta_1 X_{1i} + \beta_2 X_{2i} + \epsilon_i]}} \tag{13.12}$$

Take a close look at Equation 13.12. What is the largest $\hat{D}_i$ can be? Well, if $\hat{\beta}_0 + \hat{\beta}_1 X_{1i} + \hat{\beta}_2 X_{2i}$ equals infinity, then:

$$\hat{D}_i = \frac{1}{1 + e^{-\infty}} = \frac{1}{1} = 1 \tag{13.13}$$

4. *Odds* refers to the ratio of the number of times a choice will be made divided by the number of times it will not. In today's world, odds are used most frequently with respect to sporting events, such as horse races, on which bets are made.

5. Those interested in this proof should see Exercise 4.

because e to the minus infinity equals zero. What's the smallest that $\hat{D}_i$ can be? If $\hat{\beta}_0 + \hat{\beta}_1 X_{1i} + \hat{\beta}_2 X_{2i}$ equals minus infinity, then:

$$\hat{D}_i = \frac{1}{1 + e^{\infty}} = \frac{1}{\infty} = 0 \qquad (13.14)$$

Thus, $\hat{D}_i$ is bounded by one and zero. As can be seen in Figure 13.2, $\hat{D}_i$ approaches one and zero very slowly (asymptotically). The binomial logit model therefore avoids the major problem that the linear probability model encounters in dealing with dummy dependent variables. In addition, the logit is quite satisfying to most researchers because it turns out that real-world data often are distributed in S-shape patterns like that in Figure 13.2.

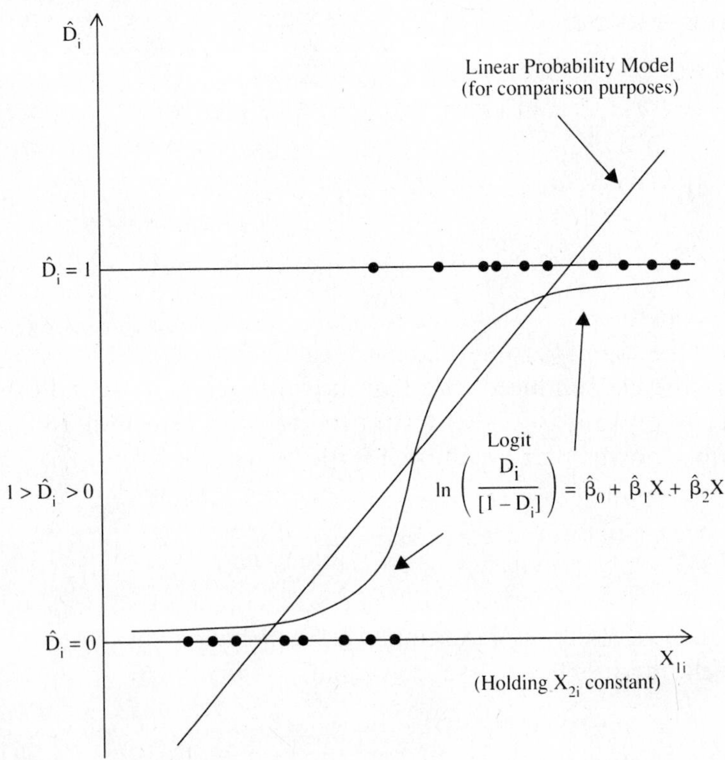

Figure 13.2 $\hat{D}_i$ IS BOUNDED BY ZERO AND ONE IN A BINOMIAL LOGIT MODEL

In a binomial logit model, $\hat{D}_i$ is nonlinearly related to X_{1i}, so even exceptionally large or small values of X_{1i}, holding X_{2i} constant, will not produce values of $\hat{D}_i$ outside the meaningful range of zero to one.

We can estimate a binomial logit with a maximum likelihood iterative process similar to the one discussed in Section 7.6. The maximum likelihood computer program is applied to a logit that has been solved for D_i (Equation 13.12), not to a logit solved for the log of the odds (Equation 13.9). This distinction is necessary because, as shown earlier, the left-hand side of Equation 13.9 can be observed only as infinity and negative infinity. Such infinite values make calculations quite difficult. With extremely large samples, maximum likelihood has the added advantage of producing normally distributed coefficient estimates, allowing the use of normal distribution hypothesis testing techniques.

Because maximum likelihood estimation of the logit works best with large samples, and because large samples allow the otherwise bimodal distribution of the error term to approach the normal distribution, minimum sample sizes for logit analysis should be substantially larger than for linear regression. Some researchers aim for samples of 500 or more.

It's also important to make sure that a logit sample contains a reasonable representation of both alternative choices. For instance, if 98 percent of a sample chooses alternative A and 2 percent chooses B, a random sample of 500 would have only 10 observations that choose B. In such a situation, our estimated coefficients would be overly reliant on the characteristics of those ten observations. A better technique would be to disproportionately sample from those who choose B. It turns out that using different sampling rates for subgroups within the sample does not cause bias in the slope coefficients of a logit model,[6] even though it might do so in a linear regression.

Once the binomial logit has been estimated, hypothesis testing and econometric analysis can be undertaken in much the same way as for linear equations. When interpreting coefficients, however, be careful to recall that they represent the impact of a one-unit change in the independent variable in question, holding the other explanatory variables constant, on the log of the odds of a given choice, not on the probability itself (as was the case with the linear probability model).

Measuring the overall fit, however, is not quite as straightforward. Recall from Chapter 7 that since the functional form of the dependent

6. The constant term, however, needs to be adjusted. Multiply $\hat{\beta}_0$ by $[\ln(p_1) - \ln(p_2)]$, where p_1 is the proportion of the observations chosen if $D_i = 1$ and p_2 is the proportion of the observations chosen if $D_i = 0$. See G. S. Maddala, *Limited-Dependent and Qualitative Variables in Econometrics* (Cambridge: Cambridge University Press, 1983), pp. 90–91.

variable has been changed, $\bar{R}^2$ cannot be used to compare the fit of a logit with an otherwise comparable linear probability model. One way around this difficulty is to use the quasi-R^2 approach of Section 7.2.1 (a nonlinear estimate of R^2) to compare the two fits. However, this quasi-R_2 shares the general faults inherent in using $\bar{R}^2$ with equations with dummy dependent variables. A better approach might be to use the percentage of correct predictions, R^2_p, from Equation 13.4. In this section, we'll calculate both measures.

To allow a fairly simple comparison between the logit and the linear probability model, let's estimate a logit on the same women's labor force participation data that we used in the previous section. The OLS estimate of that model, Equation 13.6, was:

$$\hat{D}_i = -0.28 - \underset{(0.15)}{0.38M_i} + \underset{(0.03)}{0.09S_i} \qquad (13.6)$$
$$n = 30 \qquad \bar{R}^2 = .32 \qquad R^2_p = .80$$

where: D_i = 1 if the *i*th woman is in the labor force, 0 otherwise
M_i = 1 if the *i*th woman is married, 0 otherwise
S_i = the number of years of schooling of the *i*th woman

If we estimate a logit on the same data (from Table 13.1) and the same independent variables, we obtain:[7]

$$\widehat{\ln\left(\frac{D_i}{[1 - D_i]}\right)} = -5.89 - \underset{\substack{(1.18)\\ t = -2.19}}{2.59M_i} + \underset{\substack{(0.31)\\ 2.19}}{0.69S_i} \qquad (13.15)$$
$$n = 30 \qquad \text{quasi-}R^2 = .40 \qquad R^2_p = .80 \qquad \text{iterations} = 5$$

Let's compare Equations 13.6 and 13.15. As expected, the signs and general significance of the slope coefficients are the same. Note, however, that the actual sizes of the coefficients are quite different because the dependent variable is different. The coefficient of M changes from −0.38 to −2.59! Despite these differences, the overall fits are roughly comparable, especially after taking account of the different dependent variables and estimation techniques. In this example, then, the two

7. Equation 13.15 has the log of the odds as its dependent variable, but the maximum likelihood computer estimation program that produces the β estimates uses a functional form with D_i as the dependent variable (similar to Equation 13.12).

estimation procedures differ mainly in that the logit does not produce $\hat{D}_i$s outside the range of zero and one.

However, if the size of the sample in this example is too small for a linear probability model, it certainly is too small for a logit, making any in-depth analysis of Equation 13.15 problematic. Instead, we're better off finding an example with a much larger sample.

13.2.2 An Example of the Use of the Binomial Logit

For a more complete example of the binomial logit, let's look at a model of the probability of passing the California State Department of Motor Vehicles drivers' license test. To obtain a license, each driver must pass a written and a behind-the-wheel test. Even though the tests are scored from 0 to 100, all that matters is that you pass and get your license.

Since the test requires some boning up on traffic and safety laws, driving students have to decide how much time to spend studying. If they don't study enough, they waste time because the time it takes to retake the test exceeds the extra studying that would have been needed to pass the test in the first place. If they study too much, however, they also waste time, because there's no bonus for scoring above the minimum, especially since there is no evidence that doing well on the test has much to do with driving well after the test (this, of course, might be worth its own econometric study).

Recently, two students decided to collect data on test takers in order to build an equation explaining whether someone passed the Department of Motor Vehicles test. They hoped that the model, and in particular the estimated coefficient of study time, would help them decide how much time to spend studying for the test. (Of course, it took more time to collect the data and run the model than it would have taken to memorize the entire traffic code, but that's another story.)

After reviewing the literature, choosing variables, and hypothesizing signs, the students realized that the appropriate functional form was a binomial logit because their dependent variable was a dummy variable:

$$D_i = \begin{cases} 1 \text{ if the } i\text{th test taker passed the test on the first try} \\ 0 \text{ if the } i\text{th test taker failed the test on the first try} \end{cases}$$

The rest of their hypothesized equation was:

$$D_i = f(\overset{+}{A_i}, \overset{+}{H_i}, \overset{+}{E_i}, \overset{+}{C_i})$$

where: A_i = the age of the *i*th test taker
H_i = the number of hours the *i*th test taker studied (usually less than one hour!)
E_i = a dummy variable equal to 1 if the *i*th test taker's primary language is English, 0 otherwise
C_i = a dummy variable equal to 1 if the *i*th test taker has any college experience, 0 otherwise

After collecting data from 480 test takers, the students estimated the following equation:

$$\widehat{\ln\left(\frac{D_i}{[1 - D_i]}\right)} = -1.18 + \underset{\substack{(0.009)\\ t = 1.23}}{0.011A_i} + \underset{\substack{(0.54)\\ 4.97}}{2.70H_i} + \underset{\substack{(0.34)\\ 4.65}}{1.62E_i} + \underset{\substack{(0.99)\\ 4.00}}{3.97C_i}$$

$$n = 480 \quad \text{quasi-}R^2 = .18 \quad R^2_p = .74 \quad \text{iterations} = 5 \quad LR = 90.59 \tag{13.16}$$

Note how similar this result looks to a typical linear regression result. All the estimated coefficients have the expected signs, and all but one are significantly different from zero. Remember, though, that the coefficient estimates have different meanings than in a linear regression model. For example, 2.70 is the impact of an extra hour of studying on the log of the odds of passing the test, holding constant the other three independent variables.

The quasi-R^2 of .18 might seem fairly low, but it really isn't that bad since D_i takes on only two values, both of which are at the extremes of the range of the equation. As evidence of this, note that R^2_p is .74, indicating that the equation correctly "predicted" whether almost three quarters of the sample passed the test based on nothing but the four variables in Equation 13.16.

And what about the two students? Did the equation help them? How much did they end up deciding to study? They found that given their ages, their college experience, and their English-speaking backgrounds, the expected value of $\hat{D}_i$ for each of them was quite high, even if H_i was set equal to zero. So what did they actually do? They studied for a half hour "just to be on the safe side" and passed with flying colors, having devoted more time to passing the test than anyone else in the history of the state.

13.3 Other Dummy Dependent Variable Techniques

While the binomial logit is the most frequently used estimation technique for equations with dummy dependent variables, it's by no means the only one. In this brief section, we'll mention two alternatives, the binomial probit and the multinomial logit, that are useful in particular circumstances. Our main goal is to briefly describe these estimation techniques, not to cover them in any detail.[8]

13.3.1 The Binomial Probit Model

The **binomial probit model** is an estimation technique for equations with dummy dependent variables that avoids the unboundedness problem of the linear probability model by using a variant of the cumulative normal distribution:

$$P_i = \frac{1}{\sqrt{2\pi}} \int_{-\infty}^{Z_i} e^{-s^2/2}\, ds \qquad (13.17)$$

where: P_i = the probability that the dummy variable $D_i = 1$
$Z_i = \beta_0 + \beta_1 X_{1i} + \beta_2 X_{2i} + \epsilon_i$
s = a standardized normal variable

As different as this probit looks from the logit that we examined in the previous section, it can be rewritten to look quite familiar:

$$Z_i = F^{-1}(P_i) = \beta_0 + \beta_1 X_{1i} + \beta_2 X_{2i} + \epsilon_i \qquad (13.18)$$

where F^{-1} is the inverse of the normal cumulative distribution function. Probit models typically are estimated by applying maximum likelihood techniques to the model in the form of Equation 13.17, but the results often are presented in the format of Equation 13.18.

The fact that both the logit and the probit are cumulative distributive functions means that the two have similar properties. For example, a graph of the probit looks almost exactly like the logit in

8. For more, see G. S. Maddala, *Limited Dependent Variables and Qualitative Variables in Econometrics* (Cambridge: Cambridge University Press, 1983) and T. Amemiya, "Qualitative Response Models: A Survey," *Journal of Economic Literature*, 1981, pp. 1483–1536. These surveys also cover additional techniques, like the Tobit model, that are useful with bounded dependent variables or other special situations.

Figure 13.2. In addition, the probit has the same requirement of a fairly large sample before hypothesis testing becomes meaningful. Finally, $\bar{R}^2$ continues to be of questionable value as a measure of overall fit.

From a researcher's point of view, the biggest differences between the two models are that the probit is based on the cumulative normal distribution and that the probit estimation procedure uses considerably more computer time than does the logit. As computer programs are improved, and as computer time continues to fall in price, this latter difference may eventually disappear. Since the probit is similar to the logit and is more expensive to run, why would you ever estimate one? The answer is that since the probit is based on the normal distribution, it's quite theoretically appealing (because many economic variables are normally distributed). With extremely large samples, this advantage falls away, since maximum likelihood procedures can be shown to be asymptotically normal under fairly general conditions.

For an example of a probit, let's estimate one on the same women's labor force participation data we used in the previous logit and linear probability examples (standard errors in parentheses):

$$\hat{Z}_i = \widehat{F^{-1}(P_i)} = -3.44 - \underset{(0.62)}{1.44M_i} + \underset{(0.17)}{0.40S_i} \qquad (13.19)$$

$$n = 30 \qquad \text{quasi-}R^2 = .40 \qquad R_p^2 = .80 \qquad \text{iterations} = 5$$

Compare this result with Equation 13.15 from the previous section. Note that except for a slight difference in the scale of the coefficients, the logit and probit models provide virtually identical results in this example.

13.3.2 The Multinomial Logit Model

In many cases, there are more than two qualitative choices available. In some cities, for instance, a commuter has a choice of car, bus, or subway for the trip to work. How could we build and estimate a model of choosing from more than two different alternatives?

One answer is to hypothesize that choices are made sequentially and to model a multichoice decision as a series of binary decisions. For example, we might hypothesize that the commuter would first decide whether or not to drive to work, and we could build a binary model of car vs. public transportation. For those commuters who choose public transportation, the next step would be to choose whether to take the bus or the subway, and we could build a second binary model of that choice. This method, called a **sequential binary logit,** or a *nested logit,* is cumbersome and at times unrealistic, but it does allow a researcher to use a binary technique to model an inherently multichoice decision.

If a decision between multiple alternatives is truly made simultaneously, a better approach is to build a multinomial logit model of the decision. A **multinomial logit model** is an extension of the binomial logit technique that allows several discrete alternatives to be considered at the same time. If there are n different alternatives, we need n − 1 dummy variables to describe the choice, with each dummy equalling one only when that particular alternative is chosen. For example, D_{1i} would equal one if the *i*th person chose alternative #1 and would equal zero otherwise. As before, the probability that D_{1i} is equal to one, P_{1i}, cannot be observed.

In a multinomial logit, one alternative is selected as the "base" alternative, and then each other possible choice is compared to this base alternative with a logit equation. A key distinction is that the dependent variable of these equations is the log of the odds of the *i*th alternative being chosen *compared to the base alternative:*

$$\ln\left(\frac{P_{1i}}{P_{bi}}\right)$$

where: P_{1i} = the probability of the *i*th person choosing the 1st alternative

P_{bi} = the probability of the *i*th person choosing the base alternative

If there are n alternatives, there should be n − 1 different logit equations in the multinomial logit model system, because the coefficients of the last equation can be calculated from the coefficients of the first n − 1 equations. (If you know that A/C = 6 and B/C = 2, then you can calculate that A/B = 3.) For example, if n = 3, as in the commuter-work-trip example cited above, and the base alternative is taking the bus, then a multinomial logit model would have a system of two equations:

$$\ln\left(\frac{P_{si}}{P_{bi}}\right) = \alpha_0 + \alpha_1 X_{1i} + \alpha_2 X_{2i} + \epsilon_{si} \tag{13.20}$$

$$\ln\left(\frac{P_{ci}}{P_{bi}}\right) = \beta_0 + \beta_1 X_{1i} + \beta_2 X_{3i} + \epsilon_{ci} \tag{13.21}$$

where s = subway, c = car, and b = bus

The definitions of the independent variables (and therefore the meanings of their coefficients) are unusual in a multinomial logit. Some of the Xs are characteristics of the decision maker (like the income of the *i*th commuter). The coefficients of these variables represent the *difference* between the impact of income on the probability of choosing one mode and the impact of income on the probability of choosing the other mode considered in the equation. For example, in Equation 13.20, if X_1 is income, the coefficient α_1 is the impact of an extra dollar of income on the probability of taking the subway to work *minus* the impact of an extra dollar of income on the probability of taking the bus to work (holding X_2 constant).

Xs that aren't characteristics of the decision-maker are usually characteristics of the alternative (like travel time for one of the possible modes of travel). A variable that measures a characteristic of an alternative in a multinomial logit model should be *defined* as the difference between the characteristic for the two modes. For example, if the second independent variable in our model is travel time to work, X_2 should be defined as the travel time to work by subway *minus* the travel time to work by bus, while X_3 is measured as the travel time to work by car minus the travel time to work by bus. The coefficients of such characteristics of the alternatives measure the impact of a unit of time on the ratio of the probabilities (holding X_1 constant). For practice with the meanings of the independent variables and their coefficients in a multinomial logit model, see Exercise 11.

The multinominal logit system has all the basic properties of the binomial logit but with two additional complications in estimation. First, Equations 13.20 and 13.21 are estimated simultaneously,[9] so the iterative nonlinear maximum likelihood procedure used to estimate the system is more costly than for the binomial logit. Second, the relationship between the error terms in the equations (ϵ_{si} and ϵ_{ci}) must be strictly accounted for by using a GLS procedure, a factor that also complicates the estimation procedure.[10]

9. As with the binomial logit, the maximum likelihood computer package doesn't estimate these precise equations. Instead, it estimates versions of Equations 13.20 and 13.21 that are similar to Equation 13.12.

10. For an interesting and yet accessible example of the estimation of a multinomial logit model, see Kang H. Park and Peter M. Kerr, "Determinants of Academic Performance: A Multinomial Logit Approach," *The Journal of Economic Education*, 1990, pp. 101–111.

13.4 Summary

1. A linear probability model is a linear-in-the-coefficients equation used to explain a dummy dependent variable (D_i). The expected value of D_i is the probability that D_i equals one (P_i).

2. The estimation of a linear probability model with OLS encounters four major problems:
   a. The error term is not normally distributed.
   b. The error term is inherently heteroskedastic.
   c. $\bar{R}^2$ is not an accurate measure of overall fit.
   d. The expected value of D_i is not limited by 0 and 1.

3. When measuring the overall fit of equations with dummy dependent variables, an alternative to $\bar{R}^2$ is R_p^2, the percentage of the observations in the sample that a particular estimated equation would have explained correctly.

4. The binomial logit is an estimation technique for equations with dummy dependent variables that avoids the unboundedness problem of the linear probability model by using a variant of the cumulative logistic function:

$$\ln\left(\frac{D_i}{[1 - D_i]}\right) = \beta_0 + \beta_1 X_{1i} + \beta_2 X_{2i} + \epsilon_i$$

5. The binomial logit is best estimated using the maximum likelihood technique and a large sample. A slope coefficient from a logit measures the impact of a one-unit change of the independent variable in question (holding the other explanatory variables constant) on the log of the odds of a given choice.

6. The binomial probit model is an estimation technique for equations with dummy dependent variables that uses the cumulative normal distribution function. The binomial probit has properties quite similar to the binomial logit except that it takes more computer time to estimate than a logit and is based on the normal distribution.

7. The multinomial logit model is an extension of the binomial logit that allows more than two discrete alternatives to be considered simultaneously. One alternative is chosen as a base alternative, and then each other possible choice is compared to that base alternative with a logit equation.

Exercises

(Answers to even-numbered exercises are in Appendix A.)

1. Write the meaning of each of the following terms without reference to the book (or your notes), and compare your definition with the version in the text for each:
   a. linear probability model
   b. R_p^2
   c. binomial logit model
   d. log of the odds
   e. binomial probit model
   f. sequential binary model
   g. multinomial logit model

2. On graph paper, plot each of the following linear probability models. For what range of X_i is $1 < D_i$? How about $D_i < 0$?
   a. $D_i = 0.3 + 0.1X_i$
   b. $D_i = 3.0 - 0.2X_i$
   c. $D_i = -1.0 + 0.3X_i$

3. Bond ratings are letter ratings (AAA = best) assigned to firms that issue debt. These ratings measure the quality of the firm from the point of view of the likelihood of repayment of the bond. Suppose you've been hired by an arbitrage house that wants to predict *Moody's Bond Ratings* before they're published in order to buy bonds whose ratings are going to improve. In particular, suppose your firm wants to distinguish between A rated bonds (high quality) and B rated bonds (medium quality) and has collected a data set of 200 bonds with which to estimate a model. As you arrive on the job, your boss is about to buy bonds based on the results of the following model (standard errors in parentheses):

$$\hat{Y}_i = 0.70 + \underset{(0.05)}{0.05P_i} + \underset{(0.02)}{0.05PV_i} - \underset{(0.002)}{0.020D_i}$$

$$\bar{R}^2 = .69 \qquad DW = 0.50 \qquad n = 200$$

where: Y_i = 1 if the rating of the *i*th bond = A, 0 otherwise
P_i = profit rate of the firm that issued the *i*th bond
PV_i = standard deviation of P_i over the last 5 years
D_i = the ratio of debt to total capitalization of the firm that issued the *i*th bond

a. What econometric problems, if any, exist in this equation?
b. What suggestions would you have for a re-run of this equation with a different specification?
c. Suppose that your boss rejects your suggestions, saying "This is the real world, and I'm sure that my model will forecast bond ratings just as well as yours will." How would you respond? [Hint: Saying, "OK, boss, you win," is sure to keep you your job, but it won't get much credit on this question.]

4. Show that the logistic function, $D = 1/(1 + e^{-Z})$ is indeed equivalent to the binomial logit model, $\ln[D/(1 - D)] = Z$, where $Z = \beta_0 + \beta_1X_1 + \beta_2X_2 + \epsilon$.

5. Plot each of the following binomial logit models. For what range of X_i is $1 < D_i$? How about $D_i < 0$? [Hint: When you finish, compare your answers to those for Exercise #2 above.]
a. $\ln[D_i/(1 - D_i)] = 0.3 + 0.1X_i$
b. $\ln[D_i/(1 - D_i)] = 3.0 - 0.2X_i$
c. $\ln[D_i/(1 - D_i)] = -1.0 + 0.3X_i$

6. Amatya[11] estimated the following logit model of birth control for 1145 continuously married women aged 35 to 44 in Nepal:

$$\widehat{\ln\left(\frac{D_i}{[1 - D_i]}\right)} = -4.47 + \underset{\substack{(0.36) \\ t = 5.64}}{2.03WN_i} + \underset{\substack{(0.14) \\ 10.36}}{1.45ME_i}$$

where: D_i = 1 if the ith woman has ever used a recognized form of birth control, 0 otherwise
WN_i = 1 if the ith woman wants no more children, 0 otherwise
ME_i = number of methods of birth control known to the ith woman

a. Explain the theoretical meaning of the coefficients for WN and ME. How would your answer differ if this were a linear probability model?

11. Ramesh Amatya, "Supply-Demand Analysis of Differences in Contraceptive Use in Seven Asian Nations, Late 1970s." Presented at the Meetings of the Western Economic Association, 1988, Los Angeles.

b. Do the signs, sizes, and significance of the estimated slope coefficients meet your expectations? Why or why not?
c. What is the theoretical significance of the constant term in this equation?
d. If you could make one change in the specification of this equation, what would it be? Explain your reasoning.

7. What happens if we define a dummy dependent variable over a range other than zero to one? For example, suppose that in the research cited above, Amatya had defined D_i as being equal to 2 if the *i*th woman had ever used birth control and zero otherwise.
   a. What would happen to the size and theoretical meaning of the estimated logit coefficients? Would they stay the same? Would they change? (If so, how?)
   b. How would your answers to part a above change if Amatya had estimated a linear probability model instead of a binomial logit?

8. Return to our data on women's labor force participation and consider the possibility of adding A_i, the age of the *i*th woman, to the equation. Be careful when you develop your expected sign and functional form because the expected impact of age on labor force participation is difficult to pin down. For instance, some women drop out of the labor force when they get married, while others continue working even while they're raising their children. Still others work until they get married, stay at home to have children, and then return to the workforce once the children reach school age. The Cohen-Rea-Lerman study, for example, found the age of a woman to be relatively unimportant in determining labor force participation, except for women who were 65 and older and were likely to have retired.[12] The net result for our model is that age appears to be a theoretically irrelevant variable. A possible exception, however, is a dummy variable equal to one if the *i*th woman is 65 or over and zero otherwise.
   a. Look over the data set in Table 13.1. What problems do you see with adding an independent variable equal to one if the *i*th woman is 65 or older and zero otherwise?
   b. If you go ahead and add the dummy implied above to Equation 13.15 and re-estimate the model, you obtain the equation be-

12. Cohen, et al., *Ibid*, p. 212. (See footnote #3.)

low. Which equation do you prefer, Equation 13.15 or the one below? Explain your answer.

$$\widehat{\ln\left(\frac{D_i}{[1 - D_i]}\right)} = -5.89 - \underset{(1.18)}{2.59M_i} + \underset{(0.31)}{0.69S_i} - \underset{(0.30)}{0.03AD_i}$$
$$t = -2.19 \qquad 2.19 \qquad -0.01$$
$$n = 30 \qquad \text{quasi-}R^2 = .40 \qquad R_p^2 = .80 \qquad \text{iterations} = 5$$

where $AD_i = 1$ if the age of the *i*th woman is > 65, 0 otherwise

9. To get practice in actually estimating your own linear probability, logit, and probit equations, test the possibility that age (A_i) is actually a relevant variable in our women's labor force participation model. That is, take the data from Table 13.1 and estimate each of the following equations. Then use our specification criteria to compare your equation with the parallel version in the text (without A_i). Explain why you do or do not think that age is a relevant variable.

a. the linear probability model D = f(M,A,S)
b. the logit D = f(M,A,S)
c. the probit D = f(M,A,S)

10. An article published in Kouskoulaf and Lytle[13] presents coefficients from an estimated logit model of the choice between the car and public transportation for the trip to work in Boston. All three public transportation modes in Boston (bus, subway, and train, of which train is the most preferred) were lumped together as a single alternative to the car in a binomial logit model. The dependent variable was the log of the odds of taking public transportation for the trip to work, so the first coefficient implies that as income rises, the log of the odds of taking public transportation falls, etc.

Independent Variable	*Coefficient*
Family income (9 categories with 1 = low and 9 = high	−0.12
Number employed in the family	−1.09
Out-of-pocket costs (cents)	−3.16

13. "The Use of the Multinomial Logit in Transportation Analysis," in A. Kouskoulaf and B. Lytle, *Urban Housing and Transportation* (Detroit: Wayne State University, 1975), pp. 87–90.

Wait time (tenths of minutes)	0.18
Walk time (tenths of minutes)	−0.03
In-vehicle travel time (tenths of minutes)	−0.01

The last four variables are defined as the difference between the value of the variable for taking public transportation and its value for taking the car.

a. Do the signs of the estimated coefficients agree with your prior expectations? Which one(s) differ?
b. The transportation literature hypothesizes that people would rather spend time traveling in a vehicle than waiting for or walking to that vehicle. Do the sizes of the estimated coefficients of time support this hypothesis?
c. Since trains run relatively infrequently, the researchers set wait time for train riders fairly high. Most trains run on known schedules, however, so the average commuter learns that schedule and attempts to hold down wait time. Does this fact explain any of the unusual results indicated in your answers to parts a and b above?

11. Suppose that you want to build a multinomial logit model of how students choose which college to attend. For the sake of simplicity, let's assume that there are only four colleges to choose from: your college (c), the state university (u), the local junior college (j), and the nearby private liberal arts college (a). Further assume that everyone agrees that the important variables in such a model are the family income (Y) of each student, the average SAT scores of each college (SAT), and the tuition (T) of each college.
    a. How many equations should there be in such a multinomial logit system?
    b. If your college is the base, write out the definition of the dependent variable for each equation.
    c. Carefully write out the definitions of all the independent variables in each equation.
    d. Carefully write out the meanings of all the slope coefficients in each equation.

14

Simultaneous Equations

Unfortunately, the single-equation models we've covered to date ignore much of the interdependence that characterizes the modern world. Most econometric applications are inherently interdependent or simultaneous in nature, and the best approach to understanding this simultaneity is to explicitly acknowledge it with feedback loops in our models. This means specifying and estimating simultaneous equations systems instead of looking at just one equation at a time.

The most important models in economics and business are simultaneous in nature. Supply and demand, for example, is obviously simultaneous. To study the demand for chicken without also looking at the supply of chicken is to take a chance on missing important linkages and thus making significant mistakes. Virtually all the major approaches to macroeconomics, from Keynesian aggregate demand models to more recent rational expectations schemes, are inherently simultaneous. Even models that appear to be inherently single equation in nature often turn out to be much more simultaneous than you might think. Passbook deposits in S&Ls, for instance, are affected dramatically by the level of economic activity, the prevailing rate of interest in alternative assets, and a number of other simultaneously determined variables.

All this wouldn't mean much to econometricians if it weren't for the fact that the estimation of simultaneous equations systems with OLS causes a number of difficulties that aren't encountered with single equations. Most important, Classical Assumption III, which states that all explanatory variables should be uncorrelated with the error term, is violated in simultaneous models. Mainly because of this, OLS coefficient estimates are biased in simultaneous models. As a result, an alternative estimation procedure called Two-Stage Least Squares is usually employed in such models instead of OLS.

You're probably wondering why we've waited until now to discuss simultaneous equations if they're so important in economics and if OLS encounters bias when estimating them. The answer is that the simultaneous estimation of an equation changes every time the specification of any equation in the entire system is changed, so a researcher must be well equipped to deal with specification problems like those of the previous eight chapters. As a result, it does not make sense to learn how to estimate a simultaneous system until you are fairly adept at estimating a single equation.

14.1 Structural and Reduced-Form Equations

Before we can study the problems encountered in the estimation of simultaneous equations, we need to introduce a few concepts. Readers well versed in the subject are encouraged to skip to Section 14.1.2.

14.1.1 The Nature of Simultaneous Equations Systems

Which came first, the chicken or the egg? This question is impossible to answer satisfactorily because chickens and eggs are *jointly determined;* the more eggs you have, the more chickens you'll get, but the more chickens you have, the more eggs you'll get.[1] More realistically, the economic world is full of the kind of *feedback effects* and *dual*

1. This also depends on how hungry you are, which is a function of how hard you're working, which depends on how many chickens you have to take care of. (While this chicken/egg example is simultaneous in an annual model, it would not be truly simultaneous in a quarterly or monthly model because of the time lags involved.)

causality that require the application of simultaneous equations. Besides the supply and demand and simple macroeconomic model examples mentioned above, we could talk about the dual causality of population size and food supply, the joint determination of wages and prices, or the interaction between foreign exchange rates and international trade and capital flows. In terms of a typical econometric equation:

$$Y_t = \beta_0 + \beta_1 X_{1t} + \beta_2 X_{2t} + \epsilon_t \qquad (14.1)$$

a simultaneous system is one in which Y clearly has an effect on at least one of the Xs in addition to the effect that the Xs have on Y.

Such topics are usually modeled by distinguishing between variables that are simultaneously determined (the Ys, called **endogenous variables**) and those that are not (the Xs, called **exogenous variables**):

$$Y_{1t} = \alpha_0 + \alpha_1 Y_{2t} + \alpha_2 X_{1t} + \alpha_3 X_{2t} + \epsilon_{1t} \qquad (14.2)$$

$$Y_{2t} = \beta_0 + \beta_1 Y_{1t} + \beta_2 X_{3t} + \beta_3 X_{2t} + \epsilon_{2t} \qquad (14.3)$$

For example, Y_1 and Y_2 might be the quantity and price of chicken (respectively), X_1 the income of consumers, X_2 the price of beef (beef is a substitute for chicken in both consumption and production), and X_3 the price of chicken feed. With these definitions, Equation 14.2 would characterize the behavior of consumers of chickens and Equation 14.3 the behavior of suppliers of chickens. These behavioral equations are also called *structural equations.* **Structural equations** characterize the underlying economic theory behind each endogenous variable by expressing it in terms of both endogenous and exogenous variables. Researchers must view them as an entire system in order to see all the feedback loops involved. For example, the Ys are jointly determined, so a change in Y_1 will cause a change in Y_2, which will in turn cause Y_1 to change *again.* Contrast this feedback with a change in X_1, which will not eventually loop back and cause X_1 to change again. The αs and the βs in the equations are *structural coefficients,* and hypotheses should be made about their signs just as we did with the regression coefficients of single equations.

Note that a variable is endogenous because it is jointly determined, not just because it appears in both equations. That is, X_2, which is the price of beef but could be another factor beyond our control, is in both equations but is still exogenous in nature, because it is not simultaneously determined within the chicken market. In a large general equilibrium model of the entire economy, however, such a price variable would also likely be endogenous. How do you decide whether a particular variable should be endogenous or exogenous? Some variables are almost always exogenous (the weather, for example), but most others can be considered either endogenous or exogenous depending on the number and characteristics of the other equations in the system. Thus the distinction between endogenous and exogenous variables usually depends on how the researcher defines the scope of the research project.

Sometimes, lagged endogenous variables appear in simultaneous systems, usually when the equations involved are distributed lag equations (described in Chapter 12). To avoid confusion, **predetermined variables** are defined to include all exogenous and lagged endogenous variables. "Predetermined" implies that exogenous and lagged endogenous variables are determined outside the system of specified equations or prior to the current period. Endogenous variables that are not lagged are not predetermined, because they are jointly determined by the system. Therefore, econometricians tend to speak in terms of endogenous and predetermined variables when discussing simultaneous equations systems.

Let's look at the specification of a simple supply and demand model, say for the "cola" soft-drink industry:

$$Q_{Dt} = \alpha_0 + \alpha_1 P_t + \alpha_2 X_{1t} + \alpha_3 X_{2t} + \epsilon_{Dt} \tag{14.4}$$

$$Q_{St} = \beta_0 + \beta_1 P_t + \beta_2 X_{3t} + \epsilon_{St} \tag{14.5}$$

$$Q_{St} = Q_{Dt} \quad \text{(equilibrium condition)}$$

where: Q_{Dt} = the quantity of cola demanded in time period t
Q_{St} = the quantity of cola supplied in time period t
P_t = the price of cola in time period t
X_{1t} = dollars of advertising for cola in time period t
X_{2t} = another "demand-side" exogenous variable (e.g., income or the prices or advertising of other drinks)
X_{3t} = a "supply-side" exogenous variable (e.g., the price of artificial flavors or other factors of production)
ϵ_t = classical error terms (each equation has its own error term, subscripted "D" and "S" for demand and supply)

In this case, price and quantity are simultaneously determined, but price, one of the endogenous variables, is not on the left side of any of the equations. It's incorrect to assume automatically that the endogenous variables are those that appear on the left side of at least one equation; in this case, we could have just as easily written Equation 14.5 with price on the left side and quantity supplied on the right side, as we did in the chicken example in Equations 14.2 and 14.3. While the estimated coefficients would be different, the underlying relations would not. Note also that there must be as many equations as there are endogenous variables. In this case, the three endogenous variables are Q_D, Q_S, and P.

What would be the expected signs for the coefficients of the price variables in Equations 14.4 and 14.5? We'd expect price to enter negatively in the demand equation but to enter positively in the supply equation. The higher the price, after all, the less quantity will be demanded, but the more quantity will be supplied. These signs would result in the typical supply and demand diagram (Figure 14.1) that we're all used to. Look at Equations 14.4 and 14.5 again, however, and note that they would be identical but for the different predetermined variables. What would happen if we accidentally put a supply-side predetermined variable in the demand equation or vice versa?

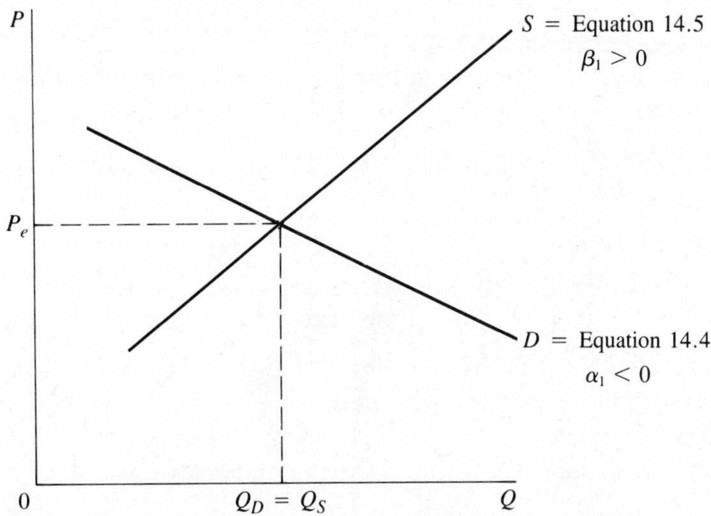

Figure 14.1 SUPPLY AND DEMAND SIMULTANEOUS EQUATIONS
An example of simultaneous equations that jointly determine two endogenous variables is the supply and demand for a product. In this case, Equation 14.4, the downward-sloping demand function, and Equation 14.5, the upward-sloping supply function, intersect at the equilibrium price and quantity for this market.

We'd have a very difficult time identifying which equation was which, and the expected signs for the coefficients of the endogenous variable P would become ambiguous. As a result, we must take care when specifying the structural equations in a system.

14.1.2 Simultaneous Equations Systems Violate Classical Assumption III

Recall from Chapter 4 that Classical Assumption III states that the error term and each explanatory variable must be independent of each other. If there is such a correlation, then the OLS regression estimation program is likely to attribute to the particular explanatory variable any variations in the dependent variable that are actually being caused by variations in the error term. The result will be biased estimates.

To see why simultaneous equations violate the assumption of independence between the error term and the explanatory variables, look again at a simultaneous system, Equations 14.2 and 14.3 (repeated with directional arrows):

$$\overset{\uparrow}{Y_{1t}} = \alpha_0 + \alpha_1 \overset{\uparrow}{Y_{2t}} + \alpha_2 X_{1t} + \alpha_3 X_{2t} + \overset{\uparrow}{\epsilon_{1t}} \quad (14.2)$$

$$\overset{\uparrow}{Y_{2t}} = \beta_0 + \beta_1 \overset{\uparrow}{Y_{1t}} + \beta_2 X_{3t} + \beta_3 X_{2t} + \epsilon_{2t} \quad (14.3)$$

Let's work through the system and see what happens when one of the error terms increases, holding everything else in the equations constant:

1. If ϵ_1 increases in a particular time period, Y_1 will also increase due to Equation 14.2.
2. If Y_1 increases, Y_2 will also rise[2] due to Equation 14.3.
3. But if Y_2 increases in Equation 14.3, it also increases in Equation 14.2 where it is an explanatory variable.

Thus an increase in the error term of an equation causes an increase in an explanatory variable in that same equation: If ϵ_1 increases, Y_1 increases, and then Y_2 increases, violating the assumption of independence between the error term and the explanatory variables.

2. This assumes that β_1 is positive. If β_1 is negative, Y_2 will decrease and there will be a negative correlation between ϵ_1 and Y_2, but this negative correlation will still violate Classical Assumption III. Also note that both Equations 14.2 and 14.3 could have Y_{1t} on the left side; if two variables are jointly determined, it doesn't matter which variable is considered dependent and which explanatory, because they are actually mutually dependent. We used this kind of simultaneous system in the cola model portrayed in Equations 14.4 and 14.5.

This is not an isolated result that depends on the particular equations involved. Indeed, as you'll find in Exercise 3, this result works for other error terms, equations, and simultaneous systems. All that is required for the violation of Classical Assumption III is that there be endogenous variables that are jointly determined in a system of simultaneous equations.

14.1.3 Reduced-Form Equations

An alternative way of expressing a simultaneous equations system is through the use of **reduced-form equations,** equations that express a particular endogenous variable solely in terms of an error term and all the predetermined (exogenous plus lagged endogenous) variables in the simultaneous system.

The reduced-form equations for the structural Equations 14.2 and 14.3 would thus be:

$$Y_{1t} = \pi_0 + \pi_1 X_{1t} + \pi_2 X_{2t} + \pi_3 X_{3t} + v_{1t} \tag{14.6}$$

$$Y_{2t} = \pi_4 + \pi_5 X_{1t} + \pi_6 X_{2t} + \pi_7 X_{3t} + v_{2t} \tag{14.7}$$

where the vs are stochastic error terms and the πs are called **reduced-form coefficients** because they are the coefficients of the predetermined variables in the reduced-form equations. Note that each equation includes only one endogenous variable, the dependent variable, and that each equation has exactly the same set of predetermined variables. The reduced-form coefficients, such as π_1 and π_5, are known as **impact multipliers** because they measure the impact on the endogenous variable of a one-unit change in the value of the predetermined variable, after allowing for the feedback effects from the entire simultaneous system.

There are at least four reasons for using reduced-form equations:

1. Since the reduced-form equations have no inherent simultaneity, they do not violate Classical Assumption III. Therefore, they can be estimated with OLS without encountering the problems discussed in this chapter.
2. The reduced-form coefficients estimated in this way can sometimes be mathematically manipulated to allow the estimation of the structural coefficients. That is, estimates of the πs of Equations 14.6 and 14.7 can be used to solve for the αs and βs of Equations 14.2 and 14.3. This method of calculating estimates of the structural coefficients from estimates of the reduced-form coefficients is called **Indirect Least Squares** (ILS). Unfortunately, ILS turns out to be

useful only in very limited situations. For more on ILS, see Exercise 4.

3. The interpretation of the reduced-form coefficients as impact multipliers means that they have economic meaning and useful applications of their own. For example, if you wanted to compare a government spending increase with a tax cut in terms of the per dollar impact in the first year, estimates of the impact multipliers (reduced-form coefficients or πs) would allow such a comparison.
4. Perhaps most importantly, reduced-form equations play an important role in the estimation technique most frequently used for simultaneous equations. This technique, Two-Stage Least Squares, will be explained in Section 14.3.

To conclude, let's return to the cola supply and demand model and specify the reduced-form equations for that model. (To test yourself, flip back to Equations 14.4 and 14.5 and see if you can get the right answer before going on.) Since the equilibrium condition forces Q_D to be equal to Q_S, we need only two reduced-form equations:

$$Q_t = \pi_0 + \pi_1 X_{1t} + \pi_2 X_{2t} + \pi_3 X_{3t} + v_{1t} \tag{14.8}$$

$$P_t = \pi_4 + \pi_5 X_{1t} + \pi_6 X_{2t} + \pi_7 X_{3t} + v_{2t} \tag{14.9}$$

Even though P never appears on the left side of a structural equation, it's an endogenous variable and should be treated as such.

14.2 The Bias of Ordinary Least Squares (OLS)

All the classical assumptions must be met for OLS estimates to be BLUE; when an assumption is violated, we must determine which of the properties no longer holds. It turns out that applying OLS directly to the structural equations of a simultaneous system, called *Direct Least Squares,* produces biased estimates of the coefficients. Such bias is called simultaneous equations bias or simultaneity bias.

14.2.1 Understanding Simultaneity Bias

Simultaneity bias refers to the fact that in a simultaneous system, the expected values of the OLS-estimated structural coefficients ($\hat{\beta}$s) are not equal to the true βs. These estimated coefficients are also inconsistent. That is, the expected values of the $\hat{\beta}$s do not approach the true βs even if the sample size gets quite large. We are therefore faced with the problem that in a simultaneous system:

$$E(\hat{\beta}) \neq \beta \tag{14.10}$$

Why does this simultaneity bias exist? Recall from Section 14.1.2 that in simultaneous equations systems, the error terms (the ϵs) tend to be correlated with the endogenous variables (the Ys) whenever the Ys appear as explanatory variables. Let's follow through what this correlation means (assuming positive coefficients for simplicity) in typical structural equations like 14.11 and 14.12:

$$Y_{1t} = \beta_0 + \beta_1 Y_{2t} + \beta_2 X_t + \epsilon_{1t} \tag{14.11}$$

$$Y_{2t} = \alpha_0 + \alpha_1 Y_{1t} + \alpha_2 Z_t + \epsilon_{2t} \tag{14.12}$$

Since we cannot observe the error term (ϵ_1) and don't know when ϵ_{1t} is above average, it will appear that if every time Y_1 is above average, so too is Y_2. As a result, the OLS estimation program will tend to attribute increases in Y_1 caused by the error term ϵ_1 to Y_2, thus overestimating β_1. This overestimation is simultaneity bias. If the error term is abnormally negative, Y_{1t} is less than it would have been otherwise, causing Y_{2t} to be less than it would have been otherwise, and the computer program will attribute the decrease in Y_1 to Y_2, once again causing us to overestimate β_1 (that is, induce upward bias).

Recall that the causation between Y_1 and Y_2 runs in both directions because the two variables are interdependent. As a result, β_1, when estimated by OLS, can no longer be interpreted as the impact of Y_2 on Y_1, holding X constant. Instead, $\hat{\beta}_1$ now measures some mix of the effects of the two endogenous variables on each other! In addition, consider β_2. It's supposed to be the effect of X on Y_1 holding Y_2 constant, but how can we expect Y_2 to be held constant when a change in Y_1 takes place? As a result, there is potential bias in all the estimated coefficients in a simultaneous system.

What does this bias look like? It's possible to derive an equation[3] for the expected value of the regression coefficients in a simultaneous

3. For Equation 14.11, the expected value of $\hat{\beta}_1$ simplifies to:

$$E(\hat{\beta}_1) = \beta_1 + E[\Sigma(Y_{2t} - \overline{Y}_2)(\epsilon_{1t})/\Sigma(Y_{2t} - \overline{Y}_2)^2]$$

In a nonsimultaneous equation, where Y_2 and ϵ_1 are not correlated, the expected value of $\hat{\beta}_1$ equals the true β_1 because the expected value of the term $\Sigma(Y_{2t} - \overline{Y}_2)(\epsilon_{1t})$ is zero. If Y_2 and ϵ_1 are positively correlated, as would be true in most simultaneous systems in economics, then the expected value of $\hat{\beta}_1$ is greater than the true β_1 because the expected value of $\Sigma(Y_{2t} - \overline{Y}_2)(\epsilon_{1t})$ is positive. In the less likely case that Y_2 and ϵ_1 are negatively correlated, the expected value of $\hat{\beta}_1$ is less than the true β_1.

system that is estimated by OLS. This equation shows that as long as the error term and any of the explanatory variables in the equation are correlated, then the coefficient estimates will be biased. In addition, it also shows that the bias will have the same sign as the correlation between the error term and the endogenous variable that appears as an explanatory variable in that error term's equation. Since that correlation is usually positive[4] in economic and business examples, so is the bias of OLS. The violation of Classical Asumption III will almost always mean bias in the estimation of β_1. In addition, this bias will usually be positive in economic applications, although the direction of the bias in any given situation will depend on the specific details of the structural equations and the model's underlying theory.

This does not mean that every coefficient from a simultaneous system estimated with OLS will be a bad approximation of the true population coefficient; indeed, most researchers use OLS to estimate equations in simultaneous systems under a number of circumstances. Instead, it's vital at least to consider an alternative to OLS whenever simultaneous equations systems are being estimated. Before we investigate the alternative estimation technique most frequently used (Two-Stage Least Squares), let's look at an example of simultaneity bias.

14.2.2 An Example of Simultaneity Bias

To show how the application of OLS to simultaneous equations estimation causes bias, we generated an example of such biased estimates. Since it's impossible to know whether any bias exists unless you also know the true βs, we picked a set of coefficients to be arbitrarily considered true, stochastically generated data-sets based on these coefficients, and obtained repeated OLS estimates of these coefficients from the generated data-sets. The expected value of these estimates turned out to be quite different from the true coefficient values, thus exemplifying the bias in OLS estimates of coefficients in simultaneous systems.

We used a supply and demand model as the basis for our example:

$$Q_t = \beta_0 + \beta_1 P_t + \beta_2 X_t + \epsilon_{Dt} \tag{14.13}$$

$$Q_t = \alpha_0 + \alpha_1 P_t + \alpha_2 Z_t + \epsilon_{St} \tag{14.14}$$

4. See Exercise 5 to examine this general statement in more detail for various economic models.

where: Q_t = the quantity demanded and supplied in time period t
P_t = the price in time period t
X_t = a "demand-side" exogenous variable, such as income
Z_t = a "supply-side" exogenous variable, such as weather
ϵ_t = classical errors (different for each equation)

The first step was to choose a set of true coefficient values that corresponded to our expectations for this model:

$$\beta_1 = -1 \qquad \beta_2 = +1 \qquad \alpha_1 = +1 \qquad \alpha_2 = +1$$

In other words, we have a negative relationship between price and quantity demanded, a positive relationship between price and quantity supplied, and positive relationships between the exogenous variables and their respective dependent variables.

The next step was to randomly generate a number of data-sets based on the true values. This also meant specifying some other characteristics of the data[5] before generating the different data sets (5000 in this case).

The final step was to apply OLS to the generated data-sets and to calculate the estimated coefficients of the demand equation (14.13). (Similar results were obtained for the supply equation.) The arithmetic means of the results for the 5000 regressions were:

$$\hat{Q}_{Dt} = -0.37P_t + 1.84X_t \tag{14.15}$$

In other words, the expected value of $\hat{\beta}_1$ should have been -1.00, but instead it was -0.37; the expected value of $\hat{\beta}_2$ should have been $+1.00$, but instead it was 1.84:

$$E(\hat{\beta}_1) = -0.37 \neq -1.00$$

$$E(\hat{\beta}_2) = 1.84 \neq 1.00$$

This is simultaneity bias! As the diagram of the sampling distributions of the $\hat{\beta}$s in Figure 14.2 shows, the OLS estimates of β_1 were almost

5. Other assumptions included a normal distribution for the error term, $\beta_0 = 0$, $\alpha_0 = 0$, $\sigma_s^2 = 3$, $\sigma_D^2 = 2$, $r_{xz}^2 = 0.4$, and $n = 20$. In addition, it was assumed that the error terms of the two equations were not correlated. This is another example of a Monte Carlo experiment.

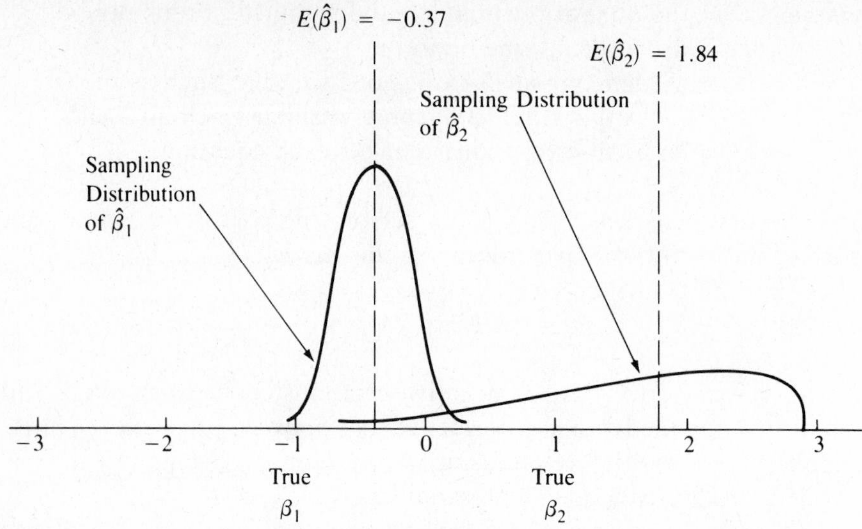

Figure 14.2 SAMPLING DISTRIBUTIONS SHOWING SIMULTANEITY BIAS OF OLS ESTIMATES
In the experiment in Section 14.2.2, simultaneity bias is evident in the distribution of the estimates of β_1, which had a mean value of -0.37 compared with a true value of -1.00, and in the estimates of β_2, which had a mean value of 1.84 compared with a true value of 1.00.

never very close to -1.00, while the OLS estimates of β_2 were distributed over a wide range of values.

The biased estimation in this example did not cause incorrect *signs* for the majority of the estimates, but any kind of bias is worth avoiding if it is at all possible. The most frequently used method of reducing the simultaneity bias is a technique called Two-Stage Least Squares (2SLS).

14.3 Two-Stage Least Squares (2SLS)

While there are a number of econometric estimation techniques available that help mitigate the bias and avoid the inconsistency inherent in the application of OLS to simultaneous equations systems, the most frequently used alternative to OLS is called Two-Stage Least Squares (2SLS).

14.3.1 What Is Two-Stage Least Squares?

OLS encounters bias in the estimation of simultaneous equations mainly because such equations violate Classical Assumption III, so one solution to the problem is to explore ways to avoid violating that assumption. We could do this if we could find a variable that is:

1. a good proxy for the endogenous variable
2. uncorrelated with the error term

If we then substitute this new variable for the endogenous variable where it appears as an explanatory variable, our new explanatory variable will be uncorrelated with the error term, and Classical Assumption III will be met. This general approach is called *instrumental variables.*

That is, consider Equation 14.16 in the following system:

$$Y_{1t} = \beta_0 + \beta_1 Y_{2t} + \beta_2 X_t + \epsilon_{1t} \tag{14.16}$$

$$Y_{2t} = \alpha_0 + \alpha_2 Y_{1t} + \alpha_2 Z_t + \epsilon_{2t} \tag{14.17}$$

If we could find a variable that was highly correlated with Y_2 but that was uncorrelated with ϵ_1, then we could substitute this new variable for Y_2 on the right side of Equation 14.16, and we'd conform to Classical Assumption III. An **instrumental variable** replaces an endogenous variable (when it is an explanatory variable); it is a good proxy for the endogenous variable and is independent of the error term.

Since there is no joint causality between the instrumental variable and any endogenous variable, the use of the instrumental variable avoids the violation of Classical Assumption III. The job of finding such a variable is another story, though. How do we go about finding variables with these qualifications? For simultaneous equations systems, 2SLS provides an approximate answer.

Two-Stage Least Squares (2SLS) is a method of systematically creating instrumental variables to replace the endogenous variables where they appear as explanatory variables in simultaneous equations systems. 2SLS does this by running a regression on the reduced form of the right side endogenous variables in need of replacement and then using the $\hat{Y}$s (or fitted values) from those reduced-form regressions as the instrumental variables. More specifically, the two-step procedure consists of:

Stage One: *Run OLS on the reduced-form equations for each of the endogenous variables that appear as explanatory variables in the structural equations in the system.*

Since the predetermined (exogenous plus lagged endogenous) variables are uncorrelated with the reduced-form error term, the OLS estimates of the reduced-form coefficients (the $\hat{\pi}$s) are unbiased. These $\hat{\pi}$s can then be used to calculate estimates of the endogenous variables:[6]

$$\hat{Y}_{1t} = \hat{\pi}_0 + \hat{\pi}_1 X_t + \hat{\pi}_2 Z_t \quad (14.18)$$

$$\hat{Y}_{2t} = \hat{\pi}_3 + \hat{\pi}_4 X_t + \hat{\pi}_5 Z_t \quad (14.19)$$

These $\hat{Y}$s are the instrumental variables that will be used as proxies in the structural equations of the simultaneous system.

Stage Two: *Substitute the reduced-form $\hat{Y}$s (instrumental variables) for the Ys that appear on the right side (only) of the structural equations, and then estimate these revised structural equations with OLS.*

That is, Stage Two consists of estimating the following equations with OLS:

$$Y_{1t} = \beta_0 + \beta_1 \hat{Y}_{2t} + \beta_2 X_t + u_{1t} \quad (14.20)$$

$$Y_{2t} = \alpha_0 + \alpha_2 \hat{Y}_{1t} + \alpha_2 Z_t + u_{2t} \quad (14.21)$$

Note that the dependent variables are still the original endogenous variables and that the substitutions are only for the endogenous variables where they appear on the right-hand side of the structural equations.

This description of 2SLS can be generalized to *m* different simultaneous structural equations with *m* reduced-form equations for each of the *m* endogenous variables. Each reduced-form equation has as explanatory variables every predetermined variable in the entire system

6. Because the πs are not uncorrelated with the ϵs, this procedure produces only approximate instrumental variables that provide consistent (for large samples) but biased (for small samples) estimates of the coefficients of the structural equation (the $\hat{\beta}$s).

of equations. The OLS estimates of the reduced-form equations are used to compute the estimated vaues of all the endogenous variables that appear as explanatory variables in the *m* structural equations. After substituting these fitted values for the original values of the endogenous independent variables, OLS is applied to each stochastic equation in the set of structural equations.

14.3.2 The Properties of Two-Stage Least Squares

1. *2SLS estimates are still biased, but they are now consistent.* That is, for small samples, the expected value of a $\hat{\beta}$ produced by 2SLS is still not equal to the true β,[7] but as the sample size gets larger, the expected value of the $\hat{\beta}$ approaches the true β. As the sample size gets bigger, the variances of both the OLS and the 2SLS estimates decrease. OLS estimates become very precise estimates of the wrong number, and 2SLS estimates become very precise estimates of the correct number. As a result, the larger the sample size, the better a technique 2SLS is.

   To illustrate, let's look again at the example of Section 14.2. We returned to that example and expanded the data-set from 5000 different samples of size 20 each to 5000 different samples of 50 observations each. As expected, the average $\hat{\beta}_1$ for 2SLS moved from -1.25 to -1.06 compared to the "true" value of -1.00. By contrast, the OLS average estimate went from -0.37 to -0.44. Such results are typical; large sample sizes will produce unbiased estimates for 2SLS but biased estimates for OLS.

2. *The bias in 2SLS for small samples is of the opposite sign of the bias in OLS.* Recall that the bias in OLS was positive, indicating that a $\hat{\beta}$ produced by OLS for a simultaneous system is likely to be greater than the true β. For 2SLS, the expected bias is negative, and thus a $\hat{\beta}$ produced by 2SLS is likely to be less than the true β. For any given set of data, the 2SLS estimate can be larger than the OLS estimate, but it can be shown that the majority of 2SLS estimates are likely to be less than the corresponding OLS estimates. For large samples, there is little bias in 2SLS.

   Return to the example of Section 14.2. Compared to the true value of -1.00 for β_1, the small sample 2SLS average estimate was -1.25, as mentioned above. This means that the 2SLS estimates

7. This bias is caused by remaining correlation between the $\hat{Y}$s produced by the first stage reduced-form regressions and the ϵs, the error terms of the structural equations. The effect of the correlation tends to decrease as the sample size increases. Even for small samples, though, it's worth noting that the expected bias due to 2SLS usually is smaller than the expected bias due to OLS.

showed negative bias. The OLS estimates, on the other hand, averaged −0.37; since −0.37 is more positive than −1.00, the OLS estimates exhibited positive bias. Thus the observed bias in the example due to OLS was opposite the observed bias due to 2SLS.

3. *If the fit of the reduced-form equation is quite poor, then 2SLS will not work very well.* Recall that the instrumental variable is supposed to be a good proxy for the endogenous variable. To the extent that the fit (as measured by R^2) of the reduced-form equation is poor, then the instrumental variable is no longer highly correlated with the original endogenous variable, and there is no reason to expect 2SLS to be effective. As the R^2 of the reduced-form equation increases, the usefulness of 2SLS will increase.

4. *If the predetermined variables are highly correlated, 2SLS will not work very well.* The first stage of 2SLS includes explanatory variables from different structural equations in the same reduced-form equation. As a result, severe multicollinearity between explanatory variables from different structural equations is possible in the reduced-form equations. When this happens, a $\hat{Y}$ produced by a reduced-form equation can be highly correlated with the exogenous variables in the structural equation. Consequently, the second stage of 2SLS will also show a high degree of multicollinearity, and the variances of the estimated coefficients will be high. Thus the higher the simple correlation coefficient between predetermined variables (or the higher the variance inflation factors), the less precise 2SLS estimates will be.

5. *The use of the* t*-test for hypothesis testing is far more accurate using 2SLS estimators than it is using OLS estimators.* The *t*-test is not exact for the 2SLS estimators, but it is accurate enough in most circumstances. By contrast, the biasedness of OLS estimators in simultaneous systems implies that its t-statistics are not accurate enough to be relied upon for testing purposes.[8] This means that it may be appropriate to use 2SLS even when the predetermined variables are highly correlated.

On balance, then, 2SLS will almost always be a better estimator of the coefficients of a simultaneous system than OLS will be. The major exception to this general rule is when the fit of the reduced-form equation in question is quite poor for a small sample.

8. In our experiments, OLS estimators rejected the correct null hypothesis more than eight times as often as would have been expected from an unbiased procedure. In contrast, the 2SLS estimators were found to reject a correct null hypothesis only twice as frequently as would have been expected.

14.3.3 An Example of Two-Stage Least Squares

Let's work through an example of the application of 2SLS to a naive linear Keynesian macroeconomic model of the U.S. economy. We'll specify the following system:

$$Y_t = C_t + I_t + G_t + NX_t \tag{14.22}$$

$$C_t = \beta_0 + \beta_1 YD_t + \beta_2 C_{t-1} + \epsilon_{1t} \tag{14.23}$$

$$YD_t = Y_t - T_t \tag{14.24}$$

$$I_t = \beta_3 + \beta_4 Y_t + \beta_5 ra_t + \epsilon_{2t} \tag{14.25}$$

$$ra_t = \frac{r_t + r_{t-1}}{2} \tag{14.26}$$

$$r_t = \beta_6 + \beta_7 Y_t + \beta_8 M1_t + \epsilon_{3t} \tag{14.27}$$

G_t, NX_t, T_t, and $M1_t$ are exogenous

where: Y_t = GNP in year t
C_t = total personal consumption in year t
I_t = total gross private domestic investment in year t
G_t = government purchases of goods and services in year t
NX_t = net exports of goods and services (exports minus imports) in year t
T_t = taxes (actually equal to taxes, depreciation, corporate profits, government transfers, and other adjustments necessary to convert GNP to disposable income) in year t
r_t = the interest rate (on Aaa-rated corporate bonds) in year t
$M1_t$ = the money supply (narrowly defined) in year t
YD_t = disposable income in year t
ra_t = the mean of r_t and r_{t-1} (in effect, interest rates lagged by six months)

All variables are in real terms (measured in billions of 1982 dollars) except the interest rate variables, which are measured in nominal percent. The data for this example are from 1964 through 1988 and are presented in Table 14.1. (To calculate lagged variables, it's necessary to know that C in 1963 was 1108.3 and r in 1963 was 4.26.)

TABLE 14.1 DATA FOR THE SMALL MACROMODEL

Year	C	I	r	Y	YD	M1	G	NX
1964	1170.7	325.9	4.40	1973.3	1291.0	160.4	470.8	5.9
1965	1236.3	367.0	4.49	2087.6	1365.7	167.9	487.0	−2.7
1966	1298.9	390.5	5.13	2208.3	1431.3	172.1	532.6	−13.7
1967	1337.7	374.4	5.51	2271.4	1493.2	183.3	576.2	−16.9
1968	1405.9	391.8	6.18	2365.6	1551.3	197.5	597.6	−29.7
1969	1456.7	410.3	7.03	2423.3	1599.8	204.0	591.2	−34.9
1970	1492.1	381.5	8.04	2416.2	1668.1	214.5	572.6	−30.0
1971	1538.8	419.3	7.39	2484.8	1728.4	228.4	566.5	−39.8
1972	1621.8	465.4	7.21	2608.5	1797.4	249.4	570.7	−49.4
1973	1689.5	520.8	7.44	2744.1	1916.3	263.0	565.3	−31.5
1974	1674.0	481.3	8.57	2729.3	1896.6	274.4	573.2	0.8
1975	1711.9	383.3	8.83	2695.0	1931.7	287.6	580.9	18.9
1976	1803.9	453.5	8.43	2826.7	2001.0	306.5	580.3	−11.0
1977	1883.7	521.3	8.02	2958.6	2066.6	331.4	589.1	−35.5
1978	1961.0	576.9	8.73	3115.2	2167.4	358.7	604.1	−26.8
1979	2004.5	575.2	9.63	3192.4	2212.6	386.1	609.1	3.6
1980	2000.3	509.3	11.94	3187.1	2214.3	412.2	620.5	57.0
1981	2024.2	545.5	14.17	3248.8	2248.6	439.1	629.7	49.4
1982	2050.7	447.3	13.79	3166.0	2261.5	476.4	641.7	26.3
1983	2146.0	504.0	12.04	3279.1	2331.9	522.1	649.0	−19.9
1984	2249.3	658.4	12.71	3501.4	2469.8	551.9	677.7	−84.0
1985	2354.8	637.0	11.37	3618.7	2542.8	620.1	731.2	−104.3
1986	2455.2	643.5	9.02	3721.7	2640.9	725.4	760.5	−137.5
1987	2520.9	674.8	9.38	3847.0	2686.3	744.2	780.2	−128.9
1988	2592.2	721.8	9.71	3996.1	2788.3	776.0	782.3	−100.2

Source: *The Economic Report of the President, 1989*

Equations 14.22 through 14.27 are the structural equations of the system, but only equations 14.23, 14.25, and 14.27 are stochastic (behavioral) and need to be estimated. The endogenous variables are those jointly determined by the system, namely Y_t, C_t, YD_t, I_t, ra_t, and r_t. To see why all six of these variables are jointly determined, try holding one of them constant and letting another change. For instance, if Y_t changes, so too must C_t, YD_t, I_t, ra_t, and r_t. On the other hand, G_t, NX_t, T_t, and $M1_t$ are all exogenous and therefore predetermined. In addition, C_{t-1} and r_{t-1} are lagged endogenous variables and are also predetermined. To sum, then, we have six structural equations, six endogenous variables, and six predetermined variables.

What is the economic content of the stochastic structural equations? The consumption function, Equation 14.23, is a Koyck distributed lag consumption function of the kind we discussed in Chapter

12. We discuss this exact equation in Section 12.1, going so far as to estimate Equation 14.23 with OLS on data from Table 14.1, and the reader is encouraged to reread that analysis.

The investment function, Equation 14.25, includes simplified multiplier and cost of capital components. The multiplier term β_4 measures the stimulus to investment that is generated by an increase in GNP. In a Keynesian model, β_4 thus would be expected to be positive. On the other hand, the higher the cost of capital, the less investment we'd expect to be undertaken (holding multiplier effects constant), mainly because the expected rate of return on marginal capital investments is no longer sufficient to cover the higher cost of capital. Thus β_5 is expected to be negative. It takes time to plan and start up investment projects, though, so the interest rate is lagged six months[9] (by using ra_t, calculated in Equation 14.26 as the average of this year's and last year's interest rate).

The interest rate equation is a liquidity preference function solved for the interest rate under the assumption of equilibrium in the money market. In such a situation, an increase in GNP with the money supply held constant would increase the transactions demand for money, pushing up interest rates, so we'd expect β_7 to be positive. If the money supply increased with GNP held constant, we'd expect interest rates to fall, so β_8 should be negative. Recall that a naive Keynesian model has constant prices by assumption.

We're now ready to apply 2SLS to our model.

Stage One: Even though there are six endogenous variables, only three of them appear on the right-hand side of stochastic equations, so only three reduced-form equations need to be estimated to apply 2SLS. These reduced-form equations are estimated automatically by all 2SLS

9. This investment equation is a simplified mix of the accelerator and the neoclassical theories of the investment function. The former emphasizes that changes in the level of output are the key determinant of investment, while the latter emphasizes that user cost of capital (the opportunity cost that the firm incurs as a consequence of owning an asset) is the key. Note also that ra_t doesn't *really* lag interest rates six months, it just approximates it. For an introduction to the determinants of consumption and investment, see any intermediate macroeconomics textbook. For traditional and modern overviews of the estimation of consumption and investment equations in simultaneous macroeconomic models, see Otto Eckstein, *The DRI Model of the U.S. Economy* (New York: McGraw-Hill, 1983); Michael K. Evans, *Macroeconomic Activity* (New York: Harper and Row, 1969); and Thomas M. Havrilesky, *Modern Concepts in Macroeconomics* (Arlington Heights, Illinois: Harlan Davidson, Inc., 1985).

computer estimation programs, but it's instructive to take a look at one anyway:

$$\widehat{YD}_t = 232.6 - 0.60G_t \quad - 0.73NX_t + 0.08T_t$$
$$\qquad (0.24) \qquad (0.17) \qquad (0.15)$$
$$t = -2.5 \qquad -4.3 \qquad 0.6$$
$$+ 1.17C_{t-1} + 9.80r_{t-1} - 0.27M1_t \tag{14.28}$$
$$\qquad (0.07) \qquad (3.57) \qquad (0.13)$$
$$t = 17.8 \qquad 2.7 \qquad -2.1$$
$$\bar{R}^2 = .998 \qquad n = 25 \qquad DW = 2.51$$

This reduced form has an excellent overall fit but is almost surely suffering from severe multicollinearity. Note that we don't test any hypotheses on reduced forms, nor do we consider dropping a variable (like T_t) that is statistically and theoretically irrelevant. The whole purpose of stage one of 2SLS is not to generate meaningful reduced-form estimated equations but rather to generate useful instruments ($\hat{Y}$s) to use as substitutes for endogenous variables in the second stage. To do that, we calculate the $\hat{Y}_t$s, $\widehat{YD}_t$s, and $\widehat{ra}_t$s for all 25 observations by plugging the actual values of all 6 predetermined variables into reduced form equations like Equation 14.28.

Stage Two: We then substitute these $\hat{Y}_t$s, $\widehat{YD}_t$s, and $\widehat{ra}_t$s for the endogenous variables where they appear on the right sides of Equations 14.23, 14.25, and 14.27. For example, the $\widehat{YD}_t$ from Equation 14.28 would be substituted into Equation 14.23, resulting in:

$$C_t = \beta_0 + \beta_1 \widehat{YD}_t + \beta_2 C_{t-1} + \epsilon_{1t} \tag{14.29}$$

If we use OLS to estimate Equation 14.29 and the other second stage equations given the data in Table 14.1, we obtain the following 2SLS[10] results:

10. A few notes about 2SLS estimation and this model are in order. The 2SLS estimates in Equations 14.30 through 14.32 are correct, but if you were to estimate those equations with OLS (using as instruments $\hat{Y}$s and $\widehat{YD}$s generated like Equation 14.28) you would obtain the same coefficient estimates but a different set of estimates of the standard errors (and t-scores). This difference comes about because running OLS on the second stage alone ignores the fact that the first stage was run at all. To get accurate estimated standard errors and t-scores, the estimation should be done on a complete 2SLS program. Most computer regression packages, including ECSTAT, include such a 2SLS program. Finally, note that because there are two stages involved, rounding errors are likely to be encountered if the regression results from different computer estimation packages are compared.

$$\hat{C}_t = -52.2 + 0.61\widehat{YD}_t + 0.37C_{t-1} \qquad (14.30)$$
$$(0.22) \qquad (0.23)$$
$$t = \quad 2.8 \qquad 1.6$$
$$\bar{R}^2 = .996 \qquad DW = 0.71$$

$$\hat{I}_t = -72.0 + 0.24\hat{Y}_t - 13.9\widehat{ra}_t \qquad (14.31)$$
$$(0.018) \qquad (3.7)$$
$$t = \quad 13.4 \qquad -3.8$$
$$\bar{R}^2 = .924 \qquad DW = 1.74$$

$$\hat{r}_t = -10.2 + 0.008\hat{Y}_t - 0.015M1_t \qquad (14.32)$$
$$(0.002) \qquad (0.007)$$
$$t = \quad 3.4 \qquad -2.0$$
$$\bar{R}^2 = .585 \qquad DW = 0.49$$

Compare these results with what we would have obtained if, instead of using 2SLS, we had estimated the equations with OLS alone:

$$\hat{C}_t = -55.3 + 0.64YD_t + 0.34C_{t-1} \qquad (14.33)$$
$$(0.17) \qquad (0.18)$$
$$t = 3.8 \qquad 1.9$$
$$\bar{R}^2 = .996 \qquad DW = 0.69$$

$$\hat{I}_t = -70.4 + 0.23Y_t - 13.1ra_t \qquad (14.34)$$
$$(0.018) \qquad (3.6)$$
$$t = 13.4 \qquad -3.6$$
$$\bar{R}^2 = .924, \qquad DW = 1.77$$

$$\hat{r}_t = -9.8 + 0.0082Y_t - 0.014M1_t \qquad (14.35)$$
$$(0.0025) \qquad (0.007)$$
$$3.33 \qquad -2.0$$
$$\bar{R}^2 = .585 \qquad DW = 0.48$$

Let's compare the OLS and 2SLS results. First, there doesn't seem to be much difference between them. If OLS is biased, how could this occur? When the fit of the Stage One reduced-form equations is excellent, as in Equation 14.28, then Y and $\hat{Y}$ are virtually identical, and the second stage of 2SLS is quite similar to the OLS estimate. Second, we'd expect positive bias in the OLS estimation and smaller negative bias in the 2SLS estimation, but the differences between OLS and 2SLS appear to be in the expected direction only about half the time. This might have been caused by the extreme multicollinearity in the 2SLS estimations as well as by the superb fit of the reduced forms mentioned above.

Finally, take a look at the Durbin-Watson d statistics. In the consumption and interest equations, d is well below the d_L of 1.21 (5 percent one-sided significance, n = 25, k′ = 2) despite d's bias towards 2 in the consumption equation (because it's a Koyck distributed lag). Consequently, positive serial correlation is likely to exist in the residuals of these two equations. This is a serious problem in the consumption function because, as mentioned in Section 12.2, serial correlation in an equation with a lagged dependent variable causes bias. One solution to this problem, running GLS *and* 2SLS, is discussed in Exercise 12.

This model provides us with a complete example of the use of 2SLS to estimate a simultaneous system. However, the application of 2SLS requires that the equation being estimated be "identified," so before we can conclude our study of simultaneous systems, we need to address the problem of identification.

14.4 The Identification Problem

Two-Stage Least Squares cannot be applied to an equation unless that equation is *identified*. Before estimating any equation in a simultaneous system, the researcher must address the identification problem. Once an equation is found to be identified, then it can be estimated with 2SLS, but if an equation is not identified *(underidentified)*, then 2SLS cannot be used no matter how large the sample. It's important to point out that an equation being identified (and therefore capable of being estimated with 2SLS) does not ensure that the resulting 2SLS estimates will be good ones. The question being asked is not how good the 2SLS estimates will be but whether the 2SLS estimates can be obtained at all.

14.4.1 What Is the Identification Problem?

Identification is a precondition for the application of 2SLS to equations in simultaneous systems; a structural equation is identified only when enough of the system's predetermined variables are omitted from the equation in question to allow that equation to be distinguished from all the others in the system. Note that one equation in a simultaneous system might be identified while another might not.

How is it possible that we could have equations that we could not identify? To see this, let's consider a supply and demand simultaneous system where only price and quantity are specified:

$$Q_{Dt} = \alpha_0 + \alpha_1 P_t + \epsilon_{Dt} \qquad \text{(demand)} \qquad (14.36)$$

$$Q_{St} = \beta_0 + \beta_1 P_t + \epsilon_{St} \qquad \text{(supply)} \qquad (14.37)$$

Although we've labeled one equation as the demand equation and the other as the supply equation, the computer will not be able to identify them from the data because the right-side and the left-side variables are exactly the same in both equations; without some predetermined variables included to distinguish between the two equations, it would be impossible to distinguish supply from demand.

What if we added a predetermined variable to one of the equations, say the supply equation? Then, Equation 14.37 would become:

$$Q_{St} = \beta_0 + \beta_1 P_t + \beta_2 Z_t + \epsilon_{St} \qquad (14.38)$$

In such a circumstance, every time Z changed, the supply curve would shift, but the demand curve would not, so that eventually we would be able to collect a good picture of what the demand curve looked like.

Figure 14.3 demonstrates this. Given four different values of Z, we get four different supply curves, each of which intersects with the

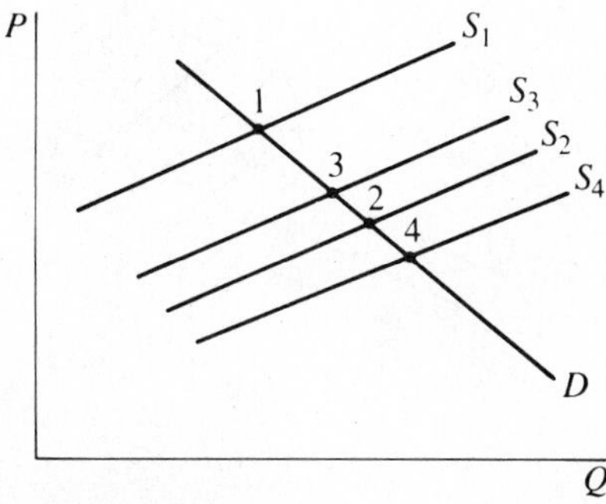

Figure 14.3 A SHIFTING SUPPLY CURVE ALLOWS THE IDENTIFICATION OF THE DEMAND CURVE

If the supply curve shifts but the demand curve does not, then we move along the demand curve, allowing us to identify and estimate the demand curve (but not the supply curve).

constant demand curve at a different equilibrium price and quantity (intersections 1-4). These equilibria are the data that we would be able to observe in the real world and are all that we could feed into the computer. As a result, we would be able to identify the demand curve because we left out at least one predetermined variable; when this predetermined variable changed, but the demand curve didn't, the supply curve shifted so that quantity demanded moved along the demand curve and we gathered enough information to estimate the coefficients of the demand curve. The supply curve, on the other hand, remains as much a mystery as ever because its shifts give us no clue whatsoever about its shape. In essence, the demand curve was identified by the predetermined variable that was included in the system but excluded from the demand equation. The supply curve is not identified because there is no such excluded predetermined variable for it.

Even if we added Z to the demand curve as well, that would not identify the supply curve. In fact, if we had Z in both equations, the two would be identical again, and while both would shift when Z changed, those shifts would give us no information about either curve! As illustrated in Figure 14.4, the observed equilibrium prices and quantities would be almost random intersections describing neither the demand nor the supply curve. That is, the shifts in the supply curve are the same as before, but now the demand curve also shifts with Z

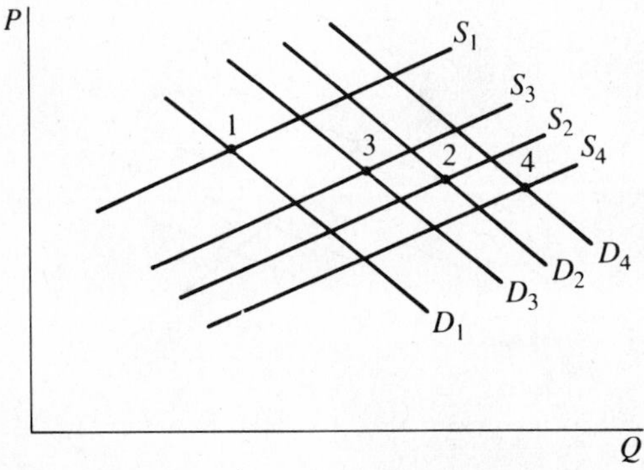

Figure 14.4 IF BOTH THE SUPPLY CURVE AND THE DEMAND CURVE SHIFT, NEITHER CURVE IS IDENTIFIED

If both the supply curve and the demand curve shift in response to the same variable, then we move from one equilibrium to another, and the resulting data points identify neither curve. To allow such an identification, at least one exogenous factor must cause one curve to shift while allowing the other to remain constant.

In this case, it's not possible to identify either the demand curve or the supply curve.[11]

The way to identify both curves is to have at least one predetermined variable in each equation that is not in the other, as in:

$$Q_{Dt} = \alpha_0 + \alpha_1 P_t + \alpha_2 X_t + \epsilon_{Dt} \quad (14.39)$$

$$Q_{St} = \beta_0 + \beta_1 P_t + \beta_2 Z_t + \epsilon_{St} \quad (14.40)$$

Now when Z changes, the supply curve shifts, and we can identify the demand curve from the data on equilibrium prices and quantities. When X changes, the demand curve shifts, and we can identify the supply curve from the data. Of course, if X and Z are highly correlated, we still will have problems of estimation, as mentioned in the previous section.

To sum, identification is a precondition for the application of 2SLS to equations in simultaneous systems. A structural equation is identified only when the predetermined variables are arranged within the system so as to allow us to use the observed equilibrium points to distinguish the shape of the equation in question. Most systems are quite a bit more complicated than the ones above, however, so econometricians need a general method by which to determine whether equations are identified. The method typically used is the *order condition* of identification.

14.4.2 The Order Condition of Identification

The **order condition** is a systematic method of determining whether a particular equation in a simultaneous system has the potential to be identified. If an equation can meet the order condition, then it is identified in all but a very small number of cases. We thus say that the order condition is a necessary but not sufficient condition of identification.[12]

11. An exception would be if you knew the relative magnitudes of the true coefficients of Z in the two equations, but such knowledge is unlikely.

12. A sufficient condition for an equation to be identified is called the *rank condition,* but this is so difficult to apply by hand to large systems of simultaneous equations that most researchers examine just the order condition before estimating an equation with 2SLS. These researchers let the computer estimation procedure tell them whether the rank condition has been met (by its ability to apply 2SLS to the equation). A rapid increase in the cost of computer time might cause econometricians to reconsider this usage, but for the present a beginning researcher needs to understand only the order condition. Those interested in the rank condition are encouraged to consult an advanced econometrics text.

What is the order condition? Recall that we have used the phrases endogenous and predetermined to refer to the two kinds of variables in a simultaneous system. Endogenous variables are those that are jointly determined in the system in the current time period. Predetermined variables are exogenous variables plus any lagged endogenous variables that might be in the model. For each equation in the system, we need to determine:

1. The number of predetermined (exogenous plus lagged endogenous) variables in the entire simultaneous system.
2. The number of slope coefficients estimated in the equation in question.

The Order Condition: *A necessary condition for an equation to be identified is that the number of predetermined (exogenous plus lagged endogenous) variables in the system be greater than or equal to the number of slope coefficients in the equation of interest.*

In equation form, a structural equation meets the order condition[13] if:

$$\underset{\text{(in the simultaneous system)}}{\text{The \# of predetermined variables}} \geq \underset{\text{(in the equation)}}{\text{The \# of slope coefficients}}$$

14.4.3 Two Examples of the Application of the Order Condition

Let's apply the order condition to some of the simultaneous equations systems encountered in this chapter. For example, consider once again the cola supply and demand model of Section 14.1.1:

$$Q_{Dt} = \alpha_0 + \alpha_1 P_t + \alpha_2 X_{1t} + \alpha_3 X_{2t} + \epsilon_{Dt} \quad (14.4)$$

$$Q_{St} = \beta_0 + \beta_1 P_t + \beta_2 X_{3t} + \epsilon_{St} \quad (14.5)$$

$$Q_{St} = Q_{Dt}$$

Equation 14.4 is identified by the order condition because the number of predetermined variables in the system (three, X_1, X_2, and X_3) is equal to the number of slope coefficients in the equation (three, α_1, α_2,

13. A more popular but harder to remember way of stating this condition is that the number of predetermined variables in the system that are excluded from the equation must be greater than or equal to the number of endogenous variables included in the equation minus one.

and α_3). This particular result (equality) implies that Equation 14.4 is *exactly identified* by the order condition. Equation 14.5 is also identified by the order condition because there still are three predetermined variables in the system, but there are only two slope coefficients in the equation; this condition implies that Equation 14.5 is *overidentified.* 2SLS can be applied to equations that are identified (which includes exactly identified and overidentified) but not to equations that are underidentified.

A more complicated example is the small macroeconomic model of Section 14.3.3:

$$Y_t = C_t + I_t + G_t + NX_t \quad (14.22)$$

$$C_t = \beta_0 + \beta_1 YD_t + \beta_2 C_{t-1} + \epsilon_{1t} \quad (14.23)$$

$$YD_t = Y_t - T_t \quad (14.24)$$

$$I_t = \beta_3 + \beta_4 Y_t + \beta_5 ra_t + \epsilon_{2t} \quad (14.25)$$

$$ra_t = \frac{r_t + r_{t-1}}{2} \quad (14.26)$$

$$r_t = \beta_6 + \beta_7 Y_t + \beta_8 M1_t + \epsilon_{3t} \quad (14.27)$$

As we've noted, there are six predetermined variables (exogenous plus lagged endogenous) in this system (G_t, NX_t, T_t, $M1_t$, C_{t-1}, and r_{t-1}). Equation 14.23 has two slope coefficients (β_1 and β_2) so this equation is overidentified ($6 > 2$) and meets the order condition of identification. As the reader can verify, Equations 14.25 and 14.27 also turn out to be overidentified. Since the 2SLS computer program did indeed come up with estimates of the βs in the model, we knew this already. Note that Equations 14.22, 14.24, and 14.26 are identities and are not estimated, so we're not concerned with their identification properties.

14.5 Summary

1. Most economic and business models are inherently simultaneous because of the dual causality, feedback loops, or joint determination of particular variables. These simultaneously determined variables are called endogenous, while nonsimultaneously determined variables are called exogenous.

2. A structural equation characterizes the theory underlying a particular variable and is the kind of equation we have used to date in this text. A reduced-form equation expresses a particular endogenous variable solely in terms of an error term and all the predetermined (exogenous and lagged endogenous) variables in the simultaneous system.

3. Simultaneous equations models violate the classical assumption of independence between the error term and the explanatory variables because of the feedback effects of the endogenous variables. For example, an unusually high observation of an equation's error term works through the simultaneous system and eventually causes a high value for the endogenous variables that appear as explanatory variables in the equation in question, thus violating the assumption of no correlation (Classical Assumption III).

4. If OLS is applied to the coefficients of a simultaneous system, the resulting estimates are biased and inconsistent. This occurs mainly because of the violation of Classical Assumption III; the OLS regression package attributes to explanatory variables changes in the dependent variable actually caused by the error term (with which the explanatory variables are correlated).

5. Two-Stage Least Squares is a method of decreasing the amount of bias in the estimation of simultaneous equations systems. It works by systematically using the reduced-form equations of the system to create proxies for the endogenous variables that are independent of the error terms (called instrumental variables). It then runs OLS on the structural equations of the system with the instrumental variables replacing the endogenous variables where they appear as explanatory variables.

6. Two-Stage Least Squares estimates are biased (with a sign opposite that of the OLS bias) but consistent (becoming more unbiased and closer to zero variance as the sample size gets larger). If the fit of the reduced-form equations is poor or if the predetermined variables are highly correlated, then 2SLS will not work very well. The larger the sample size, the better it is to use 2SLS.

7. 2SLS cannot be applied to an equation that's not identified. A necessary (but not sufficient) requirement for identification is the order condition, which requires that the number of predetermined variables in the system be greater than or equal to the number of

slope coefficients in the equation of interest. Sufficiency is usually determined by the ability of 2SLS to estimate the coefficients.

Exercises

(Answers to even-numbered exercises are in Appendix A.)

1. Write the meaning of each of the following terms without reference to the book (or your notes), and compare your definition with the version in the text for each:
   a. endogenous variables
   b. predetermined variables
   c. structural equations
   d. reduced-form equations
   e. simultaneity bias
   f. Two-Stage Least Squares
   g. identification
   h. order condition for identification

2. Which of the equations in the following systems are simultaneous, and which allow the "endogenous" variables to be determined sequentially (also called *recursive*)? Be sure to specify which variables are endogenous and which are predetermined (exogenous or lagged endogenous):
   a. $Y_{1t} = f(Y_{2t}, X_{1t}, X_{2t-1})$
      $Y_{2t} = f(Y_{3t}, X_{3t}, X_{4t})$
      $Y_{3t} = f(Y_{1t}, X_{1t-1}, X_{4t-1})$
   b. $Z_t = g(X_t, Y_t, H_t)$
      $X_t = g(Z_t, P_{t-1})$
      $H_t = g(Z_t, B_t, C_t, D_t)$
   c. $Y_{1t} = f(Y_{2t}, X_{1t}, X_{2t})$
      $Y_{2t} = f(Y_{3t}, X_{5t})$

3. Section 14.1.2 works through Equations 14.2 and 14.3 to show the violation of Classical Assumption III by an unexpected increase in ϵ_1. Show the violation of Classical Assumption III by working through the following examples:
   a. a decrease in ϵ_2 in Equation 14.3
   b. an increase in ϵ_D in Equation 14.4
   c. an increase in ϵ_1 in Equation 14.23

4. As mentioned in Section 14.1.3, Indirect Least Squares (ILS) is a method of calculating the coefficients of the structural equations directly from the coefficients of the reduced-form equations without the necessity of a second-stage regression. This technique, which requires that the equations be exactly identified, can be used because the coefficients of the structural equations can be expressed in terms of the coefficients of the reduced-form equations. These expressions can be derived by solving the structural equations for one of the endogenous variables in terms of the predetermined variables and then substituting.
   a. Return to Equations 14.4 and 14.5 and confirm that the reduced-form equations for that system are Equations 14.8 and 14.9.
   b. By mathematically manipulating the expressions, express the coefficients of Equation 14.4 in terms of the πs of the reduced-form equations.
   c. What disadvantages does the method of ILS have? Why isn't it used frequently?
   d. Think through the application of ILS to an overidentified equation. What problem would that estimation encounter?

5. Section 14.2.1 makes the statement that the correlation between the ϵs and the Ys (where they appear as explanatory variables) is usually positive in economic examples. To see if this is true, investigate the sign of the error term/explanatory variable correlation in the following cases:
   a. the three examples in Exercise 3 above
   b. the more general case of all the equations in a typical supply and demand model (for instance, the model for cola in Section 14.1)
   c. the more general case of all the equations in a simple macroeconomic model (for instance, the small macroeconomic model in Section 14.3.3)

6. Determine the identification properties of the following equations. In particular, be sure to note the number of predetermined variables in the system, the number of slope coefficients in the equation, and whether the equation is underidentified, overidentified, or exactly identified.
   a. Equations 14.2-3
   b. Equations 14.13-14
   c. Part a of Exercise 2 above (assume all equations are stochastic)
   d. Part b of Exercise 2 above (assume all equations are stochastic)

7. Determine the identification properties of the following equations. In particular, be sure to note the number of predetermined variables in the system, the number of slope coefficients in the equation, and whether the equation is underidentified, overidentified or exactly identified. (Assume all equations are stochastic unless specified otherwise.)

a. $A_t = f(B_t, C_t, D_t)$

$B_t = f(A_t, C_t)$

b. $Y_{1t} = f(Y_{2t}, X_{1t}, X_{2t}, X_{3t})$

$Y_{2t} = f(X_{2t})$

$X_{2t} = f(Y_{1t}, X_{4t}, X_{3t})$

c. $C_t = f(Y_t)$

$I_t = f(Y_t, R_t, E_t, D_t)$

$R_t = f(M_t, R_{t-1}, Y_t - Y_{t-1})$

$Y_t = C_t + I_t + G_t$ (nonstochastic)

8. Return to the supply and demand example for cola in Section 14.1 and explain exactly how 2SLS would estimate the αs and βs of Equations 14.4 and 14.5. Write out the equations to be estimated in both stages, and indicate precisely what, if any, substitutions would be made in the second stage.

9. As an exercise to gain familiarity with the 2SLS program on your computer system, take the data provided for the simple Keynesian model in Section 14.3.3, and:

a. Estimate the investment function with OLS.

b. Estimate the reduced-form for Y with OLS.

c. Substitute the $\hat{Y}$ from your reduced-form into the investment function and run the second stage yourself with OLS.

d. Estimate the investment function with your computer system's 2SLS program (if there is one) and compare the results with those obtained in part c above.

10. Suppose that one of your friends recently estimated a simultaneous equation research project and found the OLS results to be virtually identical to the 2SLS results. How would you respond if he or she said, "What a waste of time! I shouldn't have bothered with 2SLS in the first place! Besides, this proves that there wasn't any bias in my model anyway."

a. What is the value of 2SLS in such a case?

b. Does the similarity between the 2SLS and OLS estimates indicate a lack of bias?

11. Think over the problem of building a model for the supply of and demand for labor (measured in hours worked) as a function of the wage and other variables.
    a. Completely specify labor supply and labor demand equations and hypothesize the expected signs of the coefficients of your variables.
    b. Is this system simultaneous or not? That is, is there likely to be feedback between the wage and hours demanded and supplied? Why or why not?
    c. Is your system likely to encounter biased estimates? Why?
    d. What sort of estimation procedure would you use to obtain your coefficient estimates? (Hint: Be sure to determine the identification properties of your equations.)

12. Let's analyze the problem of serial correlation in simultaneous models. For instance, recall that in our small macroeconomic model, the 2SLS version of the consumption function, Equation 14.30, was:

$$\hat{C}_t = -52.2 + 0.61\widehat{YD}_t + 0.37C_{t-1} \qquad (14.30)$$
$$(0.22) \qquad (0.23)$$
$$t = \quad 2.8 \qquad 1.6$$
$$\bar{R}^2 = .996 \quad n = 25 \quad DW = 0.71$$

where C is consumption and YD is disposable income.
    a. Test Equation 14.30 to confirm that we do indeed have a serial correlation problem. (Hint: This should seem familiar.)
    b. Equation 14.30 will encounter both simultaneity bias and bias due to serial correlation with a lagged endogenous variable. If you could only solve one of these two problems, which would you choose? Why? (Hint: Compare Equation 14.30 with the OLS version of the consumption function, Equation 14.33.)
    c. Suppose you wanted to solve both problems? Can you think of a way to adjust for both serial correlation and simultaneity bias at the same time? Would it make more sense to run GLS first and then 2SLS, or would you rather run 2SLS first and then GLS? Could they be run simultaneously?

13. Suppose that the recent fad for oats (resulting from the announcement of the health benefits of oat bran) has made you toy with the idea of becoming a broker in the oat market. Before spending your money, you decide to build a simple model of supply and demand (identical to those in Sections 14.1 and 14.2) of the market for oats:

$$Q_{Dt} = \beta_0 + \beta_1 P_t + \beta_2 YD_t + \epsilon_{Dt}$$
$$Q_{St} = \alpha_0 + \alpha_1 P_t + \alpha_2 W_t + \epsilon_{St}$$
$$Q_{Dt} = Q_{St}$$

where: Q_{Dt} = the quantity of oats demanded in time period t
Q_{St} = the quantity of oats supplied in time period t
P_t = the price of oats in time period t
W_t = average oat-farmer wages in time period t
YD_t = disposable income in time period t

a. You notice that no left-hand side variable appears on the right side of either of your stochastic simultaneous equations. Does this mean that OLS estimation will encounter no simultaneity bias? Why or why not?
b. You expect that when P_t goes up, Q_{Dt} will fall. Does this mean that if you encounter simultaneity bias in the demand equation, it will be negative instead of the positive bias we typically associate with OLS estimation of simultaneous equations? Explain your answer.
c. Carefully outline how you would apply 2SLS to this system. How many equations (including reduced forms) would you have to estimate? Specify precisely which variables would be in each equation.
d. Given the following hypothetical data,[14] estimate OLS and 2SLS versions of your oat supply and demand equations.

year	Q	P	W	YD
1	50	10	100	15
2	54	12	102	12
3	65	9	105	11
4	84	15	107	17
5	75	14	110	19
6	85	15	111	30
7	90	16	111	28
8	60	14	113	25
9	40	17	117	23
10	70	19	120	35

14. These data are from the excellent course materials that Professors Bruce Gensemer and James Keeler prepared to supplement the use of the first edition of this text at Kenyon College.

e. Compare your OLS and 2SLS estimates. How do they compare with your prior expectations? Which equation do you prefer? Why?

14. In a recent study, James Ragan[15] examined the effects of unemployment insurance (hereafter UI) eligibility standards on unemployment rates and the rate at which workers quit their jobs. Ragan used a pooled data set that contained observations from a number of different states from four different years (requirements for UI eligibility differ by state). His results are as follows (t-scores in parentheses):

$$\begin{aligned}\widehat{QU}_i = 7.00 &+ \underset{(0.10)}{0.089UR_i} - \underset{(-0.63)}{0.063UN_i} - \underset{(-1.98)}{2.83RE_i} - \underset{(-0.73)}{0.032MX_i} \\ &+ \underset{(0.01)}{0.003IL_i} - \underset{(-0.52)}{0.25QM_i} + \cdots\end{aligned}$$

$$\begin{aligned}\widehat{UR}_i = -0.54 &+ \underset{(1.01)}{0.44QU_i} + \underset{(3.29)}{0.13UN_i} + \underset{(1.71)}{0.049MX_i} \\ &+ \underset{(2.03)}{0.56IL_i} + \underset{(2.05)}{0.63QM_i} + \cdots\end{aligned}$$

where: QU_i = the quit rate (quits per 100 employees) in the *i*th state

UR_i = the unemployment rate in the *i*th state

UN_i = union membership as a percentage of nonagricultural employment in the *i*th state

RE_i = average hourly earnings in the *i*th state relative to the average hourly earnings for the U.S.

IL_i = dummy variable equal to 1 if workers in the *i*th state are eligible for UI if they are forced to quit a job because of illness, 0 otherwise

QM_i = dummy variable equal to 1 if the *i*th state maintains full UI benefits for the quitter (rather than lowering benefits), 0 otherwise

MX_i = maximum weekly UI benefits relative to average hourly earnings in the *i*th state

15. James F. Ragan, Jr., "The Voluntary Leaver Provisions of Unemployment Insurance and Their Effect on Quit and Unemployment Rates," *Southern Economic Journal,* July 1984, pp. 135–146.

a. Hypothesize the expected signs for the coefficients of each of the explanatory variables in the system. Use economic theory to justify your answers. Which estimated coefficients are different from your expectations?
b. Ragan felt that these two equations would encounter simultaneity bias if they were estimated with OLS. Do you agree? Explain your answer. (Hint: Start by deciding which variables are endogenous and why.)
c. The actual equations included a number of variables not documented above, but the only predetermined variable in the system that was included in the QU equation but not the UR equation was RE. What does this information tell you about the identification properties of the QU equation? The UR equation?
d. What are the implications of the lack of significance of the endogenous variables where they appear on the right-hand side of the equations?
e. What, if any, policy recommendations do these results suggest?

Appendix 14.6: Errors in the Variables

Until now, we have implicitly assumed that our data were measured accurately. That is, while the stochastic error term was defined as including measurement error, we never explicitly discussed what the existence of such measurement error did to the coefficient estimates. Unfortunately, in the real world, errors of measurement are common. Mismeasurement might result from the data being based on a sample, as are almost all national aggregate statistics, or simply because the data were reported incorrectly. Whatever the cause, these **errors in the variables** are mistakes in the measurement of the dependent and/or one or more of the independent variables that are large enough to have potential impacts on the estimation of the coefficients. Such errors in the variables might be better called "measurement errors in the data." We will tackle this subject by first examining errors in the dependent variable and then moving on to look at the more serious problem of errors in an independent variable. We assume a single equation model. The reason we have included this section here is that errors in explanatory variables give rise to biased OLS estimates very similar to simultaneity bias.

14.6.1 Measurement Errors in the Data for the Dependent Variable

Suppose that the true regression model is

$$Y_i = \beta_0 + \beta_1 X_i + \epsilon_i \tag{14.41}$$

and further suppose that the dependent variable, Y_i, is measured incorrectly, so that Y_i^* is observed instead of Y_i, where

$$Y_i^* = Y_i + v_i \tag{14.42}$$

and where v_i is an error of measurement that has all the properties of a classical error term. What does this mismeasurement do to the estimation of Equation 14.41?

To see what happens when $Y_i^* = Y_i + v_i$, let's add v_i to both sides of Equation 14.41, obtaining

$$Y_i + v_i = \beta_0 + \beta_1 X_i + \epsilon_i + v_i \tag{14.43}$$

which is the same as

$$Y_i^* = \beta_0 + \beta_1 X_i + \epsilon_i^* \tag{14.44}$$

where $\epsilon_i^* = (\epsilon_i + v_i)$. That is, we estimate equation 14.44 when in reality we want to estimate Equation 14.41. Take another look at Equation 14.44. When v_i changes, both the dependent variable and the error term ϵ_i^* move together. This is no cause for alarm, however, since the dependent variable is always correlated with the error term. While the extra movement will increase the variability of Y and therefore be likely to decrease the overall statistical fit of the equation, an error of measurement in the dependent variable does not cause any bias in the estimates of the βs.

14.6.2 Measurement Errors in the Data for an Independent Variable

This is not the case when the mismeasurement is in the data for one or more of the independent variables. Unfortunately, such errors in the independent variables cause bias that is quite similar in nature (and in remedy) to simultaneity bias. To see this, once again suppose that the true regression model is Equation 14.41:

$$Y_i = \beta_0 + \beta_1 X_i + \epsilon_i \tag{14.41}$$

and now suppose that the independent variable, X_i, is measured incorrectly, so that X_i^* is observed instead of X_i, where

$$X_i^* = X_i + u_i \tag{14.45}$$

and where u_i is an error of measurement just like v_i above. To see what this mismeasurement does to the estimation of Equation 14.41, let's add the term $0 = (\beta_1 u_i - \beta_1 u_i)$ to Equation 14.41, obtaining

$$Y_i = \beta_0 + \beta_1 X_i + \epsilon_i + (\beta_1 u_i - \beta_1 u_i) \tag{14.46}$$

which can be rewritten as

$$Y_i = \beta_0 + \beta_1(X_i + u_i) + (\epsilon_i - \beta_1 u_i) \tag{14.47}$$

or

$$Y_i = \beta_0 + \beta_1 X_i^* + \epsilon_i^{**} \tag{14.48}$$

where $\epsilon_i^{**} = (\epsilon_i - \beta_1 u_i)$. In this case, we estimate Equation 14.48 when we should be trying to estimate Equation 14.41. Notice what happens to Equation 14.48 when u_i changes, however. When u_i changes the stochastic error term ϵ_i^{**} and the independent variable X_i^* move in opposite directions; they are correlated. Such a correlation is a direct violation of Classical Assumption III in a way that is remarkably similar to the violation (described in Section 14.1) of the same assumption in simultaneous equations. Not surprisingly, this violation causes the same problem, bias, for errors-in-the-variables models that it causes for simultaneous equations. That is, because of the measurement error in the independent variable, the OLS estimates of the coefficients of Equation 14.48 are *biased*.

A frequently used technique to rid an equation of the bias caused by measurement errors in the data for one or more of the independent variables is *instrumental variables,* the same technique used to alleviate simultaneity bias. A proxy for X is chosen that is highly correlated with X but is uncorrelated with ϵ. Recall that 2SLS is an instrumental variables technique. Such techniques are applied only rarely to errors in the variables problems, however, because while we may suspect that there are errors in the variables, it's unusual to know positively that they exist, and it's difficult to find an instrumental variable that satisfies both conditions. As a result, X^* is about as good a proxy for X as we usually can find, and no action is taken. If the mismeasurement in X were known to be large, however, some remedy would be required.

To sum, an error of measurement in one or more of the independent variables will cause the error term of Equation 14.48 to be correlated with the independent variable, causing bias analogous to simultaneity bias.[16] While instrumental variables is a possible remedy for this problem, more often than not, corrective steps are not taken.

16. If errors exist in the data for the dependent variable and one or more of the independent variables, then both decreased overall statistical fit and bias in the estimated coefficients will result.

15

Forecasting

Of the uses of econometrics outlined in Chapter 1, we have discussed forecasting the least. Accurate forecasting is vital to successful planning, so it's the primary goal of many business and governmental uses of econometrics. For example, manufacturing firms need sales forecasts, banks need interest rate forecasts, and governments need unemployment and inflation rate forecasts.

To many business and government leaders, the words "econometrics" and "forecasting" mean the same thing. Such a simplification gives econometrics a bad name because some consulting econometricians overestimate their ability to produce accurate forecasts, resulting in unrealistic claims and unhappy clients. Some of these clients would probably applaud the 19th century New York law (luckily unenforced but apparently also unrepealed) that provides that persons "pretending to forecast the future" shall be liable to a $250 fine and/or six months in prison.[1] While many econometricians might wish that such consultants would call themselves "futurists" or "soothsayers," it's

1. Section 899 of the N.Y. State Criminal Code: the law does not apply to "ecclesiastical bodies acting in good faith and without personal fees."

impossible to ignore the importance of econometrics in forecasting in today's world.

The ways in which the prediction of future events is accomplished are quite varied. At one extreme, some forecasters use models with hundreds of equations.[2] At the other extreme, quite accurate forecasts can be created with nothing more than a good imagination and a healthy dose of self-confidence.

Unfortunately, it's unrealistic to think we can cover even a small portion of the topic of forecasting in one short chapter. Indeed, there are a number of excellent books on this subject alone.[3] Instead, this chapter is meant to be a brief introduction to the use of econometrics in forecasting. We will begin by using simple linear equations and then move on to investigate a few more complex forecasting situations. The chapter concludes with an introduction to a technique, called ARIMA or time-series analysis, that calculates forecasts entirely from past movements of the dependent variable without the use of any independent variables at all.

15.1 What Is Forecasting?

In general, forecasting is the act of predicting the future; in econometrics, **forecasting** is the estimation of the expected value of a dependent variable for observations that are not part of the sample data-set. In most forecasts, the values being predicted are for time periods in the future, but cross-sectional predictions of values for countries or people not in the sample are also common. To simplify terminology, the words prediction and forecast will be used interchangeably in this chapter. (Some authors limit the use of the word forecast to out-of-sample prediction for a time-series.)

We've already encountered an example of a forecasting equation. Think back to the weight/height example of Section 1.3 and recall that the purpose of that model was to guess the weight of a male customer

2. For example, Data Resources Incorporated, one of the leading econometric forecasting firms in the U.S., has an 800-equation model. A number of other macroeconomic models are of similar size. For more on this, see Otto Eckstein, *The DRI Model of the U.S. Economy* (New York: McGraw-Hill, 1983). For an interesting approach to the comparison of such models, see Ray C. Fair and Robert J. Shiller, "Comparing Information in Forecasts from Econometric Models," *American Economic Review,* June 1990, pp. 375–389.

3. See, for example, C. W. J. Granger, *Forecasting in Business and Economics* (New York: Academic Press, 1980), or Hans Levenbach and James P. Cleary, *The Modern Forecaster* (Belmont, California: Wadsworth, Inc., 1984).

based on his height. In that example, the first step in building a forecast was to estimate Equation 1.21:

$$\text{Estimated Weight}_i = 103.4 + 6.38 \cdot \text{Height}_i \text{ (inches over five feet)} \quad (1.21)$$

That is, we estimated that a customer's weight on average equaled a base of 103.4 pounds plus 6.38 pounds for each inch over 5 feet. To actually make the prediction, all we had to do was to substitute the height of the individual whose weight we were trying to predict into the estimated equation. For a male who is 6′1″ tall, for example, we'd calculate:

$$\text{Predicted Weight} = 103.4 + 6.38 \cdot \text{(13 inches over five feet)} \quad (15.1)$$

or 103.4 + 82.9 = 186.3 pounds.

The weight-guessing equation is a specific example of using a single linear equation to predict or forecast. Our use of such an equation to make a forecast can be summarized in two steps:

1. *Specify and estimate an equation that has as its dependent variable the item that we wish to forecast.* We obtain a forecasting equation by specifying and estimating an equation for the variable we want to predict:

$$\hat{Y}_t = \hat{\beta}_0 + \hat{\beta}_1 X_{1t} + \hat{\beta}_2 X_{2t} \qquad (t = 1,2,\ldots,T) \qquad (15.2)$$

Such specification and estimation has been the topic of the first 14 chapters of this book. The use of t = (1,2,...,T) to denote the sample size is fairly standard for time-series forecasts (t stands for "time").

2. *Obtain values for each of the independent variables for the observations for which we want a forecast and substitute them into our forecasting equation.* To calculate a forecast for Equation 15.2, this would mean finding values for period T + 1 (for a sample of size T) for X_1 and X_2 and substituting them into the equation:

$$\hat{Y}_{T+1} = \hat{\beta}_0 + \hat{\beta}_1 X_{1T+1} + \hat{\beta}_2 X_{2T+1} \qquad (15.3)$$

What is the meaning of this $\hat{Y}_{T+1}$? It is a prediction of the value that Y will take in observation T + 1 (outside the sample) based upon our values of X_{1T+1} and X_{2T+1} and based upon the particular specification and estimation that produced Equation 15.2.

To understand these steps more clearly, look at two applications of this forecasting approach.

Forecasting Chicken Consumption: Let's return to the chicken demand model, Equation 6.5 of Section 6.1, to see how well that equation forecasts aggregate per capita chicken consumption:

$$\hat{Y}_t = -60.5 - \underset{(0.07)}{0.45PC_t} + \underset{(0.05)}{0.12PB_t} + \underset{(1.2)}{12.2LYD_t} \quad (6.5)$$
$$t = \quad -6.4 \qquad 2.5 \qquad 10.6$$
$$\bar{R}^2 = .984 \quad n = 35 \quad DW = 0.94$$

where Y is pounds of chicken consumption per capita, PC and PB are the prices of chicken and beef, respectively, in cents per pound, and LYD is the log of per capita U.S. disposable income (in dollars).

To forecast with this model, we would obtain values for the three independent variables and substitute them into Equation 6.5. For example, in 1985,

$$\hat{Y}_{85} = -60.5 - 0.45(14.8) + 0.12(53.7) + 12.2(\ln 11861) = 53.7 \quad (15.4)$$

Continuing on through 1988, we'd end up with:

Year	Forecast	Actual	Percent Error
1985	53.7	57.6	6.8
1986	55.3	58.7	5.8
1987	57.6	62.7	8.1
1988	59.9	64.6	7.3

How does the model do? Well, forecasting accuracy, like beauty, is in the eye of the beholder. While the equation consistently underforecasts[4] (see Figure 15.1), it predicts the consumption of chicken to within eight percent four years beyond the sample period.

4. The data for this example are in Table 6.3; the actual values of the independent variables for the three other forecast years are PC(12.5, 11.0, 9.2); PB(52.6, 61.1, 66.6); and YD(12496, 13157, 14123). A simplistic way to avoid such consistent over (or under) forecasting in some cases is to add a dummy variable to the equation (before it is estimated) which equals one in the last time period in the sample and throughout the forecast period but zero otherwise. Such a forecasting dummy constrains the forecasts to link up to the actual data at the beginning of the forecast period, but it is a "one-time dummy" that uses up a degree of freedom, so it should be used with caution.

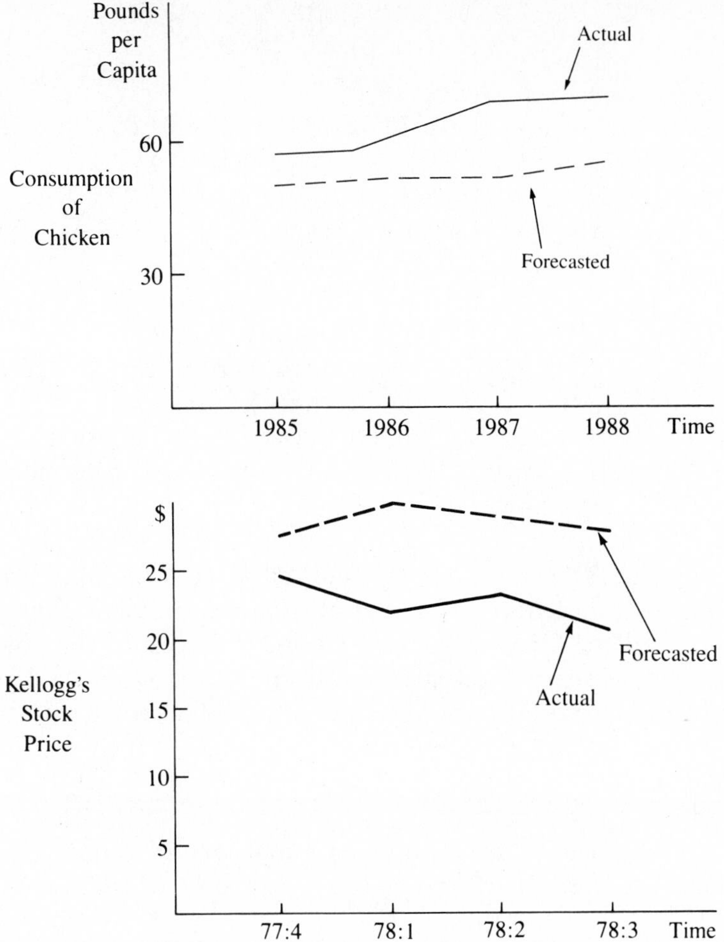

Figure 15.1 FORECASTING EXAMPLES
In the chicken consumption example, the equation underforecasted slightly. For the stock price model, even actual values for the independent variables and an excellent fit within the sample could not produce an accurate forecast.

Forecasting Stock Prices: Some students react to the previous example by wanting to build a model to forecast stock prices and make a killing on the stock market. "If we could predict the price of a stock four years from now to within eight percent," they reason, "we'd know which stocks to buy." To see how such a forecast might work, let's look at a simplified model of the quarterly price of a particular individual stock, that of the J. L. Kellogg Company (maker of breakfast cereals and other products):

$$\widehat{PK_t} = -7.80 + \underset{\substack{(0.0024)\\ t = 3.91}}{0.0096 DJIA_t} + \underset{\substack{(2.83)\\ 0.95}}{2.68 KEG_t}$$

$$+ \underset{\substack{(22.70)\\ t = 0.71}}{16.18 DIV_t} + \underset{\substack{(1.47)\\ 3.29}}{4.84 BVPS_t} \quad (15.5)$$

$$\bar{R}^2 = .95 \quad n = 35 \text{ (Quarterly 69:1 to 77:3)} \quad DW = 1.88$$

where: PK_t = the dollar price of Kellogg's stock in quarter t
$DJIA_t$ = the Dow-Jones Industrial Average in quarter t
KEG_t = Kellogg's earnings growth (percent change in annual earnings over the previous five years)
DIV_t = Kellogg's declared dividends (in dollars) that quarter
$BVPS_t$ = per-share book value of the Kellogg corporation that quarter

The signs of the estimated coefficients all agree with those hypothesized before the regression was run, $\bar{R}^2$ indicates a reasonably good overall fit, and the Durbin-Watson d statistic indicates that the hypothesis of no positive serial correlation cannot be rejected. The low t-scores for KEG and DIV are caused by multicollinearity ($r = .985$), but both variables are left in the equation because of their theoretical importance.

In order to forecast with Equation 15.5, we obtained actual values for all of the independent variables for the next four quarters and substituted them into the right side of the equation, obtaining:

Quarter	Forecast	Actual	Percent Error
77:4	$26.32	$24.38	8.0
78:1	27.37	22.38	22.3
78:2	27.19	23.00	18.2
78:3	27.13	21.88	24.0

How did our forecasting model do? Even though the $\bar{R}^2$ within the sample was .95, even though we used actual values for the independent variables, and even though we forecasted only four quarters beyond our sample, the model was something like 20 percent off. If we had decided to buy Kellogg's stock based on our forecast, we'd have *lost* money! Since other attempts to forecast stock prices have also encountered difficulties, this doesn't seem a reasonable use for econometric forecasting. Individual stock prices (and many other items) are simply too variable and depend on too many nonquantifiable items to consistently forecast accurately, even if the forecasting equation has an excellent fit! The reason for this apparent contradiction is that equations that worked in the past may or may not work well in the future.

15.2 More Complex Forecasting Problems

The forecasts generated above are unrealistically simple, however, and most actual forecasting involves one or more additional questions. For example:

1. *Unknown Xs:* It's unrealistic to expect always to know the values for the independent variables outside the sample. For instance, we'll almost never know what the Dow-Jones Industrial Average will be in the future when we are making forecasts of the price of a given stock, and yet we assumed that knowledge when making our Kellogg price forecasts. What happens when we don't know the values of the independent variables for the forecast period?
2. *Serial Correlation:* If there is serial correlation involved, the forecasting equation may be estimated with GLS. How should predictions be adjusted when forecasting equations are estimated with GLS?
3. *Confidence Intervals:* All the forecasts above were single values, but such single values are almost never exactly right. Wouldn't it be more helpful if we forecasted an interval within which we were confident that the actual value would fall a certain percentage of the time? How can we develop these confidence intervals?
4. *Simultaneous Equations Models:* As we saw in Chapter 14, many economic and business equations are part of simultaneous models. How can we use an independent variable to forecast a dependent variable when we know that a change in value of the dependent variable will change, in turn, the value of the independent variable that we used to make the forecast?

Even a few questions like these should be enough to convince you that forecasting involves issues that are more complex than implied by Section 15.1.

15.2.1 Conditional Forecasting (Unknown X Values for the Forecast Period)

A forecast in which all values of the independent variables are known with certainty can be called an **unconditional forecast,** but, as mentioned above, the situations in which one can make such unconditional forecasts are rare. More likely, we will have to make a **conditional forecast,** for which actual values of one or more of the independent variables are *not* known. We are forced to obtain forecasts for the independent variables before we can use our equation to forecast the dependent variable, making our forecast of Y conditional on our forecast of the Xs.

One key to an accurate conditional forecast is accurate forecasting of the independent variables. If the forecasts of the independent variables are unbiased, using a conditional forecast will not introduce bias into the forecast of the dependent variable. Anything but a perfect forecast of the independent variables will contain some amount of forecast error, however, and so the expected error variance associated with conditional forecasting typically will be larger than that associated with unconditional forecasting. Thus, one should try to find unbiased, minimum variance forecasts of the independent variables when using conditional forecasting.

To get good forecasts of the independent variables, take the forecastability of potential independent variables into consideration when making specification choices. For instance, when you choose which of two redundant variables to include in an equation to be used for forecasting, you should choose the one that is easier to forecast accurately. When you can, you should choose an independent variable that is regularly forecasted by someone else (an econometric forecasting firm, for example) so that you don't have to forecast X yourself.

The careful selection of independent variables can sometimes help you avoid the need for conditional forecasting in the first place. This opportunity can arise when the dependent variable can be expressed as a function of leading indicators. A **leading indicator** is an independent variable whose movements anticipate movements in the dependent variable. For instance, the impact of interest rates on investment typically is not felt until two or three quarters after interest rates have changed. To see this, let's look at a quarterly version of the investment function of the small macroeconomic model of Section 14.3.3:

$$I_t = \beta_0 + \beta_1 Y_t + \beta_2 r_{t-2} + \epsilon_t \qquad (15.6)$$

where I equals gross investment, Y equals GNP, and r equals the interest rate. In this equation, actual values of r can be used to help forecast I_{T+1} and I_{T+2}. Note, however, that to predict I_{T+3}, we need to forecast r. Thus leading indicators like r help avoid conditional forecasting for only a time period or two. For long-range predictions, a conditional forecast is usually necessary.

15.2.2 Forecasting with Serially Correlated Error Terms

Recall from Chapter 9 that pure first-order serial correlation implies that the current observation of the error term ϵ_t is affected by the previous error term and an autocorrelation coefficient, ρ:

$$\epsilon_t = \rho\epsilon_{t-1} + u_t$$

where u_t is a nonserially-correlated error term. Also recall that when serial correlation is severe, one remedy is to run Generalized Least Squares (GLS) as noted in Equation 9.18:

$$Y_t - \rho Y_{t-1} = \beta_0(1 - \rho) + \beta_1(X_t - \rho X_{t-1}) + u_t \qquad (9.18)$$

Unfortunately, whenever the use of GLS is required to rid an equation of pure first-order serial correlation, the procedures used to forecast with that equation become a bit more complex. To see why this is necessary, note that if Equation 9.18 is estimated, the dependent variable will be:

$$Y_t^* = Y_t - \hat{\rho} Y_{t-1} \qquad (15.7)$$

Thus, if a GLS equation is used for forecasting, it will produce predictions of Y_{T+1}^* rather than of Y_{T+1}. Such predictions thus will be of the wrong variable.

If forecasts are to be made with a GLS equation, Equation 9.18 should first be rewritten to be solved for Y_t, before forecasting is attempted:

$$Y_t = \rho Y_{t-1} + \beta_0(1 - \rho) + \beta_1(X_t - \rho X_{t-1}) + u_t \qquad (15.8)$$

We now can forecast with Equation 15.8 as we would with any other. If we substitute subscript T + 1 for t (to forecast time period T + 1) and insert estimates for the coefficients, ρs and Xs into the right side of the equation, we obtain:

$$\hat{Y}_{T+1} = \hat{\rho} Y_T + \hat{\beta}_0(1 - \hat{\rho}) + \hat{\beta}_1(\hat{X}_{T+1} - \hat{\rho} X_T) \qquad (15.9)$$

Equation 15.9 thus should be used for forecasting when an equation has been estimated with GLS to correct for serial correlation.

We now turn to an example of such forecasting with serially correlated error terms. In particular, recall from Chapter 9 that the Durbin-Watson statistic of the chicken demand equation used as an example in Section 15.1 was 0.94, indicating significant positive first-order serial correlation. As a result, we estimated the chicken demand equation with GLS, obtaining Equation 9.23:

$$\hat{Y}_t = -67.2 - \underset{(0.09)}{0.29}PC_t + \underset{(0.05)}{0.13}PB_t + \underset{(1.3)}{12.7}LYD_t \quad (9.23)$$
$$t = \quad -3.3 \qquad 2.6 \qquad 9.6$$
$$\bar{R}^2 = .989 \quad n = 35 \quad \hat{\rho} = 0.69$$

Since Equation 9.23 was estimated with GLS, Y is actually Y_t^*, which equals $(Y_t - \hat{\rho}Y_{t-1})$, PC_t is actually PC_t^*, which equals $PC_t - \hat{\rho}PC_{t-1}$, and so on. Thus, to forecast with Equation 9.23, we have to convert it to the form of Equation 15.9, or:

$$\begin{aligned}\hat{Y}_{T+1} = 0.69Y_T &- 67.2(1 - 0.69) - 0.29(PC_{T+1} - 0.69PC_T)\\ &+ 0.13(PB_{T+1} - 0.69PB_T) \qquad (15.10)\\ &+ 12.7(LYD_{T+1} - 0.69LYD_T)\end{aligned}$$

Substituting the actual values for the independent variables into Equation 15.10, we obtain:

Year	Forecast	Actual	Percent Error
1985	55.7	57.6	3.3
1986	57.9	58.7	1.4
1987	60.0	62.7	4.3
1988	63.4	64.6	1.9

Note that these forecasts compare favorably to those that were calculated without taking serial correlation into consideration in Section 15.1. Indeed, GLS often will provide superior forecasting performances to OLS in the presence of serial correlation.

Whether to use GLS is not the topic of this section, however. Instead the point is that if GLS is used to estimate the coefficients of an equation, then Equation 15.9 must be used to forecast with the GLS estimates.

15.2.3 Forecasting Confidence Intervals

Until now, the emphasis in this text has been on obtaining point (or single value) estimates. This has been true whether we have been estimating coefficient values or estimating forecasts. Recall, though, that a point estimate is only one of a whole range of such estimates that could have been obtained from different samples (for coefficient estimates) or different independent variable values or coefficients (for forecasts). The usefulness of such point estimates is improved if we can also generate some idea of the variability of our forecasts. The measure of variability typically used is the **confidence interval,** which is defined as the range of values within which the actual value of the item being estimated is likely to fall some percentage of the time (called the level of confidence). This is the easiest way to warn forecast users that a sampling distribution exists.

Suppose you are trying to decide how many hot dogs to order for your city's July 4th fireworks show and that the best point forecast is that you'll sell 24,000 hot dogs. How many hot dogs should you order? If you order 24,000, you're likely to run out about half the time! This is because a point forecast is usually the mean of the distribution of possible sales figures; you will sell more than 24,000 about as frequently as less than 24,000. It would be easier to decide how many dogs to order if you also had a confidence interval that told you the range within which hot dog sales would fall 95 percent of the time. This is because the usefulness of the 24,000 hot dog forecast changes dramatically depending on the confidence interval; an interval of 22,000 to 26,000 would pin down the likely sales, while an interval of 4,000 to 44,000 would leave you virtually in the dark about what to do.[5]

The same techniques we use to test hypotheses can also be adapted to create confidence intervals. Given a point forecast, $\hat{Y}_{T+1}$, all we need to generate a confidence interval around that forecast are t_c, the critical t value (for the desired level of confidence) and S_F, the estimated standard error of the forecast:

$$\text{Confidence Interval} = \hat{Y}_{T+1} \pm S_F t_c \tag{15.11}$$

or, equivalently,

$$\hat{Y}_{T+1} - S_F t_c \leq Y_{T+1} \leq \hat{Y}_{T+1} + S_F t_c \tag{15.12}$$

5. The decision as to how many hot dogs to order would also depend on the costs of having the wrong number. These may not be the same per hot dog for overestimates as they are for underestimates. For example, if you don't order enough, then you lose the entire retail price of the hot dog minus the wholesale price of the dog (and bun) because your other costs, like hiring employees and building hot dog stands, are essentially fixed. On the other hand, if you order too many, you lose the wholesale cost of the dog and bun minus whatever salvage price you might be able to get for day-old buns, etc. As a result, the right number to order would depend on your profit margin and the importance of nonreturnable inputs in your total cost picture.

The critical t-value, t_c, can be found in Statistical Table B-1 (for a two-tailed test with T − K − 1 degrees of freedom), while the standard error of the forecast, S_F, for an equation with just one independent variable, equals the square root of the forecast error variance:

$$S_F = \sqrt{s^2[1 + 1/T + (\hat{X}_{T+1} - \overline{X})^2 / \sum_{t=1}^{T} (X_t - \overline{X})^2]} \quad (15.13)$$

where s^2 is the estimated variance of the error term, T is the number of observations in the sample, $\hat{X}_{T+1}$ is the forecasted value of the single independent variable, and $\overline{X}$ is the arithmetic mean of the observed X's in the sample.[6]

Note that Equation 15.13 implies that the forecast error variance decreases the larger the sample, the more X varies within the sample, and the closer $\hat{X}$ is to its within-sample mean. An important implication is that the farther the X used to forecast Y is from the within-sample mean of the Xs, the wider the confidence interval around the $\hat{Y}$ is going to be. This can be seen in Figure 15.2, in which the confidence interval actually gets wider as $\hat{X}_{T+1}$ is father from $\overline{X}$. Since forecasting outside the sample range is common, researchers should be aware of this phenomenon. Also note that Equation 15.13 is for unconditional forecasting. If there is any forecast error in $\hat{X}_{T+1}$, then the confidence interval is larger and more complicated to calculate.

As mentioned above, Equation 15.13 assumes that there is only one independent variable; the equation to be used with more than one variable is similar but more complicated. Some researchers who must deal with more than one independent variable avoid this complexity by ignoring Equation 15.13 and estimating a confidence interval equal to $\hat{Y}_{T+1} \pm st_c$, where s is the standard error of the equation. Compare this shortcut with Equation 15.13; note that the shortcut confidence interval will work well for large samples and $\hat{X}$s near the within-sample means but will provide only a rough estimate of the confidence interval in other cases. At a minimum in such cases, Equation 15.13 should be modified with a $(\hat{X}_{T+1} - \overline{X})^2/\Sigma(X_t - \overline{X})^2$ term for each X.

Let's look at an example of building a forecast confidence interval by returning to the weight/height example. In particular, let's create a

6. Equation 15.13 is valid whether Y_t is in the sample period or outside the sample period, but it applies only to point forecasts of individual Y_ts. If a confidence interval for the expected value of Y, $E(Y_t)$, is desired, then the correct equation to use is

$$S_F = \sqrt{s^2[1/T + (\hat{X}_{T+1} - \overline{X})^2 / \Sigma(X_t - \overline{X})^2]}.$$

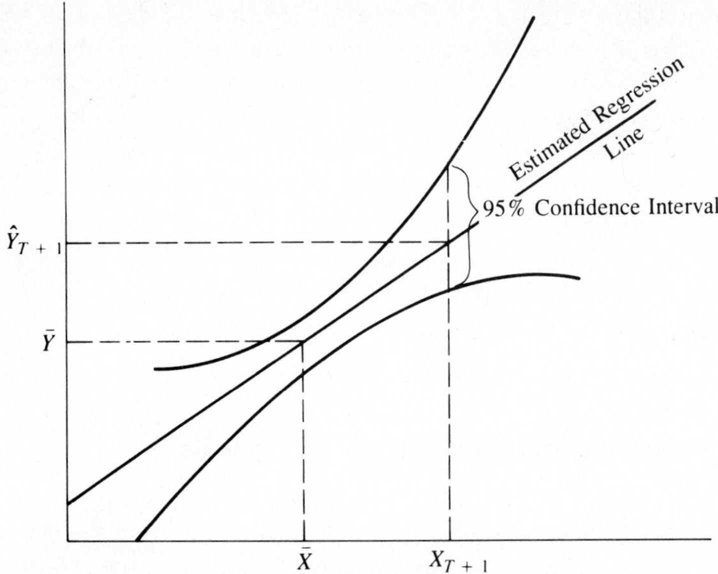

Figure 15.2 A CONFIDENCE INTERVAL FOR $\hat{Y}_{T+1}$
A 95 percent confidence interval for $\hat{Y}_{T+1}$ includes the range of values within which the actual Y_{T+1} will fall 95 percent of the time. Note that the confidence interval widens as X_{T+1} differs more from its within-sample mean, $\overline{X}$.

95 percent confidence interval around the forecast for a 6′1″ male calculated in Equation 15.1 (repeated for convenience):

$$\text{Predicted Weight}_i = 103.4 + 6.38 \cdot (\text{13 inches over five feet}) \quad (15.1)$$

for a predicted weight of 103.4 + 82.9 or 186.3 pounds. In order to calculate a 95 percent confidence interval around this prediction, we substitute Equation 15.13 into Equation 15.11, obtaining a confidence interval of:

$$186.3 \pm \left(\sqrt{s^2[1 + 1/T + (\hat{X}_{T+1} - \overline{X})^2 / \sum_{t=1}^{T} (X_t - \overline{X})^2]}\right) t_c \quad (15.14)$$

We then substitute the actual figures into Equation 15.14. From the data-set for the example, we find that T = 20, the mean X = 10.35, the summed square deviations of X around its mean is 92.50, and s^2 = 65.05. From Statistical Table B-1, we obtain the 95 percent, two-tailed critical t-value for 18 degrees of freedom of 2.101. If we now combine this with the information that our $\hat{X}$ is 13, we obtain:

$$186.3 \pm \left(\sqrt{65.05[1 + 1/20 + (13.0 - 10.35)^2/92.50]}\right)t_c \quad (15.15)$$

$$186.3 \pm 8.558(2.101) = 186.3 \pm 18.0 \quad (15.16)$$

In other words, our 95 percent confidence interval for a 6′1″ college-age male is from 168.3 to 204.3 pounds. Ask around; are 19 out of 20 of your male friends that tall within that range?

15.2.4 Forecasting with Simultaneous Equations Systems

As we learned in Chapter 14, most economic and business models are actually simultaneous in nature; for example, the investment equation used in Section 15.2.1 was estimated with 2SLS as a part of our small simultaneous macromodel in Chapter 14. Since GNP is one of the independent variables in the investment equation, when investment rises, so will GNP, causing a feedback effect that is not captured if we just forecast with a single equation. How should forecasting be done in the context of a simultaneous model? There are two approaches to answering this question, depending on whether there are lagged endogenous variables on the right side of any of the equations in the system.

If there are no lagged endogenous variables in the system, then the reduced-form equation for the particular endogenous variable can be used for forecasting because it represents the simultaneous solution of the system for the endogenous variable being forecasted. Since the reduced-form equation is the endogenous variable expressed entirely in terms of the predetermined variables in the system, it allows the forecasting of the endogenous variable without any feedback or simultaneity impacts. This result explains why some researchers forecast potentially simultaneous dependent variables with single equations that appear to combine supply-side and demand-side predetermined variables; they are actually using modified reduced-form equations to make their forecasts.

If there are lagged endogenous variables in the system, then the approach must be altered to take into account the dynamic interaction caused by the lagged endogenous variables. For simple models, this sometimes can be done by substituting for the lagged endogenous variables where they appear in the reduced-form equations. If such a manipulation is difficult, however, then a technique called simulation analysis can be used. *Simulation* involves forecasting for the first postsample period by using the reduced-form equations to forecast all endogenous variables and by using sample values for the lagged endogenous variables where they appear in the reduced-form equations. The forecast for the second postsample period, however, uses the endogenous variable *forecasts* from the last period as lagged values for any endogenous variables that have one-period lags while continuing to use sample values for endogenous variables that have lags of two or more periods. This process continues until all forecasting is done with reduced-form equations that use as data for lagged endogenous variables the forecasts from previous time periods. While such dynamic analyses are beyond the scope of this chapter, they're important to remember when considering forecasting with a simultaneous system.[7]

15.3 Forecasting with ARIMA, or Time-Series Analysis

The forecasting techniques of the previous two sections are applications of familiar regression models. We use linear regression equations to forecast the dependent variable by plugging likely values of the independent variables into the estimated equations and calculating a predicted value for Y; this bases the prediction of the dependent variable on the independent variables (and on their estimated coefficients).

ARIMA, or time-series analysis, is an increasingly popular forecasting technique that completely ignores independent variables in making forecasts. **ARIMA** is a highly refined curve-fitting device that uses current and past values of the dependent variable to produce often accurate short-term forecasts of that variable. Examples of such forecasts are stock market price predictions created by brokerage analysts (called "chartists" or technicians") based entirely on past patterns of movement of the stock prices.

Any forecasting technique that ignores independent variables also essentially ignores all potential underlying theories except those that hypothesize repeating patterns in the variable under study. Since we have emphasized the advantages of developing the theoretical under-

7. For more on this topic, see pp. 723–731 in Jan Kmenta, *Elements of Econometrics* (New York: Macmillan, 1985); chapters 11–13 in Robert S. Pindyck and Daniel L. Rubinfeld, *Econometric Models and Economic Forecasts* (New York: McGraw-Hill, 1991); or W. J. Baumol, *Economic Dynamics* (New York: Macmillan, 1970).

pinnings of particular equations before estimating them, why would we advocate using ARIMA? The answer is that the use of ARIMA is appropriate when little or nothing is known about the dependent variable being forecasted, when the independent variables known to be important really cannot be forecasted effectively, or when all that is needed is a one- or two-period forecast. In these cases, ARIMA has the potential to provide short-term forecasts that are superior to more theoretically satisfying regression models. In addition, ARIMA can sometimes produce better explanations of the residuals from an existing regression equation (in particular, one with known omitted variables or other problems). In other circumstances, the use of ARIMA is not recommended. This introduction to ARIMA is intentionally brief; a more complete coverage of the topic can be obtained from a number of other sources.[8]

15.3.1 The ARIMA Approach

The ARIMA approach combines two different specifications (called *processes*) into one equation. The first specification is an *autoregressive* process (hence the AR in ARIMA), and the second specification is a *moving average* process (hence the MA).

An **autoregressive process** expresses a dependent variable Y_t as a function of past values of the dependent variable, as in:

$$Y_t = f(Y_{t-1}, Y_{t-2}, \ldots, Y_{t-p}) \tag{15.17}$$

where Y_t is the variable being forecasted and p is the number of past values used. This equation is similar to the serial correlation error term function of Chapter 9 and to the distributed lag equation of Chapter 12. Since there are p different lagged values of Y in this equation, it is often referred to as a "*p*th-order" autoregressive process.

A **moving-average process** expresses a dependent variable Y_t as a function of past values of the error term, as in:

$$Y_t = f(\epsilon_{t-1}, \epsilon_{t-2}, \ldots, \epsilon_{t-q}) \tag{15.18}$$

8. See, for example, T. M. O'Donovan, *Short-Term Forecasting* (New York: Wiley, 1983); C. W. J. Granger and Paul Newbold, *Forecasting Economic Time Series* (New York: Academic Press, 1977); Walter Vandaele, *Applied Time Series and Box-Jenkins Models* (New York: Academic Press, 1983); and Chapters 14–19 in Robert S. Pindyck and Daniel L. Rubinfeld, *Econometric Models and Economic Forecasts* (New York: McGraw-Hill, 1991).

where ϵ_t is the error term associated with Y_t and q is the number of past values of the error term used. Such a function is a moving average of past error terms that can be added to the mean of Y to obtain a moving average of past values of Y. Such an equation would be a *q*th-order moving-average process.

To create an ARIMA model, we begin with an econometric equation with no independent variables ($Y_t = \beta_0 + \epsilon_t$) and add to it both the autoregressive and moving average processes:

$$Y_t = \beta_0 + \overbrace{\theta_1 Y_{t-1} + \theta_2 Y_{t-2} + \cdots + \theta_p Y_{t-p}}^{\text{autoregressive process}} + \epsilon_t \\ + \underbrace{\phi_1 \epsilon_{t-1} + \phi_2 \epsilon_{t-2} + \cdots + \phi_q \epsilon_{t-q}}_{\text{moving-average process}} \tag{15.19}$$

where the θs and the ϕs are the coefficients of the autoregressive and moving-average processes, respectively.

Before this equation can be applied to a time-series, however, it must be assured that the time-series is *stationary.* A **stationary series** is a time-series in which the dependent variable has a constant mean and variance over time. A **nonstationary series** is a time-series that exhibits some sort of upward or downward trend (or a nonconstant variance) over time. Figure 15.3 graphically depicts a stationary and a nonstationary series. If a series is plotted with respect to time and determined to be nonstationary, then steps must be taken to convert the series into a stationary one before the ARIMA technique can be applied. For example, a nonstationary series can often be converted into a stationary one by taking the first difference of the variable in question:

$$Y_t^* = \Delta Y_t = Y_t - Y_{t-1} \tag{15.20}$$

If the first differences do not produce a stationary series, then first differences of this first-differenced series can be taken. The resulting series is a second-difference transformation:

$$Y_t^{**} = (\Delta Y_t^*) = Y_t^* - Y_{t-1}^* = \Delta Y_t - \Delta Y_{t-1} \tag{15.21}$$

In general, successive differences are taken until the series is stationary. The number of differences required to be taken before a series becomes stationary is denoted with the letter d. For example, suppose that GNP is increasing by a fairly consistent amount each year. A plot of GNP with respect to time would depict a nonstationary series, but a plot of

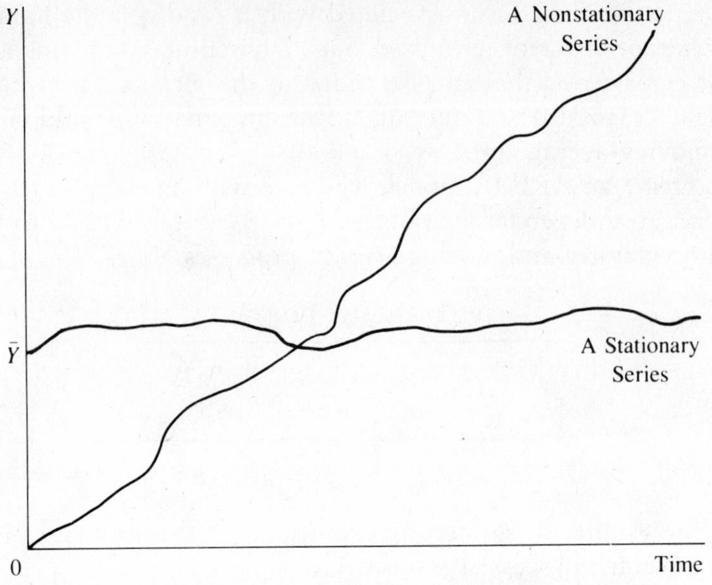

Figure 15.3 STATIONARY VS. NONSTATIONARY SERIES
Whether a series is stationary or nonstationary usually can be determined by plotting Y with respect to time. If the series appears to have a fairly constant mean and variance, then it is stationary. If it exhibits an upward or downward trend over time, it is nonstationary.

the first differences of GNP might depict a fairly stationary series. In such a case, d would be equal to one because one first difference was necessary to convert the nonstationary series into a stationary one.

The dependent variable in Equation 15.19 must be stationary, and so the Y in that equation may be Y, Y* or even Y** depending on the variable in question.[9] If a forecast of Y* or Y** is made, then it must be converted back into Y terms before its use; for example, if d = 1, then

$$\hat{Y}_{T+1} = Y_T + \hat{Y}^*_{T+1} \qquad (15.22)$$

This conversion process is similar to integration in mathematics, so the "I" in ARIMA stands for "integrated." ARIMA thus stands for

9. If Y in Equation 15.19 is Y*, then β_0 represents the coefficient of the linear trend in the original series, and if Y is Y**, the β_0 represents the coefficient of the second-difference trend in the original series. In such cases, for example Equation 15.23, it's not always necessary that β_0 be in the model.

AutoRegressive Integrated Moving Average. (If the original series is stationary and d therefore equals 0, this is sometimes shortened to ARMA.)

As a shorthand, an ARIMA model with p, d, and q specified is usually denoted as ARIMA(p,d,q) with the specific integers chosen inserted for p, d, and q as in ARIMA(2,1,1). ARIMA (2,1,1) would indicate a model with two autoregressive terms, one first difference, and one moving average term:

$$\text{ARIMA}(2,1,1)\text{: } Y_t^* = \beta_0 + \theta_1 Y_{t-1}^* + \theta_2 Y_{t-2}^* + \epsilon_t + \phi_1 \epsilon_{t-1} \tag{15.23}$$

where $Y_t^* = Y_t - Y_{t-1}$.

15.3.2 Estimating ARIMA Models

The first task in estimating an ARIMA model is to specify p (the number of autoregressive terms), d (the number of first-differences or other transformations) and q (the number of moving-average terms). Generally, the higher the p, d, and q, the better the fit, but the lower the degrees of freedom. Once the parameters p, d, and q have been chosen, the job of specifying the ARIMA equation is complete, and the computer estimation package will then calculate estimates of the appropriate coefficients. Because the error terms in the moving average process are of course not observable, a nonlinear estimation technique must be used instead of OLS.

How are the values for p, d, and q to be chosen? While experience with choosing p, d, and q will make these choices seem fairly routine, there are some diagnostic tests that can be used to determine good (integer) values for these parameters:

Choosing d: The number of first differences usually is found by examining the plot of the series. If Y shows growth over time, then calculate Y^*; if Y^* shows growth over time, then use Y^{**}. It's rare to find in economic examples a situation calling for $d > 2$. If the correct d has been chosen, the simple correlation coefficient between the dependent variable and lagged values of the dependent variable should approach zero as the number of lags increases. Such a simple correlation coefficient is called the autocorrelation function (ACF), and is often included as output in ARIMA packages. **Autocorrelation functions (ACFs)** are simple correlation coefficients[10] between a variable

10. Like simple correlation coefficients, autocorrelation coefficients (ACFs) vary between -1 and $+1$, with zero indicating no correlation.

and the same variable lagged a number of time periods. If the ACFs approach zero as the number of lags increases, then the series is stationary; a nonstationary series will show little tendency for the ACFs to decrease in size as the number of lags increases. After an examination of the plot and/or the ACFs makes the researcher feel comfortable that enough transformations have been applied to make the resulting series stationary, the next step is to choose integer values for p and q.

Choosing p and q: The number of autoregressive terms (p) and moving-average terms (q) to be included typically are determined at the same time. These are chosen by finding the lowest p and q for which the residuals of the estimated equation are devoid of autoregressive and moving-average components. This is done by:

1. Choosing an initial (p, q) set (making them as small as is reasonable),
2. Estimating Equation 15.19 for that (p, q) and the d chosen above, and
3. Testing the residuals of the estimated equation to see if they are free of autocorrelations. If they are not, then either p or q is increased by one, and the process is begun again.

To complete step 1, choosing an initial (p, q) set, an ARIMA(0,d,0) is run and the residuals from this estimate are analyzed with two statistical measures. These are the ACF, this time applied to the residuals instead of to the variable itself, and a new measure, the partial ACF (PACF), which is similar to the ACF (applied to the residuals) except that it holds the effects of other lagged residuals constant. That is, the **partial autocorrelation function (PACF)** for the *k*th lag is the correlation coefficient between e_t and e_{t-k}, holding constant all other residuals. (Another way to picture a PACF for the *k*th lag is as the coefficient of e_{t-k} in a regression of e_t on $e_{t-1}, e_{t-2}, \ldots, e_{t-k}$.) Almost every different ARIMA model has a unique ACF/PACF combination; the theoretical ACF and PACF patterns are different for different ARIMA(p,d,q) specifications. In particular, the last lag before the PACF tends to zero is typically a good value for p, and the last lag before the ACF tends to zero is typically a good value for q. That is, as can be seen in the examples in Table 15.1, the ACF and PACF for Y can be used as guides[11] to the selection of p and q.

11. The phrase "tends to zero" in Table 15.1 typically means that the ACF or PACF moves toward zero with an exponential decay, but for ARIMA(2,0,0), a damped sine wave is also possible. In addition, note that the stability of estimated ARIMA models (referred to as stationarity and invertability) depends on the values of the estimated parameters. For example, in ARIMA(1,0,0), θ must be between -1 and $+1$ for the model to be stable.

TABLE 15.1 UNIQUE ACFs AND PACFs FOR THE MAJOR ARIMA MODELS

Model	Match Up Estimated ACF and PACF for Y_t to: Theoretical ACF		Theoretical PACF
Nonstationary	Different from zero	or	Different from zero
ARIMA(0,0,0) (white noise)	All are zero	and	All are zero
ARIMA(1,0,0) $Y_t = \theta_0 + \theta_1 Y_{t-1} + \epsilon_t$	Tends to zero	and	Zero after 1 lag
ARIMA(2,0,0) $Y_t = \theta_0 + \theta_1 Y_{t-1} + \theta_2 Y_{t-2} + \epsilon_t$	Tends to zero	and	Zero after 2 lags
ARIMA(0,0,1) $Y_t = \theta_0 + \epsilon_t + \phi_1 \epsilon_{t-1}$	Zero after 1 lag	and	Tends to zero
ARIMA(0,0,2) $Y_t = \theta_0 + \epsilon_t + \phi_1 \epsilon_{t-1} + \phi_2 \epsilon_{t-2}$	Zero after 2 lags	and	Tends to zero
ARIMA(1,0,1) $Y_t = \theta_0 + \theta_1 Y_{t-1} + \epsilon_t + \phi_1 \epsilon_{t-1}$	Zero after 1 lag	and	Zero after 1 lag

Once the (p,d,q) parameters have been chosen, the equation is estimated by the computer's ARIMA estimation package, producing p different $\hat{\theta}$s, q different $\hat{\phi}$s, and $\hat{\beta}_0$ (if appropriate). Since the moving-average process contains error terms that cannot actually be observed, an ARIMA estimation with $q > 0$ requires the use of a nonlinear estimation procedure rather than OLS. If $q = 0$, OLS can be used.

After estimating an initial (p,d,q) combination, calculate and inspect the ACFs and PACFs of the residuals. If these ACFs and PACFs are all insignificantly different from zero (use the *t*-test if in doubt) then the equation can be considered the final specification (free from autoregressive and moving-average components).[12] If one or more of the PACFs or ACFs are significantly different from zero, then increase

12. As a final check, some researchers compare the variance of ARIMA(p,d,q) with those of ARIMA(p+1,d,q) and ARIMA(p,d,q+1). If ARIMA(p,d,q) has the lowest variance, then it should be considered the final specification. Also useful is the *Q statistic,* a measure of whether the first k ACFs are (jointly) significantly different from zero. If the Q statistic, which is usually printed out by ARIMA computer packages, is less than the critical chi-square value, then the model can be considered free from autoregressive and moving-average components.

p (if a PACF is significant) or q (if an ACF is significant) by one and re-estimate the model. This process continues until an ARIMA model whose residuals have no autoregressive or moving-average components can be found. With experience, you'll learn to choose the best (p,d,q) from the first set of residuals, thus avoiding the testing and re-estimation process of this paragraph.

15.3.3 Forecasting with ARIMA

To forecast with an estimated ARIMA (or time-series) model, plug past values of Y and e into the estimated version of Equation 15.19, obtaining a forecast for Y. That is, once p, d, and q have been selected and ARIMA(p,d,q) has been estimated, ARIMA forecasts for Y_t can be made using

$$\begin{aligned} Y_t = \beta_0 + \theta_1 Y_{t-1} + \theta_2 Y_{t-2} + \cdots + \theta_p Y_{t-p} + e_t \\ + \phi_1 e_{t-1} + \phi_2 e_{t-2} + \cdots + \phi_q e_{t-q} \end{aligned} \qquad (15.24)$$

This is done by substituting current and past values of the (possibly transformed) dependent variable and the *residuals* into Equation 15.24. Note that in Equation 15.24, the residuals are used as proxies for the error terms specified in Equation 15.19. After the first value of the dependent variable is forecasted, the second-forecast-period value of Y is computed by using the first forecasted value of Y on the right side of the equation and by assuming that the error terms outside the sample of Equation 15.24 are zero (based on their expected values). In other words, the forecasts are made sequentially, with one forecast value being used to forecast subsequent values.

To illustrate, assume that an ARIMA(2,0,1) has been estimated using the procedures of Sectrion 15.3.2 and that the subscript T denotes the last data point of the sample period. An ARIMA(2,0,1) equation is:

$$Y_t = \beta_0 + \theta_1 Y_{t-1} + \theta_2 Y_{t-2} + \epsilon_t + \phi_1 \epsilon_{t-1} \qquad (15.25)$$

To forecast this equation for time period T + 1, we would estimate Equation 15.25, substitute residuals for the error terms, and substitute (T + 1) for t where it appears in Equation 15.25, producing

$$\hat{Y}_{T+1} = \hat{\beta}_0 + \hat{\theta}_1 Y_T + \hat{\theta}_2 Y_{T-1} + \hat{\phi}_1 e_T \qquad (15.26)$$

where the residual from time period T + 1 is set equal to zero. To forecast for time period T + 2, add one to the subscripts of Equation 15.26, being careful to use the value of Y_{T+1} just computed and to set the residual from time period T + 1 equal to zero, obtaining:

$$\hat{Y}_{T+2} = \hat{\beta}_0 + \hat{\theta}_1(\hat{Y}_{T+1}) + \hat{\theta}_2 Y_T \tag{15.27}$$

Note that the moving-average process has disappeared. Further forecasted values of Y can be computed in essentially the same sequential (or recursive) way.

15.4 Summary

1. Forecasting is the estimation of the expected value of a dependent variable for observations that are not part of the sample data-set. Forecasts are generated (via regressions) by estimating an equation for the dependent variable to be forecasted, and substituting values for each of the independent variables (for the observations to be forecasted) into the equation.

2. An excellent fit within the sample period for a forecasting equation does not guarantee that the equation will forecast well outside the sample period.

3. A forecast in which all the values of the independent variables are known with certainty is called an unconditional forecast, but if one or more of the independent variables have to be forecasted, it is a conditional forecast. Conditional forecasting introduces no bias into the prediction of Y (as long as the X forecasts are unbiased), but increased forecast error variance is virtually unavoidable with conditional forecasting.

4. If the coefficients of an equation have been estimated with GLS (to correct for pure first-order serial correlation), then the forecasting equation is:

   $$\hat{Y}_{T+1} = \hat{\rho}Y_T + \hat{\beta}_0(1 - \hat{\rho}) + \hat{\beta}_1(X_{T+1} - \hat{\rho}X_T)$$

   where ρ is the coefficient of autocorrelation, rho.

5. Forecasts are often more useful if they are accompanied by a confidence interval, which is a range within which the actual value of

the dependent variable should fall a given percentage of the time (the level of confidence). This is:

$$\hat{Y}_{T+1} \pm S_F t_c$$

where S_F is the estimated standard error of the forecast and t_c is the critical two-tailed t-value for the desired level of confidence.

6. ARIMA, or time-series analysis, is a highly refined curve-fitting technique that uses current and past values of the dependent variable (and only the dependent variable) to produce often accurate short-term forecasts of that variable. The first step in using ARIMA is to make the dependent variable series stationary by taking d first-differences until the resulting transformed variable has a constant mean and variance. The ARIMA(p,d,q) approach then combines an autoregressive process (with $\theta_1 Y_{t-1}$ terms) of order p with a moving-average process (with $\phi_1 \epsilon_{t-1}$ terms) of order q to explain the *d*th differenced dependent variable.

7. Specifying an ARIMA model involves choosing the correct p, d, and q. The best d value is the smallest number of first differences (or other transformations) that will produce a stationary series. The best p and q values are the smallest integer values that produce an equation with residuals that are free of autoregressive and moving-average components. The basic tools used to make such specification choices are the simple (ACF) and partial (PACF) autocorrelation functions, which are correlation coefficients between a variable (or a residual) and itself lagged a number of time periods.

Exercises

(Answers to even-numbered exercises are in Appendix A.)

1. Write the meaning of each of the following terms without reference to the book (or your notes), and compare your definition with the version in the text for each:
   a. conditional forecast
   b. leading indicator
   c. confidence interval
   d. autoregressive process
   e. moving-average process

f. ARIMA(p,d,q)
g. stationary series
h. simple and partial autocorrelation functions

2. Calculate the following unconditional forecasts:
   a. Bond prices given the simplified equation in Exercise 5 in Chapter 1 and the following data for the Federal funds rate: 5.82, 5.04, 5.54, 7.93.
   b. The expected level of check volume at three possible future sites for new Woody's restaurants, given Equation 3.14 and the following data. If you could only build one new eatery, in which of these three sites would you build (all else equal)?

Site	*Competition*	*Population*	*Income*
Richburgh	6	58,000	38,000
Nowheresville	1	14,000	27,000
Slick City	9	190,000	15,000

   c. Per capita consumption of fish in the U.S. for 1971–74 given Equation 8.27 and the following data:

Year	*PF*	*PB*	*YD*
1971	130.2	116.7	2679
1972	141.9	129.2	2767
1973	162.8	161.1	2934
1974	187.7	164.1	2871

   (Hint: Reread Section 8.5.2 before attempting this forecast.)

3. Forecast values for the appropriate independent variables, and then calculate the following conditional forecasts:
   a. The weights of the next three males you see, using Equation 1.21 and your *estimates* of their heights. (Ask for actual values after finishing.)
   b. S & L passbook deposits for all four quarters of 1980 given the equation you chose in the interactive exercise in Section 11.4.
   c. S & L passbook deposits for all four quarters of 1980 given the estimates in Table 11.5 and the forecasts of the Xs you generated for part b. (Hint: Note that this regression was estimated with GLS.) Compare your part b forecasts with your GLS forecasts. Which seem more likely to be accurate? Why?

4. Calculate 95 percent confidence interval forecasts for the following:
   a. the weight of a male who is 5′9″ tall (Hint: Modify Equation 15.15.)
   b. Next month's sales of ice cream cones at the Campus Cooler given an expected price of 60 cents per cone and:

$$\hat{C}_t = 2{,}000 - \underset{(5.0)}{20.0P_t} \qquad \bar{R}^2 = .80$$
$$t = -4.0 \qquad T = 30 \qquad \bar{P} = 50$$

where: C_t = number of ice cream cones sold in month t
P_t = the price of the Cooler's ice cream cones (in cents) in month t
$s^2 = 25{,}000$, and $\Sigma(P_t - \bar{P})^2 = 1000$

   c. Your forecast for first quarter 1980 passbook deposits from Question 3b above. (Hint: Use the actual data set in Section 11.4.)

5. Build your own (nonARIMA) forecasting model from scratch: Pick a dependent variable, specify your equation, hypothesize signs, find a data-set, estimate your model (leaving a couple of the most current observations out of the sample) and forecast your dependent variable. Now comes the "fun" of comparing your forecast with the actual Ys. How did you do?

6. For each of the following series, calculate and plot Y_t, $Y_t^* = \Delta Y_t$, and $Y_t^{**} = \Delta Y_t^*$, describe the stationarity properties of the series, and choose an appropriate value for d.
   a. 2, 3, 4, 5, 6, 7, 8, 9, 10, 11, 12, 13
   b. 2, 2, 3, 4, 5, 6, 8, 10, 12, 15, 19, 24
   c. 2, 3, 6, 3, 4, 2, 3, 5, 1, 4, 4, 6

7. Take the three Y_t^* series you calculated as part of your answer to Question 6 above and check to see if they are correct by calculating backwards and seeing if you can derive the original three Y_t series from your three Y_t^* series. (Hint: Equation 15.22 can be adapted for this "integration" purpose.)

8. Suppose you have been given two different ARIMA(1,0,0) fitted time-series models of the variable Y_t:

$$\text{Model A: } Y_t = 15.0 + 0.5Y_{t-1} + \epsilon_t$$

$$\text{Model T: } Y_t = 45.0 - 0.5Y_{t-1} + \epsilon_t$$

where ϵ_t is a normally-distributed error term with mean zero and standard deviation equal to one.

a. The final observation in the sample (time period 86) is $Y_{86} = 31$. Determine forecasts for periods 87, 88, and 89 for both models.
b. Suppose you now find out that the actual Y_{87} was equal to 33. Revise your forecasts for periods 88 and 89 to take the new information into account.
c. Based on the fitted time-series and your two forecasts, which model (Model A or Model T) do you expect to exhibit smoother behavior? Explain your reasoning.

9. Suppose you have been given an ARIMA(1,0,1) fitted time-series model:

$$Y_t = 0.0 + 1.0Y_{t-1} + \epsilon_t - 0.5\epsilon_{t-1}$$

where ϵ_t is a normally-distributed error term with mean zero and standard deviation equal to one and where $T = 99$, $Y_{99} = 27$, and where $\hat{Y}_{99} = 27.5$.

a. Calculate e_{99}.
b. Calculate forecasts for Y_{100}, Y_{101}, and Y_{102}. (Hint: Use your answer to part a.)

10. The actual data used to estimate Equation 15.5 for the Kellogg stock forecasting problem are given below. Calculate an ARIMA forecast for this data-set by:

a. Choosing p, d, and q
b. Estimating the coefficients of your ARIMA(p, d, q) model
c. Forecasting the Kellogg's stock price four quarters into the future
d. Calculating percent forecast errors and comparing your results with those in Section 15.1. Did you do better or worse than the regression-generated forecasts? Why do you think this happened?

DATA (The dollar price of J. L. Kellogg's stock quarterly from 1969 through third quarter 1977):

Year	First Quarter	Second Quarter	Third Quarter	Fourth Quarter
1969:	10.50	10.44	9.94	10.25
1970:	11.00	9.88	10.50	12.00
1971:	13.94	12.25	12.61	13.50
1972:	13.44	12.44	13.50	15.39
1973:	15.75	13.88	14.50	15.50
1974:	16.13	14.75	11.75	15.25
1975:	17.13	20.50	19.00	21.50
1976:	20.25	25.63	26.88	27.63
1977:	23.88	26.38	24.00	—

11. You've been hired to forecast *Sports Illustrated* subscriptions (S) using the following function of GNP (Y) and a classical error term (ϵ):

$$S_t = \beta_0 + \beta_1 Y_t + \beta_2 S_{t-1} + \epsilon_t$$

Explain how you would forecast (out two time periods) with this equation in the following cases:

a. If future values of Y are known. [Hint: Be sure to comment on the functional form of this relationship.]
b. If future values of Y are unknown and *Sports Illustrated* subscriptions are small in comparison to GNP.
c. If *Sports Illustrated* subscriptions are about half of GNP (obviously a sports-lover's heaven!) and all other components of GNP are known to be stochastic functions of time.

Appendix A

Answers to Even-Numbered Exercises

Chapter One

1-2. **a.** positive, **b.** negative, **c.** positive, **d.** negative, **e.** ambiguous, **f.** negative.

1-4. **a.** **b.** and **c.** coefficients, **d.**, **e.** and **f.** neither, **g.** both.

1-6. **a.** Customers #3, 4 and 20; no.
b. Weight is determined by more than just height.
c. People who decide to play the weight-guessing game may feel they have a weight that is hard to guess.

1-8. **a.** The coefficient of L_i represents the change in the percentage chance of making a putt when the length of the putt increases by one foot. In this case, the percentage chance of making the putt decreases by 4.1 for each foot longer the putt is.
b. 42.6 percent, yes; 79.5 percent, no (too low); −18.9 percent, no (negative!).
c. One problem is that the theoretical relationship between the length of the putt and the percentage of putts made is almost surely nonlinear in the variables; we'll discuss models appropriate to this problem in Chapter 7. A second problem is that the actual dependent variable is limited by zero and one while the regression estimate is not; we'll discuss models appropriate to this problem in Chapter 13.
d. The equations are identical. To convert one to the other, you need to know that $\hat{P}_i = P_i - e_i$, which is true because $e_i = P_i - \hat{P}_i$ (or more generally, $e_i = Y_i - \hat{Y}_i$).

1-10. a. 21.4: A \$1 billion increase in GNP will be associated with an increase of \$21.40 in the average price of a new house. 4363.1: Technically, the constant term equals the value of the dependent variable when all the independent variables equal zero, but in most cases (as in this one) such a definition has little economic meaning. As we'll learn in Chapters 4 and 7, estimates of the constant term should not be relied on for inference.

b. It doesn't matter what letters we use as symbols for the dependent and independent variables.

c. You could measure both P_t and Y_t in real terms by dividing each observation by the GNP deflator (or the CPI) for that year (and multiplying by 100).

d. The price of houses is determined by the forces of supply and demand, and we won't discuss the estimation of simultaneous equations until Chapter 14. In a demand-oriented sense, GNP is probably measuring buying power, which is better represented by disposable income. In a supply-oriented sense, GNP might be standing for costs like wages and the price of materials.

Chapter Two

2-2. Because of Equation 2.7, $\hat{\beta}_0$ and $\hat{\beta}_1$ tend to compensate for each other in a two-variable model. Thus if $\hat{\beta}_1$ is too high, then $\hat{\beta}_0$ is likely to be too low.

2-4. a. $\hat{\beta}_1 = -0.5477$, $\hat{\beta}_0 = 12.289$.

b. $R^2 = .465$, $\bar{R}^2 = .398$, $r = -0.682$ (rounding will cause slight differences).

c. $\widehat{\text{Income}} = 12.289 - 0.5477(8) = 7.907$.

2-6. Start with $R^2 = ESS/TSS = \Sigma(\hat{Y} - \bar{Y})^2/\Sigma(Y - \bar{Y})^2$, and substitute $\bar{Y} - \hat{\beta}_1\bar{X} + \hat{\beta}_1 X$ for $\hat{Y}$. Now factor out $\hat{\beta}_1^2$, simplify, and rearrange, obtaining:

$$R^2 = \frac{\hat{\beta}_1^2\Sigma(X - \bar{X})^2}{\Sigma(Y - \bar{Y})^2}$$

Now substitute $\Sigma(X - \bar{X})(Y - \bar{Y})/\Sigma(X - \bar{X})^2$ for $\hat{\beta}_1$ above and cancel out one $\Sigma(X - \bar{X})^2$. The remaining equation is the square of the simple correlation coefficient between X and Y.

2-8. a. Yes, **b.** at first glance, perhaps, but see below.
c. Three dissertations, since $(489 \times 3 = \$1467) > (\$230)$ or $(120 \times 2 = \$240)$.
d. The coefficient of D seems to be too high; perhaps it is absorbing the impact of an independent variable that has been omitted from the regression. For example, students may choose a dissertation advisor on the basis of reputation, a variable not in the equation.

2-10. a. V_i: positive
H_i: negative, (although some would argue that in a world of perfect information, drivers would take fewer risks if they knew the state had few hospitals.)
C_i: ambiguous because a high rate of driving citations could indicate risky driving (raising fatalities) *or* zealous police citations policies (reducing risky driving and therefore fatalities).
b. No, because the coefficient differences are small and the data will differ from year to year. We'd be more concerned if the coefficients differed by orders of magnitude or changed sign.
c. Since the equation for 1982 has more degrees of freedom but a lower R^2, no calculation is needed to know that the equation for 1981 has a higher $\bar{R}^2$. Just to be sure, we calculated $\bar{R}^2$ and obtained .652 for 1981 and .565 for 1982.

Chapter Three

3-2. a. D = 1 if graduate student and D = 0 if undergraduate.
b. Yes, for example E = how many exercises (such as this) the student did.
c. If D is defined as in answer a, then its coefficient's sign would be expected to be positive. If D is defined as 0 if graduate student, 1 if undergraduate, then the expected sign would be negative.
d. A coefficient with a value of .5 indicates that, all else equal, a graduate student would be expected to earn half a grade point higher than an undergraduate. If there were only graduate students or only undergraduates in class, the coefficient of D could not be estimated.

3-4. a. There are many possible omitted explanatory variables; for example, the number of parking spaces near the restaurant.

c. This calculation gives the ratio of checks to population without taking the other variables into consideration. The regression coefficient is an estimate of the impact of a change in population on check volume, holding constant the other variables in the equation.

3-6. a. New P = Old P/1000, so $\hat{\beta}$ goes from 0.3547 to 354.7.
b. 286.3.
c. no.

3-8. a. The Midwest (the fourth region of the country).
b. Including the omitted condition as a variable will cause the dummies to sum to a constant (1.0). This constant will be perfectly collinear with the constant term, and the computer will not be able to estimate the equation.
c. Positive.
d. Most correct = III, least correct = I.

3-10. a. PP_t − ; PB_t + ; YD_t +.
b. Yes.
c. The United States consumes 0.9 pounds less pork per person in quarter one than in quarter four, holding all other included explanatory variables constant (analysis is similar for other quarters).
d. Real price = nominal price $(CPI_{\text{base year}}/CPI_{\text{current year}})$.

3-12. a. All positive except for the coefficient of F_i, which in today's male-dominated movie industry probably has a negative expected sign.
b. Arnold, because \$500,000 < (\$4,000,000 − \$3,027,000).
c. Yes, since 200 × 15.4 = \$3,080,000 > \$1,200,000.
d. Yes, since \$1,770,000 > \$1,000,000.
e. Yes, the unexpected sign of the coefficient of B_i.

Chapter Four

4-2. a. β_1.
b. yes; double-log production function.

4-4. c. definitely violates Classical Assumption VI, and a. might for some samples.

4-6. $Z_i = (X_i - \mu)\sigma = (1.0 - 0.0)/\sqrt{0.5} = 1.414$; for this Z_i, Table B-7 gives 0.0787, which is the probability of observing an X greater than + 1. To also include the probability of an X less than − 1, we need to double 0.0787, obtaining a final answer of 0.1574.

4-8. We know that: $\Sigma e_i^2 = \Sigma(Y_i - \hat{Y}_i)^2 = \Sigma(Y_i - \hat{\beta}_0 - \hat{\beta}_1 X_i)^2$. To find the minimum, differentiate Σe_i^2 with respect to $\hat{\beta}_0$ and $\hat{\beta}_1$ and set each derivative equal to zero (these are the "normal equations"):

$$\delta(\Sigma e_i^2)/\delta\hat{\beta}_0 = -2[\Sigma(Y_i - \hat{\beta}_0 - \hat{\beta}_1 X_i)] = 0$$
$$\text{or } \Sigma Y_i = n(\hat{\beta}_0) + \hat{\beta}_1(\Sigma X_i)$$

$$\delta(\Sigma e_i^2)/\delta\hat{\beta}_1 = -2[\Sigma(Y_i - \hat{\beta}_0 - \hat{\beta}_1 X_i)X_i] = 0$$
$$\text{or } \Sigma Y_i X_i = \hat{\beta}_0(\Sigma X_i) + \hat{\beta}_1(\Sigma X_i^2)$$

Solve the two equations simultaneously and rearrange:

$$\hat{\beta}_1 = [n(\Sigma Y_i X_i) - \Sigma Y_i X_i]/[n(\Sigma X_i^2) - (\Sigma X_i)^2]$$
$$= \Sigma(X_i - \overline{X})(Y_i - \overline{Y})/\Sigma(X_i - \overline{X})^2 = \Sigma x_i y_i/\Sigma x_i^2$$

where $x_i = (X_i - \overline{X})$ and $y_i = (Y_i - \overline{Y})$

$$\hat{\beta}_0 = [\Sigma X_i^2 \Sigma Y_i - \Sigma X_i \Sigma X_i Y_i]/[n(\Sigma X_i^2) - (\Sigma X_i)^2] = \overline{Y} - \hat{\beta}_1\overline{X}$$

To prove linearity: $\hat{\beta}_1 = \Sigma x_i y_i/\Sigma x_i^2 = \Sigma x_i(Y_i - \overline{Y})/\Sigma x_i^2$

$$= \Sigma x_i Y_i/\Sigma x_i^2 - \Sigma x_i(\overline{Y})/\Sigma x_i^2$$
$$= \Sigma x_i(Y_i)/\Sigma x_i^2 - \overline{Y}\Sigma x_i/\Sigma x_i^2$$
$$= \Sigma x_i(Y_i)/\Sigma x_i^2 \text{ since } \Sigma x_i = 0$$
$$= \Sigma k_i Y_i \text{ where } k_i = x_i/\Sigma x_i^2$$

$\hat{\beta}_1$ is a linear function of Y, since this is how a linear function is defined. It is also a linear function of the βs and ε, which is the basic interpretation of linearity. $\hat{\beta}_1 = \beta_0\Sigma k_i + \beta_1\Sigma k_i x_i + \Sigma k_i\epsilon_i$. $\hat{\beta}_0 = \overline{Y} - \hat{\beta}_1(\overline{X})$ where $\overline{Y} = \hat{\beta}_0 + \hat{\beta}_1(\overline{X})$, which is also a linear equation.

To prove unbiasedness: $\hat{\beta}_1 = \Sigma k_i Y_i = \Sigma k_i(\beta_0 + \beta_1 X_i + \epsilon_i)$

$$= \Sigma k_i\beta_0 + \Sigma k_i\beta_1 X_i + \Sigma k_i\epsilon_i$$

Since $k_i = x_i/\Sigma x_i^2 = (X_i - \overline{X})/\Sigma(X_i - \overline{X})^2$,

then $\Sigma k_i = 0$, $\Sigma k_i^2 = 1/\Sigma x_i^2$, $\Sigma k_i x_i = \Sigma k_i X_i = 1$.

So, $\hat{\beta}_1 = \beta_1 + \Sigma k_i \epsilon_i$ and given the assumptions of ϵ_i,

$E(\hat{\beta}_1) = \beta_1 + \Sigma k_i E(\epsilon_i) = \beta_1$, proving $\hat{\beta}_1$ is unbiased.

To prove minimum variance (of all linear unbiased estimators): $\hat{\beta}_1 = \Sigma k_i Y_i$. Since $k_i = x_i/\Sigma x_i^2 = (X_i - \overline{X})/\Sigma(X_i - \overline{X})^2$, $\hat{\beta}_1$ is a weighted average of the Ys, and the k_i are the weights. To write an expression for any linear estimator, substitute w_i for k_i, which are also weights but not necessarily equal to k_i:

$$\beta_1^* = \Sigma w_i Y_i, \text{ so } E(\beta_1^*) = \Sigma x_i E(Y_i) = \Sigma w_i (\beta_0 + \beta_1 X_i)$$
$$= \beta_0 \Sigma w_i + \beta_1 \Sigma w_i X_i$$

In order for β_1^* to be unbiased, $\Sigma w_i = 0$ and $\Sigma w_i X_i = 1$. The variance of β_1^*:

$$\text{var}(\beta_1^*) = \text{var}\Sigma w_i Y_i = \Sigma w_i \text{var} Y_i = \sigma^2 \Sigma w_i^2$$
$$[\text{var}(Y_i) = \text{var}(\epsilon_i) = \sigma^2]$$
$$= \sigma^2 \Sigma(w_i - x_i/\Sigma x_i^2 + x_i/\Sigma x_i^2)^2$$
$$= \sigma^2 \Sigma(w_i - x_i/\Sigma x_i^2)^2 + \sigma^2 \Sigma x_i/(\Sigma x_i^2)^2$$
$$+ 2\sigma^2 \Sigma(w_i - x_i/\Sigma x_i^2)(x_i/\Sigma x_i^2)$$
$$= \sigma^2 \Sigma(w_i - x_i/\Sigma x_i^2)^2 + \sigma^2/(\Sigma x_i^2)$$

The last term in this equation is a constant, so the variance of β_1^* can be minimized only by manipulating the first term. The first term is minimized only by letting $w_i = x_i/\Sigma x_i^2$, then

$$\text{var}(\beta_1^*) = \sigma^2/\Sigma x_i^2 = \text{var}(\hat{\beta}_1).$$

When the least-squares weights, k_i, equal w_i, the variance of the linear estimator β_1 is equal to the variance of the least-squares estimator, $\hat{\beta}_1$. When they are not equal, $\text{var}(\beta_1^*) > \text{var}(\hat{\beta}_1)$ Q.E.D.

4-10. a. Most experienced econometricians would prefer an unbiased nonminimum variance estimate.

b. Yes; an unbiased estimate with an extremely large variance has a high probability of being far from the true value. In such a case, a slightly biased estimate with a very small variance would be better.

c. The most frequently used possibility is to minimize the Mean Square Error (MSE), to be discussed in Section 6.7.

Chapter Five

5-2. a. H_0: $\beta_1 \leq 0$, H_A: $\beta_1 > 0$

b. H_0: $\beta_1 \geq 0$, H_A: $\beta_1 < 0$; H_0: $\beta_2 \leq 0$, H_A: $\beta_2 > 0$; H_0: $\beta_3 \leq 0$, H_A: $\beta_3 > 0$ (The hypothesis for β_3 assumes that it is never too hot to go jogging.)

c. H_0: $\beta_1 \leq 0$, H_A: $\beta_1 > 0$; H_0: $\beta_2 \leq 0$, H_A: $\beta_2 > 0$; H_0: $\beta_3 \geq 0$, H_A: $\beta_3 < 0$ (The hypothesis for β_3 assumes you're not breaking the speed limit.)

d. H_0: $\beta_G = 0$; H_A: $\beta_G \neq 0$ (G for grunt.)

5-4. a. $t_c = 1.363$; reject H_0 for β_1, cannot reject H_0 for β_2 and β_3.

b. $t_c = 1.318$; reject H_0 for β_1, cannot reject H_0 for β_2 and β_3.

c. $t_c = 3.143$; cannot reject the null hypothesis for β_1, β_2, and β_3.

5-6. a. $t_2 = (200 - 160)/25.0 = 1.6$; $t_c = 2.052$; therefore cannot reject H_0. (Notice the violation of the strawman approach here.)

b. $t_3 = 2.37$; $t_c = 2.756$; therefore cannot reject the null hypothesis.

c. $t_2 = 5.6$; $t_c = 2.447$; therefore reject H_0 if it is formulated as in the exercise, but this poses a problem because the original hypothesized sign of the coefficient was negative. Thus the alternative hypothesis ought to have been stated: H_A: $\beta_2 < 0$, and H_0 cannot be rejected.

5-8. a. $F = [R^2/(K)]/[(1 - R^2)/(n - K - 1)]$.

b. F is a statistical measure of fit while R^2 is a qualitative measure; printing both saves the reader time and avoids (human) computation errors.

5-10. a. $t = 8.509$; with $t_c = 1.746$, reject H_0 of no collinearity.

b. $t = 16.703$; with $t_c = 2.060$, reject H_0.

c. $t = 3.216$; with $t_c = 3.365$, cannot reject H_0 of no collinearity.

d. $t = -7.237$; with $t_c = -1.303$, reject H_0.
e. $t = 3.213$; with $t_c = 2.048$, reject H_0.

5-12. $F = (3764.99/1)/(1305.43/18) = 51.91$; the estimated equation is significant since $51.91 > 4.41$, the critical F-value at the 5 percent level.

5-14. a. T: H_0: $\beta_T \leq 0$, H_A: $\beta_T > 0$. Reject H_0 since $|+5.57| > 1.711$ and $+5.57$ has the sign of H_A.
P: H_0: $\beta_P \geq 0$, H_A: $\beta_P < 0$. Reject H_0 since $|-2.35| > 1.711$ and -2.35 has the sign of H_A.
A: H_0: $\beta_A \leq 0$, H_A: $\beta_A > 0$. Cannot reject H_0 since $|+1.244| < 1.711$.
C: H_0: $\beta_C \leq 0$, H_A: $\beta_C > 0$. Cannot reject H_0 since $|-1.10| < 1.711$ or since -1.10 does not have the sign of H_A.
b. Reject all null hypotheses.
c. Reject null hypotheses for T, J, and S. Cannot reject null hypotheses for F (since $|-1.47| < 1.645$) or B (since $|-1.33| < 1.645$ or since -1.33 does not have the sign of H_A).

Chapter Six

6-2. Expected bias in $\hat{\beta} = (\beta_{\text{omitted}}) \cdot f(r_{\text{omitted, included}})$
a. Expected bias $= (-) \cdot (+) = (-) =$ negative bias.
b. $(+) \cdot (+) = (+) =$ positive bias; this bias will be potentially large since age and experience are highly correlated.
c. $(+) \cdot (+) = (+) =$ positive bias.
d. $(-) \cdot (0) = 0 =$ no bias; it may seem as though it rains more on the weekends, but there is no theoretical relationship between the two.

6-4. Yes; you could run a regression that includes variables for the risk and taxability of the bonds as well as maturity date and interest rate.

6-6. a.

Coefficient:	β_1	β_2	β_3	β_4
Hypothesized sign:	+	+	+	−
Calculated t-score:	5.0	1.0	10.0	3.0
$t_c = 2.485$ (1% level), so:	signif.	insig.	signif.	unexpected sign

b. The significant unexpected sign of $\hat{\beta}_4$ is evidence of a possible omitted variable that is exerting positive bias. The omitted variable must either be correlated positively with X_4 *and* have a positive expected coefficient or else be correlated negatively with X_4 *and* have a negative expected coefficient. The fairly low calculated t-score for β_2 is not strong evidence of a specification error.

c. A second run might add an independent variable that is theoretically sound and that could have caused positive bias in $\hat{\beta}_4$. For example, X_5 = the number of "attractions" like movie theaters or shopping malls in the area would have a positive expected coefficient and be positively correlated with the number of nearby competing stores.

6-8. a. Nothing is certain, but the best guess is: X_1 = # of students, X_2 = chain price, X_3 = temperature, X_4 = Cooler price.

b. X_4 has the only negative coefficient, and Cooler price has the only negative expected sign. # of students (in thousands) should be the most significant and have the largest coefficient. Weather should be the least significant and also have a small coefficient (since that variable can be the largest in size). X_2 = chain price by elimination.

c. Note that developing hypotheses includes determining the desired level of significance. A possible rerun would be to drop (or reformulate to absolute degrees difference from optimal hamburger-eating range, if there is such a thing) the weather variable. If there is omitted variable bias, it is positive on $\hat{\beta}_4$ (advertising?).

6-10. a. Consumers and producers can react differently to changes in the same variable. For example, price: a rise in price causes consumers to demand a lower quantity and producers to supply a greater quantity.

b. Include only variables affecting demand ("demand-side variables") in demand equations and only variables affecting supply ("supply-side variables") in supply equations.

c. Review the literature, decide whether the equation you wish to estimate is a supply or a demand equation, and, when specifying the model, think carefully about whether an independent variable is appropriate for a demand or supply equation.

6-12. a. No bias $(+ \cdot 0)$ unless weather patterns indicate a correlation between rainfall and temperature. If it tends to rain more when it's cold, then there would be a small negative bias $(+ \cdot -)$.

b. Positive bias $(+ \cdot +)$.

c. Positive bias $(+ \cdot +)$.

d. Negative bias $(+ \cdot -)$ given a likely negative correlation between hours studied for the test and hours slept.

6-14. a.

Coefficient:	β_P	β_L	β_A	β_N
Hypothesized sign:	+	+	+	+
Calculated t-score:	3.3	1.5	−0.6	13.5

$t_c = 1.677$ (approx.), so H_0 can be rejected only for β_P and β_N.

b. The inclusion of N means that the other variables probably should be aimed at differences between the situations of farm women and rural nonfarm women, but they are not, probably because such data are unavailable. Thus the equation certainly has omitted variables. If they can be found, then L, A, and possibly even P might be irrelevant.

c. Add a variable measuring the average income of farm women in the *i*th state, if possible as a ratio to the average income of rural nonfarm women in the state. Note that this is a sample of states, so suggesting adding a variable relating to specific women would create a mismatch in the data-set.

d. Theory: if L is the best proxy available for the relative income of farm women, then it has a strong theoretical basis until the preferred variable can be found.

t-score: insignificant at the 5 percent level (but significant at the 10 percent level)

$\bar{R}^2$: $\bar{R}^2$ is not given, but it turns out that the deletion of any variable with a t-score greater than one in absolute value will lower $\bar{R}^2$, in this case not by much.

Bias: none of the coefficients change significantly.

Thus the four criteria are inconclusive. As long as relative income data are unavailable, L probably should be retained in the equation.

Chapter Seven

7-2. a. Semi-log [where $Y = f(\ln X)$]; as income increases, the sales of shoes will increase, but at a declining rate.

b. Linear (intercept dummy); there is little theory for any other form.

c. Semi-log (as in a. above) or linear are both justifiable.

d. Inverse function [where $Y = f(1/X)$]; as the interest rate gets higher, the quantity of money demanded will decrease, but even at very high interest rates there still will be some money held to allow for transactions.

e. Quadratic function [where $Y = f(X,X^2)$]; as output levels are increased, we will encounter diminishing returns to scale.

f. While functional form should be chosen on the basis of theory, one outlier is capable of shifting an estimated quadratic unreasonably; in such cases, a double-log function might avoid the problem.

7-4. a.

Coefficient:	β_1	β_2
Hypothesized sign:	+	+
t-value:	4.0	2.20

$t_c = 1.708$ at the 5%level, so H_0: $\beta \leq 0$ can be rejected for both.

b. It is the sum of the constant effect of omitted independent variables and the nonzero mean of the sample error term observations; it does not mean that salaries (logged) could be negative.

c. For this semi-log function, the elasticities are $\beta_1 ED_i$ and $\beta_2 EXP_i$ and the slopes are $\beta_1 SAL_i$ and $\beta_2 SAL_i$, which both increase as the Xs rise. This implies that a one-unit change in ED_i will cause a β_1 *percent* change in SAL_i, which makes sense for salaries.

d. The $\bar{R}^2$s cannot be compared because the dependent variables are different. To do so, you would need to calculate a "quasi-R^2."

7-6. a. To check your answer, compare your R^2s with those below.

b. The R^2 for the linear equation = .982, and the R^2 for the double-log equation = .971, but since the equations have different dependent variables, they are not directly compatible.

c. "Quasi-R^2" for the double-log equation = $1 - (541.9/8787.8) = .938$.

7-8. Let PCI_i = per capita income in the *i*th period
GR_i = rate of growth in the *i*th period (ϵ_i = a classical error term)
a. $GR_i = \beta_0 + \beta_1 PCI_i + \beta_2 D_i + \beta_3 D_i PCI_i + \epsilon_i$ where $D_i = 0$ if $PCI_i \leq \$2{,}000$ and $D_i = 1$ if $PCI_i > \$2{,}000$.
b. $GR_i = \alpha_0 + \alpha_1 PCI_i + \alpha_2 PCI_i^2 + \epsilon_i$ where we'd expect $\alpha_1 > 0$ and $\alpha_2 < 0$.
c. A semi-log function alone cannot change from positive to negative slope, so it is not appropriate.

7-10. a. The expected signs are β_1, + or ?; β_2, + ; β_3, + ; β_4, + .
b. AD_i/SA_i: the inverse form implies that the larger sales are, the smaller will be the impact of advertising on profits. CAP_i, ES_i, DG_i: the semi-log functional form implies that as each of these variables increases (holding all others constant), PR increases at a decreasing rate.
c. β_2, β_3, and β_4 all have positive expected signs, so $(+) \cdot (+) = (+)$ = positive expected bias on β_1 if one of the other Xs were omitted.

7-12. a. The estimated coefficients all are in the expected direction, and those for A and S are significant. $\bar{R}^2$ seems fairly low, even for a cross-sectional data-set of this nature.
b. It implies that wages rise and then fall with respect to age but does not imply perfect collinearity.
c. With a semi-log functional form (lnY), a slope coefficient represents the percentage change in the dependent variable caused by a one-unit change in the independent variable (holding constant all the other independent variables). Since pay raises are often discussed in percentage terms, such a functional form frequently is used to model wage rates and salaries.
d. It's a good habit to ignore $\hat{\beta}_0$ (except to make sure that one exists) even if it looks too large or too small.
e. The poor fit and the insignificant estimated coefficient of union membership are all reasons for being extremely cautious about using this regression to draw any conclusions about union membership.

7-14. a. 43.56, 3.673

c. The semi-log version has a lower RSS, 0.062133 to 0.203649, even though the linear version has a higher R^2.

d. No. The Box-Cox test only tells you which functional form fits better, not which form to use. In some cases, a functional form that makes little or no sense will fit better than the theoretically sound one because of the influence of a few datapoints.

Chapter Eight

8-2. a., c.

8-4. Likely dominant variables = a and d. In **a.** # of games won = # of games played (which is a constant) − # of games lost, while in **d.**, # of autos = (# of tires bought)/(# of tires per car, which = 4 if no spare is sold with the cars or = 5 if a spare is included).

8-6. a.

Coefficient:	β_F	β_S	β_A
Hypothesized sign:	+	+	+
t-value:	2.90	−1.07	5.97
t_c = 1.699 at the 5% level, so:	signif.	insig. unexp. sign	signif.

b. All three are possibilities.

c. Multicollinearity is a stronger possibility.

d. Yes; the distribution of the $\hat{\beta}$s is wider with multicollinearity.

8-8. a. Don't change your regression just because a fellow student says you are going to have a problem; in particular, even if you do have multicollinearity, you may well end up doing nothing about it.

b. There is a relatively high $\bar{R}^2$ ($\bar{R}^2$ = .75) while all the estimated coefficients are insignificant at the 10 percent level. Furthermore, the simple correlation coefficient between HR and RBI of .90 is significant at the 1 percent level ($t = 5.06 > t_c = 3.143$).

c. Since multicollinearity is a sample problem, the best solution here would be to try to increase the sample size (more than eight baseball players received MVP votes).

8-10. a.

Coefficient:	β_C	β_P	β_E
Hypothesized sign:	+	+	+
t-value:	31.15	−0.07	−0.85
$t_c = 1.69$ at the 5% level, so:	signif.	insig.	insig.
		unexpected signs	

b. From the information given, omitted variables, irrelevant variables, and multicollinearity are all possible problems in this equation.

c. Yes; with the high $\bar{R}^2$, the low t-values for $\hat{\beta}_P$ and $\hat{\beta}_E$, and the high simple correlation coefficient between the two, there is definite multicollinearity in the equation. The high correlation coefficient between the dependent variable and C_i would not be evidence of multicollinearity, but it is cause for a reexamination of the definitions of the two variables just to make sure that they are not tautologically related.

d. Since contracts often last for several years and can vary greatly in how much they are worth, C_i does not seem to be a dominant variable. The payroll for defense workers and the number of civilians employed in defense industries are redundant, however; they measure the same thing. As a result, one or the other should be dropped.

8-12. a. 2.35, 2.50, 1.18

b. 9.12, 1.88, 10.53

8-14. a.

Coefficient:	β_Y	β_{Y^2}	β_H	β_A
Hypothesized sign:	+	−	+	+
Calculated t-score:	3.00	−0.80	6.50	−1.00
$t_c = 1.282$, so:	signif.	insig.	signif.	insig. unexpected sign

b. The functional form appears reasonable. The coefficient of Y can be greater than 1.0 since Y^2 is in the equation with a negative coefficient.

c. A and H seem potentially redundant.

d. The high VIFs strengthen the answer.

e. Either drop A or, if the purpose behind A was to measure the differential eating habits of children, change the two variables to A and (H − A).

Hints for Section 8.7.2: The SAT Interactive Regression Learning Exercise:

1. Severe multicollinearity between APMATH and APENG is the only possible problem in this regression. You should switch to the AP linear combination immediately.

2. An omitted variable is a distinct possibility, but be sure to choose the one to add on the basis of theory.

3. Either an omitted or irrelevant variable is a possibility. In this case, theory seems more important than any mild statistical insignificance.

4. On balance, this is a reasonable regression. We see no reason to worry about theoretically sound variables that have slightly insignificant coefficients with expected signs. We're concerned that the coefficient of GEND seems larger in absolute size than those reported in the literature, but none of the specification alternatives seems remotely likely to remedy this problem.

5. An omitted variable is a possibility, but there are no signs of bias and this is a fairly reasonable equation already.

6. We'd prefer not to add PREP (since many students take prep courses because they did poorly on their first shots at the SAT) or RACE (because of its redundancy with ESL and the lack of real diversity at Arcadia High). If you make a specification change, be sure to evaluate the change with our four specification criteria.

7. Either an omitted or irrelevant variable is a possibility, although GEND seems theoretically and statistically strong.

8. The unexpected sign makes us concerned with the possibility that an omitted variable is causing bias or that PREP is irrelevant. If PREP is relevant, what omission could have caused this result? How strong is the theory behind PREP?

9. This is a case of imperfect multicollinearity. Even though the VIFs are only between 3.8 and 4.0, the definitions of ESL and RACE (and the high simple correlation coefficient between them) make

them seem like redundant variables. Remember to use theory (and not statistical fit) to decide which one to drop.

10. An omitted variable or irrelevant variable is a possibility, but there are no signs of bias and this is a fairly reasonable equation already.

11. Despite the switch to the AP linear combination, we still have an unexpected sign, so we're still concerned with the possibility that an omitted variable is causing bias or that PREP is irrelevant. If PREP is relevant, what omission could have caused this result? How strong is the theory behind PREP?

12. All of the choices would improve this equation except switching to the AP linear combination. If you make a specification change, be sure to evaluate the change with our four specification criteria.

13. To get to this result, you had to have made at least three suspect specification decisions, and you're running the risk of bias due to a sequential specification search. Our advice is to stop, take a break, review Chapters 6–8, and then try this interactive exercise again.

14. We'd prefer not to add PREP (since many students take prep courses because they did poorly on their first shots at the SAT) or ESL (because of its redundancy with RACE and the lack of real diversity at Arcadia High). If you make a specification change, be sure to evaluate the change with our four specification criteria.

15. Unless you drop one of the redundant variables, you're going to continue to have severe multicollinearity.

16. From theory and from the results, it seems as if the decision to switch to the AP linear combination was a waste of a regression run. Even if there were severe collinearity between APMATH and APENG (which there isn't), the original coefficients are significant enough in the expected direction to suggest taking no action to offset any multicollinearity.

17. On reflection, PREP probably should not have been chosen in the first place. Many students take prep courses only because they did poorly on their first shots at the SAT or because they anticipate doing poorly. Thus, even if the PREP courses improve SAT scores, which they probably do, the students who think they need to take them were otherwise going to score worse than their colleagues

(holding the other variables in the equation constant). The two effects seem likely to offset each other, making PREP an irrelevant variable. If you make a specification change, be sure to evaluate the change with our four specification criteria.

18. Either adding GEND or dropping PREP would be a good choice, and it's hard to choose between the two. If you make a specification change, be sure to evaluate the change with our four specification criteria.

19. On balance, this is a reasonable regression. We'd prefer not to add PREP (since many students take prep courses because they did poorly on their first shots at the SAT), but the theoretical case for ESL (or RACE) seems strong. We're concerned that the coefficient of GEND seems larger in absolute size than those reported in the literature, but none of the specification alternatives seem remotely likely to remedy this problem. If you make a specification change, be sure to evaluate the change with our four specification criteria.

Chapter Nine

9-2. a. Reject H_0 of no positive serial correlation ($d < d_L = 1.03$).

b. Cannot reject H_0 of no positive serial correlation ($d > d_U = 1.25$).

c. Inconclusive ($d_L = 1.07 < d < 1.83 = d_U$).

d. Inconclusive ($4 - d_U = 4 - 1.63 = 2.37 < d < 4 - d_L = 4 - 1.13 = 2.87$).

e. Cannot reject H_0 of no positive serial correlation ($d > d_U = 1.57$).

f. Reject H_0 of no serial correlation ($d < d_L = 1.04$).

g. Inconclusive ($d_L = 0.90 < d < 1.99 = d_U$).

9-4. a. $\beta_0^* = \beta_0(1 - \hat{\rho})$, so to get β_0, divide β_0^* by $(1 - \hat{\rho})$.

b. To account for the fact that the equation was estimated with GLS.

c. $\hat{\beta}_0 = -67.2/(1 - 0.69) = -216.8$.

d. $\hat{\beta}_0 = 1.48/(1 - 0.762) = 6.22$.

e. The equations are inherently different, and different equations can have drastically different constant terms, because β_0 acts as a "garbage collector" for the equation it is in. As a result, we should not analyze the estimated values of the constant term.

9-6. The same test applies, but the inconclusive region has expanded because of the small sample size and the large number of explanatory variables. As a result, even if the DW d = 2, you cannot conclude that there is no positive serial correlation.

9-8. a. Except for the first and last observations in the sample, the DW test's ability to detect first-order serial correlation is unchanged.

b. GLS can be applied mechanically to correct for serial correlation, but this procedure generally does not make sense; this time's error term is now hypothesized to be a function of *next* time's error term.

c. First-order serial correlation in data that have been entered in reverse chronological order means that this time's error term is a function of next time's error term. This might occur if, for example, the decision makers accurately predict and adjust to future random events before they occur, which would be the case in a world of rational expectations and perfect information.

9-10. a.

$$\hat{F}_t = 13.99 - \underset{\substack{(1.840) \\ t = -0.38}}{0.700RP_t} + \underset{\substack{(0.250) \\ 3.41}}{0.854D_t} \qquad \bar{R}^2 = .288 \quad n = 25 \quad DW = 1.247$$

b. The relative price coefficient is now insignificant, and the dummy variable is now significant in the unanticipated direction (that is, the Pope's decision significantly *increased* the fish consumption). In addition the DW is inconclusive in testing for serial correlation, but the DW of 1.247 is quite close to the d_L of 1.21 (for a 10 percent two-sided level of significance). Thus the omitted variable has not only caused bias, it also has moved the DW d just about into the positive serial correlation range.

c. This exercise is a good example of why it makes sense to search for specification errors before adjusting for serial correlation.

9-12. a.

Coefficient:	$\beta_{\ln Y}$	β_{PB}	β_{PP}	β_D
Hypothesized sign:	+	−	+	−
t_k:	6.6	−2.6	2.7	−3.7

at 5 percent $t_c = 1.714$, so all four are significantly different from zero in the expected direction.

b. With a 5 percent, one-sided test and n = 28, k′ = 4, the critical values are d_L = 1.10 and d_U = 1.75. Since d = 0.94 < 1.10, we can reject the null hypothesis of no positive serial correlation.
c. The probable serial correlation suggests GLS.
d. We prefer the GLS equation because we've rid the equation of much of the serial correlation while retaining estimated coefficients that make economic sense. Note that the dependent variables in the two equations are different, so an improved fit is not evidence of a better equation.

9-14. a. With a 1 percent, one-sided test and n = 19, k′ = 1, the critical values are d_L = 0.93 and d_U = 1.13. Since d = 0.48 < 0.93, we can reject the null hypothesis of no positive serial correlation. (Impure serial correlation caused by an incorrect functional form tends to be positive.)
b. See the answer to Exercise 1-8c.
c. 1.22 > 1.13, so we can't reject the null hypothesis.
d. 9.31, but, as we'll learn in Chapter 13, neither equation is perfect because the $\hat{P}$s are not limited by zero and one even though in theory they should be.

Chapter Ten

10-2. a. LIKELY: the number of professors, the number of undergraduates.
b. LIKELY: aggregate gross investment, population.
c. LIKELY: U.S. disposable income, population, and, less likely but still possible, U.S. per capita disposable income.

10-4. a. At the 1 percent level, t_c = 2.787; reject the null hypothesis of homoskedasticity.
b. At the 1 percent level, t_c = 4.032; t = 1.30, cannot reject null hypothesis of homoskedasticity.
c. It depends on the underlying theory that led you to choose Z as a good proportionality factor. If you believe that the absolute value of Z is what makes the variance of ϵ large, then there is no difference between −200 and +200. On the other hand, if you believe that the *relative* value of Z is important, then you are forced to add a constant (greater than 200) to each Z (which changes the nature of Z) and run the Park test.

d. At the 1 percent level, $t_c = 3.012$; $t = 0.666$, and we cannot reject the null hypothesis of homoskedasticity.

10-6. $\epsilon_i = u_i Z_i$, so $VAR(\epsilon_i) = VAR(u_i Z_i) = E[u_i Z_i - E(u_i Z_i)]^2$. Since u_i is a classical error term, $E(u_i) = 0$ and u_i is independent of Z_i, so $E(u_i Z_i) = 0$ and $VAR(\epsilon_i) = E(u_i Z_i)^2 = E[(u_i^2)(Z_i^2)] = Z_i^2 E(u^2_i) = \sigma^2 Z_i^2$ (since Z_i is constant with respect to ϵ_i.).

10-8. a. $GQ = 57.97 > 6.51 = F_c$, so reject H_0 of homoskedasticity.

b. $L = 95.37 > 11.34$ = critical Chi-square value, so reject H_0 of homoskedasticity.

c. $nR^2 = 33.226 > 15.09$ = critical Chi-square value, so reject H_0 of homoskedasticity. Thus all three tests agree with the Park test result.

10-10. a.

Coefficient:	β_1	β_2	β_3	β_4
Hypothesized sign:	+	+	+	+
t-value:	7.62	2.19	3.21	7.62
$t_c = 1.645$ (5% level) so:	signif.	signif.	signif.	signif.

b. Some authors suggest the use of a double-log equation to avoid heteroskedasticity because the double-log functional form compresses the scales on which the variables are measured, reducing a tenfold difference between two values to a twofold difference.

c. A reformulation of the equation in terms of output per acre (well, stremmata) would likely produce homoskedastic error terms.

d. Assuming the heteroskedastic error term is $\epsilon_i = Z_i u_i$, where u_i is a homoskedastic error term, Z_i is the proportionality factor, and $Z_i = X_{1i}$, then the equation to estimate is:

$$Y_i/X_{1i} = \beta_0/X_{1i} + \beta_1 + \beta_2 X_{2i}/X_{1i} + \beta_3 X_{3i}/X_{1i} + \beta_4 X_{4i}/X_{1i} + u_i.$$

10-12. a. $Y_i/\sqrt{X_{1i}} = \alpha_0 + \beta_0/\sqrt{X_{1i}} + \beta_1\sqrt{X_{1i}} + \beta_2 X_{2i}/\sqrt{X_{1i}} + u_i.$

b. $Y_i/X_{3i} = \alpha_0 + \beta_0/X_{3i} + \beta_1 X_{1i}/X_{3i} + \beta_2 X_{2i}/X_{3i} + u_i.$

c. $Y_i/\hat{Y}_i = \alpha + \beta_0/\hat{Y}_i + \beta_1 X_{1i}/\hat{Y}_i + \beta_2 X_{2i}/\hat{Y}_i + u_i.$

10-14. a. Heteroskedasticity is still a theoretical possibility. Young pigs are much more likely to grow at a high percentage rate than are old ones, so the variance of the error terms for young pigs might be greater than that of the error terms for old pigs.

b. Yes, $|-6.31|$ is greater than the two-tailed one percent t_c of 2.576.

c. An analysis of the sign of the coefficient can be useful in deciding how to correct any heteroskedasticity. In this case, the variance of the error term *decreases* as the proportionality factor increases, so dividing the equation again by weight wouldn't accomplish much.

d. One possibility would be to regroup the sample into three subsamples by age and rerun the WLS specification on each. This is an unusual solution, but since the sample is so large, it's a feasible method of obtaining more homogeneous groups of pigs.

Chapter Eleven

Hints for Section 11.4.2: The Passbook Deposits Interactive Exercise:

1. Serial correlation, omitted variables, and irrelevant variables are all possible, but the obvious problem is severe multicollinearity between QYDUS and QYPERM (the coefficients are much less significant than would have been expected for such important variables, and the simple correlation coefficient is .999). Any expected signs in such circumstances are merely the result of chance correlations.

2. Serial correlation and irrelevant variables are both possible, but the unexpected sign for QRDPASS raises the question of an omitted variable or a switch to the spread interest rate formulation. Does economic theory argue in favor of or against the spread formulation? Are there any potential omitted variables that could cause negative bias?

3. If you did anything but drop QYPERM (you're required to keep QYDUS because of the way the example is set up), you probably will not be very happy with your results, and you will have wasted a regression run.

4. Serial correlation, omitted variables, and irrelevant variables are all possible, but the insignificant interest rate variables and the unexpected sign for QRDPASS raise the question of a switch to

the spread interest rate formulation. Does economic theory argue in favor of or against such a switch?

5. Dropping EXPINF is not a terrible mistake, but it is too soon to tell, given the potential omitted variable problems. The switch to SPREAD seems much more pressing at this point.

6. Adding BRANCH is a poor idea, since leaving it out is unlikely to have caused the unexpected sign (+ · + = + bias), but adding EXPINF at least could potentially solve the unexpected sign on the coefficient of QRDPASS (+ · − = − bias). The switch to SPREAD also seems likely to get at that problem, but this particular choice is not obvious from the regression results. It really comes down to your prior economic thinking: which of the two variables (EXPINF or SPREAD) makes more sense?

7. Serial correlation, omitted variables, and irrelevant variables are all possible, but if there is serial correlation, it almost *has* to be at least partly impure serial correlation because the DW is virtually zero. This means we should focus on an omitted variable or a switch to the spread interest rate formulation. Does economic theory argue in favor of or against the spread formulation? Are there any potential omitted variables that could cause negative bias?

8. Adding MMCDUM seems to be the best route from a theoretical point of view; the empirical results support this choice, since the omission of a dummy variable that equals zero for the first part of the sample and one thereafter would be quite likely to cause the kind of positive (impure) serial correlation indicated by the Durbin-Watson statistic.

9. Adding BRANCH is a poor idea, since the theory behind it is weaker than the theory behind SPREAD or MMCDUM, and adding MMCDUM could potentially solve the apparent negative bias on the coefficient of QRDPASS (+ · − = − bias). The switch to SPREAD also seems likely to get at that problem, but this particular choice is not obvious from the regression results. It really comes down to your prior economic thinking: which of the two variables (MMCDUM or SPREAD) makes more sense?

10. If one of the variables is irrelevant, dropping it will not fix any bias problem. Adding MMCDUM could potentially solve the ap-

parent negative bias of the coefficient of QRDPASS (+ · − = − bias). The switch to SPREAD also seems likely to get at that problem, but this particular choice is not obvious from the regression results. It really comes down to your prior economic thinking: which of the two variables (MMCDUM or SPREAD) makes more sense?

11. Dropping BRANCH makes more sense than dropping EXPINF because the theory is weaker for BRANCH. While it is true that the coefficient of BRANCH has an unexpected sign, that result is insignificant and should not make you more likely to drop BRANCH than EXPINF for that reason alone. Extra branches would help an individual savings and loan association compete with other associations, but the overall impact might be expected to net out to zero because variations in return and income seem far more important than ease of deposit in choosing a portfolio. In addition, having more branches also makes it easier to withdraw funds as well as to deposit them!

12. Serial correlation, omitted variables, and irrelevant variables are all possible, but if there is serial correlation, it almost *has* to be at least partly impure serial correlation because the DW is virtually zero. This means we should focus on an omitted variable instead of worrying about an irrelevant variable right now.

13. Serial correlation and omitted variables are both possible.

14. Serial correlation and omitted variables are both possible, but if there is serial correlation, it almost *has* to be at least partly impure serial correlation because the DW is virtually zero. This means we should focus on an omitted variable; which one should you consider adding?

15. It is unclear whether EXPINF belongs in the equation. In the final analysis, it depends on whether the expectation of inflation will make asset holders more sensitive to interest rate differentials and therefore less likely to leave a portion of their assets in relatively low-earning passbook accounts. The choice between these two equations becomes a matter of the strength of your prior belief in EXPINF; there is not always a "best" answer in the art of econometrics!

16. It is vital that you drop MMCDUM immediately; as long as you leave it in the equation, you will have higher R^2s and a significant negative sign.

17. Serial correlation, irrelevant variables, and omitted variables are all possible problems.

18. Dropping BRANCH makes a lot of sense, since the theory behind it is weak. While extra branches would help an individual savings and loan association compete with other associations, the overall impact might be expected to net out to zero because variations in return and income seem far more important than ease of deposit in choosing a portfolio. In addition, having more branches also makes it easier to withdraw funds as well as to deposit them!

Hints for Section 11.6.2: The Demand for Pork Interactive Exercise:

19. Serial correlation and omitted variables are both possible.

20. Serial correlation and irrelevant variables are both possible.

21. Omitted variables should always be tackled before serial correlation because of the possibility of impure serial correlation.

22. Omitted variables and irrelevant variables are both possible.

23. Serial correlation, omitted variables, and irrelevant variables are all possible.

24. Income could be irrelevant, or its coefficient could be negatively biased due to an omitted variable. Before you go too far, though, review the theory behind PROPK.

25. An omitted variable is possible; what is it?

26. Serial correlation may well be the biggest remaining problem.

27. Deciding what to add is not easy, but if you added PROPK you should carefully review the theory behind that variable.

28. Income could be irrelevant, or its coefficient could be negatively biased due to an omitted variable. Before you go too far, though, review the theory behind PROPK. How should the two insignificant seasonal dummy variables be handled?

29. The theory behind choosing PROPK instead of an income variable needs to be reviewed carefully. Does supply create its own demand *without* price changes?

30. Note that DW > 2 indicates negative serial correlation, which is usually a sign of specification error. In addition, the theory behind choosing PROPK instead of an income variable needs to be reviewed carefully. Does supply create its own demand *without* price changes?

Hints for Section 11.7.2: The Housing Price Interactive Exercise:

The biggest problem most students have with this interactive exercise is that they run far too many different specifications "just to see" what the results look like. In our opinion, all but one or two of the specification decisions involved in this exercise should be made before the first regression is estimated, so one measure of the quality of your work is the number of different equations you estimated. Typically, the fewer the better.

As to which specification to run, most of the decisions involved are matters of personal choice and experience. Our favorite model on theoretical grounds is:

$$P = f(\overset{+}{S}, \overset{-}{N}, \overset{-}{A}, \overset{+}{A^2}, \overset{+}{Y}, \overset{+}{C})$$

We think that BE and BA are redundant with S. In addition, we can justify both positive and negative coefficients for P, giving it an ambiguous expected sign, so we'd avoid including it. We would not quibble with someone who preferred a linear functional form for A to our quadratic. In addition, we recognize that C is quite insignificant for this sample, but we'd retain it, at least in part because it gets quite hot in Monrovia in the summer.

As to interactive variables, the only one we can justify is between S and N. Note, however, that the proper variable is not $S \cdot N$ but instead is $S \cdot (5 - N)$—or something similar—to account for the different expected signs. This variable turns out to improve the fit while being quite collinear (redundant) with N and S.

In none of our specifications did we find evidence of serial correlation or heteroskedasticity, although the latter is certainly a possibility in such cross-sectional data.

Chapter Twelve

12-2. a. $\hat{Y}_t \approx 13.0 + 12.0X_t + 0.48X_{t-1} + 0.02X_{t-2}$
(smoothly decreasing impact)

b. $\hat{Y}_t \approx 13.0 + 12.0X_t + 0.96X_{t-1} + 0.08X_{t-2} + 0.01X_{t-3}$
(smoothly decreasing impact)

c. $\hat{Y}_t \approx 13.0 + 12.0X_t + 24.0X_{t-1} + 48.0X_{t-2} + \cdots$
(explosively positive impact)

d. $\hat{Y}_t \approx 13.0 + 12.0X_t - 4.8X_{t-1} + 1.92X_{t-2} - \cdots$
(damped oscillating impact)

e. $0 < \lambda < 1$

12-4. a.

Coefficient:	β_{Pt}	β_{Pt-1}	β_U
Hypothesized sign:	+	+	−
t-value:	4.55	0.06	−3.89
$t_c = 1.341$, so	signif.	insig.	signif.

b. The hypothesis being tested here is that the impact of a change in price on wages is distributed over time rather than instantaneous. Such a distributed lag (in this case ad hoc) could occur because of long-term contracts, slowly adapting expectations, and do forth. P_{t-1} is extremely insignificant in explaining W, but it's not obvious that it should be dropped from the equation. Collinearity might be the culprit, or the lag involved may be more or less than a year. In the latter case, it would not be a good idea to test many different lags on the same data-set, but if another data-set could be developed, such tests (scans) would probably be useful.

c. The equation would no longer be an ad hoc distributed lag equation.

12-6. a. $$Y_t = a + \beta_0 X_t + \alpha_0(X_{t-1} + X_{t-2} + X_{t-3}) + \alpha_1(X_{t-1} + 2X_{t-2} + 3X_{t-3}) + \alpha_2(X_{t-1} + 4X_{t-2} + 9X_{t-3}) + \epsilon_t$$

where: $b = \hat{a}$, $\beta_0 = \hat{\alpha}_0$, $\beta_1 = \hat{\alpha}_0 + \hat{\alpha}_1 + \hat{\alpha}_2$,
$\beta_2 = \hat{\alpha}_0 + 2\hat{\alpha}_1 + 4\hat{\alpha}_2$, $\beta_3 = \hat{\alpha}_0 + 3\hat{\alpha}_1 + 9\hat{\alpha}_2$

b. $Y_t = a + \beta_0 X_t + \alpha_0(X_{t-1} + X_{t-2} + X_{t-3} + X_{t-4})$
$+ \alpha_1(X_{t-1} + 2X_{t-2} + 3X_{t-3} + 4X_{t-4})$
$+ \alpha_2(X_{t-1} + 4X_{t-2} + 9X_{t-3} + 16X_{t-4})$
$+ \alpha_3(X_{t-1} + 8X_{t-2} + 27X_{t-3} + 64X_{t-4}) + \epsilon_t$

where: $b = \hat{a}$, $\beta_0 = \hat{\alpha}_0$, $\beta_1 = \hat{\alpha}_0 + \hat{\alpha}_1 + \hat{\alpha}_2 + \hat{\alpha}_3$,
$\beta_2 = \hat{\alpha}_0 + 2\hat{\alpha}_1 + 4\hat{\alpha}_2 + 8\hat{\alpha}_3$,
$\beta_3 = \hat{\alpha}_0 + 3\hat{\alpha}_1 + 9\hat{\alpha}_2 + 27\hat{\alpha}_3$,
$\beta_4 = \hat{\alpha}_0 + 4\hat{\alpha}_1 + 16\hat{\alpha}_2 + 64\hat{\alpha}_3$

c. $Y_t = a + \beta_0 X_t$
$+ \alpha_0(X_{t-1} + X_{t-2} + X_{t-3} + X_{t-4} + X_{t-5})$
$+ \alpha_1(X_{t-1} + 2X_{t-2} + 3X_{t-3} + 4X_{t-4} + 5X_{t-5})$
$+ \alpha_2(X_{t-1} + 4X_{t-2} + 9X_{t-3} + 16X_{t-4} + 25X_{t-5})$
$+ \alpha_3(X_{t-1} + 8X_{t-2} + 27X_{t-3} + 64X_{t-4} + 125X_{t-5}) + \epsilon_t$

where: $b = \hat{a}$, $\beta_0 = \hat{\alpha}_0$, $\beta_1 = \hat{\alpha}_0 + \hat{\alpha}_1 + \hat{\alpha}_2, + \hat{\alpha}_3$,
$\beta_2 = \hat{\alpha}_0 + 2\hat{\alpha}_1 + 4\hat{\alpha}_2 + 8\hat{\alpha}_3$,
$\beta_3 = \hat{\alpha}_0 + 3\hat{\alpha}_1 + 9\hat{\alpha}_2 + 27\hat{\alpha}_3$,
$\beta_4 = \hat{\alpha}_0 + 4\hat{\alpha}_1 + 16\hat{\alpha}_2 + 64\hat{\alpha}_3$,
$\beta_5 = \hat{\alpha}_0 + 5\hat{\alpha}_1 + 25\hat{\alpha}_2 + 125\hat{\alpha}_3$

12-8. a. 6,2
b. 5,3

12-10. a. $h = 0.0372 < 1.96 = t_c$ (5 percent two-tailed), so we cannot reject the null hypothesis of no serial correlation. (Specific h values vary because the Durbin-Watson d is so close to 2.0, but H_0 can never be rejected.)
b. $LM = 0.135 < 3.84 = 5$ percent critical Chi-square value with one degree of freedom, so we cannot reject the null hypothesis of no serial correlation.

12-12. I Granger-causes Y: $F = 3.16 < 3.26 = F_c$ (5 percent) with 4 and 12 degrees of freedom, so we can't reject the null hypothesis that the coefficients of the lagged Is are jointly zero.

Y Granger-causes I: $F = 5.92 > 3.26 = F_c$, so we can reject the null hypothesis that the coefficients of the lagged Ys jointly equal zero.

These two results allow us to conclude that GNP does indeed Granger-cause investment, at least in this sample.

Chapter Thirteen

13-2. a. $D_i > 1$ if $X_i > 7$ and $D_i < 0$ if $X_i < -3$
b. $D_i > 1$ if $X_i < 10$ and $D_i < 0$ if $X_i > 15$
c. $D_i > 1$ if $X_i > 6.67$ and $D_i < 0$ if $X_i < 3.33$

13-4. Start with $\ln[D/(1-D)] = Z$ and take the anti-log, obtaining $D/(1-D) = e^Z$. Then cross-multiply and multiply out, which gives $D = e^Z - De^Z$. Then solve for $D = e^Z/(1 + e^Z)$. Finally, multiply the right-hand side by e^{-Z}/e^{-Z}, obtaining $D = 1/(1 + e^{-Z})$.

13-6. a. WN: The log of the odds that a woman has used a recognized form of birth control is 2.03 higher if she wants no more children than it is if she wants more children, holding ME constant.

ME: A one-unit increase in the number of methods of birth control known to a woman increases the log of the odds that she has used a recognized form of birth control by 1.45, holding WN constant.

LPM: If the model were a linear probability model, then each individual slope coefficient would represent the impact of a one-unit change in the independent variable on the probability that the *i*th woman had ever used a recognized form of birth control, holding the other independent variable constant.

b. Yes, although we did not expect $\hat{\beta}_{ME}$ to be more significant than $\hat{\beta}_{WN}$.
c. β_0 has no theoretical importance. It's fair to say, however, that in this particular case the two positive variable-coefficient pairs make it very unlikely indeed that we would observe a positive intercept.
d. We'd add one of a number of potentially relevant variables, for instance the educational level of the *i*th woman, whether or not the *i*th woman lives in a rural area, etc.

13-8. a. There are only two women in the sample over 65.
b. We prefer Equation 13.15 because AD gives every appearance of being an irrelevant variable, at least as measured by the four criteria developed in Chapter 6.

13-10. a. All signs meet expectations except that of wait time.
b. The fact that the estimated coefficient of walk time is larger in absolute value than that of travel time supports this hypothesis, but the large positive coefficient for wait time does not.
c. Yes, if train commuters know train schedules and actually adjust their station arrival to minimize wait time, then setting the wait time for trains high allowed wait time to become a proxy for being the preferred mode of travel in Boston.

Chapter Fourteen

14-2. a. All three equations are simultaneous.
Endogenous variables $= Y_{1t}, Y_{2t}, Y_{3t}$
Predetermined variables: $X_{1t}, X_{1t-1}, X_{2t-1}, X_{3t}, X_{4t}, X_{4t-1}$
b. All three equations are simultaneous.
Endogenous variables $= Z_t, X_t, H_t$
Predetermined variables: $Y_t, P_{t-1}, B_t, C_t, D_t$
c. The equations are recursive; solve for Y_2 first and use it to get Y_1.

14-4. a. $P_t = \pi_0 + \pi_1 X_{1t} + \pi_2 X_{2t} + \pi_3 X_{3t} + v_{1t}$

$Q_{St} = Q_{Dt} = \pi_4 + \pi_5 X_{1t} + \pi_6 X_{2t} + \pi_7 X_{3t} + v_{2t}$

b. Step one: Set the two structural quantity equations equal to each other and solve for P_t:

$$\alpha_0 + \alpha_1 P_t + \alpha_2 X_{1t} + \alpha_3 X_{2t} + \epsilon_{Dt} = \beta_0 + \beta_1 P_t + \beta_2 X_{3t} + \epsilon_{St}$$

$$P_t = (\beta_0 - \alpha_0)/(\alpha_1 - \beta_1) - [\alpha_2/(\alpha_1 - \beta_1)]\, X_{1t} - [\alpha_3/(\alpha_1 - \beta_1)]X_{2t} + [\beta_2/(\alpha_1 - \beta_1)]\, X_{3t} + [(\epsilon_{St} - \epsilon_{Dt})/(\alpha_1 - \beta_1)]$$

Step two: compare this equation with the first reduced-form equation in part a:

$$\pi_0 = (\beta_0 - \alpha_0)/(\alpha_1 - \beta_1);$$
$$\pi_1 = -\alpha_2/(\alpha_1 - \beta_1);$$
$$\pi_2 = -\alpha_3/(\alpha_1 - \beta_1);$$
$$\pi_3 = \beta_2/(\alpha_1 - \beta_1);$$
$$v_{1t} = (\epsilon_{St} - \epsilon_{Dt})/(\alpha_1 - \beta_1).$$

Step three: Substitute P_t into the structural Q_D equation, combine like terms, and compare this equation with the second reduced-form equation in part a:

$$\pi_4 = (\alpha_1\beta_0 - \alpha_0\beta_1)/(\alpha_1 - \beta_1);$$
$$\pi_5 = -\alpha_2\beta_1/(\alpha_1 - \beta_1);$$
$$\pi_6 = -\alpha_3\beta_1/(\alpha_1 - \beta_1);$$
$$\pi_7 = \alpha_1\beta_2/(\alpha_1 - \beta_1);$$
$$v_{2t} = (\alpha_1\epsilon_{St} - \beta_1\epsilon_{Dt})/(\alpha_1 - \beta_1).$$

Step four: Rearrange and solve simultaneously for the αs:

$$\alpha_0 = \pi_4 - \pi_0\alpha_1;$$
$$\alpha_1 = \pi_7/\pi_3;$$
$$\alpha_2 = \pi_5 - \pi_1\alpha_1;$$
$$\alpha_3 = \pi_6 - \pi_2\alpha_1$$

c. First, the equation needs to be exactly identified; to see why this is so, try to solve for the βs of the overidentified Equation 14.5. Second, for equations with more than one or two slope coefficients, it is very awkward and time-consuming to use indirect least squares. Third, 2SLS gives the same estimates in this case, and 2SLS is much easier to apply.

14-6. a. There are 3 predetermined variables in the system, and both equations have 3 slope coefficients, so both equations are exactly identified. (If the model specified that the price of beef was determined jointly with the price and quantity of chicken, then it would not be predetermined, and the equations would be underidentified.)

b. There are 2 predetermined variables in the system, and both equations have 2 slope coefficients, so both equations are exactly identified.

c. There are 6 predetermined variables in the system, and there are 3 slope coefficients in each equation, so all 3 equations are overidentified.

d. There are 5 predetermined variables in the system, and there are 3, 2, and 4 slope coefficients in the first, second, and third equations, respectively, so all 3 equations are overidentified.

14-8. Stage one: Apply OLS to the second of the reduced-form equations:

$$Q_{St} = Q_{Dt} = \pi_0 + \pi_1 X_{1t} + \pi_2 X_{2t} + \pi_3 X_{3t} + v_{1t}$$
$$P_t = \pi_4 + \pi_5 X_{1t} + \pi_6 X_{2t} + \pi_7 X_{3t} + v_{2t}.$$

Stage two: Substitute the reduced-form estimates of the endogenous variables for the endogenous variables that appear on the right side of the structural equations. This would give:

$$Q_{Dt} = \alpha_0 + \alpha_1 \hat{P}_t + \alpha_2 X_{1t} + \alpha_3 X_{2t} + u_{Dt}$$
$$Q_{St} = \beta_0 + \beta_1 \hat{P}_t + \beta_2 X_{3t} + u_{St}$$

To complete stage two, estimate these revised structural equations with OLS.

14-10. a. You don't know that OLS and 2SLS will be the same until the system is estimated with both.

b. Not necessarily. It indicates only that the fit of the reduced-form equation from stage one is excellent and that $\hat{Y}$ and Y are virtually identical. Since bias is only a general tendency, it does not show up in every single estimate; indeed, it is possible to have estimated coefficients in the opposite direction. That is, even though positive bias exists with OLS, an estimated coefficient less than the true coefficient can be produced.

14-12. a. The serial correlation is so severe that it can be detected by the Durbin-Watson d-test even though that statistic is biased towards two. $DW = 0.75 < 1.21 = d_L$ for $n = 25$, $k' = 2$ at a 5 percent level of significance. This circumstance is especially useful, because Durbin's h is undefined for Equation 14.30.

b. Since the OLS and 2SLS estimates of this equation are similar, and since the serial correlation is quite severe, we'd choose to correct for serial correlation if we could only correct for one problem.

c. We prefer a modification of a procedure first suggested by Fair.[1] This involves estimating a reduced form for YD_t that includes C_{t-2} and YD_{t-1} on the right-hand side, and then substituting $\widehat{YD}_t$ into a GLS equation. This approach might be called "2SLS/GLS" since the 2SLS portion of the procedure is carried out before the GLS portion.

1. R. C. Fair, "The Estimation of Simultaneous Equation Models with Lagged Endogenous Variables and First-Order Serially Correlated Errors," *Econometrica*, 1970, pp. 507–516.

14-14. a. QU: −, −, −, +, +, +
UR: +, +, +, +, +

b. Yes, since UR and QU are jointly determined in this system.

c. This tells us that the UR equation is exactly identified but tells us nothing about the identification properties of the QU equation.

d. The lack of significance makes us wonder if UR and QU are indeed simultaneously determined. We should be hesitant to jump to this conclusion, however, because: one, the theory indicates simultaneity, two, multicollinearity or other specification problems may be causing the insignificance, and three, the pooled cross-section/time-series data-set makes it difficult to draw inferences.

e. Given the above reservations, we should be cautious. However, the results tend to confirm the theory that states interested in lowering their unemployment rates and lowering their budget deficits might consider lowering their unemployment benefits.

Chapter Fifteen

15-2. a. 73.58; 77.31; 74.92; 63.49

b. 117,259; 132,859; 107,230; Nowheresville

c. 14.11; 14.23; 14.57; 14.24

15-4. a. 160.82 ± 17.53

b. 800 ± 344.73

15-6. a. $Y_t^* = 1, 1, 1, 1, 1, 1, 1, 1, 1, 1, 1$
$Y_t^{**} = 0, 0, 0, 0, 0, 0, 0, 0, 0, 0$ $(d = 1)$

b. $Y_t^* = 0, 1, 1, 1, 1, 2, 2, 2, 3, 4, 5$
$Y_t^{**} = 1, 0, 0, 0, 1, 0, 0, 1, 1, 1$ $(d = 2)$

c. $Y_t^* = 1, 3, -3, 1, -2, 1, 2, -4, 3, 0, 2$
$Y_t^{**} = 2, -6, 4, -3, 3, 1, -6, 7, -3, 2,$ $(d = 0)$

15-8.

		Model A	*Model T*
a.	1987	30.50	29.50
	1988	30.25	30.25
	1989	30.13	29.87
b.	1988	31.50	28.50
	1989	30.75	30.75

c. Model A should exhibit smoother behavior because of the negative coefficient in Model T.

15-10. a. Before we can forecast with ARIMA, we must specify the model by choosing p, d, and q and then estimate ARIMA (p,d,q).

The first step in specifying the model is to choose d by checking for stationarity. Based on the upward trend of the stock price data and the fact that the ACF of the series does not decline geometrically to zero, the series appears to be nonstationary. If first differences are taken, then the resulting PK* series does indeed appear to be stationary, indicating that d = 1.

To choose p and q, we examine the ACFs and PACFs of the first-differenced series PK*. The PACFs are all statistically insignificant, which implies p = 0. The ACFs have spikes which are significantly different from zero (at the 5 percent level) for lags 3 and 4, indicating that q = 4. Thus one reasonable specification is ARIMA(0,1,4). (Another might be to suppress the MA coefficients for one and two lags, but such a specification is beyond this text.)

If we estimate ARIMA(0,1,4) on the data, we obtain:

$$\widehat{PK}_t^* = 0.0199e_{t-1} - 0.0084e_{t-2} - 0.1725e_{t-3} + 0.6856e_{t-4}$$

If we now forecast with this equation, we obtain:

	Forecast	*Actual*	*Percent Error*
1977:4	24.36	24.38	0.1
1978:1	23.04	22.38	2.9
1978:2	23.73	23.00	3.2
1978:3	21.86	21.88	0.1

These results are significantly better than the regression-produced forecasts, indicating the potential of ARIMA for short-term forecasts of variables with little or no strong underlying theory. Not all ARIMA stock-market forecasts are this accurate, of course.

Appendix B

Statistical Tables

The following tables present the critical values of various statistics used primarily for hypothesis testing. The primary applications of each statistic are explained and illustrated. The tables are:

Table B-1: The t-Distribution

The t-distribution is used in regression analysis to test whether an estimated slope coefficient (say $\hat{\beta}_k$) is significantly different from a hypothesized value (such as β_{H0}). The t-statistic is computed as

$$t_k = (\hat{\beta}_k - \beta_{H0})/SE(\hat{\beta}_k)$$

where $\hat{\beta}_k$ is the estimated slope coefficient and $SE(\hat{\beta}_k)$ is the estimated standard error of $\hat{\beta}_k$. To test the one-sided hypothesis:

$$H_0\colon \beta_k \leq \beta_{H0}$$

$$H_A\colon \beta_k > \beta_{H0}$$

the computed t-value is compared with a critical t-value t_c, found in the t-table on the opposite page in the column with the desired level of significance for a one-sided test (usually 5 or 10 percent) and the row with $n-K-1$ degrees of freedom, where n is the number of observations and K is the number of explanatory variables. If $|t_k| > t_c$ and if t_k has the sign implied by the alternative hypothesis, then reject H_0; otherwise, do not reject H_0. In most econometric applications, β_{H0} is zero and most computer regression programs will calculate t_k for $\beta_{H0} = 0$. For example, for a 5 percent one-sided test with 15 degrees of freedom, $t_c = 1.753$, so any positive t_k larger than 1.753 would lead us to reject H_0 and declare that $\hat{\beta}_k$ is statistically significant in the hypothesized direction at the 95 percent level of confidence.

For a two-sided test, $H_0\colon \beta_k = \beta_{H0}$ and $H_A\colon \beta_k \neq \beta_{H0}$, the procedure is identical except that the column corresponding to the two-sided level of significance is used. For example, for a 5 percent two-sided test with 15 degrees of freedom, $t_c = 2.131$, so any t_k larger in absolute value than 2.131 would lead us to reject H_0 and declare that $\hat{\beta}_k$ is significantly different from β_{H0} at the 95 percent level of confidence.

Another use of the *t*-test is to determine whether a simple correlation coefficient (r) between two variables is statistically significant. That is, the null hypothesis of no correlation between two variables can be tested with:

$$t_r = r\sqrt{(n-2)}/\sqrt{(1-r^2)}$$

where n is the number of observations. This t_r is then compared with the appropriate t_c ($n-2$ degrees of freedom) using the methods outlined above. For more on the *t*-test, see Chapter 5.

TABLE B-1 CRITICAL VALUES OF THE t-DISTRIBUTION

Degrees of Freedom	Level of Significance				
One Sided:	10%	5%	2.5%	1%	0.5%
Two Sided:	20%	10%	5%	2%	1%
1	3.078	6.314	12.706	31.821	63.657
2	1.886	2.920	4.303	6.965	9.925
3	1.638	2.353	3.182	4.541	5.841
4	1.533	2.132	2.776	3.747	4.604
5	1.476	2.015	2.571	3.365	4.032
6	1.440	1.943	2.447	3.143	3.707
7	1.415	1.895	2.365	2.998	3.499
8	1.397	1.860	2.306	2.896	3.355
9	1.383	1.833	2.262	2.821	3.250
10	1.372	1.812	2.228	2.764	3.169
11	1.363	1.796	2.201	2.718	3.106
12	1.356	1.782	2.179	2.681	3.055
13	1.350	1.771	2.160	2.650	3.012
14	1.345	1.761	2.145	2.624	2.977
15	1.341	1.753	2.131	2.602	2.947
16	1.337	1.746	2.120	2.583	2.921
17	1.333	1.740	2.110	2.567	2.898
18	1.330	1.734	2.101	2.552	2.878
19	1.328	1.729	2.093	2.539	2.861
20	1.325	1.725	2.086	2.528	2.845
21	1.323	1.721	2.080	2.518	2.831
22	1.321	1.717	2.074	2.508	2.819
23	1.319	1.714	2.069	2.500	2.807
24	1.318	1.711	2.064	2.492	2.797
25	1.316	1.708	2.060	2.485	2.787
26	1.315	1.706	2.056	2.479	2.779
27	1.314	1.703	2.052	2.473	2.771
28	1.313	1.701	2.048	2.467	2.763
29	1.311	1.699	2.045	2.462	2.756
30	1.310	1.697	2.042	2.457	2.750
40	1.303	1.684	2.021	2.423	2.704
60	1.296	1.671	2.000	2.390	2.660
120	1.289	1.658	1.980	2.358	2.617
(Normal) ∞	1.282	1.645	1.960	2.326	2.576

Source: Reprinted from Table IV in Sir Ronald A. Fisher, *Statistical Methods for Research Workers,* 14th ed. (copyright © 1970, University of Adelaide) with permission of Hafner, a Division of the Macmillan Publishing Company, Inc.

Table B-2: The F-Distribution

The F-distribution is used in regression analysis to test two-sided hypotheses about more than one regression coefficient at a time. To test the most typical joint hypothesis (a test of the overall significance of the regression):

$$H_0: \beta_1 = \beta_2 = \cdots = \beta_K = 0$$

$$H_A: H_0 \text{ is not true}$$

the computed F-value is compared with a critical F-value, found in one of the two tables that follow. The F-statistic has two types of degrees of freedom, one for the numerator (columns) and one for the denominator (rows). For the null and alternative hypotheses above, there are K numerator (the number of restrictions implied by the null hypothesis) and $n - K - 1$ denominator degrees of freedom, where n is the number of observations and K is the number of explanatory variables in the equation. This particular F-statistic is printed out by most computer regression programs. For example, if $K = 5$ and $n = 30$, there are 5 numerator and 24 denominator degrees of freedom, and the critical F-value for a 5 percent level of significance (Table B-2) is 2.62. A computed F-value greater than 2.62 would lead us to reject the null hypothesis and declare that the equation is statistically significant at the 95 percent level of confidence. For more on the F-test, see Sections 5.6 and 5.8.

TABLE B-2 CRITICAL VALUES OF THE F-STATISTIC: 5 PERCENT LEVEL OF SIGNIFICANCE

v_1 = Degrees of Freedom for Numerator

v_2 = degrees of freedom for denominator	1	2	3	4	5	6	7	8	10	12	20
1	161	200	216	225	230	234	237	239	242	244	248
2	18.5	19.0	19.2	19.2	19.3	19.3	19.4	19.4	19.4	19.4	19.4
3	10.1	9.55	9.28	9.12	9.01	8.94	8.89	8.85	8.79	8.74	8.66
4	7.71	6.94	6.59	6.39	6.26	6.16	6.09	6.04	5.96	5.91	5.80
5	6.61	5.79	5.41	5.19	5.05	4.95	4.88	4.82	4.74	4.68	4.56
6	5.99	5.14	4.76	4.53	4.39	4.28	4.21	4.15	4.06	4.00	3.87
7	5.59	4.74	4.35	4.12	3.97	3.87	3.79	3.73	3.64	3.57	3.44
8	5.32	4.46	4.07	3.84	3.69	3.58	3.50	3.44	3.35	3.28	3.15
9	5.12	4.26	3.86	3.63	3.48	3.37	3.29	3.23	3.14	3.07	2.94
10	4.96	4.10	3.71	3.48	3.33	3.22	3.14	3.07	2.98	2.91	2.77
11	4.84	3.98	3.59	3.36	3.20	3.09	3.01	2.95	2.85	2.79	2.65
12	4.75	3.89	3.49	3.26	3.11	3.00	2.91	2.85	2.75	2.69	2.54
13	4.67	3.81	3.41	3.18	3.03	2.92	2.83	2.77	2.67	2.60	2.46
14	4.60	3.74	3.34	3.11	2.96	2.85	2.76	2.70	2.60	2.53	2.39
15	4.54	3.68	3.29	3.06	2.90	2.79	2.71	2.64	2.54	2.48	2.33
16	4.49	3.63	3.24	3.01	2.85	2.74	2.66	2.59	2.49	2.42	2.28
17	4.45	3.59	3.20	2.96	2.81	2.70	2.61	2.55	2.45	2.38	2.23
18	4.41	3.55	3.16	2.93	2.77	2.66	2.58	2.51	2.41	2.34	2.19
19	4.38	3.52	3.13	2.90	2.74	2.63	2.54	2.48	2.38	2.31	2.16
20	4.35	3.49	3.10	2.87	2.71	2.60	2.51	2.45	2.35	2.28	2.12
21	4.32	3.47	3.07	2.84	2.68	2.57	2.49	2.42	2.32	2.25	2.10
22	4.30	3.44	3.05	2.82	2.66	2.55	2.46	2.40	2.30	2.23	2.07
23	4.28	3.42	3.03	2.80	2.64	2.53	2.44	2.37	2.27	2.20	2.05
24	4.26	3.40	3.01	2.78	2.62	2.51	2.42	2.36	2.25	2.18	2.03
25	4.24	3.39	2.99	2.76	2.60	2.49	2.40	2.34	2.24	2.16	2.01
30	4.17	3.32	2.92	2.69	2.53	2.42	2.33	2.27	2.16	2.09	1.93
40	4.08	3.23	2.84	2.61	2.45	2.34	2.25	2.18	2.08	2.00	1.84
60	4.00	3.15	2.76	2.53	2.37	2.25	2.17	2.10	1.99	1.92	1.75
120	3.92	3.07	2.68	2.45	2.29	2.18	2.09	2.02	1.91	1.83	1.66
∞	3.84	3.00	2.60	2.37	2.21	2.10	2.01	1.94	1.83	1.75	1.57

Abridged from M. Merrington and C. M. Thompson, "Tables of percentage points of the inverted beta (F) distribution," *Biometrika*, Vol. 33, 1943, p. 73. By permission of the *Biometrika* trustees.

Table B-3: The F-Distribution

The F-distribution is used in regression analysis to test two-sided hypotheses about more than one regression coefficient at a time. To test the most typical joint hypothesis (a test of the overall significance of the regression):

$$H_0\colon \beta_1 = \beta_2 = \cdots = \beta_K = 0$$

$$H_A\colon H_0 \text{ is not true}$$

the computed F-value is compared with a critical F-value, found in Tables B-2 and B-3. The F-statistic has two types of degrees of freedom, one for the numerator (columns) and one for the denominator (rows). For the null and alternative hypotheses above, there are K numerator (the number of restrictions implied by the null hypothesis) and $n - K - 1$ denominator degrees of freedom, where n is the number of observations and K is the number of explanatory variables in the equation. This particular F-statistic is printed out by most computer regression programs. For example, if $K = 5$ and $n = 30$, there are 5 numerator and 24 denominator degrees of freedom, and the critical F-value for a 1 percent level of significance (Table B-3) is 3.90. A computed F-value greater than 3.90 would lead us to reject the null hypothesis and declare that the equation is statistically significant at the 99 percent level of confidence. For more on the F-test, see Sections 5.6 and 5.8.

TABLE B-3 CRITICAL VALUES OF THE F-STATISTIC: 1 PERCENT LEVEL OF SIGNIFICANCE

v_2 = degrees of freedom for denominator	v_1 = Degrees of Freedom for Numerator										
	1	**2**	**3**	**4**	**5**	**6**	**7**	**8**	**10**	**12**	**20**
1	4052	5000	5403	5625	5764	5859	5928	5982	6056	6106	6209
2	98.5	99.0	99.2	99.2	99.3	99.3	99.4	99.4	99.4	99.4	99.4
3	34.1	30.8	29.5	28.7	28.2	27.9	27.7	27.5	27.2	27.1	26.7
4	21.2	18.0	16.7	16.0	15.5	15.2	15.0	14.8	14.5	14.4	14.0
5	16.3	13.3	12.1	11.4	11.0	10.7	10.5	10.3	10.1	9.89	9.55
6	13.7	10.9	9.78	9.15	8.75	8.47	8.26	8.10	7.87	7.72	7.40
7	12.2	9.55	8.45	7.85	7.46	7.19	6.99	6.84	6.62	6.47	6.16
8	11.3	8.65	7.59	7.01	6.63	6.37	6.18	6.03	5.81	5.67	5.36
9	10.6	8.02	6.99	6.42	6.06	5.80	5.61	5.47	5.26	5.11	4.81
10	10.0	7.56	6.55	5.99	5.64	5.39	5.20	5.06	4.85	4.71	4.41
11	9.65	7.21	6.22	5.67	5.32	5.07	4.89	4.74	4.54	4.40	4.10
12	9.33	6.93	5.95	5.41	5.06	4.82	4.64	4.50	4.30	4.16	3.86
13	9.07	6.70	5.74	5.21	4.86	4.62	4.44	4.30	4.10	3.96	3.66
14	8.86	6.51	5.56	5.04	4.70	4.46	4.28	4.14	3.94	3.80	3.51
15	8.68	6.36	5.42	4.89	4.56	4.32	4.14	4.00	3.80	3.67	3.37
16	8.53	6.23	5.29	4.77	4.44	4.20	4.03	3.89	3.69	3.55	3.26
17	8.40	6.11	5.19	4.67	4.34	4.10	3.93	3.79	3.59	3.46	3.16
18	8.29	6.01	5.09	4.58	4.25	4.01	3.84	3.71	3.51	3.37	3.08
19	8.19	5.93	5.01	4.50	4.17	3.94	3.77	3.63	3.43	3.30	3.00
20	8.10	5.85	4.94	4.43	4.10	3.87	3.70	3.56	3.37	3.23	2.94
21	8.02	5.78	4.87	4.37	4.04	3.81	3.64	3.51	3.31	3.17	2.88
22	7.95	5.72	4.82	4.31	3.99	3.76	3.59	3.45	3.26	3.12	2.83
23	7.88	5.66	4.76	4.26	3.94	3.71	3.54	3.41	3.21	3.07	2.78
24	7.82	5.61	4.72	4.22	3.90	3.67	3.50	3.36	3.17	3.03	2.74
25	7.77	5.57	4.68	4.18	3.86	3.63	3.46	3.32	3.13	2.99	2.70
30	7.56	5.39	4.51	4.02	3.70	3.47	3.30	3.17	2.98	2.84	2.55
40	7.31	5.18	4.31	3.83	3.51	3.29	3.12	2.99	2.80	2.66	2.37
60	7.08	4.98	4.13	3.65	3.34	3.12	2.95	2.82	2.63	2.50	2.20
120	6.85	4.79	3.95	3.48	3.17	2.96	2.79	2.66	2.47	2.34	2.03
∞	6.63	4.61	3.78	3.32	3.02	2.80	2.64	2.51	2.32	2.18	1.88

Abridged from M. Merrington and C. M. Thompson, "Tables of percentage points of the inverted beta (F) distribution," *Biometrika,* Vol. 33, 1943, p. 73. By permission of the *Biometrika* trustees.

Tables B4, B5, and B6: The Durbin-Watson d Statistic

The Durbin-Watson d statistic is used to test for first-order serial correlation in the residuals. First-order serial correlation is characterized by $\epsilon_t = \rho\epsilon_{t-1} + u_t$, where ϵ_t is the error term found in the regression equation and u_t is a classical (nonserially correlated) error term. Since $\rho = 0$ implies no serial correlation, and since most economic and business models imply positive serial correlation if any pure serial correlation exists, the typical hypotheses are:

$$H_0: \rho \leq 0$$

$$H_A: \rho > 0$$

To test the null hypothesis of no positive serial correlation, the Durbin-Watson d statistic must be compared to two different critical d-values, d_L and d_U found in the tables that follow depending on the level of significance, the number of explanatory variables (k′) and the number of observations (n). For example, with 2 explanatory variables and 30 observations, the 1 percent one-tailed critical values are $d_L = 1.07$ and $d_U = 1.34$, so any computed Durbin-Watson statistic less than 1.07 would lead to the rejection of the null hypothesis. For computed DW d-values between 1.07 and 1.34, the test is inconclusive, and for values greater than 1.34, we can say that there is no evidence of positive serial correlation at the 99 percent level of confidence. These ranges are illustrated in the diagram below:

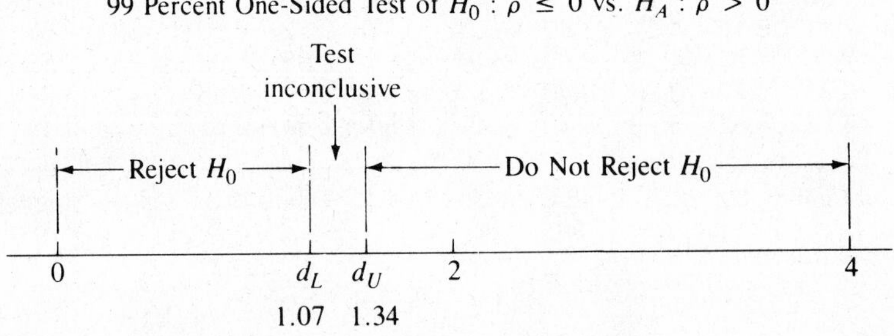

Two-sided tests are done similarly, with $4 - d_U$ and $4 - d_L$ being the critical DW d-values between 2 and 4. For more on this, see Chapter 9. Tables B-5 and B-6 (for 2.5 and 1 percent levels of significance in a one-sided test) go only up to five explanatory variables, so extrapolation for more variables (and interpolation for observations between listed points) is often in order.

TABLE B-4 CRITICAL VALUES OF THE DURBIN-WATSON TEST STATISTICS d_L AND d_U: 5 PERCENT ONE-SIDED LEVEL OF SIGNIFICANCE (10 PERCENT, TWO-SIDED LEVEL OF SIGNIFICANCE)

n	k′ = 1		k′ = 2		k′ = 3		k′ = 4		k′ = 5		k′ = 6		k′ = 7	
	d_L	d_U	d_L	d_U	d_L	d_U	d_L	d_U	d_L	d_U	d_L	d_U	d_L	d_U
15	1.08	1.36	0.95	1.54	0.81	1.75	0.69	1.97	0.56	2.21	0.45	2.47	0.34	2.73
16	1.11	1.37	0.98	1.54	0.86	1.73	0.73	1.93	0.62	2.15	0.50	2.39	0.40	2.62
17	1.13	1.38	1.02	1.54	0.90	1.71	0.78	1.90	0.66	2.10	0.55	2.32	0.45	2.54
18	1.16	1.39	1.05	1.53	0.93	1.69	0.82	1.87	0.71	2.06	0.60	2.26	0.50	2.46
19	1.18	1.40	1.07	1.53	0.97	1.68	0.86	1.85	0.75	2.02	0.65	2.21	0.55	2.40
20	1.20	1.41	1.10	1.54	1.00	1.68	0.89	1.83	0.79	1.99	0.69	2.16	0.60	2.34
21	1.22	1.42	1.13	1.54	1.03	1.67	0.93	1.81	0.83	1.96	0.73	2.12	0.64	2.29
22	1.24	1.43	1.15	1.54	1.05	1.66	0.96	1.80	0.86	1.94	0.77	2.09	0.68	2.25
23	1.26	1.44	1.17	1.54	1.08	1.66	0.99	1.79	0.90	1.92	0.80	2.06	0.72	2.21
24	1.27	1.45	1.19	1.55	1.10	1.66	1.01	1.78	0.93	1.90	0.84	2.04	0.75	2.17
25	1.29	1.45	1.21	1.55	1.12	1.66	1.04	1.77	0.95	1.89	0.87	2.01	0.78	2.14
26	1.30	1.46	1.22	1.55	1.14	1.65	1.06	1.76	0.98	1.88	0.90	1.99	0.82	2.12
27	1.32	1.47	1.24	1.56	1.16	1.65	1.08	1.76	1.00	1.86	0.93	1.97	0.85	2.09
28	1.33	1.48	1.26	1.56	1.18	1.65	1.10	1.75	1.03	1.85	0.95	1.96	0.87	2.07
29	1.34	1.48	1.27	1.56	1.20	1.65	1.12	1.74	1.05	1.84	0.98	1.94	0.90	2.05
30	1.35	1.49	1.28	1.57	1.21	1.65	1.14	1.74	1.07	1.83	1.00	1.93	0.93	2.03
31	1.36	1.50	1.30	1.57	1.23	1.65	1.16	1.74	1.09	1.83	1.02	1.92	0.95	2.02
32	1.37	1.50	1.31	1.57	1.24	1.65	1.18	1.73	1.11	1.82	1.04	1.91	0.97	2.00
33	1.38	1.51	1.32	1.58	1.26	1.65	1.19	1.73	1.13	1.81	1.06	1.90	0.99	1.99
34	1.39	1.51	1.33	1.58	1.27	1.65	1.21	1.73	1.14	1.81	1.08	1.89	1.02	1.98
35	1.40	1.52	1.34	1.58	1.28	1.65	1.22	1.73	1.16	1.80	1.10	1.88	1.03	1.97
36	1.41	1.52	1.35	1.59	1.30	1.65	1.24	1.73	1.18	1.80	1.11	1.88	1.05	1.96
37	1.42	1.53	1.36	1.59	1.31	1.66	1.25	1.72	1.19	1.80	1.13	1.87	1.07	1.95
38	1.43	1.54	1.37	1.59	1.32	1.66	1.26	1.72	1.20	1.79	1.15	1.86	1.09	1.94
39	1.43	1.54	1.38	1.60	1.33	1.66	1.27	1.72	1.22	1.79	1.16	1.86	1.10	1.93
40	1.44	1.54	1.39	1.60	1.34	1.66	1.29	1.72	1.23	1.79	1.18	1.85	1.12	1.93
45	1.48	1.57	1.43	1.62	1.38	1.67	1.34	1.72	1.29	1.78	1.24	1.84	1.19	1.90
50	1.50	1.59	1.46	1.63	1.42	1.67	1.38	1.72	1.34	1.77	1.29	1.82	1.25	1.88
55	1.53	1.60	1.49	1.64	1.45	1.68	1.41	1.72	1.37	1.77	1.33	1.81	1.29	1.86
60	1.55	1.62	1.51	1.65	1.48	1.69	1.44	1.73	1.41	1.77	1.37	1.81	1.34	1.85
65	1.57	1.63	1.54	1.66	1.50	1.70	1.47	1.73	1.44	1.77	1.40	1.81	1.37	1.84
70	1.58	1.64	1.55	1.67	1.53	1.70	1.49	1.74	1.46	1.77	1.43	1.80	1.40	1.84
75	1.60	1.65	1.57	1.68	1.54	1.71	1.52	1.74	1.49	1.77	1.46	1.80	1.43	1.83
80	1.61	1.66	1.59	1.69	1.56	1.72	1.53	1.74	1.51	1.77	1.48	1.80	1.45	1.83
85	1.62	1.67	1.60	1.70	1.58	1.72	1.55	1.75	1.53	1.77	1.50	1.80	1.47	1.83
90	1.63	1.68	1.61	1.70	1.59	1.73	1.57	1.75	1.54	1.78	1.52	1.80	1.49	1.83
95	1.64	1.69	1.62	1.71	1.60	1.73	1.58	1.75	1.56	1.78	1.54	1.80	1.51	1.83
100	1.65	1.69	1.63	1.72	1.61	1.74	1.59	1.76	1.57	1.78	1.55	1.80	1.53	1.83

NOTE: n = number of observations; k′ = number of explanatory variables, excluding the constant term. It is assumed that the equation contains a constant term and no lagged dependent variables (if so, see Table B-7). Source: N. E. Savin and Kenneth J. White, "The Durbin-Watson Test for Serial Correlation with Extreme Sample Sizes or Many Regressors," *Econometrica,* Nov. 1977, p. 1994. Reprinted with permission.

TABLE B-5 CRITICAL VALUES OF THE DURBIN-WATSON TEST STATISTICS OF d_L AND d_U: 2.5 PERCENT ONE-SIDED LEVEL OF SIGNIFICANCE (5 PERCENT, TWO-SIDED LEVEL OF SIGNIFICANCE)

n	k′ = 1		k′ = 2		k′ = 3		k′ = 4		k′ = 5	
	d_L	d_U	d_L	d_U	d_L	d_U	d_L	d_U	d_L	d_U
15	0.95	1.23	0.83	1.40	0.71	1.61	0.59	1.84	0.48	2.09
16	0.98	1.24	0.86	1.40	0.75	1.59	0.64	1.80	0.53	2.03
17	1.01	1.25	0.90	1.40	0.79	1.58	0.68	1.77	0.57	1.98
18	1.03	1.26	0.93	1.40	0.82	1.56	0.72	1.74	0.62	1.93
19	1.06	1.28	0.96	1.41	0.86	1.55	0.76	1.72	0.66	1.90
20	1.08	1.28	0.99	1.41	0.89	1.55	0.79	1.70	0.70	1.87
21	1.10	1.30	1.01	1.41	0.92	1.54	0.83	1.69	0.73	1.84
22	1.12	1.31	1.04	1.42	0.95	1.54	0.86	1.68	0.77	1.82
23	1.14	1.32	1.06	1.42	0.97	1.54	0.89	1.67	0.80	1.80
24	1.16	1.33	1.08	1.43	1.00	1.54	0.91	1.66	0.83	1.79
25	1.18	1.34	1.10	1.43	1.02	1.54	0.94	1.65	0.86	1.77
26	1.19	1.35	1.12	1.44	1.04	1.54	0.96	1.65	0.88	1.76
27	1.21	1.36	1.13	1.44	1.06	1.54	0.99	1.64	0.91	1.75
28	1.22	1.37	1.15	1.45	1.08	1.54	1.01	1.64	0.93	1.74
29	1.24	1.38	1.17	1.45	1.10	1.54	1.03	1.63	0.96	1.73
30	1.25	1.38	1.18	1.46	1.12	1.54	1.05	1.63	0.98	1.73
31	1.26	1.39	1.20	1.47	1.13	1.55	1.07	1.63	1.00	1.72
32	1.27	1.40	1.21	1.47	1.15	1.55	1.08	1.63	1.02	1.71
33	1.28	1.41	1.22	1.48	1.16	1.55	1.10	1.63	1.04	1.71
34	1.29	1.41	1.24	1.48	1.17	1.55	1.12	1.63	1.06	1.70
35	1.30	1.42	1.25	1.48	1.19	1.55	1.13	1.63	1.07	1.70
36	1.31	1.43	1.26	1.49	1.20	1.56	1.15	1.63	1.09	1.70
37	1.32	1.43	1.27	1.49	1.21	1.56	1.16	1.62	1.10	1.70
38	1.33	1.44	1.28	1.50	1.23	1.56	1.17	1.62	1.12	1.70
39	1.34	1.44	1.29	1.50	1.24	1.56	1.19	1.63	1.13	1.69
40	1.35	1.45	1.30	1.51	1.25	1.57	1.20	1.63	1.15	1.69
45	1.39	1.48	1.34	1.53	1.30	1.58	1.25	1.63	1.21	1.69
50	1.42	1.50	1.38	1.54	1.34	1.59	1.30	1.64	1.26	1.69
55	1.45	1.52	1.41	1.56	1.37	1.60	1.33	1.64	1.30	1.69
60	1.47	1.54	1.44	1.57	1.40	1.61	1.37	1.65	1.33	1.69
65	1.49	1.55	1.46	1.59	1.43	1.62	1.40	1.66	1.36	1.69
70	1.51	1.57	1.48	1.60	1.45	1.63	1.42	1.66	1.39	1.70
75	1.53	1.58	1.50	1.61	1.47	1.64	1.45	1.67	1.42	1.70
80	1.54	1.59	1.52	1.62	1.49	1.65	1.47	1.67	1.44	1.70
85	1.56	1.60	1.53	1.63	1.51	1.65	1.49	1.68	1.46	1.71
90	1.57	1.61	1.55	1.64	1.53	1.66	1.50	1.69	1.48	1.71
95	1.58	1.62	1.56	1.65	1.54	1.67	1.52	1.69	1.50	1.71
100	1.59	1.63	1.57	1.65	1.55	1.67	1.53	1.70	1.51	1.72

NOTE: n = number of observations; k′ = number of explanatory variables, excluding the constant term. It is assumed that the equation contains a constant term and no lagged dependent variables (if so, see Table B-7). Source: J. Durbin and G. S. Watson, "Testing for Serial Correlation in Least Squares Regression," *Biometrika,* vol. 38, 1951, pp. 159–77. Reprinted with permission of the *Biometrika* trustees.

TABLE B-6 CRITICAL VALUES OF THE DURBIN-WATSON TEST STATISTICS d_L AND d_U: 1 PERCENT ONE-SIDED LEVEL OF SIGNIFICANCE (2 PERCENT, TWO-SIDED LEVEL OF SIGNIFICANCE)

n	k′ = 1		k′ = 2		k′ = 3		k′ = 4		k′ = 5	
	d_L	d_U	d_L	d_U	d_L	d_U	d_L	d_U	d_L	d_U
15	0.81	1.07	0.70	1.25	0.59	1.46	0.49	1.70	0.39	1.96
16	0.84	1.09	0.74	1.25	0.63	1.44	0.53	1.66	0.44	1.90
17	0.87	1.10	0.77	1.25	0.67	1.43	0.57	1.63	0.48	1.85
18	0.90	1.12	0.80	1.26	0.71	1.42	0.61	1.60	0.52	1.80
19	0.93	1.13	0.83	1.26	0.74	1.41	0.65	1.58	0.56	1.77
20	0.95	1.15	0.86	1.27	0.77	1.41	0.68	1.57	0.60	1.74
21	0.97	1.16	0.89	1.27	0.80	1.41	0.72	1.55	0.63	1.71
22	1.00	1.17	0.91	1.28	0.83	1.40	0.75	1.54	0.66	1.69
23	1.02	1.19	0.94	1.29	0.86	1.40	0.77	1.53	0.70	1.67
24	1.04	1.20	0.96	1.30	0.88	1.41	0.80	1.53	0.72	1.66
25	1.05	1.21	0.98	1.30	0.90	1.41	0.83	1.52	0.75	1.65
26	1.07	1.22	1.00	1.31	0.93	1.41	0.85	1.52	0.78	1.64
27	1.09	1.23	1.02	1.32	0.95	1.41	0.88	1.51	0.81	1.63
28	1.10	1.24	1.04	1.32	0.97	1.41	0.90	1.51	0.83	1.62
29	1.12	1.25	1.05	1.33	0.99	1.42	0.92	1.51	0.85	1.61
30	1.13	1.26	1.07	1.34	1.01	1.42	0.94	1.51	0.88	1.61
31	1.15	1.27	1.08	1.34	1.02	1.42	0.96	1.51	0.90	1.60
32	1.16	1.28	1.10	1.35	1.04	1.43	0.98	1.51	0.92	1.60
33	1.17	1.29	1.11	1.36	1.05	1.43	1.00	1.51	0.94	1.59
34	1.18	1.30	1.13	1.36	1.07	1.43	1.01	1.51	0.95	1.59
35	1.19	1.31	1.14	1.37	1.08	1.44	1.03	1.51	0.97	1.59
36	1.21	1.32	1.15	1.38	1.10	1.44	1.04	1.51	0.99	1.59
37	1.22	1.32	1.16	1.38	1.11	1.45	1.06	1.51	1.00	1.59
38	1.23	1.33	1.18	1.39	1.12	1.45	1.07	1.52	1.02	1.58
39	1.24	1.34	1.19	1.39	1.14	1.45	1.09	1.52	1.03	1.58
40	1.25	1.34	1.20	1.40	1.15	1.46	1.10	1.52	1.05	1.58
45	1.29	1.38	1.24	1.42	1.20	1.48	1.16	1.53	1.11	1.58
50	1.32	1.40	1.28	1.45	1.24	1.49	1.20	1.54	1.16	1.59
55	1.36	1.43	1.32	1.47	1.28	1.51	1.25	1.55	1.21	1.59
60	1.38	1.45	1.35	1.48	1.32	1.52	1.28	1.56	1.25	1.60
65	1.41	1.47	1.38	1.50	1.35	1.53	1.31	1.57	1.28	1.61
70	1.43	1.49	1.40	1.52	1.37	1.55	1.34	1.58	1.31	1.61
75	1.45	1.50	1.42	1.53	1.39	1.56	1.37	1.59	1.34	1.62
80	1.47	1.52	1.44	1.54	1.42	1.57	1.39	1.60	1.36	1.62
85	1.48	1.53	1.46	1.55	1.43	1.58	1.41	1.60	1.39	1.63
90	1.50	1.54	1.47	1.56	1.45	1.59	1.43	1.61	1.41	1.64
95	1.51	1.55	1.49	1.57	1.47	1.60	1.45	1.62	1.42	1.64
100	1.52	1.56	1.50	1.58	1.48	1.60	1.46	1.63	1.44	1.65

Source and Notes: See previous table.

Table B-7: The Normal Distribution

The normal distribution is usually assumed for the error term in a regression equation. Table B-7 indicates the probability that a randomly drawn number from the standardized normal distribution (mean = 0 and variance = 1) will be greater than or equal to the number identified in the side tabs, called Z. For a normally distributed variable ϵ with mean μ and variance σ^2, $Z = (\epsilon - \mu)/\sigma$. The row tab gives Z to the first decimal place, and the column tab adds the second decimal place of Z.

The normal distribution is referred to infrequently in the text, but it does come in handy in a number of advanced settings. For instance, testing for serial correlation when there is a lagged dependent variable in the equation (distributed lags) is done with a normally distributed statistic, Durbin's h Statistic:

$$h = (1 - 0.5DW)\sqrt{n/(1 - n \cdot s_\lambda^2)}$$

where DW is the Durbin-Watson d Statistic, n is the number of observations, and s_λ^2 is the estimated variance of the estimated coefficient of the lagged dependent variable (Y_{t-1}). The h-statistic is asymptotically distributed as a standard normal variable. To test a one-sided null hypothesis of no positive serial correlation:

$$H_0: \rho \leq 0$$

$$H_A: \rho > 0$$

calculate h and compare it to a critical h-value for the desired level of significance. For a one-sided 2.5 percent test, for example, the critical h-value is 1.96 as shown in the accompanying graph. If we observed a computed h higher than 1.96, we would reject the null hypothesis of no positive serial correlation at the 97.5 percent level of confidence.

TABLE B-7 THE NORMAL DISTRIBUTION

$$Z = \frac{\epsilon - \mu}{\sigma} \quad \text{(Standardized normal)}$$

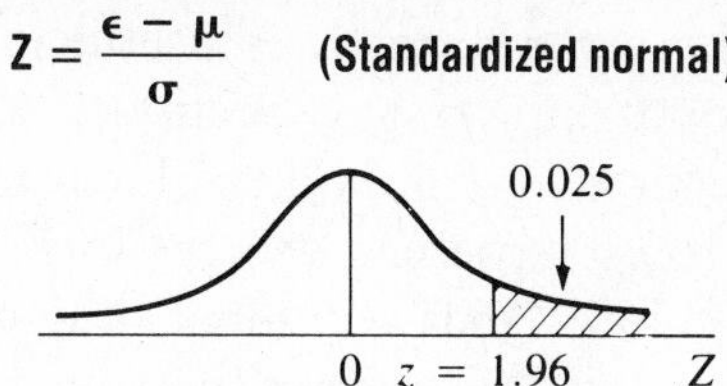

z	.00	.01	.02	.03	.04	.05	.06	.07	.08	.09
0.0	.5000	.4960	.4920	.4880	.4840	.4801	.4761	.4721	.4681	.4641
0.1	.4602	.4562	.4522	.4483	.4443	.4404	.4364	.4325	.4686	.4247
0.2	.4207	.4168	.4129	.4090	.4052	.4013	.3974	.3936	.3897	.3859
0.3	.3821	.3873	.3745	.3707	.3669	.3632	.3594	.3557	.3520	.3483
0.4	.3446	.3409	.3372	.3336	.3300	.3264	.3228	.3192	.3156	.3121
0.5	.3085	.3050	.3015	.2981	.2946	.2912	.2877	.2843	.2810	.2776
0.6	.2743	.2709	.2676	.2643	.2611	.2578	.2546	.2514	.2483	.2451
0.7	.2420	.2389	.2358	.2327	.2296	.2266	.2236	.2206	.2217	.2148
0.8	.2119	.2090	.2061	.2033	.2005	.1977	.1949	.1922	.1894	.1867
0.9	.1841	.1814	.1788	.1762	.1736	.1711	.1685	.1660	.1635	.1611
1.0	.1587	.1562	.1539	.1515	.1492	.1469	.1446	.1423	.1401	.1379
1.1	.1357	.1335	.1314	.1292	.1271	.1251	.1230	.1210	.1190	.1170
1.2	.1151	.1131	.1112	.1093	.1075	.1056	.1038	.1020	.1003	.0985
1.3	.0968	.0951	.0934	.0918	.0901	.0885	.0869	.0853	.0838	.0823
1.4	.0808	.0793	.0778	.0764	.0749	.0735	.0721	.0708	.0694	.0681
1.5	.0668	.0655	.0643	.0630	.0618	.0606	.0594	.0582	.0571	.0559
1.6	.0548	.0537	.0526	.0516	.0505	.0495	.0485	.0475	.0465	.0455
1.7	.0446	.0436	.0427	.0418	.0409	.0401	.0392	.0384	.0375	.0367
1.8	.0359	.0351	.0344	.0366	.0329	.0322	.0314	.0307	.0301	.0294
1.9	.0287	.0281	.0274	.0268	.0262	.0256	.0250	.0244	.0239	.0233
2.0	.0228	.0222	.0217	.0212	.0207	.0202	.0197	.0192	.0188	.0183
2.1	.0179	.0174	.0170	.0166	.0162	.0158	.0154	.0150	.0146	.0143
2.2	.0139	.0136	.0132	.0129	.0125	.0122	.0119	.0116	.0113	.0110
2.3	.0107	.0104	.0102	.0099	.0096	.0094	.0091	.0089	.0087	.0084
2.4	.0082	.0080	.0078	.0075	.0073	.0071	.0069	.0068	.0066	.0064
2.5	.0062	.0060	.0059	.0057	.0055	.0054	.0052	.0051	.0049	.0048
2.6	.0047	.0045	.0044	.0043	.0041	.0040	.0039	.0038	.0037	.0036
2.7	.0035	.0034	.0033	.0032	.0031	.0030	.0029	.0028	.0027	.0026
2.8	.0026	.0025	.0024	.0023	.0023	.0022	.0021	.0020	.0020	.0019
2.9	.0019	.0018	.0018	.0017	.0016	.0016	.0015	.0015	.0014	.0014
3.0	.0013	.0013	.0013	.0012	.0012	.0011	.0011	.0011	.0011	.0010

Source: Based on *Biometrika Tables for Statisticians,* Vol. 1, 3rd ed. (1966), with the permission of the *Biometrika* trustees.
NOTE: The table plots the cumulative probability $Z > z$.

Table B-8: The Chi-Square Distribution

The chi-square distribution describes the distribution of the estimate of the variance of the error term. It is useful in a number of tests, including the likelihood ratio test of Section 7.6, the Breusch-Pagan and White tests of Section 10.3.4, and the Lagrange Multiplier test of Section 12.2.2. The rows represent degrees of freedom, and the columns denote the probability that a number drawn randomly from the chi-square distribution will be greater than or equal to the number shown in the body of the table. For example, the probability is 10 percent that a number drawn randomly from any chi-square distribution will be greater than or equal to 22.3 for 15 degrees of freedom.

The steps for the likelihood ratio test of a significant difference between the fit of a constrained equation and an unconstrained equation are:

1. Compute LL $= -2\ln(\text{LR})$, where LR is the likelihood ratio, equal to the likelihood function of the constrained equation divided by the likelihood function of the unconstrained equation.

2. Compare LL to the Chi-square value found in Table B-8 for the number of degrees of freedom equal to the number of parameters constrained by the null hypothesis at a specific level of significance; if LL $>$ the critical Chi-square value, then reject the null hypothesis. For example, with two degrees of freedom and 10 percent level of significance, the critical Chi-square value is 4.61. If the observed LL is greater than 4.61, then we can reject the null hypothesis with 90 percent confidence.

TABLE B-8 THE CHI-SQUARE DISTRIBUTION

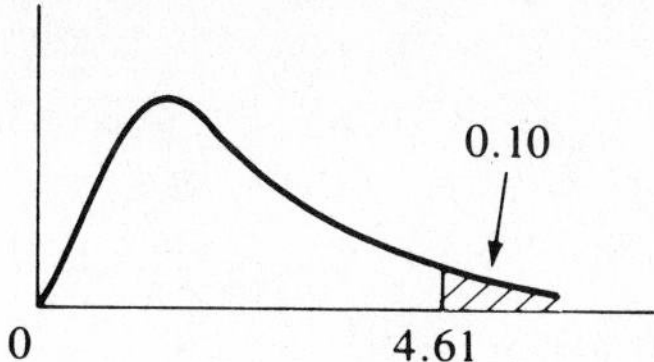

Degrees of Freedom	Level of Significance (Probability of a Value of at Least as Large as the Table Entry)			
	10%	5%	2.5%	1%
1	2.71	3.84	5.02	6.63
2	4.61	5.99	7.38	9.21
3	6.25	7.81	9.35	11.34
4	7.78	9.49	11.14	13.28
5	9.24	11.07	12.83	15.09
6	10.64	12.59	14.45	16.81
7	12.02	14.07	16.01	18.48
8	13.36	15.51	17.53	20.1
9	14.68	16.92	19.02	21.7
10	15.99	18.31	20.5	23.2
11	17.28	19.68	21.9	24.7
12	18.55	21.0	23.3	26.2
13	19.81	22.4	24.7	27.7
14	21.1	23.7	26.1	29.1
15	22.3	25.0	27.5	30.6
16	23.5	26.3	28.8	32.0
17	24.8	27.6	30.2	33.4
18	26.0	28.9	31.5	34.8
19	27.2	30.1	32.9	36.2
20	28.4	31.4	34.2	37.6

Source: See previous table.

Index